Nematology – Advances and Perspectives

Volume II: Nematode Management
and Utilization

Editors: Z. X. Chen
S. Y. Chen
D. W. Dickson

Tsinghua University Press **CABI Publishing**

CABI Publishing is a division of CAB International

CABI Publishing
CAB International
Wallingford
Oxfordshire OX10 8DE
UK

Tel: +44 (0)1491 832111
Fax: +44 (0)1491 833508
E-mail: cabi@cabi.org
Website: www.cabi-publishing.org

CABI Publishing
875 Massachusetts Avenue
7th Floor
Cambridge, MA 02139
USA

Tel: +1 617 395 4056
Fax: +1 617 354 6875
E-mail: cabi-nao@cabi.org

A catalogue record for this book is available from the British Library, London, UK.

Library of Congress Cataloging-in-Publication Data

Nematology : advances and perspectives / editors, Z.X. Chen, S.Y. Chen, D.W. Dickson.
p. cm.
Includes bibliographical references.
ISBN 0-85199-645-0 (alk. paper)
1. Nematoda. 2. Plant nematodes. 3. Soil nematodes. I. Chen, Z. X. II. Chen S. Y. III. Dickson, Donald W. (Donald Ward)

QL391.N4N43 2003
592'.57--dc21

2003048906

CABI ISBN 0 85199 646 9

Tsinghua University Press ISBN 7-302-06817-8

Published in China by Tsinghua University Press, Xue Yan Building, Tsinghua University, Beijing, 100084, PR China.

Printed and bound in the UK from copy supplied by Tsinghua University Press by Biddles Ltd, King's Lynn.

Preface

Nematology, Advances and Perspectives

During the 20th century, science and technology have developed in an unparalleled manner. This is especially true for nematology. Since Cobb first recognized nematology as an independent discipline during the early part of the century, nematology has made unparalleled advances and become an integral part of the biological sciences. The development of nematology is largely attributed to the discovery of the importance of nematodes in agricultural ecosystems and their impact on society. Nematodes are the most abundant and diversified group in the animal kingdom, and four out of five animals on earth are nematodes. Marine nematology has become an independent discipline and it has been suggested that the secret of the natural history of our planet may lie in the nematodes dwelling deep in oceans. Animal-parasitic nematodes have had great impact on human heath and society throughout history. Soil nematodes play important roles in organic degradation, mineralization, and food webs in soil ecosystems. While most nematode species in soil are beneficial, some species are important pathogens of plants and cause severe damage to crops. Worldwide crop yield losses to nematodes have been estimated at approximately $ 78 billion annually. The advances in nematology and related disciplines, however, have drastically strengthened our ability to fight this unseen enemy. Thus, over the past century, billions of dollars of crop losses from nematode damages have been prevented.

One of the most exciting and important new fields of nematology includes recent advances made in the use of nematodes as model organisms for basic biological studies, especially in developmental biology, genetics, and cellular and molecular biology. Using *Caenorhabditis elegans* as a model, scientists have unraveled for the first time in history the complete genomic sequence of DNA from a multicellular organism. Also, they have documented cellular development from a single fertilized cell to a fully developed body.

Nematology is one of the most dynamic and exciting disciplines in the biological sciences. It is impossible to cover every aspect of nematology in this book. Furthermore, several topics in nematology reached their climax during the past century, but interest has now ebbed. The major focus of this book is on topics that have made remarkable advances in the latter part of the century and currently are of primary research emphasis. These topics will relate mainly to free-living, plant-parasitic, and entomopathogenic nematodes.

The following lists of chapters evolved after lengthy discussion and consultation with fellow nematologists. We are pleased that so many renowned nematologists with international reputations and experiences have responded positively to the book proposal and agreed to make authoritative contributions. It is overwhelming to have nematologists around the world making a collaborative effort to produce a quality book. The book is to be published in 2004. Therefore, we shall consider it a memorial to the achievements in nematology in the 20th century and as a beacon for future developments in nematology during the new century. For many reasons, we believe that nematology will become a primary focus in the agricultural sciences during the new century. With the rapid growth of the world's population, food and fiber demands in the next century will increase tremendously. But the task for nematologists is daunting in that funding, personnel in nematology, and graduate education has been sharply curtailed. Also, with the suspensions and phase-outs of many soil fumigants and nonfumigants and the threat for even more suspensions in the foreseeable future, the challenges will become even greater. Increased endeavors in nematode research will be essential to achieve the sustained growth of agriculture in the 21st century.

The objective of this book is to summarize advances in nematology that have been made during the 20th century and to provide perspectives for the development of nematology in the next century. It is aimed at researcher and graduate student communities. The international representation of the science will be a key component of this book. It is our hope that the book will provide a road map to the most important aspects of the science at this time and that it will provide critical thoughts and ideas for researchers and graduate students to carry forward into the next millennium.

The editors are indebted to the authors of the respective chapters for their excellent contributions. We are grateful to the large number of scientists who have provided valuable comments and suggestions during book proposal discussion and reviews of each individual chapter manuscripts. Finally, we express our gratitude to the Tsinghua University Press and CAB International for fulfilling this endeavor.

Editors:

Zhongxiao Chen
National Food Safety and Toxicology Center
Michigan State University
East Lansing, Michigan

Senyu Chen
Southern Research and Outreach Center
University of Minnesota
Waseca, Minnesota

Donald W. Dickson
Entomology and Nematology Department
University of Florida
Gainesville, Florida

This Book Is Dedicated To

Pinsan Chen
For his 50 years of career in nematology

Pinsan Chen was born on 3 December 1927 in Mengcheng County, Anhui Province, China. He received elementary, middle and high school education in his hometown and began college education in 1947 in the Division of Plant Pathology, Department of Agronomy, College of Agriculture, Nanjing University, China. After graduation in 1952, he taught botany and plant pathology in Agronomy Department, Beijing Mechanized Agricultural College.

In 1954, Pinsan accepted a research position at the Northern China Institute of Agricultural Sciences (later the Chinese Academy of Agricultural Sciences) focused on plant nematology, an area received little attention then in China. Without formal training in nematology, he was a self-starter and quick learner. With his insightful observations, he demonstrated that the causal organism of millet nematode disease was *Aphelenchoides* sp. rather than *Tylenchus* sp. as reported in literature.

In 1963, Pinsan designed first contemporary nematology laboratory in China based on literature from UK, Russia, Japan, and USA. Soon, he examined samples collected from crop fields, pastures, mountains, and wetlands. Based on the preliminary observations, he speculated that all reported most important plant-parasitic nematodes might exist in China, and proposed extensive research and management. In the following years, his research was focused on *Aphelenchoides* sp. on millet, *A. besseyi* on rice, *Anguina tritici* on wheat, *Ditylenchus* sp. on sweet potato, and root-knot nematodes (*Meloidogyne* spp.) on peanut and vegetables.

In late 1978 after the country emerged from the Cultural Revolution, with encouragement and support from senior scientists he restored the nematology laboratory, which was almost destroyed during the Cultural Revolution. Two years later, Pinsan recruited the first Master of Science degree student in his laboratory. He designated his laboratory focusing on the biology and management of root-knot and soybean cyst nematodes. In collaboration with scientists from other institutes and laboratories, Pinsan developed soybean cyst nematode-resistant cultivars "Kangxian No. 1", "Kangxian No. 2", and "Qingdao No. 1"; and developed chemical control using Avermictin in nursery soil treatment and Methyl-isofenphus in field crops. During 1980–1996, Pinsan and co-workers identified a number of important nematodes that included eleven new records in China and three new species: *Heterodera sinensis, Meloidogyne fanzhiensis,* and *Pratylenchus artemisiae.*

In his extensive career, Mr. Chen published more than 30 papers, 50 popular articles, and seven books. He advised and co-advised eight Master of Science and two Ph.D. graduate students. He won one National Technology Invention Award, one National Science Advance Award, and six Provincial Science and Technology Advance Awards. Mr. Chen is currently a Professor Emeritus at the Institute of Plant Protection, Chinese Academy of Agricultural Sciences, Beijing, China.

Contents

List of Contributors

Rosa H. Manzanilla-López
Kenneth Evans
Nematode Interactions Unit, IACR-Rothamsted, Harpenden, Hertfordshire, AL5 2JQ, UK
Email: ken. evans@ bbsrc. ac. uk

John Bridge
Tropical Plant Nematology Advisor, CABI Bioscience (UK Center), Bakeham Lane, Egham, Surrey, TW20 9TY, UK
Email: j. bridge@ cabi. org

Derek J. F. Brown
Professor, Nematology Department, Scottish Crop Research Institute, Invergowrie, Dundee DD2 5DA, Scotland, UK
Email: djf. brown@ scri. sari. ac. uk

JingWu Zheng and Xueping Zhou
Associate Professor and Professor, College of Agriculture and Biotechnology, Huajiachi Campus, Zhejiang University, Hangzhou 100029, Peoples Republic of China

Haddish Melakeberhan
Department of Entomology, Michigan State University, East Lansing, MI 48824-1115, USA
Email: melakebe@ msu. edu

Khuong B. Nguyen and Grover C. Smart, Jr.
Entomology and Nematology Department, University of Florida, Gainesville, Florida, USA
Email: kbn@ gnv. ifas. ufl. edu
gcs@ gnv. ifas. ufl. edu

J. L. Starr
Department of Plant Pathology & Microbiology, Texas A&M University, College Station, TX 77843-2132, USA
Email: starr@ ppserver. tamu. edu

P. A. Roberts
Department of Nematology, University of California, Riverside, CA 92521-0415, USA
philip. roberts@ ucr. edu

John M. Halbrendt Associate Professor of Plant Pathology, The Pennsylvania State University, Fruit Research & Extension Center, 290 University Drive, Biglerville, PA 17307
Email: jmh23@psu.edu

James A. LaMondia Plant Pathologist/Nematologist, Connecticut Agricultural Experiment Station, 153 Cook Hill Road, Windsor, CT 06095
Email: lamondia@caes.state.ct.us

Silamar Ferraz Professor, Universidade Federal de Viçosa-36571-000-Viçosa, MG-Brazil
Email: silamar@mail.ufv.br

Leandro Grassi de Freitas Associate Professor, Universidade Federal de Viçosa-36571-000-Viçosa, MG-Brazil
Email: leandro@mail.ufv.br

Senyu Chen Southern Research and Outreach Center, University of Minnesota, Waseca, MN 56093, USA
Email: Chenx099@tc.umn.edu

Donald W. Dickson Entomology and Nematology Department, University of Florida, Gainesville, FL 32611, USA
Email: dwd@ufl.edu

Z. X. Chen National Food Safety and Toxicology Center, Michigan State University, East Lansing, MI 48824, USA
Email: chenzho@msu.edu

Harry K. Kaya Professor, Department of Nematology, University of California, Davis, California 95616, USA
Email: hkkaya@ucdavis.edu

Albrecht M. Koppenhöfer Extension Specialist, Department of Entomology, Rutgers University, New Brunswick, New Jersey 08901, USA
Email: koppenhofer@AESOP.RUTGERS.EDU

Paul S. Lehman Nematologist, Florida Department of Agriculture and Consumer Services, Division of Plant Industry, Nematology Section, Gainesville, FL 32614, USA
Email: lehmanp@doacs.state.fl.us

J. R. Rich Professor of Nematology, University of Florida, 30 Research Road, Quincy, Florida 32351, USA
Email: jrich@mail.ifas.ufl.edu

R. A. Dunn Professor of Nematology Emeritus, University of Florida, 1045 Irish Road, Scottsville, Virginia 24590, USA

J. W. Noling Associate Professor of Nematology, University of Florida, 700 Experiment Station Road, Lake Alfred, Florida 33850, USA

13 Plant Diseases Caused by Nematodes

Rosa H. Manzanilla-López, Kenneth Evans and John Bridge

13.1 Introduction

The need for increased food and fiber production to meet the demands of a growing world population and the realization that meeting those demands will depend largely upon plant pest and disease control is now very clear. It is also challenging, considering the "declining agricultural base" that has been inherited from the 20th century (Sequeira, 2000). Browning (1998) considered that we are in the "Age of Plants" in which virtually every plant and ecosystem has actual or potential value for food, fiber, aesthetic purposes or support services.

Worldwide losses from biotic causes in various crops are estimated at: rice 51.4%, potatoes 41.4%, coffee 40.0%, maize 38.3%, cotton 37.7%, wheat 34.0%, soybeans 32.4%, barley 29.4%. Globally, these losses average 42.1% of attainable yield and the figure would be 69.8% "if no physical, biological, or chemical measures were used to protect crops" (Browning, 1998; Oerke, et al., 1994).

The development of genetically engineered crop cultivars has caused concern and also has led to a re-consideration of existing agroecosystems, resulting in growing fears over the "health" of ecosystems and the "quality of food" (e.g., pesticide-free and organic food production are seen as highly desirable). Both of these concerns are re-shaping our way of dealing with plant diseases, parasites, and pathogens. Traditional knowledge and agricultural practices are being recalled and reviewed. New knowledge and ways of exploiting new technologies are required as well as an emphasis on a more holistic and ecological understanding of the nature of plant disease.

The development of genetics, cellular and molecular biology, and molecular engineering during the 20th century has opened previously unimaginable possibilities for manipulating biological processes at levels never before conceived. However, the concept of simply inserting genes taken from one organism into the genome of another may lull us into too comfortable a concept of how we may deal with plant diseases. This may lead us to disregard the need for deeper knowledge of the mechanisms involved in the interactions between plants and pathogens that might lead to identification of novel bioactive compounds with potential for nematode management, such as systemically acquired resistance ("SAR") elicitors. This may, in turn, lead us to ignore the problems of how to deal with artificially engineered organisms in such systems and ensure their safe use in the future. Engineering plants for resistance against nematodes by exploiting points of weakness in the host/parasite relationship is already under way. Such developments should not stop us from enquiring into and understanding the "nature of disease", rather they should encourage us to understand better how diseases, both old and new, were and are produced, treated and managed. However, the research life of any single nematologist could never be long enough to decipher all the enigmas related to specific patterns or processes of disease!

In this chapter we will detail some of the main events that led to the recognition of nematodes as disease agents and their place in agriculture, the nature of diseases caused by plant-parasitic nematodes, and terminology related to parasitism and disease. Diseases caused by nematodes will be divided into sub-groups according to their habitat and parasitic habit (sedentary, migratory, endo- and ecto-parasitic nematodes). The symptomatology associated with the most important diseases and their impact on agricultural crops will also be discussed.

13.2 The Development of Nematology and Recognition of its Economic Importance

Nematology, as in other disciplines, started with a taxonomic approach but research soon shifted to development of an understanding of the biology, physiology, biochemistry, host-parasite interactions and, eventually, molecular biology of nematodes. It is important to note that findings in disciplines such as bacteriology, mycology, physics, virology, along with discoveries and inventions in industry, had profound impacts on the routes that nematology followed at the end of the 19th century and through the 20th century. These discoveries aided our understanding of how nematodes cause disease, and this has been further helped by new technologies in a step-by-step process.

The *Caenorhabditis elegans* genome project (http://elegans.swmed.edu/) has greatly increased public awareness of nematodes and research efforts on their biology (Bird and Opperman, 1998). In a similar way to its sister disciplines (plant pathology and crop protection), research on diseases caused by nematodes has been shaped by ideas, findings and technologies developed through the centuries in different scientific disciplines. These started with simple comparisons between healthy and diseased plants, moved on to observations of symptoms with the naked eye and then powerful electron microscopes, and eventually reached the molecular level of the mechanisms responsible for causing disease. Thus, the old and new paradigms of pest management in agriculture have been followed by nematologists in a search for better and sustainable ways of managing agroecosystems for the production of food and fiber, for the benefit of both present and future generations.

Because of their size, plant-parasitic nematodes have remained a little-known group. The first formal reference to a plant-parasitic nematode was during the 18th century by Turbevill Needham, who discovered, in 1743, the riddle of the "cockle" when he crushed a shrunken, blackened wheat grain and examined a portion of it under his primitive microscope. In the cockle he discovered aquatic animals, which he referred to as worms, eels or serpents. The worms were *Vibrio tritici* = *Anguina tritici.* The 19th century saw more important landmarks in the science of nematology and the recognition of the economic potential of plant-parasitic nematodes. In 1859, Schacht described a serious disease caused by a nematode that threatened the sugar industry of Germany. The sugar beet nematode was named *Heterodera schachtii* by Schmidt in 1871, and investigations on the control of this species dominated the nematological scene in Europe from 1870 to 1910. Studies on its life history, habits, distribution, etiology and methods of control were undertaken and probably the first soil fumigation for nematode control was by Kühn, when he applied carbon disulfide in sugar beet nematode infested fields (Thorne, 1961).

Cobb (1893), while investigating a disease of banana trees that occurred in Fiji, found in the soil around the roots of affected trees a new species of nematode that he named *Tylenchus similis* (the burrowing nematode = *Radopholus similis*). This species was found later in roots of sugarcane from the Hawaiian Islands and diseased rhizomes of banana trees from Jamaica and other countries. It also caused yellow disease of pepper (*Piper nigrum* L.) on the island of Bangka in Indonesia (Van den Vecht, 1950). This disease became an epidemic and resulted in reduction of the number of black pepper vines on the island from 22 million to 2 million in about 20 years (Hubert, 1957; Williams and Bridge, 1983). A similar situation occurred in Central America in the late 1960s and 1970s when the *Fusarium* wilt-susceptible Gros Michel banana was replaced by wilt-resistant Cavendish varieties, which proved more susceptible to the nematode (O'Bannon, 1977).

The first specific mention of root-knot nematodes was by Cornu (1879), when he described *Anguillula marioni* = *Meloidogyne marioni,* the causal agent of galls on the roots of *Onobrychis sativa* Scop. Some years later, Goeldi (1887; published 1892) described the production of galls on roots of coffee in Brazil by *Meloidogyne exigua,* in the process erecting the genus *Meloidogyne.* Root-knot nematodes of the genus *Meloidogyne* are more widely distributed than any other group of plant-parasitic nematodes and rank high on the list of factors affecting the production of plants (Sasser, 1977).

The acceptance of revolutionary new ideas in the 19th century brought within grasp the potential to control disease in man, animals, and plants, and thereby the opportunity to enhance the quality of life for all (Kelman, 1995). The first reference to the pathogen concept was in 1876 in bacteriology. The original definition was of a parasitic organism that causes disease in another organism (Mountain, 1960a). Robert Koch and Louis Pasteur were key figures in establishing the germ theory of disease. In 1882, Robert Koch postulated his laws of pathogenicity and gave to pathology a sound and logical procedure for investigating the bacterial diseases of man and animals (Koch, 1884). Just after the turn of the century, Erwin F. Smith adapted these postulates to plant pathology and they are still of fundamental importance in plant pathology and nematology (Mountain, 1960a). However, early 20th century leaders in science and medicine did not all accept the possibility that the micro-organisms present in the diseased tissue of plants and animals could be the causal agents of the disease but, eventually, it was established that disease in plants could be caused by parasitic micro-organisms. Nevertheless, it was only in the 1930s that nematodes achieved serious recognition among the problems confronting agriculture (Thorne, 1961).

13.2.1 Nematodes as Pathogens and Vectors

The frequent lack of distinct symptoms means that nematode damage has often been confused with soil structure or nutrition problems. Nematode problems have occasionally been attributed to fungi and bacteria, their true nature only being recognized later; e. g., the destructive "Stubby root" of vegetables in Florida, which disconcerted research workers for many years until Christie and Perry (1951) determined that *Trichodorus christiei* (= *Paratrichodorus minor*) was the causal agent. This was the first time that this genus was recognized as economically important (Thorne, 1961). More recent examples include *Bursaphelenchus xylophilus,* the pine wilt nematode. This was first shown to be the causal agent of a devastating pine wilt disease in Japan as recently as 1971; the disease had been documented prior to that date but had not been associated with a nematode (Weischer and Brown, 2000).

Interactions of nematodes with other plant pathogens had already been noticed by Atkinson in 1892. Infection by root-knot nematodes seemed to increase the incidence and severity of *Fusarium* wilt in cotton, laying the historical groundwork for future studies of the role of plant-parasitic nematodes in fungus diseases (Powell, 1963). Hunger (1901) showed that tomatoes were attacked by *Pseudomonas solanacearum* (= *Ralstonia solanacearum*) in nematode-infested soil, but remained healthy in nematode-free soil (Pitcher, 1963).

Xiphinema index was confirmed as the natural vector of grapevine fanleaf nepovirus in vineyards in California (Hewitt, et al., 1958). This report encouraged the search for other associations, revealing that longidorid and trichodorid nematodes often transmit nepo- and tobra-viruses, respectively (Brown, et al., 1995). As with plant virology, nematology has benefited from the advent of the transmission electron microscope, which has helped in the recognition and identification of viral particles on the mouthparts of nematodes. Developments in immunology and ELISA also had great impact in the diagnosis of viral diseases and have proved to be useful tools in nematology.

13.3 The "Nature of Disease"

Nematology has used the terms and concepts developed in plant pathology to describe the role that plant-parasitic nematodes play in plant diseases. At one time, nematology publications were not held in particularly high regard by plant pathologists, due to the misuse of plant pathological terminology by nematologists trying to define plant-nematode relationships in particular diseases (Mountain, 1960a).

Plant health describes the appearance and performance of green plants in the absence of biotic and abiotic sources of stress – the stress that prevents the plant from achieving its genetic potential over time (Browning, 1998). Within plant pathology, disease is defined as "The injurious alteration of one or more ordered processes of energy utilization in a living system, caused by the continuous irritation of a primary causal factor or factors. When a living system is altered beyond its range of easy tolerance it is diseased as opposed to healthy; if the system is pushed beyond its limits of absolute tolerance it will die" (Bateman, 1978). A key element in disease is the chronic action of the causal agent, in contrast to injury, which comes and goes suddenly.

The causes of plant disease include biotic entities, such as pathogens and parasites, non-biotic entities such as viruses (which obligatorily replicate in their host), and abiotic nutritional disorders (e.g., mineral deficiency or toxicity) and environmental factors that affect the normal physiological mechanisms of the plant and its homeostasis with the environment. Disease is also a dynamic process with a series of events, each one often leading to the next once pathogenesis has been triggered by the pathogen (Bos and Parlevliet, 1995). When plants are subject to sustained impairment by a pathogen (or parasite), the pathogen drives the disease; if the pathogen dies, the plant may recover. The severity of a disease is decided at the time of encounter between host and parasite and by environmental conditions and (for crops) man's activities. Host susceptibility and parasite aggressiveness determine whether or not a host/parasite relationship is established, the grade of susceptibility being determined by the genetics of both the host (i. e., cultivar) and parasite (i. e., biotype, strain, race, pathotype, etc.).

The triangle of disease (host, plant and environment) is well known to nematologists. With the inclusion of man this becomes a tetrahedron, which helps us to explain the distinct interactions that may lead to the onset of disease.

Crop disease usually implies economic loss. However, yield (which is of prime interest for the farmer) is not considered in most disease definitions, disease being defined mainly by biological and physiological considerations. The diseased condition, especially the reduction in vitality, undermines the plant's competitiveness. Non-specific growth reduction, early senescence, premature death, and increased (or sometimes decreased) susceptibility and sensitivity to other pathogens often result. The quality and quantity of yield may also be reduced (Bos and Parlevliet, 1995).

Plant disease may be categorized according to the type of causal agent, the tissues infected, the epidemic characteristics, tissue utilization, resistant reactions, etc., all of which reflect perspectives of disease. In the case of plant-parasitic nematodes, diseases have been categorized mainly according to above- or below-ground symptomatology and parasitic habits.

13.3.1 Symbiosis, Parasitism, and Disease

Central to the issue of considering nematodes as agents of disease lies the definition of terms such as symbiosis, parasitism, pest, pathogen, injury, damage and disease. An accurate characterization of the relationships and terminology used to define or describe the interaction between plant-parasitic nematodes and their hosts is necessary for the characterization of plant disease and its differentiation from other conditions.

The term symbiosis, in its widest sense, means the living together of two phylogenetically unrelated species, and therefore includes all degrees of parasitism, commensalism and mutualism. Parasitism is a widely used term in plant pathology, entomology and nematology, but its interpretation varies. Food deprivation dominates in most definitions: parasitism thus is the "partial or complete nutritional dependence of one organism or virus on the tissues of another living organism" (Bos and Parlevliet, 1995). Alternatively "parasitism is generally used to designate the relationship or association between organisms, usually belonging to different species, in which one party, the parasite, benefits from the other, the host" (Bateman, 1978). The common statement that the parasite is "conferring no benefit in return" or that it is to the detriment of the host is used to enable a distinction to be made between parasitic and symbiotic relationships (Bos and Parlevliet, 1995). Parasitism has a chemical basis and host and parasite exchange chemical substances, which may frequently be of mutual benefit. Small numbers of sedentary plant-parasitic nematodes can, under certain conditions, confer some benefit on their host (Pitcher, 1965). In these instances, a well-recognized host-parasite relationship is more of a mutualistic relationship, if only for a short time or within a specific range of population densities. Situations such as these justify the use of symbiosis as an all-embracing term (Bird, 1975).

The concept to which plant pathologists have made most serious objection in its application to nematodes is that of a pathogen. Well-adapted parasites do not kill their hosts (although extreme parasite burdens may) whereas some pathogens do. The term pathogen is often associated with the inoculation of toxins or other chemical substances or compounds that induce a harmful reaction in the host, which in time develops into symptoms, pathogenesis and disease (all of which can occur in nematode infections). Pathogens usually cause disease but parasites may or may not cause disease; if they do, then they are considered pathogens as well. Sometimes, the demarcation between disease and just injury or damage is so slight that many phytophagous nematodes have been called indiscriminately "plant-parasitic" or "plant-pathogenic" nematodes because they happen to have similar feeding habits. Although the habits may seem the same, subtle distinctions may exist. Nematodes that feed on plants, in common with other disease-producing agents, are often regarded as pathogens in their own right, capable of producing a single, recognizable disease. Frequently, such a concept is valid, but nematodes may also facilitate the entry and establishment of plant-pathogenic fungi, bacteria and viruses, especially in the case of "soil-borne" diseases (Pitcher, 1965).

Injury is the simple mechanical effect of a parasite or other harmful agent upon the victim and elicits no other reaction than a wound-healing response. Plants consumed by herbivores are food plants, rather than hosts. However, in closer and more prolonged relationships between invertebrates (e.g., small insects, mites, and nematodes) and plants, the victim, which this time does act as a host, also may be "irritated" into a pathological reaction and consequently become diseased. Disease is not the same as injury (from which plants may recover) and results from a continuous irritation (Bateman, 1978; Bos and Parlevliet, 1995).

Any organism that is harmful, troublesome or destructive could be, and often is, described as a pest, although the term is usually reserved for harmful insects and other invertebrates. Cook and Evans (1987) considered plant-parasitic nematodes to be pests rather than disease-causing pathogens like bacteria, fungi, mycoplasmas and viruses (Shaner, et al., 1992). Attributes of pests include the capacity to cause an epidemic disease associated with high mortality (plague), destructiveness or noxiousness (Bos and Parlevliet, 1995), examples of all of which can be found in plant-parasitic nematodes.

The onset and/or establishment of the host/parasite relationship (i. e., the aggression by the parasite) are called attack. It may entail local or systemic ingress or invasion and colonization of the host by the parasite, but does not necessarily result in disease. Attack can lead to injury or gross damage and also covers short-duration activity by insects or nematodes on food plants. The act of attacking or the state of being attacked is a consequence of aggressiveness of the parasite or phytophagous organism and of the susceptibility of the host or victim, and does not necessarily result in disease. Host-parasite relationships may be more characterized by the attack, i. e., the visible physical presence of the parasite, rather than the resultant disease.

Some authors (e. g., Shaner, et al., 1992) consider that many nematodes are primarily parasitic and secondarily pathogenic. In fact, plant-parasitic nematodes should be considered as having a dual nature that may include both states, with the balance capable of being shifted either way by biotic (e. g., host, number of nematodes) and abiotic factors (soil, environmental conditions). According to Yeates (1971), only in plant nematodes with sedentary females do we find the truly parasitic state. He regards the various migratory, ecto- and endoparasitic plant nematodes as a "highly adaptable group of plant browsers".

13.3.2 Physiology, Biochemistry, and Genetics in the Understanding of Disease

Plant defense against nematodes includes pre-infection and post-infection mechanisms. The first include the synthesis of toxic compounds (phytoalexins) and the hypersensitive reaction, with the subsequent death of cells next to the nematode (Mateille, 1994). Once the first barrier of defense is breached, pathogenesis and post-infection mechanisms proceed. Nematodes affect plant growth by breaking down cell structure, removing cell contents, disrupting physiological processes, and modifying gene expression in the host. Comparison of different feeding sites has led to the conclusion that nematodes may exaggerate normal cell function and metabolism rather than reprogramming events (Jones, 1981). Symptom expression at the cellular, tissue and whole plant level has been reviewed extensively (Dropkin, 1969; Paulson and Webster, 1970; Bird, 1974; Jones, 1981; Hussey, 1989; Mateille, 1994; Williamson and Hussey, 1996). The nature of the changes may differ according to the location of the parasite in the plant host and the influence of other organisms and physical factors. The physiological causes of loss of host biomass and yield include most of the major processes in the plant, including respiration, photosynthesis, nutrient translocation and availability, water relations and phytohormone balance, as well as having to satisfy the energy demand of the nematode (Mateille, 1994).

Photosynthesis can be affected in different ways and CO_2 fixation has been studied on plants infected with *Meloidogyne* spp. (Bird and Loveys, 1975; Loveys and Bird, 1973; Melakeberhan, et al., 1985, 1986, 1988; Wallace, 1974), or *Globodera* (Fatemy, et al., 1985; Franco, 1980). Changes in CO_2 fixation have been attributed to a reduction in gas diffusion through stomata, or to deficiencies of cytokinins or gibberellins (Mateille, 1994).

Reduction in respiratory intensity has been reported, although it can be related to a slowing of root growth rather than to actual physiological activity (Mateille, 1994). Zacheo and Bleve-Zacheo (1987) and Zacheo and Molinari (1987) identified the NADH cytochrome system as one of the metabolic pathways more modified by nematodes (Mateille, 1994).

Nematodes influence amino acid and protein synthesis (Mateille, 1994) but results from such studies do not help explain how nematodes cause these changes. Lewis and McClure (1975) and Meon et al. (1978) showed that roots infested with *Meloidogyne* had accumulations of proline in the galls. Studies on Heteroderoidea (*Meloidogyne* spp. and *Heterodera* spp.) have demonstrated DNA and RNA synthesis in nematode feeding sites (Bird, 1972; Masood and Saxena, 1980; Singh, et al., 1984; Arya and Tiagi, 1985). Raja and Dasgupta (1986) noted that mRNA *de novo* synthesis occurred in *Angola* pods infected with *Meloidogyne incognita,* with synchronous synthesis of macromolecules such as lignins, peroxidases and polyphenols. This proved that nematodes cause the synthesis of additional types of RNA and gene activation at a transcriptional level that leads to macromolecule synthesis.

Individual plants, and crops, may be regarded as assemblages of sources and sinks, the last including new and expanding tissues, short-term, mid-term, and permanent storage sites, which are commonly associated with yield. The relationships between sources and sinks vary both temporally and spatially, affecting the relationship between biomass production and yield. Whether a crop is source- or sink-limited at specific growth stages is relevant to understanding yield development and realization, and is therefore relevant to the reaction to disease (Gaunt, 1995). The use of radiotracer techniques has allowed the concept to be developed that nematodes such as *Meloidogyne* act as nutrient sinks (Bird, 1975; Jones and Northcote, 1972; McClure, 1977). At the cellular level, cells modified by nematodes can act as strong sinks, but "whether hormone-directed transport is involved is not clear" (Jones, 1981). What is certain is that nutrients taken up by nematodes can be excreted into the rhizosphere and there influence micro-organisms and even plant succession.

13.3.3 Plant-parasitic Nematodes and Yield Loss

Nematodes are major pathogens in their own right but their interactions with other disease-causing agents make it difficult to measure their impact on yield accurately, and the latest available large-scale estimates date from 1987 (Sasser and Freckman, 1987). The difficulties lie mainly in the design of suitable experiments because of the many overlapping interactions involved (Browning, 1998). Crop yield is the result of the interaction of biotic and abiotic factors over time on the physiological processes in the plant. In understanding the influence of plant-parasitic nematodes and associated organisms on host growth and yield, it also is important to consider the phenology of the host-parasite interaction. Yield loss is the culmination of many effects, and may even occur without any visible symptoms (Bos and Parlevliet, 1995). Recent approaches to understanding yield loss relationships include consideration of the effects of diseases on key physiological processes related to growth and development (Gaunt, 1995).

13.3.4 Strategies for Nematode Management

Correct diagnosis of a malady is the starting point for successful treatment, and should be followed by a well-considered prescription by a trained professional. However, the partition between biotic and abiotic causes of plant stress and the complex interactions between them are poorly understood and it is frequently difficult to effect specific diagnoses, prescribe treatments and implement management programs (Boyer, 1982; Browning, 1998). Traditionally, the basic strategies for management of diseases have been exclusion (quarantine), prevention, eradication, protection, and genetic control. All of them have been applied to nematodes with differing degrees of success. Quarantine measures first started between 1920 and 1930, when the bulb and stem nematode gained international importance because of its possible distribution through narcissus, tulip, and other bulbs, and rigorous quarantine measures against infested foreign bulbs were imposed by the USA (Thorne, 1961). At present, species such as *Globodera pallida, G. rostochiensis, Nacobbus aberrans, Anguina tritici, Bursaphelenchus xylophilus, Rhadinaphelenchus cocophilus, Radopholus similis* and others are subject to strict international quarantines.

Prevention, eradication and protection have been possible through the use of nematicides. Soil fumigants made it possible to demonstrate that elimination of plant-parasitic nematodes frequently removed certain disease symptoms formerly attributed to fungi and bacteria, and that nematodes were far more important than previously had been suggested. The nematicidal properties of chloropicrin were discovered in 1919, but it was results from Hawaii (Godfrey, et al., 1934) that demonstrated the great potential of this compound as a nematicide (Thorne, 1961). This led in turn to the development of agricultural machinery to make field-scale fumigation feasible. Later, more compounds were shown to have nematicidal properties and became extensively used, until increasing awareness of the environmental hazards that they posed led to their withdrawal and demands for safer compounds for nematode control. Until very recently, one of the easiest ways to persuade a farmer that yield reduction was due to nematodes was to apply a nematicide to a demonstration plot. Now that the use of methyl bromide and several other nematicidal compounds has been banned it is getting harder to demonstrate nematode damage and we need to develop alternative methods to persuade growers that nematodes cause disease and yield loss. Also, data obtained with, for example, fumigant nematicides are difficult to interpret because they affect other disease-causing and beneficial organisms.

13.3.5 Genetics and Breeding for Resistance

As the 20th century progressed, new ideas, concepts, and even revolutions (e.g., the Green Revolution) occurred in agriculture. The development of the gene-for-gene concept by Flor (1955) had enormous impact on resistance breeding programs and on ideas for the management of genetic attributes of crops. However, a radical change in the concept of a gene has occurred in the last 20 years and particularly in the post-genomics era. Nevertheless, gene and product names are still associated with DNA sequences, which may code for amino-acid products or have some regulatory role (Skupski, et al., 1999).

The recognition of major diseases caused by nematodes led to the search for sources of resistance in native host species. During this process, plant breeders and nematologists have had to overcome biological difficulties encountered with both host and parasite. Controlled genetic crosses with inbred populations of *Globodera rostochiensis* have demonstrated the existence of a gene-for-gene relationship between the H1 resistance gene in potato and a dominant virulence gene in the pathogen (Williamson and Hussey, 1996). Breeding for resistance in tomatoes to root-knot nematodes, *M. incognita,* started in the early 1940s. High levels of resistance were found in *Lycopersicon peruvianum* L. (Mill.) and tight linkages preventing the development of good fruit types were finally broken by Gilbert and McGuire in 1956 in Hawaii (Harlan, 1976). This isolated the *Mi* gene, which, following the advent of molecular techniques, was cloned during the 1990s and shown to mediate resistance against aphids and nematodes, the first example of a plant resistance gene active against two organisms of different phyla (Rossi, et al., 1998).

An increase in complexity of host – parasite interactions is often associated with an enhanced capacity of the parasites to regulate host genes in their favor (Sijmons, et al., 1994). Plant-parasitic nematodes had free-living ancestors and, through different adaptive strategies, evolved from fungus-feeding habits to sedentary endoparasitism. The essentially complete sequence of the *Caenorhabditis elegans* genome was published in 1998 (http://elegans.swmed.edu/) and the majority of genes identified to date in parasitic nematodes have homologues in *C. elegans* (e. g., 60% of *M. incognita* genes). According to McCarter et al. (2000), the most effective route to gene discovery for nematode genomes is the generation of expressed sequence tags (ESTs). These tags are single-pass reads from cDNA library clones. Thousands of ESTs, approximately 300 – 600 nucleotides in length, are already available for five species of plant-parasitic nematodes (McCarter, et al., 2000).

13.4 Plant Diseases Caused by Nematodes

Plant-parasitic nematodes can be categorized in a number of different ways in relation to their interactions with plants: (1) as named diseases; (2) by their feeding types or parasitic behavior; (3) by the visible indications of the nematode's presence as a disease-causing organism; (4) as pathogenic agents; (5) by their role in pathogenesis of interactions.

13.4.1 Named Plant Diseases Caused by Nematodes

Compared with other microorganisms, nematodes generally receive less recognition as disease-causing organisms, and the assignment of specific disease names to plant-nematode interactions is quite limited. Diseases are more often given a name when specific damage symptoms are produced, for example "ear cockle" of wheat or "red ring disease" of coconuts. Where the nematode causes non-specific damage symptoms (often a combination of stunted growth, chlorosis, wilting, and poor yield), disease names often have not been given, even with nematodes of major economic importance. For example, no actual disease name is given to the damage caused by the potato cyst nematodes, *Globodera* spp., although in Mexico it has the local name "liendrecilla". *Heterodera cajani,* on the other hand, produces "pearly-root disease" (Table 13.1).

Table 13.1 Named plant diseases caused by nematodes and associated organisms.

Disease	Main crops affected	Causal nematodes/organisms
Bloat	Onions	*Ditylenchus dipsaci*
Black head	Bananas	*Radopholus similis*
Cauliflower disease (of strawberries)	Strawberries	*Aphelenchoides ritzemabosi* + *Corynebacterium fascians*
Corchosis	Coffee	*Meloidogyne* spp., *Pratylenchus* spp. + *Fusarium* spp., *Verticillium* spp., *Phialophora* spp., *Gonytrichium* spp.
Crimp (of strawberries)	Strawberries	*Aphelenchoides fragariae*
Docking disorder	Sugar beet	*Trichodorus* spp. and *Longidorus* spp.
Dry rot	Yams	*Scutellonema bradys*
Ear cockle	Wheat	*Anguina tritici*
False root-knot, rosary-root, "rosario" or "jicamilla" disease	Beans, chilli pepper, potato, sugar beet, tomato	*Nacobbus aberrans*
Grapevine fanleaf	Grapes	*Xiphinema americanum* + Grapevine fanleaf virus (GFLV)
Kalahasti malady	Groundnut	*Tylenchorhynchus brevilineatus*
Mentek	Rice	*Hirschmanniella oryzae*
Mezquino de la papa	Potato	*Meloidogyne chitwoodi*
Miti-miti	Taro	*Hirschmanniella miticausa*
Molya	Wheat, barley	*Heterodera avenae*
Panama disease	Banana	*Radopholus similis* + *Fusarium oxysporum* f. sp. *cubensis*
Pearly-root (of pigeon pea)	Pigeon pea	*Heterodera cajani*
Pine wilt	Pine	*Bursaphelenchus xylophilus* (vectored by *Monochamus alternatus*)
Potato tuber rot	Potato	*Ditylenchus destructor*
Red ring	Coconut and other palms	*Rhadinaphlenchus cocophilus* (vectored by *Rhynchophorus palmarum*)

Continued

Disease	Main crops affected	Causal nematodes/organisms
Replant diseases	Various plantation crops	*Pratylenchus* spp.
Rice white tip	Rice	*Aphelenchoides besseyi*
Slow decline	Citrus	*Tylenchulus semipenetrans*
Spraing	Potato	*Paratrichodorus, Trichodorus* + tobra-virus (TRV)
Spreading decline	Citrus	*Radopholus similis*
Tulip root	Cereals	*Ditylenchus dipsaci*
Ufra disease	Rice	*Ditylenchus angustus*
Voltaic chlorosis	Groundnut	*Aphasmatylenchus straturatus*
Yellow disease	Groundnut	*Criconemella ornata*
Yellow dwarf	Soybean	*Heterodera glycines*
Yellow ear-rot ("Tundu" or "Tannan")	Wheat	*Anguina tritici* + *Clavibacter tritici*

13.4.2 Feeding Types and Parasitic Behavior

Most plant-parasitic nematodes are minute and vermiform, ranging in size from less than 0.2 mm (*Paratylenchus* spp.) to over 12 mm (*Paralongidorus* spp.). Among the anatomical structures and systems whose characters are important in revealing and identifying a nematode feeding type are the stoma and alimentary tract. Plant-parasitic nematodes possess a stylet whose characteristics can help to identify them as members of the orders Dorylaimida, Triplonchida, Tylenchida or Aphelenchida. There are two basic types of hollow stylet: stomatostyle (Tylenchida, Aphelenchida) and odontostyle (Dorylaimida), although solid stylets may occur in Trichodoridae (Triplonchida). Three basic types of esophagus are also recognized (Table 13.2). Plant-parasitic nematodes feed by inserting their stylet into plant cells and sucking out the contents. Digestion of cell contents may be partially extra-corporeal as salivary fluid can be exuded. Such fluids may also have a toxic or growth-modifying effect on the plant tissues.

Tylenchida and Aphelenchida contain most of the plant-parasitic species. Parasitic species also occur in the families Longidoridae (Dorylaimida) and Trichodoridae (Triplonchida) and these include the nematode species that act as vectors of plant viruses. Taxonomical classification may vary according to authors or when new evidence of relatedness is provided by research, but what is clear is that an "adaptive radiation" has occurred among plant-parasitic nematodes. This has allowed them to exploit the different habitats and niches

Table 13.2 Characteristics of stylet and esophagus of plant-parasitic nematodes.

Tylenchida	Aphelenchida	Dorylaimida
Stomatostyle	Stomatostyle	Odontostyle
Esophagus with three parts: corpus, isthmus, and posterior bulb. Three esophageal glands	Esophagus with three parts: corpus, isthmus, and posterior bulb. Three esophageal glands	Esophagus with two parts: corpus and postcorpus. Three to five esophageal glands
Outlet of dorsal esophageal gland in the procorpus, usually close to the knobs of the stylet	Outlet of dorsal esophageal gland in the metacorpus, anterior to the valve	Outlet of glands close to the body of the glands
Metacorpus width less than 75% of body width	Metacorpus width more than 75% of body width	Without metacorpus
Isthmus	Isthmus absent in Aphelenchoididae	No isthmus
Esophageal glands in a bulb or overlapping lobes	Esophageal glands in a bulb (Paraphelenchidae) or in an overlapping lobe	Postcorpus widened and elongated

provided by host plants and by interaction with other microorganisms, in both the soil and above-ground environments (see Bird and Bird, 2001). Different evolutionary strategies also encompass a wide range of migratory habits, ranging from ectoparasites to endoparasites (Table 13.3). Ectoparasites remain on the surface of the plant tissues and feed by inserting the stylet into cells that are within reach. Stylet length (short or moderately long) and strength determine penetration depth, according to the softness or hardness of root tissue. No specific host reaction is related to nematode feeding. Migratory endoparasites (all stages) can completely penetrate the plant tissues, retaining mobility and remaining largely vermiform as they move through and feed on the tissues. Nematodes of this type often induce a plant response, migrate between the soil and roots, and have moderately strong stylets. In sedentary endoparasites, the stylet is small and delicate; the immature female or juvenile nematodes enter the plant tissues where they develop a permanent feeding site, become immobile and swell and become obese. Semi-endoparasites only partially penetrate the roots, usually as juveniles or immature females. These nematodes become immobile at a fixed feeding site and the projecting posterior of the female body swells (e.g., *Tylenchulus*). Some nematodes behave as facultative ecto-endoparasites (*Hoplolaimus* and *Helicotylenchus*). They have a robust stylet and may act as ecto- or endoparasites in what may be a step towards leaving the soil environment and developing a greater dependence on the root environment. Endoparasites (whether migratory or sedentary) conduct

Table 13.3 Plant-parasitic nematodes classified by parasitic habit.

Habit	Genera and species
Ectoparasites	
Foliar	
Ectoparasites that feed mainly on epidermal plant cells of young leaves, stems, and flower primordia, often enclosed by other foliage. *N. B.* most of these species may also be found within tissues (i. e., endoparasitically)	*Anguina, Aphelenchoides, Ditylenchus*
Root	
1. Ectoparasites with short stylets feeding mainly on outer epidermal and cortical cells and root hairs	*Tylenchorhynchus, Trichodorus, Paratrichodorus, Helicotylenchus* (some spp.)
2. Ectoparasites with long stylets that can be inserted deep into root tissues, often at growing tip. (Some become relatively immobile for long periods)	*Belonolaimus, Cacopaurus, Criconema, Criconemella, Dolichodorus, Hemicriconemoides, Hemicycliophora, Longidorus, Paralongidorus, Paratylenchus, Xiphinema*
Migratory endoparasites	
Above ground	
Nematodes moving freely in tissues of stems, leaves, flower primordia or seeds	*Aphelenchoides, Bursaphelenchus, Ditylenchus, Rhadinaphelenchus*
Below ground	
All stages of nematodes present, moving freely within different tissues of roots, corms, bulbs, tubers, and seed (groundnuts)	*Aphasmatylenchus, Ditylenchus* (some spp.), *Helicotylenchus* (some spp.), *Hirschmanniella, Hoplolaimus, Pratylenchoides, Pratylenchus, Radopholus, Rotylenchoides, Rotylenchus* (some spp.), *Scutellonema*
Sedentary endoparasites (all below ground)	
Enlarged nematode bodies generally within tissues. Different life stages present; females become sedentary and obese. Enlargement of tissues (galling) can occur around nematodes	*Achlysiella, Globodera, Heterodera, Meloidodera, Meloidogyne, Nacobbus, Punctodera, Cactodera*
Sedentary semi-endoparasites (all below ground)	
Different stages of nematode invade and become partly embedded in root and sedentary; posterior part of mature female projects from root and becomes swollen	*Rotylenchulus, Sphaeronema, Trophotylenchulus, Tylenchulus*

most of their vital processes inside the roots (except during the stages of dispersion), their motility and feeding habits resulting in death of cells. Migration through roots characterizes the first level of obligate endoparasitism (*Pratylenchus, Radopholus*), the next stage being sessile endoparasitism. The development of females that induce nurse cells and galls allowed the sedentary habit to develop free of the necrotic reaction caused by the migratory habit. Sedentary endoparasites are the pinnacle of Tylenchida parasitism, the second stage juveniles (J2) and males being the only motile stages. Sedentary females may be transformed into a cyst containing J2 within eggs (*Heterodera, Globodera*), or the females may be non-cyst-forming (*Meloidogyne* spp.). A long co-evolution between parasites and their hosts tends to produce less virulent reactions than those resulting from new encounters. A necrotic cell reaction to an obligate parasite is regarded as a sign of an ill-balanced host-parasite relationship and, in the form of hypersensitivity, may provide a successful defense mechanism. The most complex host reactions, e. g., to *Meloidogyne* spp., are not necrotic but represent well adjusted, if ultimately harmful (when large numbers of nematodes invade), host-parasite relationships (Pitcher, 1965).

The range of cell responses to feeding by plant-parasitic nematodes provides some clues as to how the endoparasitic relationships evolved (Jones, 1981). Surface feeders (e. g., *Trichodorus*) rapidly kill the cells they feed from. Ectoparasites that cause root tip galls (e. g., *Xiphinema*) return repeatedly to the galls to feed but kill those cells fed from. Gall cells nevertheless become bi- or tri-nucleate and slightly enlarged. *Tylenchulus* induces and feeds from nurse cells for most of its life cycle without killing them. *Rotylenchulus* additionally induces limited wall breakdown to form a syncytium. *Heterodera, Nacobbus* and *Meloidogyne* induce larger volumes of cytoplasm on which they feed. Once nematodes can feed from host cells without killing them, the females enlarge enormously and become sedentary, with a secure food supply that allows increased egg production.

13.4.3 Indications of the Nematode's Presence as a Disease-causing Organism

Human needs for food and fiber are provided by only a few crops, and yield loss statistics are available only for major crops, not for those of limited or regional use. Such crops can be important staples, especially in developing countries. Thus, although plant-parasitic nematodes may be reported in relation to known diseases, their importance and potential threat remain ignored in many other cases. Nematode damage to crops can be classified under different schemes that allow, for practical purposes, unrelated taxa to be put together according to parasitic habit, affected crop, symptom expression, or major pathogenic effect. This simplifies information in, for example, diagnostics, where symptoms can be used as a rough guide to shorten a list of potential disease agents. Useful summaries of the range of symptoms caused by nematodes, accompanied by illustrations, can be found in Agrios (1997) and in Shurtleff and Averre (2000).

13.4.3.1 Disease Expression

The most universal symptom of nematode disease in a plant is a reduction in growth rate compared with that of a healthy plant, although light infestations may stimulate growth, usually of the roots (Pitcher, 1965). Nematodes may also stimulate plant callus growth in aseptic culture (Sandstedt and Schuster, 1963, 1966a, b). Study of a disease proceeds from initial recognition of symptoms and identification of the pathogenic agent to the means by which it is brought about.

Symptoms may vary according to nematode parasitic habits and host-parasite relationships, and other factors such as host age and physiological condition. Symptoms in roots can range from microscopic to extensive lesions, from shallow to deep necrotic areas, and can include gall formation, all of which can affect root architecture and some of which may eventually weaken the anchoring system severely (e.g., *Radopholus similis* on bananas). Affected plants are yellowed and stunted, being distributed in circular to oval areas of variable size in the field ("patches"), the longitudinal axis of the infested area typically being in the direction of cultivation of the field (Ayoub, 1980). Symptoms can be divided into above ground and below ground. It is usually necessary to examine roots and other plant tissues to establish a connection between damage symptoms and nematodes. To be certain of the association, nematodes have to be extracted from soil, roots or other material and identified microscopically (Shurtleff and Averre, 2000; Southey, 1986). Samples from plants within patches of poor growth should be compared with samples taken from plants outside the patches.

13.4.3.2 Above-ground Symptoms

Symptoms are essentially the same as those caused by any root damage that interferes with the physical support and water and nutrient absorption systems. Thus, they are often similar to mineral deficiencies, inadequate or excessive water supply, and generally poor soils. Symptoms are usually more pronounced if the plants are already affected by adverse growing conditions or are being attacked by other pathogens. Plants growing under highly favorable conditions may be heavily attacked by nematodes but show few, if any, symptoms above ground. Under such circumstances, nematodes reproduce better and can represent a severe and hidden threat to succeeding crops. The more important above-ground symptoms are:

1. Stunting, reduction in amount of foliage and progressive death (die-back).
2. Poor yield.
3. Yellowing (chlorosis) of leaves.
4. Wilting.
5. Early senescence.
6. Dropping of fruits and fruit malformation.
7. Leaves with dark green spots, angular or cuneiform in shape, with interveinal discoloration and necrosis (e. g., *Aphelenchoides ritzemabosi* on chrysanthemum leaves).
8. Galls in stems, leaves and seeds of cereals and grasses (*Anguina* spp.).
9. Twisting and white tips of leaves in rice (*Aphelenchoides besseyi*).
10. Twisting of leaves and raised yellow lesions on stems and leaves (*Ditylenchus dipsaci* on narcissi and onions).
11. Twisted panicles and empty grains (*Ditylenchus angustus* on rice).
12. Yellowing and collapse of palm leaves followed by rapid death; a red necrosis in the vascular bundles of the stem forming a red ring (*Rhadinaphelenchus cocophilus* in coconut and oil palm).
13. Yellowing and rapid death of pine trees (*Bursaphelenchus xylophilus*).
14. Distorted apical growth and "crimping" of leaves and inflorescence (*Aphelenchoides besseyi* and *Aphelenchoides fragariae* on strawberry).

13.4.3.3 Below-ground Symptoms

Below-ground symptoms are reflected above ground as an unspecified decline of the plants due to reduced absorption of water and nutrients by the secondary roots. Some species of root-knot nematodes induce qualitative and quantitative changes in root exudates, such as a reduced concentration of amino acids, increased concentration of sucrose, and a disappearance of glucose, threonine, serine, histidine and citric acid (Wang and Bergeson, 1974).

Water movement also is affected by plant-parasitic nematodes, thereby influencing the water potential of leaves, stomatal conductivity, transpiration and root conductivity. *Heterodera avenae, Tylenchorhynchus dubius* and *Heterodera trifolii* have all been shown to reduce total water intake (Seinhorst, 1981). Root conductivity was reduced in bean plants infected with *M. hapla* (Wilcox and Loria, 1986; Wilcox-Lee and Loria, 1987) and in potato plants infected by *Pratylenchus penetrans* (Kotcon and Loria, 1986). *Globodera rostochiensis* caused mineral deficiency (N, P, K, and Mg) by affecting ion absorption and reduced water absorption due to poor root development (Evans, et al., 1977; Fatemy and Evans, 1986a, b). Calcium concentrations may increase in response to reduced potassium uptake and dehydration (Been and Schomaker, 1986; Trudgill, 1980), or through disruption of the endodermis by the nematodes (Price and Sanderson, 1984).

In leguminous crops, nitrogen uptake and nutrition also can be affected, but it has not always been possible to relate yield loss to the reduced nodulation caused by nematodes. Soybean lightly infected by *Heterodera glycines* can compensate for reduced nodule production by increasing nodulation in uninfected parts of the root system. However, enhanced supply of nitrate does not compensate for the effects of reduced nodulation, thus suggesting that complex mechanisms are involved in nitrogen assimilation after parasite attack (Mateille, 1994). Nematodes can parasitize the nodules without damaging their structural integrity, but bacterioids do not develop and eventually the nodules atrophy (Barker and Hussey, 1976; Ko, et al., 1984; Mateille, 1994). Although nematodes are generally detrimental to the symbiotic interaction, *M. hapla* and *P. penetrans* can stimulate nodulation (Hussey and Barker, 1976).

Below-ground symptoms caused by nematodes include:

1. Root system reduced in extent, especially the secondary feeder roots.
2. Abnormal root development:
    a. Excessive branching of secondary roots (e. g., *Meloidogyne hapla*, *Pratylenchus* spp., *Nacobbus aberrans*).
    b. Overall root galling (*Meloidogyne* spp., *N. aberrans*).
    c. Roots with longitudinal necrotic areas (*Pratylenchus* spp., *Radopholus* spp., *Hirschmanniella* spp.).
    d. Swollen, hooked root tip galls (*Subanguina* spp., *Xiphinema* spp, *Meloidogyne graminicola*).
    e. Roots ending in rounded galls (*Longidorus* spp., *Hemicycliophora* spp.).
    f. Localized proliferation of lateral roots (some *Meloidogyne* spp., *Heterodera* spp.).
    g. Dirty roots caused by accumulation of soil particles and root debris on the roots (*Tylenchulus semipenetrans*).
    h. Stubby roots, suppression of root growth (*Trichodorus* spp., *Paratrichodorus* spp.).
3. Roots with white, yellow or dark brown colored cysts (*Cactodera* spp., *Heterodera* spp., *Globodera* spp., *Punctodera* spp.).
4. Internal rotting of tubers, bulbs and corms (*Ditylenchus* spp., *Pratylenchus* spp., *Scutellonema* spp.).
5. Galled or warty tubers of potato and yam (*Meloidogyne* spp.).

6. Surface cracking of tubers such as potato, sweet potato, yam (*Ditylenchus destructor, Scutellonema bradys, Pratylenchus coffeae, Meloidogyne* spp.).
7. Lesions on groundnut pods (*Pratylenchus* spp., *Criconemella* spp.).
8. Brown and shrivelled groundnut seeds (*Aphelenchoides arachidis, Ditylenchus africanus*).
9. Fewer symbiotic nodules on leguminous hosts (*H. glycines, H. goettingiana*).

13.4.3.4 Seed/Planting Material: Symptoms of Nematode Damage

The spread of nematodes in planting material can lead to serious losses in the mature crop, or in subsequent crops if the nematodes build up to damaging levels only slowly. Many nematodes are spread by this means, mainly in vegetatively propagated material (tubers, corms, bulbs) or seedlings and rootstocks. Very few nematodes are actually disseminated in true seed (Table 13.4).

Table 13.4 Detection of nematode symptoms in seed or planting material (adapted from Bridge, 1987b).

Crop(s)	Nematode(s)	Symptoms of damage
Seeds		
Wheat (*Triticum aestivum*)	*Anguina tritici*	Black, misshapen seed galls
Beans (*Vicia faba*)	*Ditylenchus dipsaci*	Shrivelled and discolored seeds, sometimes with "nematode wool" attached
Onion (*Allium cepa*)	*Ditylenchus dipsaci*	No observable symptoms; nematodes emerge from soaked seed examined microscopically
Rice (*Oryza sativa*)	*Aphelenchoides besseyi*	No observable symptoms; nematodes emerge from soaked seed examined microscopically
Groundnut (*Arachis hypogaea*)	*Aphelenchoides arachidis*, *Ditylenchus africanus*	Shrivelled seeds, dark brown testas
Tubers		
Yam (*Dioscorea* spp.)	*Scutellonema bradys*, *Pratylenchus coffeae*, *Radopholus similis*	Internal, dark brown "dry rot" in periphery of tuber, observed when tuber is cut or when epidermis is scraped away. Whole tubers spongy to touch with surface cracks
	Meloidogyne spp.	Knobbly tubers with internal necrotic spots

Continued

Crop(s)	Nematode(s)	Symptoms of damage
Potato (*Solanum tuberosum*)	*Pratylenchus* spp.	Uneven tubers pitted with dark, dry necrotic lesions
	Meloidogyne spp.	Watery tubers, swellings and lesions on surface
	Globodera spp.	Cysts on tuber or in adhering soil (difficult to see)
	Nacobbus aberrans	No observable symptoms. Lenticels with spongy tissue
Bulbs		
Onion, garlic (*Allium* spp.), Narcissi	*Ditylenchus dipsaci*	Internal necrosis or rot of sets and cloves, sometimes with "nematode wool" attached
Rhizomes		
Ginger (*Zingiber officinale*)	*Meloidogyne* spp.	Light brown, watery, necrotic areas in outer rhizome, often in surface folds
	Radopholus similis	Small, sunken water lesions on surface
Turmeric (*Curcuma domestica*)	*Radopholus similis*	Shallow, water-soaked, brownish lesions and rotting
Corms		
Banana, plantain, abaca (*Musa* spp.), enset (*Ensete* spp.)	*Radopholus similis, Pratylenchus coffeae, P. goodeyi*	Purple to dark brown necrotic lesions throughout cortex of attached roots, and similar lesions in outer tissues of corm
	Helicotylenchus multicinctus	Purple to dark brown necrotic lesions in periphery of cortex of attached roots
Taro (coco yam) (*Colocasia esculenta*)	*Hirschmanniella miticausa*	Internal red necrotic streaks and peripheral rot
Seedlings/transplants (the following symptoms may be observed after uprooting)		
Vegetables, rice, coffee, etc.	*Meloidogyne* spp.	Swelling or galls on roots
	Nacobbus aberrans	Discrete swellings or galls on roots
	Pratylenchus spp., *Hirschmanniella* spp., *Radopholus* spp.	Yellow or brown lesions on root surface
	Heterodera spp.	Spidery or distorted root growth sometimes with white specks (females) on roots of older seedlings

Continued

Crop(s)	Nematode(s)	Symptoms of damage
Rootstocks		
Grapevine, peach, etc.	*Meloidogyne* spp.	Galls on roots, sometimes accompanied by rotting
	Pratylenchus spp.	Brown lesions on root surface
Citrus	*Tylenchulus semipenetrans*	Irregular, "dirty" roots with soil adhering to surface

13.4.4 Nematodes as Pathogens

In comparatively few cases have plant-nematode relationships been studied free from the presence of all other organisms, work by Byars (1914) and Tyler (1933) being amongst the earliest examples. As Mountain (1959, 1960b) has pointed out, only in this way is it possible to be certain that a given plant reaction is wholly due to parasitism by a nematode. The development of cell and tissue culture techniques during the last century facilitated the study *in vitro* of the onset and development of alterations and symptoms related to plant diseases caused by nematodes. However, the challenge of culturing nematodes in axenic conditions remains, leading Pitcher (1965) to state that "the bulk of our present knowledge of plant-nematode relationships thus rests upon somewhat insecure foundations".

Pitcher (1963) recommended that due consideration be given to Koch's postulates in studies of nematode-induced disease, ensuring that the same standards of purity of inocula applied to other pathogenic microorganisms are applied to the nematode inoculum. "Nematodes added in the form of infested roots or as an aqueous suspension with an uncontrolled surface microflora are not simple nematodes but nematodes plus an unknown number of unknown factors" (Pitcher, 1963). No problem caused by nematodes has been fully reproduced axenically, but there is convincing evidence from other observations that many nematode species damage plants. However, important conclusions, which confirm the pathogenic role of some plant-parasitic nematodes, have been made from gnotobiotic root culture experiments (Zuckerman, 1969; Zuckerman and Brzeski, 1965). According to Oostenbrink (1969), the concept of pathogen and the relevance of Koch's postulates are disputable in diagnostic work with nematodes. Some of the practical problems include the provision of optimal conditions in an artificial environment, the use of appropriate inoculum dosage, and quantitative rather than qualitative symptom evaluation. The formal application of Koch's postulates may lead to erroneous conclusions about the causal role of the nematode. Negative conclusions have been drawn because of inadequate sampling or extraction techniques, because a damaging species occurred deeper than the level of soil sampled, or because the wrong species was selected from a mixture of nematodes. Inoculation experiments may fail for a number of reasons, such as wrong soil type, unsuitable substratum or temperature and humidity conditions. The inoculum is often insufficiently large to compensate for such factors as the use of small containers, the aggregation of nematodes along roots in natural soil, and the heavy mortality of inoculated nematodes and nematode eggs in so-called quantitative extraction. The use of Koch's postulates is regarded more often as a "finishing touch than a trustworthy guide" in the first stages of research (Oostenbrink, 1969), but this can be changed with increasing knowledge of nematode biology, better-planned experiments, and more precise methodologies and techniques.

Sterile inoculation is not the only scientific method used in determining the role of nematodes in plant disease. Oostenbrink (1969) gave general guidelines to work on an unknown growth disturbance in searching for a causal relationship. They include: (1) observation of feeding *in situ* of immobile stages and aggregation of the nematodes in or around the affected plant organ; (2) presence of specific local symptoms such as swellings, galls, lesions, stem and root malformations; (3) association of non-specific symptoms of poor growth in the field with high densities of one or more species of suspected plant-parasitic nematodes; (4) significant correlation between severity of damage and nematode densities; (5) complete cure by the use of nematicides or other chemical compounds; (6) reproduction of local disease symptoms in another soil where a particular plant was never grown before but where the nematode population has built up on other plants; (7) correlation of severity of symptoms with geometrically increasing non-sterile nematode inoculum on whole plants grown with nematode-free soil; (8) the use of aseptic monospecific inoculum of nematodes and of associated organisms, or various mixtures of inoculum, on whole plants or plant tissue grown aseptically, may help to show if the nematode itself acts as pathogen, incitant, concerter, vector, modifier, deterrent, harmless feeder, antagonist of a pathogen or growth stimulator of the plant.

13.4.5 The Role of Nematodes in Pathogenesis of Interactions

Plant-parasitic nematodes are often regarded solely as pathogens, capable of producing a single, recognizable disease. Frequently, such a concept is valid enough but nematodes also may facilitate the entry and establishment of other plant pathogens. Often the nematode partner is affected, either to its benefit or disadvantage, leading to the use of the terms "interrelationships" and "interactions" with bacteria, fungi and viruses. Pitcher (1965) proposed several mechanisms of interaction, which represent yet another classification of the role of nematodes in soil-borne diseases. All plant-parasitic nematodes wound plants to some degree, either by a simple micro-puncture or by rupturing or separating cells. They may thereby either introduce a pathogen on or within their bodies or aid the entry of a pathogen already present on the plant cell surface. Pathogens "capable of self-establishing" once in contact with the hosts use nematodes to carry them to fresh infection sites from infected tissue on the same plant or from the soil and, less frequently, from plant to plant. Occasionally, systemic effects whereby nematodes feeding on plants aid fungal penetration at remote locations have been reported (Corbett and Hide, 1971). Virus vectors move from plant to plant, being the prime example of "nematodes as vectors of pathogens incapable of self-establishment unless introduced below the epidermis", and fulfilling much the same role as arthropod vectors of viruses above ground. Root infection cannot take place without the aid of the nematode and, once established within the root, the virus multiplies, usually becoming systemic throughout the plant. Some nematodes may act as "providers of necrotic infection courts" by modifying the substrate for the weaker, unspecialized pathogens, and providing them with a "food base", which reinforces their invasive potential. In this role, nematodes are "modifiers of substrates". This process may be simple or highly sophisticated. In destroying chemicals antagonistic to pathogens, nematodes may ultimately become "breakers of disease resistance", impairing the host defense reactions by which healthy plants would otherwise repel pathogens. Some nematodes are considered to be "deterrents of plant disease", by simply "grazing" on fungal pathogens as in the case of *Aphelenchus avenae* (Mankau and Mankau, 1963) but, in other situations, physiological changes may be involved.

13.4.5.1 Nematodes as Vectors of Plant Viruses

Nematodes are vectors of several harmful plant viruses. Under very favorable circumstances a nematode can acquire a virus in feeding periods of an hour or less and transmit equally rapidly. The viruses are widely distributed, although few are recorded in the tropics. The nematode vector may remain infective for several months, the period being longer in the genera *Xiphinema* and *Trichodorus* than in *Longidorus*. Infectivity does not survive a molt and the virus is not transmitted *via* the nematode egg. Two major virus groups are carried only by nematodes in the orders Dorylaimida and Triplonchida:

1. Nematode-transmitted polyhedral viruses (Nepoviruses). Species of *Xiphinema* and *Longidorus* transmit the following viruses:

Xiphinema - Tomato ringspot (ToRSV), Cherry rasp leaf (CRLV), Peach rosette mosaic (PRMV), Grapevine yellow vein, Arabis mosaic (ArMV), Strawberry latent ringspot (SLRSV), Grapevine fanleaf (GFLV).

Longidorus - Artichoke Italian latent (AILV), Tomato blackring (TBRV), Peach Rosette mosaic (PRMV), Raspberry ringspot (RRSV), Mulberry ringspot (MRSV).

2. Nematode-transmitted tubular viruses (Tobraviruses). Species of *Trichodorus* and *Paratrichodorus* are the vectors of Tobacco rattle (TRV), Pea early-browning (PEBV), Pepper ringspot (PRV).

13.4.5.2 Interactions with Fungi and Bacteria

Disease complexes between nematodes and wilt-inducing and root-rot fungi are well documented for several fungal and nematode genera. *Fusarium oxysporum* and other *Fusarium* spp., *Verticillium* spp., *Pythium* spp., and *Cylindrocarpon* spp. are examples of fungi that interact with nematodes, mainly *Meloidogyne* spp., *Pratylenchus* spp., and *Rotylenchulus reniformis.* There is frequently a synergistic effect on plant damage, although the effects of the interacting organisms may be simply additive.

Interactions are also known to occur between the disease-causing bacteria *Clavibacter* spp., *Pseudomonas* spp., and *Agrobacterium* spp., and species of the nematode genera *Meloidogyne, Pratylenchus, Anguina* and *Ditylenchus.* Two well-known examples of nematode-bacteria interactions are that of *Meloidogyne* spp. and *Pseudomonas* (*Ralstonia*) *solanacearum* causing bacterial wilt of many crops (tobacco, potato, eggplant) and that of the ear cockle nematode, *Anguina tritici,* and *Clavibacter tritici* (Carlson and Vidaver) Davis, Gillaspie, Vidaver, and Harris causing a disease in wheat referred to as "yellow ear rot" or "tundu" in India (Gupta and Swarup, 1972).

13.5 Root Gall Forming Species

Endo-and semi-endoparasitic nematodes of Tylenchida (e. g., Heteroderidae, Meloidogynidae, Pratylenchidae, Hoplolaimidae, Tylenchulidae) and some Dorylaimida (*Longidorus, Xiphinema*) have evolved the ability to induce changes in host cells to form feeding sites and sequester nutrients from the vascular system of the host plants. However, the host also requires nutrients, and delicate strategies have been developed by the nematodes in order to maintain a metabolic balance as, if the feeding cells are damaged, the nematodes (i. e., the sessile endoparasites) may die. The introduction of substances by the nematodes, combined with the creation of a source-sink arrangement, may cause cell modifications that tip the host physiological balance in a few critical pathways (Jones, 1981). Host cell changes include enlarged nuclei and nucleoli, increased synthesis of cytoplasm and cell organelles, reduction or loss of the vacuole, stimulation of intermediary metabolism, RNA and DNA synthesis, and the enhancement of free amino acid and solute levels. Thus, large volumes of active cytoplasm are accessible to the adult nematodes. Transfer cell wall ingrowths are present when altered cells are relatively isolated from neighboring cells (i. e., few plasmodesmatal connections), but absent when plasmodesmata are numerous. The suggestion that *Meloidogyne* giant cells possess a hydrogen ion efflux pump, solutes being taken up *via* a proton-cotransport mechanism at the plasmalemma (Jones, 1981), has been confirmed and such a pump is present between the apoplast and the giant cells of *M. incognita* (Dorhout, et al., 1992). The origin of the solutes is the vascular tissues, except for *Tylenchulus,* which apparently obtains nutrients from symplastic cortical cells (Jones, 1981).

Hypertrophy (an increase in cell size) and hyperplasia (an increase in cell numbers) are two of the most striking plant responses to nematode attack that may or may not occur together in the feeding sites associated with sedentary nematodes. One or both are involved in the production of galls by different and even unrelated groups of plant-parasitic nematodes and hosts. Hyperplasia is a feature of parasitism by *Meloidogyne* but not *Heterodera.* Hyperplastic tissue constitutes the main bulk of a root-knot gall and is also found in the hooked root tips induced by *Xiphinema* spp., and root and flower galls induced by *Ditylenchus* and *Anguina* (Pitcher, 1965).

Plant growth compounds are involved in gall production, but they may originate from the nematodes. The compound is generally an auxin or a similar molecule, and the kind of auxin in the gall varies according to the plant and the nematode (Orion, 1973; Viglierchio and Yu, 1968). It may even differ between species of the same genus of nematode. Increased growth regulator levels may result from increased synthesis or decreased degradation, and no experimental distinction has been made between development and function of modified cells (giant cells or syncytia) and gall development with respect to altered regulator levels (Mateille, 1994). Cytokinins have been found in adult nematodes, egg masses, and hatched juveniles (Dimalla and van Staden, 1977), and higher levels of endogenous cytokinins have been reported in susceptible tomato roots (van Staden and Dimalla, 1977). Symptoms of fasciation, comparable to those caused by an IAA, kinetin or gibberellin hormonal imbalance, are reported on *Lilium henry* infected by *D. dipsaci* (Stumm-Tegethoff, 1986).

The induction and development of feeding structures probably result from interactions between effects of nematode secretions and their regulation of gene expression, but events linking these two phenomena have yet to be described. Common features must be present in the process of induction of feeding structures in different hosts. *Meloidogyne incognita* probably interacts with fundamental aspects of the plant cell cycle as it can induce similar giant cells in thousands of host species (Sijmons, et al., 1994).

13.5.1 Root-knot Nematodes (Overall Galling and Swelling of Roots)

The species of the genus *Meloidogyne* Goeldi 1887, commonly known as "root-knot nematodes", are obligate endoparasites of great economic importance, being among the major limiting factors in the production of field and plantation crops, predominantly in the tropics but also in Europe and North America (Jepson, 1987; Siddiqi, 2000). The nature and morphology of species of *Meloidogyne* is such that differences between species are generally very small, with certain features being influenced by the host and environment (Jepson, 1987). Identification is difficult and no single morphological approach or single life-history stage feature can be used to distinguish all species, although different approaches have been tested to resolve identification (Barker, et al., 1985; Jepson, 1987), including molecular techniques. Species identification is essential, particularly where more than one crop is grown and *Meloidogyne* control is planned using rotation of susceptible and resistant crops. Species identification is not in itself sufficient when *M. incognita* and/or *M. arenaria* are present because these species exist as races, which have different parasitic abilities. Because of their wide distribution and broad host range, the species of major economic importance are: *M. incognita, sensu lato, M. javanica, M. arenaria* and *M. hapla.* Other species that are economically important, but with more restricted distributions and host ranges, are *M. graminicola, M. naasi, M. exigua, M. acronea, M. chitwoodi, M. artiellia, M. decalineata, M. africana, M. coffeicola, M. oryzae,* and *M. thamesi* (Jepson, 1987).

13.5.1.1 Biology

The infective stage is the J2, which locates and penetrates a root to establish a feeding site, usually within the pericycle and vascular tissues. A gall is formed due to hypertrophy and hyperplasia of the root cells. Later, the J2 becomes sausage-shaped and undergoes three further molts within the cuticle of the second molt. The third (J3) and fourth (J4) stage juveniles do not have a stylet and do not feed. The stylet reappears after the fourth molt. The female feeds, becoming spheroidal, and starts laying eggs in a gelatinous matrix secreted by the rectal glands. The gelatinous matrix has lytic properties that produce a cavity inside the galls, housing and protecting many of the eggs from the external environment (Orion, et al., 1987). Eggs hatch on exposure to moisture in warm conditions. The vermiform male does not feed and females may be parthenogenetic (mitotic or meiotic). During adverse conditions most of the developing juveniles become males in many species (Siddiqi, 2000). Duration of the life cycle varies according to species and temperature conditions.

13.5.1.2 Symptomatology and Pathogenesis

Above-ground symptoms are very general and non-specific, being similar to those produced by mineral deficiency. Growth is reduced or even terminated, resulting in stunted plants, premature leaf abscission, wilting, early senescence, and reduced fruit production and yield loss. These symptoms are due to root damage and interference with synthesis or translocation of the host growth hormones, including cytokinins and gibberellins, which cause a decrease in photosynthesis. Below-ground symptoms include galls, which are initiated and produced in response to growth regulators, proteins and glycoproteins introduced into the host from the subventral esophageal glands of the feeding J2 (Jepson, 1987). Syncytia formed in the vascular system are multinucleate, being formed by repeated mitosis of a single nucleus within the same cell, the cells having dense cytoplasm and highly invaginated cell walls (Huang and Maggenti, 1969; Jepson, 1987; Jones and Payne, 1978). There is now good evidence that these cell wall modifications are due to endoglucanase enzymes of plant, rather than nematode, origin (Goellner, et al., 2001). Vascular continuity is maintained by differentiated xylem and sieve elements, and wall ingrowths facilitate solute transport from surrounding cells to the giant cells. Final giant cell dimensions may reach 600 μm in length by 200 μm in diameter. During cell enlargement, the cytoplasmic contents increase greatly relative to the cell vacuole and cell appearance is similar to that of actively growing meristematic cells. Nuclei become amoeboid and nuclear-cytoplasmic exchange increases *via* numerous nuclear pores. Golgi bodies and mitochondria are numerous, the distribution of endoplasmic reticulum is relatively sparse, and the ground cytoplasm has many ribosomes and polysomes, implying that much of the protein synthesis is directed to internal metabolism rather than export (Jones, 1981). The syncytium is highly specialized, resembling transfer cells, and its induction and maintenance depend on continuous stimulation from the nematode (Bird, 1974).

Morphological changes in the root depend on the host and nematode species involved, and include the amount, extent, and type of galling and, in selected cases, the suppression of secondary roots and development of fine lateral roots around the infected area. Difficulties in detection may arise when dealing with species that produce small galls, the general assumption being that root-knot nematodes produce obvious galls. Also, confusion can be made with legume nodules, although these may be rubbed off the root whereas galls produced by *Meloidogyne* cannot be so dislodged (Jepson, 1987). Few species (e. g., *M. incognita, sensu lato, M. javanica, M. arenaria,* and *M. hapla*) produce very large galls. Most other species produce small, discrete galls (e. g., *M. graminis* and *M. oryzae*) and some produce no galling at all (*M. aquatilis*). Galls of *M. naasi* are characteristically horseshoe-shaped or spiral (Jepson, 1987). Rice root tips infected by *M. graminicola* or *M. oryzae* produce a hooked root tip swelling (Bridge, et al., 1990). The extent of galling is related to species, the position of the female within the gall and mode of reproduction. Gall size within a single species may depend on the plant host involved. Amphimictic species have females predominantly external or only partially embedded in the gall, to allow easy access for males (e. g., *M. acronea*), whereas those that are facultatively parthenogenetic have females partially or completely embedded (e. g., *M. graminicola*). In obligatory parthenogenetic species, the females are completely embedded within large galls (e. g., *M. incognita, M. javanica,* and *M. arenaria*), and virtually inaccessible to males.

13.5.1.3 Control

There are few specific control measures but many strategies can be used to reduce damage, although there are no practical methods of completely eliminating most species once they have become established in soil. Usually, a combination of two or more strategies (i. e., rotation, fallow, resistant varieties, removal of infested roots, etc.) represents the practical approach to managing root-knot nematodes, chemicals being used only on a limited number of high-value crops. Contaminated soil and infested seedlings are the most common means of transferring root-knot nematodes to new fields. If seedlings can be kept free of nematodes by raising them in clean seedbeds, considerable yield advantage will ensue even when planted into infested fields, because early damage is a very important factor limiting yield (Bridge, 1987a).

13.5.1.4 Differential Host Test

Care must be taken when selecting crops and varieties for rotation due to the existence of races within certain species. The differential host test, developed by the International Meloidogyne Project (IMP), is well documented (Sasser and Carter, 1985). It is designed for use with only four species (*M. arenaria, M. hapla, M. incognita, sensu lato,* and *M. javanica*) to give a preliminary identification based on their usual response to cultivars of six hosts. The hosts used are tobacco (*Nicotiana tabacum* L. cv. NC95), cotton (*Gossypium hirsutum* L. cv. Deltapine 16), pepper (*Capsicum frutescens* L. cv. California Wonder), watermelon (*Citrullus vulgaris* Schrad. cv. Charleston Gray), peanut (*Arachis hypogea* L. cv. Florunner), and tomato (*Lycopersicon esculentum* Mill. cv. Rutgers). However, the test has well-known limitations and, as races cannot be distinguished morphologically, molecular diagnostic methods are under development.

13.5.1.5 Interactions

Meloidogyne spp. have been recorded in many disease associations, which makes it the most important incitant genus. *Fusarium* spp. are the most frequent partners, although each pathogen can cause disease without the other and both can interact with other genera. There appear to be features in the biology of both genera, possibly related to their common parasitism of vascular tissues and features of their biochemistry, that may make them suitable for the complex interactions involved (Pitcher, 1965).

13.5.2 False Root-knot Nematode (Discrete Overall Root Galling)

The genus *Nacobbus* is endemic to the Americas. The species of economic importance is *N. aberrans*, a polyphagous and important pest in sugar beet, beans, chilli pepper, potato, and tomato (Canto-Saenz, 1992; Costilla, 1985; Inserra, et al., 1984). Yield losses vary according to crop but can reach 60% in tomato and 80% in potato (Ramos, et al., 1998). The duration of the life cycle is strongly influenced by temperature (Inserra, et al., 1983; Prasad and Webster, 1967). All stages of development are migratory endoparasites but the adult female becomes a sedentary endoparasite and deposits eggs in a gelatinous matrix (Clark, 1967). The biology of *N. aberrans* shares some similarities with *Meloidogyne* spp. (Souza and Baldwin, 1998). Galls are induced in the root system and spindle-shaped syncytia (2 – 3 mm in length) are formed within the stele by wall digestion at pit fields and fusion by neighboring protoplasts. Mitosis accompanies syncytium development, and cells divide before being incorporated in the syncytium. Increases in cytoplasm content and secondary vacuolation accompany cell expansion. Cell nuclei are enlarged and mitochondria and Golgi bodies are numerous. Polysaccharide deposition thickens cell walls and wall fragments. The stelar organization is badly disrupted, vascular continuity being maintained by groups of xylem and phloem cells. Numerous gall cells differentiate into wound-type sieve elements. Starch granules are found in syncytia and gall cells, probably due to increased sucrose off-loading from the sieve elements. Numerous plasmodesmata are found in pit fields between sieve elements and syncytial cells (Jones, 1981).

Due to its broad host range and the existence of races, *N. aberrans* populations have to be managed by integrated crop and pest management programs that include three-year rotation schemes, trap crops, cultural practices, and fertilization or composting (Franco, et al., 1992). Nematicides have been used to reduce soil populations and to protect tubers. Removal of root debris after harvest is especially important as they are an important source of infective stages.

13.5.3 Species that Cause Galling or Swelling of Root Tips Only

According to Fortuner and Luc (1987), the development of elongated, rigid stylets has been a recurrent characteristic within obligate ectoparasites belonging to unrelated groups (e. g., *Xiphinema, Longidorus*, and *Criconemella*). Similarly, the ability to induce galls in response to nematode feeding activity occurs in phylogenetically unrelated groups, although the physiological processes involved may differ or be characteristic of the host or nematode species.

The symptoms caused by ectoparasitic nematodes such as *Xiphinema* spp. include distinctive discoloration of roots, often a blackened appearance. A marked reduction in root system size is due to destruction of feeder roots, the remaining roots bearing numerous lesions that may lead to disintegration and decay of the cortical tissue at various points (Cohn, 1975). Some *Xiphinema* species (e. g., *X. bakeri, X. diversicaudatum*, and *X. index*) feed at root tips, others (*X. americanum* and *X. brevicollum*) rarely do so. Some young roots may have slightly swollen and curved tips (e. g., when attacked by *X. americanum, X. chambersi*, and *X. brevicollum*), whilst other species (e. g., *X. diversicaudatum* on rose and other hosts) cause prominent terminal or subterminal swellings on roots, which makes the tip curl ("curly tip" or "fish-hook tip"). The size of the swellings may differ according to root age and plant species and may even be lacking. The galls contain necrotic cells and enlarged multinucleate cells produced from mitosis without cytokinesis. Root pathology is similar for *Longidorus* and *Xiphinema*, with darkening of tissues, cortical hyperplasia, lateral root proliferation and tip galling (Hunt, 1993). Both *Xiphinema* and *Longidorus* are capable of inserting their stylets to a considerable depth into the plant tissue, depending on root diameter. The stylet may reach the vascular tissues in young thin roots, whereas in older, thicker roots it is inserted into the cortical parenchyma layers. *Longidorus* invariably feeds at root tips that transform into terminal galls (Cohn, 1975; Sijmons, et al., 1994). *Paralongidorus australis* causes brown necrotic tips (sometimes hooked or curled) on primary roots of rice; secondary roots are shorter, often having a forked appearance (Bridge, et al., 1990). *Paralongidorus maximus* causes galling, distortion and stunting of the roots with necrosis and secondary invasion by microbial pathogens (see Hunt, 1993). The histopathology of roots infected by *Xiphinema* and *Longidorus* has not been studied extensively. However, cellular and tissue alterations due to feeding of *Xiphinema* spp. include groups of dark suberized cells within the cortical parenchyma, an underlying phellogen layer in the cortex, reduced size and activity of apical meristems, and a marked hyperplastic response of cortical cells (Cohn, 1975). A multinucleate condition of enlarged cortical cells also has been reported (Cohn, 1975). Substances discharged during feeding by *Longidorus* and *Xiphinema* may be responsible for triggering gall formation and other biochemical alterations, such as increased levels of amino-acids, carbohydrates, nucleic acids, phenols, and peroxidase activity (Cohn, 1975). Secretions from the dorsal esophageal glands of *Xiphinema* liquefy the cytoplasm and nucleoplasm of uninucleate cells before food withdrawal. Neighboring cells become multinucleate and then show progressive but prolonged degradation (Sijmons, et al., 1994). *Xiphinema* spp. and *Longidorus* spp. induce DNA and RNA multiplication, and binucleate cells (Griffiths and Robertson, 1984) but, unlike the responses to feeding by *Xiphinema*, the cells on which *Longidorus* has fed do not respond by undergoing mitosis without cytokinesis (Sijmons, et al., 1994).

Hemicycliophora arenaria is an ectoparasite found in large numbers in the citrus rhizosphere and feeds at root tips, causing hyperplasia and the development of spherical galls (Duncan and Cohn, 1990).

13.6 Species that Cause Gall and Lesion Formation on Stems, Leaves, and Seeds

Galls and lesions are produced on aerial plant parts by *Anguina, Aphelenchoides, Ditylenchus, Bursaphelenchus,* and *Rhadinaphelenchus*. Subterranean seeds may be affected by *Aphelenchoides* and *Ditylenchus* but also by species such as *Pratylenchus brachyurus* and *Criconemella ornata*.

Aphelenchids are predominantly free-living and mycetophagous nematodes in habit. The most abundant and widely spread genera are *Aphelenchus* and *Aphelenchoides*. Most species of *Aphelenchoides* are mycetophagous but a few parasitize higher plants. *Aphelenchoides arachidis* reproduces readily on fungi and is clearly a facultative phytoparasite, but others such as *A. ritzemabosi* usually have been regarded as obligate plant parasites. However, no injection of glandular secretions occurs in *Aphelenchus avenae* or *Aphelenchoides bicaudatus* before they ingest material from fungal hyphae (Hunt, 1993).

Aphelenchoides arachidis (the testa nematode) is a facultative endoparasite of the seed, testa, pod shells, roots, and hypocotyl of groundnuts but it may feed ectoparasitically on roots. The pods and seeds of groundnuts are produced underground and invasion of *A. arachidis* occurs from the soil. Heavily infested (freshly removed) seeds are light brown in color, with dark vascular strands within the translucent testas. Dried seeds have wrinkled testas and are darker brown. Nematodes are found mainly in the sub-epidermal parenchymatous layer, and around the tracheids of the testa. The cell walls of the sub-epidermal parenchyma are broken and the cells enlarged. The epidermal layer of the seed coat is reduced and the basal tissue, including the aleurone layer, is disorganized. Nematode-infested seeds had greater levels of fungal infection (*Rhizoctonia solani, Sclerotium rolfsii, Macrophomina phaseolina*, and *Fusarium* spp.) than nematode-free seeds (McDonald, et al., 1979).

Groundnut seeds can also be infected by *Ditylenchus africanus* (previously described as *D. destructor*), which produces symptoms very similar to those caused by *A. arachidis*. Both species are endoparasitic, feeding on tissues of the roots, pegs, hulls, and seeds. As with *A. arachidis*, developing pods are invaded after fruiting pegs have penetrated the soil. Most nematodes occur in the hulls and seed testas (Jones and de Waele, 1990; Wendt, et al., 1995).

Bursaphelenchus xylophilus (the pine wilt nematode) is the causal agent of a devastating pine wilt disease. In 1981 the annual loss of timber in Japan was estimated to be 2 million cubic meters, which approximates to some 10 million trees (Weischer and Brown, 2000). Pine wilt disease is the product of a complex series of interactions and interrelationships, the nematode being the causal agent of the disease, insects acting as dispersal agents for the nematode, and fungi providing an alternative food source for the nematode. The first disease symptoms are the cessation of resin production and transpiration. Transpiration from the needles at first declines and then ceases altogether, and the needles wilt and turn yellow. Initially, symptoms may be localized near the site of infection, but eventually they spread to the crown of the tree (Hunt, 1993; Malek and Appleby, 1984; Mamiya, 1983). Tree death can occur 30 – 40 days after infection in warm climates but takes longer in cooler areas. The major vectors are *Monochamus alternatus* Hope in Japan and *M. carolinensis* Olivier in the USA (Hunt, 1993).

Red ring disease of the coconut palm (*Cocos nucifera* L.) is caused by

Rhadinaphelenchus cocophilus, transmitted by the palm weevil (*Rhynchophorus palmarum* L.), but the nematode can also gain entry through the root system. Disease is common in trees two and a half to ten years old, with greatest incidence in those four to seven years old (Griffith and Koshy, 1990). A cross-section of the stem reveals an internal orange ring (2 – 4 cm wide), at a distance of 3 to 5 cm from the periphery. This ring normally extends the whole length of the roots and stem and into the petioles, where it takes on a crescent-like shape. Toxins are produced in the discolored area and there is also a partial breakdown of the vascular system. The root cortices become spongy in texture and assume a reddish-brown color. The lower leaves of infected trees become yellow, starting at the tips of the distal pinnae and later extending progressively downwards to the base of the pinnae and to the base of the leaf. The yellowing is succeeded by browning and eventual death of the leaf. The yellowing and browning subsequently spread to the other leaves and the tree dies within 3 – 4 months after the appearance of the first symptoms (Fenwick, 1969). The larvae of the palm weevil become infested with the nematode while feeding in infected trees, nematodes entering the haemocoel *via* the gut tract; in adult weevils, nematodes can be found in the gut, body cavity, and the ovipositor region. The nematodes are injected into the tissues of the coconut tree when the adult insect deposits its eggs in leaf axils in the crown of the tree. The nematodes first invade only parenchymatous tissue in roots, stems, and leaves as intercellular parasites, but later they can be found both intercellularly and intracellularly. Nematodes have never been found in xylem vessels and the cause of the restriction of nematodes to the narrow band or ring of necrotic tissue in stems has never been explained satisfactorily. There is no report of any tree, once affected, having recovered. Control is based on prevention and destruction of infested palm material and by the trapping and killing of the weevil vectors (Griffith and Koshy, 1990).

Mycetophagy is common in Anguinidae, but some genera are plant parasites and cause the formation of well-defined galls, usually on above-ground parts (leaves or flower structures) and exceptionally on roots (e. g., *Subanguina radicicola*), in which the adults develop. Their stylets are short and delicate (11 μm), and females may be enlarged but not globose or kidney-shaped. Anguinidae are well adapted to above-ground plant parasitism but, even if gall formation is considered an advanced stage of parasitism, some authors have considered this niche as a blind alley when compared to evolutionary tendencies in Tylenchida as a whole (Fortuner and Maggenti, 1987). *Anguina* and their close relatives show more marked adaptation to parasitism than *Ditylenchus* species.

The "ear-cockle" disease of wheat and barley is caused by *Anguina tritici*, which also vectors a bacterial disease ("yellow ear-rot"). The first symptom of infection by *A. tritici* is an enlargement of the basal stem portion, near the soil base. The emerging leaves are twisted and crinkled but straighten out later, with faint ridges on the surface. Affected plants are dwarfed with a spreading habit but may not show symptoms under very light infestations, even when a few seed galls are produced in the ears; severely infested plants may die without heading. Infested seedlings produce more tillers that grow faster than normal plants. Short and broad ears emerge 30 – 40 days earlier in diseased plants, having very small or no awns on the glumes. Nematode galls replace all or some of the grains. The production of a bright yellow slime or gum-like substance on the abortive ears as well as the leaves is a characteristic feature of yellow ear-rot disease. The infected spike is narrow and short with the wheat grain partially or completely replaced by slime (Swarup and Sosa Moss, 1990).

Soil moisture facilitates the release of juveniles from the nematode gall in the soil. Juveniles then move upward on the growing point of the germinating plant. After penetrating the flower primordia, the juveniles develop into adults, females being obese and sluggish. The life cycle is completed in 113 days, depending on temperature. Temperature, humidity, planting depth, and gall source are major determinants in symptom expression (Swarup and Sosa Moss, 1990).

Histological changes caused by *Anguina graminis* on *Festuca rubra* L. appear in the epidermis, parenchyma, and vascular bundles of the grass blades (Solov'eva and Kovalenko, 1980). Mature galls on *Poa annua* L., caused by *S. radicicola*, are characterized by a hyperplastic cortex that has five to ten more cell layers than healthy cortical parenchyma, and the nematodes are surrounded by multinucleate cells with necrotic walls (Vovlas, 1983). *Subanguina radicicola* causes root gall formation in graminaceous hosts (barley, rye, *Poa annua*, and wheat). The root tips are invaded by the juveniles, which cause the formation of galls. The galls gradually enlarge due to increase in size of cortical and endodermal cells, the cortex gradually breaks down and the cavities become filled with eggs, juveniles, and adults (Hooper and Southey, 1978).

Another ectoparasite, *Criconemella ornata*, causes the "Yellow disease" of groundnut. Roots, pods, and pegs growing in heavily infested soils can be severely discolored, having brown necrotic lesions that appear superficial and small but usually extend deep into tissues. Losses have not been well defined but infestation reduces the numbers of lateral roots and reduces pod yield by about a half. An interaction (enhancement) occurs between *C. ornata* and *Cylindrocladium crotalariae* (black rot, CBR) on CBR-susceptible Florunner but not on CBR-resistant cv. NC 3033 (Diomandé and Beute, 1981).

Considered as one of the most important pests of sugarcane in Indonesia, *Criconemella xenoplax*, an ectoparasite, feeds continuously for up to eight days from a single cortical cell, nutrients being withdrawn *via* a zone of modified cytoplasm in intimate contact with the stylet tip orifice (the only part not being covered by callose depositions). Plasmodesmata between the food cell and surrounding cells are modified to facilitate solute transport for consumption by the nematode (Sijmons, et al., 1994).

13.7 Root Lesion Forming Nematodes

Modifications of plant tissues can be extreme, as in the gall formers, but lesions, although apparently less extreme, can be very disruptive and potentially destructive. Among the Tylenchida, the Pratylenchidae include several genera with different levels of adaptation to migratory endoparasitism and that are capable of causing lesions ranging from punctiform and superficial in nature to deep, coalescent cavities. The physiology of parasitism and host-parasite relationships involves an active role for nematode secretions and host response. A transition from migratory endoparasitism, with several infective stages (J2, J3, J4, adults) that produce necrotic reactions in roots (e.g., *Pratylenchus, Radopholus*), to a female sedentary endoparasitism with syncytium induction and gall formation (i.e., *Nacobbus*) occurs within Pratylenchidae (Luc, 1987).

Pratylenchus spp. cause mechanical destruction of root cells during their migration through roots. They usually reside in roots, rhizomes or tubers but may leave plant tissues and live for some time in the soil (Brodie, et al., 1993). The nematodes feed on the parenchyma, inflicting extensive injury that it is not confined to the cortex. Lesions can be gradually enlarged by nematodes feeding at the periphery, and infestations can result in significant plant growth reduction. Affected cells lose turgor pressure, the nucleus increases in size, and cell death occurs (Zunke, 1990). Intracellular migration kills cortical and adjacent cells, tannin deposition occurs, membrane integrity is lost, and cell organelles degenerate, the effects being more pronounced in the endodermis but extending into stelar cells (Sijmons, et al., 1994; Townshend, et al., 1989). Secondary invasions by other plant pathogens may result in tissue necrosis and rotting (Christie, 1959). Many species are polyphagous and, according to Loof (1978), most species for which only one (or a small number) of hosts are known are either rare or recently described species. Host lists can be very long and *P. penetrans* is one of the most important pests in temperate agriculture, horticulture, tree and flower nurseries, bulb fields, and orchards. It has also been associated with "pre-plant disease" in Europe and America. In The Netherlands, *P. penetrans* causes a potato sickness similar to that caused by *Globodera rostochiensis*, but it only occurs where potatoes are grown for the first time. Affected plants are stunted, dull, and dark, and flowering may be abnormally intensive. Later in the season many rootlets are broken off, with dead tips. Early maturing potato cultivars may die whilst late maturing cultivars suffer retarded growth from which they may recover. Patches of diseased plants increase in size gradually over the course of a few years (Loof, 1978). *Pratylenchus* spp. can infect potato tubers, symptoms varying according to species from a scabby appearance or sunken lesions (*P. scribneri*) to wart-like protuberances (*P. penetrans*), which vary from brown to black in color at harvest and turn purple in storage (Brodie, et al., 1993). *Pratylenchus crenatus* damages grasses, cereals, vegetable crops and especially carrots, causing a kind of "carrot sickness" with infested fields showing irregular patches of poorly growing, thin, pale plants; the main root is small and often branched, the other roots are short, with dead tips and brown lesions (Loof, 1978). *Pratylenchus pratensis, P. loosi, P. brachyurus*, and *P. coffeae* are important pests of coffee. The last two species reduced the growth of *Coffea arabica* L. cv. Mundo Novo seedlings and increased soluble sugars in the leaves of infected plants (Inomoto et al., 1998). In Latin America, *Pratylenchus* spp. have been associated with a disease complex in coffee known as "corchosis" (Castillo, et al., 1992; Marbàn-Mendoza, 1989).

Radopholus similis (the burrowing nematode), has a wide host range and geographical distribution, thus not only posing a threat in tropical and subtropical countries whose national economies depend on export of agricultural products, but also to many crops that are important in world commerce (e. g., the horticultural and ornamental industries). It has more than 250 known hosts, but most research has been done on banana and citrus. Crop loss severity depends on host susceptibility and environmental conditions (O'Bannon, 1977).

Nematode parasitism of banana is a serious problem and is widely distributed throughout banana-producing areas. At least three genera and four species of lesion forming nematodes (*R. similis, Helicotylenchus multicinctus, Pratylenchus coffeae*, and *P. goodeyi*) are pathogens of considerable economic importance (Blake, 1969; Gowen and Quénéhervé, 1990). Other nematodes, such as *Meloidogyne* spp., *Rotylenchulus reniformis*, and *Hoplolaimus* are also common on banana. *Radopholus similis* is responsible for the greatest worldwide losses and can reduce banana yields from 73 to 30 t/ha per year

(Williams and Bridge, 1983). *Radopholus similis* causes extensive root necrosis, bunch weight reduction, and toppling. Fruit yield can be suppressed by as much as 50% within 3 – 4 years of planting, and the number of uprooted or blown-down plants may be increased by 60% (O'Bannon, 1977). The species is introduced to virgin soils by planting infected rhizomes; these can be disinfected before planting by paring or by immersion in hot water at 53 – 55℃ for 20 – 25 min (O'Bannon, 1977). Nematodes gain entry into the rhizome through roots, leaf attachment points, emerging buds, or directly from the soil. In roots, most nematodes enter near the root tip, taking up a feeding position between parenchyma cells in the cortex and causing contiguous cells to separate. The size of the nucleus and nucleolus in the cells surrounding the feeding nematode increases significantly but the amount of cytoplasm decreases. Ultimately, the nucleus disintegrates, the primary cell wall ruptures, a cavity forms and the nematode moves into the space. Nematodes continue enlarging these cavities (which may coalesce) by feeding on peripheral cells and tunneling in the cortex. Hyperplasia and hypertrophy are rare and necrosis usually is confined to cells lining cavities and tunnels and to epidermal cells that have been injured during invasion. Lesions and cavities extend from the cortex to the endodermis, which acts as a barrier to stelar invasion. Migration of nematodes in the cortex and colonization of the lesions by other parasites and saprophytes enlarge the lesions. The surface of the invaded tissue is black, with the advancing margin typically reddish brown. The rotting cortical tissue atrophies, and the stele is sometimes reduced to a few short root stubs at the base of the corm. There is usually no early evidence of damage on the surface of the banana roots, but reddish brown cortical lesions may be present. Later, when extensive cavities are formed in the cortex, one or more longitudinal cracks with raised margins become evident on the surface of the roots overlying the lesion. Reproduction is amphimictic and the life cycle from egg to egg spans 20 days or 25 days at 32℃ or 24℃, respectively. The males are morphologically degenerate in the esophageal region and do not normally enter roots. The incidence of Panama wilt (*Fusarium oxysporum* f. sp. *cubense*) is increased by the presence of *R. similis*.

Two races of *R. similis* are recognized, one attacking banana only (the banana race), and the other attacking both banana and citrus (the citrus race). The banana race has a more restricted host range than the citrus race. The two are morphologically similar and have even been considered as sibling species (Huettel, et al., 1984), but recent genetic and molecular studies demonstrated that the citrus and banana races of burrowing nematodes are conspecific. According to Kaplan et al. (2000) they should be considered as pathotypes of

R. similis and not races, since the genetic basis of citrus parasitism in *R. similis* remains unknown.

The citrus-parasitic burrowing nematode is only encountered on Florida's central ridge of deep sandy soils where it causes "spreading decline", a severe disease in which extensive tunneling is produced through root tissue. *Radopholus similis* infestations suppress grapefruit yields by 50 – 80% and orange yields by 40 – 70%. The rate of spread of decline in infested groves can reach 15m/year. It is generally distinguishable from other major decline diseases, such as citrus blight, in that large contiguous groups of trees are affected and expansion of the affected area is rapid. Decline trees have sparse foliage, particularly noticeable high in the canopy during the early stages of disease development. Preventive measures are the most effective and economic means of control. Management focuses on prevention of spread by the certification of planting stock, sanitation and imposition of physical barriers, cultural management practices, and the use of resistant and tolerant rootstocks and nematicides (Duncan and Cohn, 1990).

Hirschmanniella species are mainly parasites of plants with an aquatic growing phase. Some species are important parasites of rice, contributing to crop yield losses estimated at 25% (Bridge, et al., 1990). Although there are no obvious above-ground symptoms, growth retardation occurs and early growth and tillering are especially decreased. Chlorosis of rice plants occurs occasionally and flowering may be delayed. Invaded roots turn yellowish brown and rot. The nematodes produce necrotic cavities and channels through the cortex of the root (Bridge, et al., 1990). *Hirschmanniella miticausa* causes a corm taro (*Colocasia esculenta*) disease known as "miti-miti" (see section 13.10).

The family Hoplolaimidae includes genera differentiated by short tails, strong stylets, sclerotization of the cephalic region and esophageal glands overlapping the intestine dorsally or ventrally (Fortuner, 1987). They are ecto- or semi-endo- plant parasites, accomplishing part or all of their life cycle in the soil. Depending on genus, host damage may include superficial lesions or even the production of a syncytium (e. g., *Rotylenchulus reniformis*). Little is known about their feeding activities. *Helicotylenchus dihystera* and *H. varicaudatus* feed for many days on a single cell in cereal roots. The food cell, located close to the stele and surrounded by a few cells with dense cytoplasm, shows signs of increased metabolic activity without nuclear enlargement. Feeding tubes are surrounded by an extensive membranous network and have also been found in cortical cells modified by feeding of *Scutellonema brachyurus* (Sijmons, et al., 1994). Few hoplolaimid species are considered true endoparasites (e. g., *Hoplolaimus* spp., *Scutellonema bradys*, and

Aphasmatylenchus straturatus). Other hoplolaimids are the so-called "spiral nematodes" (*Helicotylenchus* and *Rotylenchus*). Although common in soil and root samples, the presence of these nematodes (even in large numbers) cannot always be linked with pathogenicity (e. g., *H. dihystera* on tea; Campos, et al., 1990), although they can be linked with stunting and root lesions (e. g., *Helicotylenchus* spp. on banana; Gowen and Quénéhervé, 1990). However, spiral nematodes, especially *H. multicinctus* on banana, are of considerable economic importance. *Helicotylenchus multicinctus* penetrates the epidermis and feeds on parenchyma cells of the cortex. The cell cytoplasm contracts, some cell walls may rupture and the nucleus may increase in size. In contrast to *R. similis, H. multicinctus* forms few cavities and the histological changes are confined to parenchyma cells beneath the epidermis (Blake, 1969).

Most *Helicotylenchus* spp. are considered mild pathogens of little or no economic importance, although hosts include sugarcane and various tubers and corms. More than 30 species of *Helicotylenchus* have been recorded from the rhizosphere of sugarcane, *H. dihystera* being the commonest and its pathogenicity being demonstrated experimentally (Rao and Swarup, 1975). They feed ectoparasitically or semi-endoparasitically in the root cortex causing brownish red lesions, distortion and collapse of the cells, and blunt and malformed primary roots with fewer lateral roots (Spaull and Cadet, 1990). *Rotylenchulus reniformis, H. erythrinae, H. dihystera*, and *S. bradys* have been reported on cassava (*Manihot esculenta* Crantz), their importance on this crop only being exceeded by *M. incognita, M. javanica*, and *P. brachyurus*. Their abundance, wide host ranges, and potential to interact with other pathogenic organisms to develop disease complexes mean that intercropping of susceptible hosts with cassava cannot be recommended (Jatala and Bridge, 1990).

Aphasmatylenchus straturatus causes groundnut yield reductions of 30 – 70%, and may cause total yield loss in pigeon pea (Germani and D'Héry, 1973). It is, therefore, a serious potential threat for groundnut and other Leguminosae (CAB International, 1999).

13.8 Cyst-forming Nematodes

Members of the family Heteroderidae are among the most important plant-parasitic nematodes in agriculture, especially in temperate zones (see Sharma, 1998). Cyst nematodes derive their common name from the swollen (cyst-like), endoparasitic female, which contains fully embryonated eggs inside its body, protected by the hardened cuticle (Luc, et al., 1988). Females can be seen on the surface of the roots and have a whitish, pearled, brown or golden color according to genus and cyst maturity. They are easily recovered from soil (Shepherd, 1986). Penetration and establishment of juveniles generally occurs in regions away from the root meristem (Endo, 1987). The first cells to be affected may be in the cortex, endodermis, pericycle or vascular parenchyma. The J2 invades the host roots and migrates intracellularly, toward the vascular cylinder. The response of host cells to feeding is the formation of a syncytium, which arises from breakdown of adjacent cell walls and not hyperplasia. A plug of material forms around the stylet where it pierces the cell wall and extends as a collar ("feeding tube") into the cell cytoplasm, sealing the stylet in place; the channel blocks after the stylet is withdrawn when the J2 changes feeding position or molts. In *H. schachtii*, the feeding tube is produced from secretions of the dorsal esophageal gland (Sijmons, et al., 1994). Once established at a feeding site, the J2 develops endoparasitically into J3, J4, adult male or female through molts. Syncytia induced by male nematodes are considerably smaller than those of females at a similar developmental stage (Sijmons, et al., 1994). After their final molt, males leave the roots but adult females continue to feed and develop inside roots. Stylet penetration after molting occurs at a new site nearby, where a new plug forms around the stylet. Increase in cytoplasmic organelles and secondary vacuolation is accompanied by wall digestion at the pit fields and fusion of neighboring cell protoplasts. The syncytium may reach 2 – 3 cm in length. Cell expansion accompanies incorporation into the syncytium, cell vacuoles are lost, nuclei and nucleoli enlarge, and numerous mitochondria and Golgi bodies occur. Wall ingrowths are laid down where a syncytium contacts vascular tissues and, as with root knot nematodes, recent evidence suggests that these ingrowths form under the influence of endoglucanase enzymes of plant rather than nematode origin (Goellner, *et al.*, 2001). As a syncytium extends, wall degradation occurs at the extremities, the enzymes responsible for wall digestion being of plant origin. As secondary thickening of the root proceeds, the cambial region that has been replaced by the syncytium is lost, and a wedge-shaped region (i.e., in transverse section) pointing to the center of the stele is often seen to be occupied by a syncytium. In heavy infections all four arcs of a tetrarch root may be deformed with severe restriction of longitudinal water movement (Jones, 1981).

The most important species for North American crops (Canada and USA) are *Globodera rostochiensis, G. pallida, G. tabacum, G. t. solanacearum, Heterodera glycines, H. schachtii, H. cruciferae, H. trifolii*, and *H. avenae* (Miller, 1985). Species of known or potential economic importance elsewhere in the world include: *Cactodera cacti* on cactus and cereals, *Punctodera punctata* on wheat, *P. chalcoensis* and *H. zeae* on maize, *H. lespedezae* on lespedeza, *H. leucilyma* on Augustin grass, *H. fici* on figs, *H. humuli* on hops, *H. oryzicola* and *H. elachista* on rice, *H. sacchari* on sugarcane and rice, and *H. carotae* on carrot.

The potato cyst nematodes, *G. pallida* and *G. rostochiensis* (see Marks and Brodie, 1998) are found in virtually all potato growing regions of the world, although *G. pallida* is notably absent from some areas, such as the USA. This latter species is the more difficult to control as there are no cultivars with complete resistance to this species and nematicides and rotation are less effective than against *G. rostochiensis*. For these reasons, *G. pallida* is becoming increasingly dominant where both species occur in intensive potato production areas (Evans and Haydock, 2000). In the UK, almost two thirds of potato production land is infested with one or other (or both) of these species, with 92% of infested land containing at least some *G. pallida*. Annual costs due to these nematodes are estimated at almost £50 million in the UK, and 300 million euros in Europe (Mulholland, et al., 1996).

Heterodera avenae affects wheat in most cereal-growing regions (Meagher, 1977). Oats are actually less tolerant than wheat but encourage the build-up of biological antagonists of the nematodes. Continuous high temperatures and soil moisture do not favor survival of *H. avenae* and the severity of damage is related to soil, climate, juvenile emergence patterns, soil temperature, long-term survival, cultural practices, and host range. Increased leaching of nutrients, particularly nitrogen, can intensify symptoms. Wheat yield losses can reach 50%. Annual loss can be equivalent to US$70 million in Australia, US$4.5 million in Europe and US$9.6 million in India (CAB International, 1999). In the USA, because of its expected economic impact among exotic pests, it ranks 16th with expected losses of US$132 million. In India and Pakistan, *H. avenae* is the most important nematode pest of wheat and barley and causes "molya disease" of wheat (Sharma, et al., 1997). Plants heavily infested with the cyst nematode are stunted with reddish yellow leaves and narrow leaf blades. The most severe damage results from nematode-fungus interactions, *Rhizoctonia solani* being the most important species (Meagher, 1977). In Australia, two eggs/g of soil represent the economic damage threshold on wheat, but the threshold may be lower or higher in other temperate areas. Pathotypes occur in *H. avenae* and serological techniques have been used in diagnostics (CAB International, 1999). Sources of resistance are available in various species and cultivars of *Hordeum, Triticum, Secale*, and *Triticale* (Rivoal and Cook, 1993). Rotation schemes (resistant and tolerant cultivars), control of grassy weeds, biological, cultural and chemical control have been used to restrain populations to sub-threshold levels

(CAB International, 1999). Heavy attack can be detected by remote sensing techniques, using both visible and infrared images, and the infrared waveband responsive to the disturbed transpiration that results from infection (CAB International, 1999; Caubel, et al., 1978; Lili, et al., 1991).

Soybean is one of the major world food sources and *Heterodera glycines*, the soybean-cyst nematode (SCN), is considered a major threat for production (Riggs and Wrather, 1992). SCN occurs as a series of "races" and at least three recessive genes and one dominant gene control resistance. Rotation schemes include one year with a non-host, one year with a resistant cultivar, and one year with a susceptible cultivar. Quarantine costs from 1956 to 1972 reached US $9.4 million in the USA (Riggs, 1977).

13.9 Root-shortening Nematodes

Belonolaimus longicaudatus ("sting nematode") is a destructive, migratory ectoparasite of numerous crops. It was first identified as a pathogen of cotton (*Gossypium hirsutum*), by Graham and Holdeman (1953). Symptoms on cotton roots include shrunken lesions along the root axis or at the root tip, but also shortened or stubby root symptoms on "DPL 5O" cotton (Crow, et al., 1997, 2000a). This species increases the severity of *Fusarium* wilt of cotton (Holdeman and Graham, 1954; Minton and Minton, 1966; Yang, et al., 1976). Soil texture greatly influences its distribution and it is found predominantly in sandy soils. If cotton production expands into this type of soil, sting nematode may become a significant problem on this crop. Experimentally, *B. longicaudatus* caused a 39% reduction in fine cotton roots with as few as 10 nematodes/130 cm^3 of soil, and 70% reduction with as few as 60 nematodes/130 cm^3 of soil. In the field, large populations (> 100 nematodes/130 cm^3) reduced yields to near zero (Crow, et al., 2000a). The economic threshold for management varies between 2 and 5 nematodes/130 cm^3 of soil, depending on the nematicide used (Crow, et al., 2000b).

Belonolaimus longicaudatus also occurs in 5% of Florida citrus orchards, damaging citrus by feeding on root tips and greatly reducing the number of fibrous roots. Relatively small populations (40 nematodes/dm^3) can cause stunted, chlorotic plants (Kaplan, 1985). Root systems of infested trees appear very coarse due to a reduction in the number of lateral roots and the presence of swollen fibrous roots that also have terminal swellings as well as multiple apices. The epidermis may slough off easily due to secondary infection. Histological examination has shown several meristematic zones at root tips with tissue disorganization that includes hyperplastic tissue, cavities, and extensive vascular formation. Cell disruption at the cavity borders results in cytoplasm leakage into these spaces and suggests that these are the sites at which the nematodes feed (Duncan and Cohn, 1990; Kaplan, 1985; Standifer and Perry, 1960). Preplant soil fumigation and post-plant nematicide treatments alleviate symptoms. Hot water treatment for 5 min at 49℃ has been proposed as an eradication method for bare-root seedlings (Duncan and Cohn, 1990; Kaplan, 1985).

13.10 Bulb, Tuber, Rhizome, and Corm Diseases

Ditylenchus spp. are important migratory endoparasites of higher plants, but some species feed on higher plants and/or fungi while the feeding habits of others are unknown. Esophagus modifications, such as gland enlargement, are possibly related to plant parasitism (Fortuner and Maggenti, 1987). Species of economic importance include *D. dipsaci* (stem nematode), *D. destructor* (potato tuber nematode), *D. myceliophagus* (which destroys mushroom mycelium), and *D. angustus* (which causes "Ufra" disease of rice).

Ditylenchus dipsaci feeds upon parenchymatous tissue in stems and bulbs, causing the breakdown of the middle lamellae of cell walls, swelling, and distortion (Reed, et al., 1979; Bleve-Zacheo, et al., 1980). The life cycle in onion plants is completed in 21 days at 15℃; there are four molts, the first within the egg. This species is able to survive, in a coiled, dry condition for several years, mainly as fourth-stage juveniles. These juveniles tend to aggregate on or just below the surface of heavily infested tissue to form clumps of "eelworm wool".

There are distinct races of *D. dipsaci*, some of which can interbreed to yield progeny that may have different host preferences. DNA probes (Palmer, et al., 1991), monoclonal antibodies (Palmer, et al., 1992), and restriction fragment length polymorphisms are used in race identification (CAB International, 1999). Important host crops for the different races include flower bulbs (narcissi, tulips, and hyacinths), onion, garlic, oats, rye, field beans, clover, lucerne, and strawberry. In narcissus bulbs, individual scales are often heavily infested, starting from the neck of the bulb. Secondary invasion of infested tissue by bacterial pathogens causes a typical "brown ring" effect when the bulb is cut across. In the growing season the leaves from infested bulbs often have small, lighter colored, swellings or "spickels". Infested bulbs rot when stored. Onion and garlic are very susceptible; seedlings quickly become infested and affected tissue is swollen and distorted with characteristic "bloat" symptoms. In oats and rye, the leaf bases of the main stem and tillers swell to give a swollen-base to the plant, a condition known as "tulip root". Heavily infested plants are usually stunted and often fail to produce panicles. Infested plants of clover and lucerne are stunted, stems and leaf petioles are often swollen, flower heads become infested, and the seed is contaminated (Hooper and Southey, 1978).

Ditylenchus destructor is a migratory endoparasite of underground parts of plants. Its host range includes at least 70 crops and weeds and it has a similar number of fungal host species. It enters potato tubers through lenticels, causing small white mealy spots below the surface that are only visible if the skin is removed. Infested areas enlarge and coalesce, and light brown lesions may be visible beneath the skin. As the infestation progresses, the tissues and the skin become dry and papery, and secondary pathogens may gain entry. *Ditylenchus myceliophagus* is a migratory parasite of economic importance because it destroys the mycelium of cultivated mushroom (*Agaricus bisporus* = *A. hortensis*).

"Miti-miti disease" is a rot of taro (*Colocasia esculenta*) corms caused by *Hirschmanniella miticausa*, which occurs in Pacific regions (Bridge, et al., 1983). Foliar symptoms include wilting of the older leaves, which eventually become chlorotic, while the new central leaf, instead of bending, remains straight. Premature death occurs as a result of corm damage. Red streaks radiating from the base of the corm are found throughout longitudinal sections of the corms. "Soft root" in corms is invariably associated with this disease, and the fungi *Corticium solani, Pythium vexans, Fusarium solani*, and *F. oxysporum* have been isolated from affected areas (Bridge, et al., 1983; Jatala and Bridge, 1990). Miti-miti renders the corm inedible and the disease can be so devastating that, in parts of the Solomon Islands, the crop has been almost entirely abandoned, particularly where it was continuously cultivated in swamp pits. It is a quarantine pest in the Pacific Island countries and in SE Asia (CAB International, 1999).

"Dry rot" disease of yam (*Dioscorea*) tubers is caused by *Scutellonema bradys*, the yam nematode, and *Pratylenchus coffeae*, both of which are migratory endoparasites. The biology of *S. bradys* on yams has been described and it is likely that the behavior of *P. coffeae* in yam tubers is generally similar as symptoms are identical. Nematodes invade young tubers through the growing points of tissues, alongside emerging roots and shoots, and through roots and cracks or other damaged areas of the tuber skin (Bridge, 1972; Jatala and Bridge, 1990). Intracellular feeding by the nematodes results in cell wall rupture, loss of cell contents, and the formation of cavities. The nematodes are mainly confined to the sub-dermal, peridermal, and underlying parenchymatous tissues in the outer 1 or 2 cm of tuber. They feed and reproduce in yams stored after harvesting. Cream and light yellow lesions below the tuber outer skin occur first, spreading later into the tuber (2 cm depth or more). Infected tissues become light brown and then turn dark brown to black, external cracks appearing in the tuber skin. Patches of dark brown, dry rot tissues are exposed when tuber parts flake off. Severe symptoms occur in mature tubers especially during storage. *Botryodiploidia theobromae, Fusarium* spp., and *Erwinia* spp. are among pathogens responsible for internal tuber decay. Nematodes and fungi or bacteria are found together in the transitional stage between dry rot and wet rot, but nematodes do not occur in the "late wet rot" stage deep in the tuber (Jatala and Bridge, 1990). Dry rot of yams causes a marked reduction in the quality, marketable value, and edible portions of tubers. Dry rot followed by wet rot in stored yams causes losses as high as 80% to 100%. Small nematode populations produce discrete internal areas of yellow necrotic tissues or dry rot. Populations exceeding 1000 nematodes/50 g of tuber peelings are necessary to produce external symptoms, but populations can exceed 300,000 nematodes/50 g of tuber peelings (Bridge, 1973). Hot water treatment (50 – 55℃) for up to 40 min gives the best control without damaging tubers. Chemical control has had some success but information on the economics is lacking for large-scale usage (Jatala and Bridge, 1990). Other nematodes reported from yam roots or tubers include *Rotylenchulus reniformis, Scutellonema clathricaudatum*, and *Helicotylenchus dihystera*.

Rots of ginger rhizomes are caused by *Meloidogyne incognita, Radopholus similis*, and *Pratylenchus coffeae* in different countries. Rhizomes infested with *M. incognita* have water-soaked areas in the outer tissues, particularly in the angles between shoots (Koshy and Bridge, 1990). These infested rhizomes serve as a source of infection and means of dissemination as nematodes remain active in the harvested crop. *Radopholus similis* and *P. coffeae* produce similar shallow, sunken, water-soaked lesions and, as migratory endoparasites, produce large infection channels or galleries within the rhizomes. Hot water treatment is one of the main recommended means of controlling nematodes in ginger rhizomes. In another spice, turmeric (*Curcuma domestica*), a similar rot of rhizomes is caused by *R. similis*.

13.11 Nematodes that Reduce General Root Growth

In inoculation trials, *Hemicriconemoides mangiferae* has been shown to be potentially damaging to mango seedlings at a population level of 6 nematodes/ cm^3 of soil. In sapodilla (*Manilkara sapota* (L.) Royen), pathogenicity was demonstrated at similar population densities (Cohn and Duncan, 1990).

Xiphinema brevicollum and *H. mangiferae* are major pests of lychee (*Litchi chinensis* Sonn.), causing a severe decline syndrome in South Africa. Above-ground symptoms include bare twigs and branches, leaf chlorosis, leaf tip burn, poor flowering, and excessive fruit drop. In some orchards up to 40% of the trees may die. Severe stubby root symptoms and root darkening occur, leading to loss of a large proportion of the total feeder root mass and consequent interference in the uptake of nutrients and water. *Xiphinema brevicollum* feeds more superficially, whilst *H. mangiferae* causes extensive destruction of the cortical tissue and is therefore a more severe pathogen (Cohn and Duncan, 1990; Milne, 1982).

Reniform nematodes (*Rotylenchulus* spp.) are sedentary root semi-endoparasites that occur largely in tropical and subtropical latitudes. *Rotylenchulus reniformis* appears to have the widest host range and has been reported to reproduce on 86% of 364 plant species (Robinson, et al., 1997). However, more than 50 crop or ornamental plants support little or no reproduction and examples include barley, maize, onion, rice, several *Crotalaria* species, and resistant cultivars of soybean. All *Rotylenchulus* species have a similar life cycle. The J2 and following juvenile stages (J3 and J4) retain the cuticle of the previous stages after molting and do not feed. Males can be rare or absent in populations of some species. Immature females partly penetrate the root cortex with the anterior third of their body and establish a permanent feeding site in the stele, becoming sedentary and laying eggs into a gelatinous matrix on the root surface. Life cycle length varies according to species and temperature. In the absence of a host, vermiform stages in the soil can remain in a state of arrested development indefinitely. When a host is present, immature females penetrate roots perpendicular to the long axis. The initial cell to be modified is usually an endodermal cell ("prosyncyte"), although occasionally it can be one in the pericycle. Expansion and wall degradation is most marked near the head of the nematode. The replacement of cell vacuoles by active cytoplasm occurs in expanded cells (a curved sheet of 100 – 200 pericycle cells). Enlarged irregular nuclei are found in the cells close to the nematode. Wall thickening occurs, but wall ingrowths do not form where altered cells contact xylem or sieve elements. A peg of material containing the stylet and a "feeding tube" projecting into the cytoplasm mark the site of stylet penetration. There is no stimulation of mitosis. Plasmodesmata are present between altered cells and neighboring cells. There are clear anatomical differences in the feeding site induced by *Rotylenchulus macrodoratus* (uninucleate giant cell) and syncytia induced by other species, which also have different parasitic abilities (Robinson, et al., 1997). In *R. macrodoratus* infection, the uninucleate giant cell develops from

Table 13.5 Distribution and crop hosts of nematode genera and species known to be important pests worldwide. Much of the information in this table can be found in CAB International Crop Protection Compendium (1999) and Eisenback (1999).

Nematode	Damage symptoms	Main crops affected	Distribution
Achlysiella			
A. williamsi (Siddiqi, 1961) Hunt, Bridge, and Machon, 1989	Root necrosis, purplish in color	Sugarcane	Australasia, Mauritius
Anguina			
A. tritici (Steinbuch, 1799) Chitwood, 1935 (Ear cockle nematode)	Stunting, chlorosis; leaves rolled or folded; spikes reduced; glumes protruding at an abnormal angle; galls	Temperate cereals, mainly wheat	Temperate: eastern Europe, N Africa, Asia, N and S America
Aphasmatylenchus			
A. straturatus Germani, 1970	Stunting and chlorosis	Groundnuts, pigeon pea	Tropical: W Africa
Aphelenchoides			
A. arachidis Bos, 1977 (Groundnut testa nematode)	Seeds wrinkled, discoloured to darker brown. Thicker and uneven testas, shrivelled seeds	Groundnut	Tropical: W Africa
A. besseyi Christie, 1942 (Rice white tip nematode)	Flagleaf shortened, twisted. Sterile panicles. New leaves with chlorotic tips. Young leaves of tillers with chlorotic areas (speckled). Leaf margin distorted	Rice	Tropical: rice growing areas worldwide
A. composticola (Mushroom nematode)	Destruction of mycelium	Mushrooms	In mushroom cultivation areas worldwide

Continued

Nematode	Damage symptoms	Main crops affected	Distribution
A. fragariae (Ritzema-Bos, 1891) Christie, 1932 (Strawberry crimp nematode)	Shoots malformed. Leaves: smaller, twisted, wrinkled, chlorotic, crinkled edges, reddening of petioles. Internodes of runners shortened, reduced flower trusses and dying crown bud	Strawberry	Temperate: Europe, N and S America (Brazil), Asia (China, India, Japan, Korea), Oceania
A. ritzemabosi (Schwartz, 1911) Steiner and Buhrer, 1932 (Leaf nematode)	Leaves: crinkled, deformed, discoloured, brown spots on veins. Reddish-yellow lesions in young lower leaves; older leaves chlorotic and necrotic. Scant foliage. Flowers: fewer, deformed	Chrysanthemum	Temperate: Europe, N and S America, Asia (China, India, Japan), E and S Africa, Australia, Oceania (Fiji and New Zealand)
Belonolaimus			
B. longicaudatus Rau, 1958 (Sting nematode)	Severe stunting, wilting, leaf chlorosis. Reduced root system with stubby side branches, terminal galling. Dark lesions at feeding and stylet penetration points	Sweet corn, vegetables, groundnut, citrus, cotton	Subtropical: N and Central Ame-rica, Caribbean
Bursaphelenchus			
B. xylophilus (Steiner and Buhrer, 1934) Nickle, 1970	Chlorosis and tree death	Pine	Temperate: Asia (Japan), N America, Europe
Criconemella			
C. onoensis (Luc, 1959) Luc and Raski, 1981	Root decay, secondary roots stunted with lesion near club-shaped root tips	Rice	Tropical: USA, W Africa, C and S America

Continued

Nematode	Damage symptoms	Main crops affected	Distribution
C. ornata (Raski, 1958) Luc and Raski, 1981	Pods and pegs discolored. Brown necrotic lesions, small (superficial) or large (coalescent, deep). Death of lateral root primordia. Extensive lesions and pits on roots (peach seedlings)	Groundnut, fruit, and nut crops	Subtropical: Africa, N and S America, Australia, Europe, Asia (India, Japan)
C. xenoplax (Raski, 1962) Luc and Raski, 1981	Lack of feeder roots. Predispose trees to cold injury and/or bacterial canker	Fruit and nut crops	Subtropical: USA
Ditylenchus			
D. africanus Wendt et al., 1995 (Groundnut pod nematode)	Greyish-black necrosis at peg connection on hulls, extending in broad bands along longitudinal veins. Shrunken seeds, brown to black micropiles; yellow to brown testa	Groundnut	Subtropical: S Africa
D. angustus (Butler, 1913) Filipjev, 1936 (Ufra nematode)	Stunting or rosetting. Leaves rolled or folded; brown stains on leaves and sheaths; white speckles at the base of young leaves. Panicle head and flag leaf distorted; panicles partially or completely enclosed within leaf sheath, with few seeds. Empty spikelets distorted and discolored. Stem bark discolored	Rice	Tropical: Asia (Bangladesh, Burma, India, Indonesia, Madagascar, Pakistan, Philippines, Thailand), Africa (Egypt, Sudan)
D. destructor Thorne, 1945	Tubers: small, white or dark spots, coalescent; woolly texture. Sunken areas, cracked and wrinkled skin. Dark to black, dry, mealy flesh	Potatoes	Temperate: Europe, N and S America (Ecuador), Asia, Africa (South Africa), Oceania

Continued

Nematode	Damage symptoms	Main crops affected	Distribution
D. dipsaci (Kühn, 1857) Filipjev, 1936 (Stem and bulb nematode)	Leaf: deformed, swollen (blister-like) areas; chlorosis, wilting; disordered growth. Bulbs: soft, concentric circles of brown scales. Stem: short internodes (lucerne); tissues swollen, deformed; reddish-black lesions (faba beans). Pods: dark brown, necrosis on petioles (faba beans). Seeds: darker, smaller, speckled. Secondary formation of tillers (cereals)	Field beans, onions, narcissi and other bulb crops, cereals	Europe, N and S America, Australasia, NE and SE Africa
D. myceliophagus Goodey, 1958	Mycelium sparse. Affected areas sunken, soggy and foul-smelling, pH raised	Mushrooms	Temperate: mushroom cultivating areas worldwide
Dolichodorus			
D. heterocephalus Cobb, 1914 (Awl nematode)	Stubby roots, severe root necrosis. Tissues necrotic at the base of hypocotyl	Vegetables (celery, tomato)	Warm temperate: USA, S Africa
Globodera			
G. pallida Stone, 1973 (White potato cyst nematode)	Reduced root systems. Early senescence. Smaller tubers with rough skin or damage	Potatoes	Temperate: potato growing areas worldwide, with some exceptions (e.g. USA)
G. rostochiensis (Wollenweber, 1923) Skarbilovich, 1959 (Golden nematode)	Reduced root systems. Early senescence. Smaller tubers with rough skin or damage	Potatoes	Temperate: potato growing areas worldwide
G. tabacum (Lownsbery and Lownsbery, 1954) Skarbilovich, 1959 (Tobacco cyst nematode)	Reduced root systems, stunting of tops and early death	Tobacco, tomato, aubergine	N America (USA, Mexico), S America (Argentina, Colombia), Europe, Asia, Africa

Continued

Nematode	Damage symptoms	Main crops affected	Distribution
Helicotylenchus			
H. multicinctus (Cobb, 1893) Golden, 1956 (Banana spiral nematode)	Toppling. Root: yellow, reddish-brown-black necrotic lesions; decay; rotten? flecks in cortex.	Bananas and plantains	Tropical/subtropical: banana growing areas worldwide
Hemicriconemoides			
H. mangiferae Siddiqi, 1961	Slow decline. Bare twigs and branches; leaf chlorosis, tip burn, poor flowering; excessive fruit drop. Stubby roots.	Tropical fruit trees	Subtropical: S Asia, Africa, S and C America, Caribbean, Oceania
Hemicycliophora			
H. arenaria Raski, 1958	Root tip galling, stubby roots (citrus), reduced root system (vegetables)	Citrus, fruit trees, vegetables, weeds	N (USA) and S America (Argentina, Brazil)
Heterodera			
H. avenae Wollenweber, 1924 (Cereal cyst nematode)	Reduced tillering; small heads, poorly filled. Leaves: reddish-yellow, leaf burn, thin and narrow blades. Roots: shortened and thickened (barley), main root elongated; excessive branching (barley) rootlet tips bunched, slight swelling or knotting (wheat)	Temperate cereals	Cereal growing areas worldwide
H. cajani Koshy, 1967 (Pigeon pea cyst nematode)	Stunting (pigeon pea); suppression of Rhizobium nodule formation (cowpea)	Pigeon pea, cowpea	India, Pakistan, Egypt

Continued

Nematode	Damage symptoms	Main crops affected	Distribution
H. ciceri Vovlas, Greco, and di Vito, 1985 (Chickpea cyst nematode)	Early senescence, limited number of flowers and pods which may be empty. Small, longitudinal necrotic spots on root. Reduction in number of nitrogen-fixing nodules	Chickpea, lentil	Mediterranean (Asia)
H. glycines Ichinohe, 1952 (Soybean cyst nematode)	Reduction of yield, stunting, and chlorosis may or may not occur together. Die-off before pods are full. Root system reduced and devoid of Rhizobium nodules	Soybean, beans	Subtropical: N and S America, Asia (China, Japan, Korea)
H. goettingiana Liebscher, 1892 (Pea cyst nematode)	Reduced root growth without proliferation of lateral roots; suppression of nitrogen-fixing nodules. Rapid yellowing and haulm death	Pea, beans, lentil	Temperate: Europe, N America, Mediterranean
H. oryzae Luc and Berdon-Brizuela, 1961 (Rice cyst nematode)	Reduced tillering. Leaf chlorosis	Rice	Tropical: Asia (Bangladesh, India, Japan, Korea, Pakistan), Africa (Côte d'Ivoire, Senegal)
H. sacchari Luc and Merny, 1963 (Sugarcane cyst nematode)	Roots necrotic and black. Reduced root system. Leaf folding. Tillering reduced	Sugarcane, rice	Tropical: W Africa, India; Caribbean (Trinidad and Tobago)
H. schachtii Schmidt, 1871 (Sugar beet cyst nematode)	Tap root destroyed, only lateral roots developing	Beets, swedes and other brassicas	Temperate/subtropics: Europe, N America, W and S Africa, Australia
H. trifolii Goffart, 1932 (Clover cyst nematode)	Poor plant growth. Reduced root length and number. Fewer, smaller Rhizobium nodules	Clover	Europe

Continued

Nematode	Damage symptoms	Main crops affected	Distribution
H. zeae Koshy, Swarup, and Sethi, 1970	Stunting, chlorotic plants with narrow leaves	Maize	Tropical: India, Pakistan, Thailand, USA
Hirschmanniella			
H. gracilis (de Man, 1880) Luc and Goodey, 1964	Early growth; decrease in tillering, occasional yellowing. Delay of flowering. Roots yellowish-brown, rot	Rice	Tropical
H. imamuri Sher, 1968	Similar to other species	Rice	Tropical
H. miticausa Bridge, Mortimer, and Jackson, 1984	Corms with red streaks, decay, and rot. Number of cormels reduced	Taro	Tropical: Pacific
H. oryzae (van Breda de Haan, 1902) Luc and Goodey, 1964 (Rice root nematode)	Retardation of growth and flowering, decrease of tillering. Root necrosis, yellowish-brown, and rot	Rice	Tropical: W Africa, N and S America, S and SE Asia
H. spinicaudata (Schuurmans Stekhoven, 1944) Luc and Goodey, 1964	Roots turn yellowish brown and rot. Root surface lesions yellowish. Retarded growth. Tillering decreases. Wilting and yellowing of older leaves, new central leaf straight.	Rice	Tropical: Africa, N and S America
Hoplolaimus			
H. columbus Sher, 1963	Feeding site yellowish brown, discoloration of tissues. Large cavities in cortex. Shorter shoot length	Cotton	Tropical: USA, Egypt

Continued

Nematode	Damage symptoms	Main crops affected	Distribution
H. seinhorsti Luc, 1958	Browning and necrosis of roots	Cotton, cowpea, rice, sugar cane, fruits, and vegetables	Tropical: Africa, S Asia, C America
Longidorus			
L. elongatus (de Man, 1876) Micoletzky, 1922	Galling of roots at or near tips, stunting of roots, fangy roots in severely attacked sugar beet	Strawberry, sugar beet	Temperate: Europe, Canada
Meloidogyne			
M. acronea Coetzee, 1956	Inconspicuous galls. Sorghum: delayed flowering, extensive root proliferation	Cotton, sorghum	Tropical: southern Africa
M. africana Whitehead, 1960	Small galls (1 – 5 mm diameter). Seedlings stunted, numerous rootlets behind the affected root tip	Coffee	Tropical: Africa
M. arabicida Lopez and Salazar, 1989	Chlorosis, defoliation, mummification, and loss of fruits, poor root system with thicke-ned roots showing cracks	Coffee	Tropical: Costa Rica
M. arenaria (Neal, 1889) Chitwood, 1949	Galls on roots, corms, tubers or groundnut pods. Reduced yield, Early senescence	Groundnut	Tropical: worldwide
M chitwoodi Golden, O'Bannon, Santo, and Finley, 1980	Reduced yield. Small galls on tuber surface; internal tissue necrotic and brownish below galls	Potatoes, sugar beet, cereals	Temperate: N America, Mexico, S Africa, Australia, Europe

Continued

Nematode	Damage symptoms	Main crops affected	Distribution
M. coffeicola Lordello and Zamith, 1960	Defoliation, general decline, death, peeling, and cracking of roots, but does not produce galls	Coffee	Tropical: S America
M. exigua Goeldi, 1892	Defoliation; uprooted trees; root hairs and rootlets destroyed. Stunting of seedlings	Coffee	Tropical: South America
M. graminicola Golden and Birchfield, 1965	Root tip swollen and hooked. Death or drowning of plants	Rice	Tropical: N (USA) and S America (Brazil) S and SE Asia
M. hapla Chitwood, 1949	Galls small and sub-spherical with proliferation of small roots. Brown spots on potato tubers	Pyrethrum, vegetables, clover	Temperate: worldwide
M. incognita (Kofoid and White, 1919) Chitwood, 1949	Peeling and cracking of cortical parts of the root tissue. Reduction of root system and tissues; defoliation and even death	Vegetables, cotton, tobacco, guava, coffee (very wide host range)	Tropical: worldwide
M. konaensis Eisenback, Bernard, and Schmitt, 1995	Significant damage and death of coffee trees	Coffee	Kona Island, Hawaii
M. javanica (Treub, 1885) Chitwood, 1949	Galls small and irregular on young feeder roots. Blister-like swellings which enlarge and may deform corms. Enlarged roots with fewer feeder roots (groundnut); Large galls (tobacco)	Vegetables, cotton, tobacco (very wide host range)	Tropical: worldwide

Continued

Nematode	Damage symptoms	Main crops affected	Distribution
M. mayaguensis Rammah and Hirschmann, 1988	Severe root galling, decline symptoms, yellowing, and defoliation	Coffee, guava, Solanaceae	Caribbean (Cuba, Puerto Rico), Africa (Senegal, S Africa)
M. naasi Franklin, 1965	Galls similar to but smaller than *M. incognita*. Hook or horseshoe-like curving of the roots	Grasses	Temperate: Europe, USA, New Zealand
M. oryzae Maas, Sanders, and Dede, 1978	Stunting, distinct (2–3 mm) galls on rice, conspicuous galls on spring wheat, relatively small swellings on Fimbristylis grass, potato, and tomato	Rice, *Fimbristylis miliacea* (L.), wheat, potato, tomato	Tropical: Surinam
Nacobbus			
N. aberrans (Thorne, 1935) Thorne and Allen, 1944 (False root-knot nematode)	Galls resembling rosary-beads	Vegetables, potato, sugar beet	Temperate/Subtropical: S, C, and N America
Paralongidorus			
P. australis Stirling and McCulloch, 1985	Root tips necrotic, hooked or galled. Secondary roots shortened, reduced root growth	Rice	Subtropical: Australia
Paratrichodorus			
P. minor (Colbran, 1956) Siddiqi 1974	Cessation of root growth, stubby roots. Discoloration of roots may occur	Vegetables	Temperate: Europe, USA
Pratylenchus			
P. brachyurus (Godfrey, 1929) Filipjev and Schuurmans Steckhoven, 1941	Defoliation, lesions brown to red on roots; girdling and root severing	Groundnuts, pineapple	Tropical: worldwide

Continued

Nematode	Damage symptoms	Main crops affected	Distribution
P. coffeae (Zimmermann, 1898) Filipjev and Schuurmans Steckhoven, 1941	Toppling. Roots: Reduction, cortex slough off, black or purple necrosis on root, epidermal and cortical tissues and corm; water-soaked lesions (tobacco); tunnels, cavities (citrus). Tubers: dry rot, deep cracks, corky appearance. Outer peel of tubers with lesions, scabs, pustules	Bananas, yam, coffee, citrus, spices (very wide host range)	Tropical: worldwide
P. goodeyi Sher and Allen, 1953	Small bunches, weak or reduced flowers, toppling. Necrotic lesions and destruction of root cortical tissues	Bananas and plantains	Subtropical: E and W Africa, Canaries
P. zeae Graham, 1951	Root: brown to red lesions in the cortex. Necrosis of stelar and cortical tissues forming cavities. Reduction of stalk length (sugar cane)	Maize, sorghum, sugarcane	Worldwide
Punctodera			
P. chalcoensis Stone, Sosa-Moss, and Mulvey, 1976 (Maize cyst nematode)	Smaller root systems, short lateral roots ("bottle brush"-like). Pale strips on the leaves	Maize, teozinte	Mexico
Radopholus			
R. citri Machon and Bridge, 1996	Cortical necrosis and root destruction. Reduction of leaf nodes	Citrus	Tropical: Indonesia

Continued

Nematode	Damage symptoms	Main crops affected	Distribution
R. similis (Cobb, 1893) Thorne, 1949 (Burrowing nematode)	Banana: toppling. Dark lesions on roots and corm, girdling. Black pepper: defoliation, cessation of growth, die-back, shedding of inflorescences. Orange to purple lesions on roots	Bananas, citrus, root and tuber crops, coconut, tea, black pepper, and other spices	Tropical: worldwide
Rhadinaphelenchus			
R. cocophilus (Cobb, 1919) Goodey, 1960 (Red ring nematode)	Tip chlorosis of oldest leaves that spreads toward their bases. Stem cross section with an orange to brick-red colored ring (2 – 4 cm wide), 3.5 cm from periphery	Coconut, oil palm	S and C America, Caribbean
Rotylenchulus			
R. parvus (Williams, 1960) Sher, 1961	Severe distortion of roots, occasional lesions. Slowing decline (sugarcane)	Pigeon pea, sweet potato, maize, sugarcane, cowpea	Tropical: Africa
R. reniformis Linford and Oliveira, 1940	Stunting and leaf curling. Root and cortical necrosis. Dark appearance of roots due to soil adhering to egg masses	Pineapple, vegetables	Tropical: worldwide
Rotylenchus			
R. robustus (de Man, 1876) Filipjev, 1936	Reduction of seedling growth (spruce)	Vegetables, tree seedlings	Temperate: Europe, N America, India

Continued

Nematode	Damage symptoms	Main crops affected	Distribution
Scutellonema			
S. bradys (Steiner and LeHew, 1933) Andrássy, 1958 (Yam nematode)	Cream, light yellow lesions below outer skin; light brown to dark tissues, external cracks and flaking off	Yams	Tropical: W Africa, Caribbean
S. cavenessi Sher, 1964	Reduction of number of lateral roots; chlorotic foliage	Groundnuts	Tropical: W Africa
Trichodorus			
T. primitivus (de Man, 1880) Micoletzky, 1922	Docking disorder, spraing in potato tubers (Tobacco Rattle Virus)	Sugar beet, potato, coniferous and deciduous trees	Temperate: Europe, N America, New Zealand
T. viruliferus Hooper, 1963	Browning of root tip tissue, cessation of root growth; vector for Pea Early Browning Virus and Tobacco Rattle Virus	Sugar beet, potato	Temperate: Europe, N America
Trophotylenchulus			
T. piperis Mohandas, Ramana, and Raski, 1985	Yellowing and die-back. Reduced root growth, dirty roots. Roots discolored	Black pepper	Southern India
T. obscurus (Colbran, 1961) Cohn and Kaplan, 1983	Damage to root cortical cells. Dark brown capsules (containing eggs and nematodes) on the root surface	Coffee	Tropical: Africa

Continued

Nematode	Damage symptoms	Main crops affected	Distribution
Tylenchorhynchus			
T. acutus Allen, 1955 (Stylet stunt nematode)	Poor root system with few feeder roots	Sorghum, soybean, tobacco, lucerne	N America (Canada, USA), Pakistan
Tylenchulus			
T. semipenetrans Cobb, 1913 (Citrus nematode)	Wilting. Smaller, chlorotic leaves. Dirty roots. Fibrous roots thicker, root lesions, cortical sloughing	Citrus	Subtropical/tropical: citrus growing areas worldwide
Xiphinema			
X. americanum Cobb, 1913	Necrosis and destruction of feeder roots	Trees, grapevines, strawberries	N America
X. diversicaudatum (Micoletzky, 1927) Thorne, 1939	Root tip galls, necrosis	Roses, grapevines, soft fruits	Temperate: Europe, N America, Australia, New Zealand
X. index Thorne and Allen, 1950	Swelling of root tips; breaks in the cortex, necrosis	Grapevines, fruit trees, rose	Temperate: Europe, S America, Mediterranean, S Africa, Australia

a cell of the endodermis, pericycle or vascular parenchyma. The cell vacuole is replaced by cytoplasm with many organelles and secondary vacuoles. Wall ingrowths typical of transfer cells are present. Root vascular organization is distorted by enlargement of the giant cell, which contacts both xylem and sieve elements and appears not to crush surrounding cells.

The altered cells or "nurse cells" at the feeding site of adult *Tylenchulus semipenetrans* are situated in the root cortex. The head of the nematode is located in an empty cortical cell, surrounded by 6 – 10 nurse cells. They are discrete cortical cells with enlarged nuclei and nucleoli and in which the central vacuole is replaced by cytoplasm. No mitotic stimulation or cell enlargement occurs.

Tylenchorhynchus dubius is among the most abundant and widespread of phytophagous nematodes in Europe and is also recorded from other areas, including North America and India. It possesses a short stylet and causes only moderate perturbance of the protoplast within the cells on which it feeds. Secretions are injected from the dorsal gland after perforation of the cell wall but before ingestion, and these may pre-digest a zone of cytoplasm around the inserted stylet tip (Sijmons, et al., 1994). It has occasionally been reported as a noxious plant parasite from countries in temperate as well as subtropical zones, and has been associated with poor growth of cotton, field beans, oats, grasses, spruce seedlings, wheat, turnip, millet, and sorghum (Bridge, 1974). In combination with *Phoma medicaginis*, it induces black stem foot disease in peas. *Tylenchorhynchus dubius* causes damage at densities that occur normally in the field and therefore is a plant-parasitic nematode of potential economic importance to which attention should be paid (Sharma, 1971).

Most plant-parasitic species that have well-known roles as disease or damage causing agents are listed in Table 13.5.

13.12 Worldwide Recognition

The worldwide distribution of economically important plant parasites means that they have played important parts in the economy of ancient civilizations, and may have been wholly or partly responsible for related rapid deterioration of soils. They may even have been responsible for land being abandoned for agricultural purposes and the consequent migration of human populations (Thorne, 1961).

Diseases caused by nematodes are usually correlated with the presence of nematologists, so the distribution of nematologists gives us clues to the importance that nematodes have had in the past and at present. However, scarcity of nematologists may result in poor or wrong diagnosis, considering the broad spectrum of symptoms that are shared with other diseases not caused by nematodes (i. e., yellowing, wilting, stunting, etc.). The importance of nematodes is often overlooked because they are small, are concealed in soil, and often do not cause specific symptoms. Lack of awareness of plant-parasitic nematodes in soil can lead to the slow, invisible build-up and spread of potentially pathogenic populations. Nematodes are frequently discarded as a first option in diagnosis of plant disease, being ranked as less important than fungi, viruses, and bacteria. Also, recognition of the importance of plant-parasitic nematodes may be related to the value and importance of the crop concerned in a particular region or country. This is partly why we have amassed much information on species like potato cyst nematodes (*Globodera rostochiensis* and *G. pallida*), soybean cyst nematode (*Heterodera glycines*), and root-knot nematodes (*Meloidogyne* spp.) in temperate zones, but not much on the nematodes in subsistence agroecosystems.

Nematode pathology studies are less numerous than in other plant pathology disciplines, the reason apparently being that plant nematologists are more focused on the management of nematode populations than on understanding pathological mechanisms or searching for the basis of susceptibility (Mateille, 1994). We still know little or nothing of the feeding habits, hosts, and pathogenicity of hundreds of species of stylet-bearing (and, therefore, potentially plant pathogenic) nematodes. Association of nematodes with numerous plant diseases is now well recognized, especially those involved in root rots of various types. These associations may present widely varying symptoms because of the many factors involved. Vigor of host, species of nematode, and cultural and climatic conditions combine with certain fungi and bacteria to create the disease complex. The doctrine of specific etiology is easier to work with in order to explain and understand plant disease, but nature does not work with pure cultures. Questions and problems related to disease etiology have been (and still are) the bane of plant pathologists and nematologists, ever since the establishment of the germ theory led to the intensive investigation of the causes of disease (Powell, 1971) and disease complexes (Powell, 1963). The "slow decline", or "dieback", of walnut, cherry, prune, apple, citrus, and other orchard trees has been almost invariably associated with large populations of *Pratylenchus, Xiphinema, Tylenchulus, Rotylenchulus, Meloidogyne, Criconemella*, and other plant-parasitic nematodes whose pathogenicity has been difficult to prove. Often these species were indigenous and merely transferred from their native hosts to the orchard trees on which they multiplied. Certain endoparasitic forms, however, have been introduced through infested nursery stock, seedling plants, tubers, bulbs, soil, and other agencies (Thorne, 1961).

13. 13 Future Perspectives

The development of a scientific method to prove the pathogenicity of plant-parasitic nematodes and their role as disease agents has been a major achievement in the discipline of plant nematology. To study, understand and control disease, different approaches have been considered: holistic and analytical, both strongly influenced by economic and scientific objectives that are also linked to scientific philosophy.

Agricultural sicknesses are caused by both biotic and abiotic sources of stress that limit the expression of the genetic potential of major world crops (Boyer, 1982). Despite our poor knowledge of how to partition the causes, there is still confidence that we can improve yields, lower production costs, and improve crop quality. This will be achieved by acknowledging soil properties and recognizing weed, pest, and disease problems. The measurement of the spatial distribution of nematode populations (and other factors) more accurately will allow us to forecast yields through geographic information systems and perhaps to use remote sensing and variable application rate technology to manage the problems (Strickland, et al., 1998). The application of site-specific management technologies seems most promising for nematode problems (Evans, et al., 2003).

At the beginning of the new millennium, many of the challenges related to the causes of disease still remain, despite the tremendous advances that immunological and molecular technologies have allowed us to make. Elucidating the nature of nematode secretions and their biological activity is one of the most intriguing and challenging areas of future research. Purification and characterization of these secretions will enhance our understanding of the molecular interactions that occur during pathogenesis (Hussey, 1989).

The generation of cDNA libraries and expressed sequence tags (ESTs) constitute a novel approach to isolate nematode parasitism genes. So far, the only putative parasitism genes from plant-parasitic nematodes have been cloned from the esophageal gland cells (Ding, et al., 1998; Lambert, et al., 1999; Rosso, et al., 1999; Smant, et al., 1998). The identification of genes encoding bioactive molecules from nematodes that initiate and maintain successful parasitic interactions with host plants is another area of active investigation, primarily by reverse-genetic approaches (Davis, et al., 2000).

The developmental regulation and the detailed characterizations conducted with genes cloned from the esophageal gland cells of plant-parasitic nematodes has made them good models for the analysis of gene structure, regulation, and function. Modifications of functional assays used for *C. elegans* genes are being explored to develop methods to determine the function(s) of isolated plant nematode parasitism genes but "proof of concept" for gene/protein function is still difficult, due to lack of a functional mutant analysis and complementation scheme (Davis, et al., 2000). However, new insights are being provided into the evolution of nematodes and their adaptation to parasitism. The discovery that a gene expressed in the esophageal gland cells of a plant-parasitic nematode is very similar to a bacterial gene lends support to the hypothesis that parasitism genes in plant nematodes may have been acquired, at least in part, by gene transfer from micro-organisms that inhabit the same parasitic niche (Davis, et al., 2000).

Few diseases caused by plant-parasitic nematodes have been subjected to such detailed study as those caused by root-knot and cyst nematodes. Increased competition for the allocation of research funds has meant that research on plant-parasitic nematodes has become increasingly restricted to the major species attacking high value crops, and our knowledge of other nematode-induced diseases remains limited. Thus, certain areas of research on plant-nematode interactions are advancing at tremendous pace (Hussey, 1989; Davis, et al., 2000), whilst others are neglected. Worldwide, the study, management, and treatment of major crop diseases shows wide variation in knowledge, and a tremendous gap between the research approaches undertaken in the developed and developing worlds. The first is deeply involved in genomics and molecular engineering, whilst accurate diagnosis and management strategies remain of prime importance for the second. Agrochemicals (i. e., nematicides) are expensive and only considered for use on very valuable crops in the developing world, so researchers are continuously faced with a pressing need to develop inexpensive, and therefore, almost by default, environmentally friendly control strategies. This is not usually because of ecological awareness, but to overcome the low incomes of farmers and to sustain production. How will we fill this gap?

Nematologists will require continuous feedback from agricultural sciences and should ideally be exposed to diagnosis of diseases in the field. Some diseases are tentatively linked with nematodes (e. g., "Platte Valley Yellows" with spiral nematodes; May, 2000) but proof of pathogenicity is still awaited. New diseases will be discovered, described, and characterized, but living systems are integrated wholes whose properties cannot necessarily be reduced to those of their constituent parts. Their essential or "systemic" properties are properties of the whole, which none of the parts have (Capra, 1997). As nematologists, we will need to recognize the necessity of a multidisciplinary, integrated systems approach, including both agronomy and social science, if we want to overcome present (and future) limitations to crop production. Genetic improvement of crops is considered the most viable approach to ensure that food production keeps pace with the anticipated growth of the human population, and should benefit from the use of molecular techniques to search the germplasm of wild species for useful genes (Browning, 1998; Tanksley and McCouch, 1997).

Diseases can have devastating effects and cause famine, and we should not forget that security of food supply is a priority for every country. Insufficient appropriate research and badly planned disease management strategies can also have devastating effects. Global warming will modify crop production and the relative importance of present crop diseases. Also, new arable land and the intensive exploitation (mono-cropping) of previously localized crops will bring new diseases into prominence. Improvement of management strategies to meet the challenges that these changes will bring requires both broader and deeper knowledge of crop disease. All involved in agriculture, not only researchers but also administrators, decision makers, and policy makers, should recognize the possible influence of plant nematodes on crop production and the contribution that a research effort on these pathogens can make to the development of the more intensive, yet sustainable, agriculture the world is going to need (Sharma, et al., 1997).

"Biodiversity" in agroecosystems, "sustainable management", and "organic farming" are recently introduced terms, but are already familiar to consumers, who require scientists and producers to accomplish the practicalities of achieving their philosophy. We must link these approaches and use them effectively in future comprehensive schemes for disease management.

"A sustainable society is one that satisfies its needs without diminishing the prospects of future generations" (Brown, 1981).

References

Agrios, G. N. 1997. Plant Pathology. San Diego, CA: Academic Press.

Arya, M., and B. Tiagi. 1985. Nucleic acid changes in total proteins in three carrot cultivars infested with *Meloidogyne incognita*. Indian Journal of Nematology 12: 398 – 400.

Atkinson, G. F. 1892. Some diseases of cotton. Alabama Polytechnic Institute Agricultural Experiment Station Bulletin No. 41: 61 – 65.

Ayoub, S. M. 1980. Plant Nematology. An Agricultural Training Aid. Sacramento, CA: NemaAid Publication.

Barker, K. R., and R. S. Hussey. 1976. Histopathology of nodular tissues infected with certain nematodes. Phytopathology 66: 851 – 855.

Barker, K. R., C. C. Carter, and J. N. Sasser. 1985. An Advanced Treatise on *Meloidogyne*, Volume II Methodology. Raleigh, NC: North Carolina State University Graphics.

Bateman, D. F. 1978. The dynamic nature of disease. Pp. 53 – 83. In J. G. Horsfall and E. B. Cowling, eds. Plant Disease an Advanced Treatise. New York: Academic Press.

Been, T. H., and C. H. Schomaker. 1986. Quantitative analysis of growth, mineral composition and ion balance of the potato cultivar Irene infested with *Globodera pallida* (Stone). Nematologica 32: 339 – 355.

Bird, A. F. 1972. Quantitative studies on the growth of syncytia induced in plants by root-knot nematodes. International Journal of Parasitology 2: 157 – 170.

Bird, A. F. 1974. Plant response to root-knot nematode. Annual Review of Phytopathology 12: 69 – 85.

Bird, A. F. 1975. Symbiotic relationships between nematodes and plants. Symposia of the Society for Experimental Biology 29: 351 – 371.

Bird, A. F., and B. R. Loveys. 1975. The incorporation of photosynthates by *Meloidogyne javanica*. Journal of Nematology 7: 111 – 113.

Bird, D. McK., and A. Bird. 2001. Plant-parasitic nematodes. Pp. 139 – 166. In M. W. Kennedy and W. Hamett, eds. Parasitic Nematodes. Wallingford, UK: CAB International.

Bird, D. McK., and C. H. Opperman. 1998. *Caenorhabditis elegans*: a genetic guide to parasitic nematode biology. Journal of Nematology 30: 299 – 308.

Blake, C. D. 1969. Nematode parasites of banana and their control. Plant nematodes - proofs of pathogenicity. Pp. 109 – 132. In J. E. Peachey, ed. Nematodes of Tropical Crops. Farnham Royal, UK: Commonwealth Agricultural Bureaux.

Bleve-Zacheo, T., G. Zacheo, M. T. Melillo, F. Lamberti, and O. Arrigoni. 1980. Cytological changes induced by the stem nematode *Ditylenchus dipsaci* in strawberry leaves. Nematologia Mediterranea 8: 153 – 163.

Bos, L., and J. E. Parlevliet. 1995. Concepts and terminology on plant/pest relationships: toward consensus in plant pathology and crop protection. Annual Review of Phytopathology 33: 69 – 102.

Boyer, J. S. 1982. Plant productivity and environment. Science 218: 443 – 448.

Bridge, J. 1972. Nematode problems with yams (*Dioscorea* spp.) in Nigeria. PANS 1: 89 – 91.

Bridge, J. 1973. Nematodes as pests of yams in Nigeria. Mededelingen van de Faculteit Landbouwwetenschappen, Rijksuniversiteit, Gent 38: 841 – 852.

Bridge, J. 1974. *Tylenchorhynchus dubius*. C. I. H. Descriptions of Plant-parasitic Nematodes, Set 4, No. 51. Farnham Royal, UK: Commonwealth Agricultural Bureaux.

Bridge, J. 1987a. Aspects of control. Pp. 8 – 11. In S. B. Jepson. Identification of Root-knot Nematodes (*Meloidogyne* species). Wallingford, UK: CAB International.

Bridge, J. 1987b. Control strategies in subsistence agriculture. Pp. 389 – 420. In R. H. Brown and B. R. Kerry, eds. Principles and Practice of Nematode Control in Crops.

Sydney, Australia: Academic Press.

Bridge, J., M. Luc, and R. A. Plowright. 1990. Nematode parasites of rice. Pp. 69 – 108. In M. Luc, R. A. Sikora, and J. Bridge, eds. Plant Parasitic Nematodes in Subtropical and Tropical Agriculture. Wallingford, UK: CAB International.

Bridge, J., J. J. Mortimer, and G. V. H. Jackson. 1983. *Hirschmanniella miticausa* n. sp. (Nematoda: Pratylenchidae) and its pathogenicity on taro (*Colocasia esculenta*). Revue de Nématologie 6: 285 – 290.

Brodie, B. B., K. Evans, and J. Franco. 1993. Nematode parasites of potatoes. Pp. 87 – 132. In K. Evans, D. L. Trudgill, and J. M. Webster, eds. Plant Parasitic Nematodes in Temperate Agriculture. Wallingford, UK: CAB International.

Brown, D. J. F., W. M. Robertson, and D. L. Trudgill. 1995. Transmission of viruses by plant nematodes. Annual Review of Phytopathology 33: 223 – 249.

Brown, L. R. 1981. Building a Sustainable Society. New York: Norton.

Browning, J. A. 1998. One phytopathologist's growth through IPM to holistic plant health: the key to approaching genetic yield potential. Annual Review of Phytopathology 36: 1 – 24.

Byars, L. P. 1914. Preliminary notes on the cultivation of the plant parasitic nematode, *Heterodera radicicola*. Phytopathology 4: 323 – 326.

CAB International. 1999. Crop Protection Compendium. Global Module (CD). Wallingford, UK: CAB International.

Campos, V. P., P. Sivapalan, and N. C. Gnanapragasam. 1990. Nematode parasites of coffee, cocoa and tea. Pp. 387 – 430. In M. Luc, R. A. Sikora, and J. Bridge, eds. Plant Parasitic Nematodes in Subtropical and Tropical Agriculture. Wallingford, UK: CAB International.

Canto-Saenz, M. 1992. Life cycle and pathogenicity of *Nacobbus aberrans* (Thorne, 1935) Thorne and Allen, 1944. Pp. 113 – 127. In F. J. Gommers and P. W. Th. Maas, eds. Nematology from Molecule to Ecosystem. Wildervank, The Netherlands: Dekker and Huisman.

Capra, F. 1997. The Web of Life. London: Flamingo, Harper Collins Publishers.

Castillo, P. G., V. E. Hernandez, O. D. Teliz, A. D. Nieto, H. J. Castillo, and B. R. Ruiz. 1992. La corchosis del cafeto en Mexico. XV Simposio Latinoamericano de Cafeticultura; Jalapa, Vera Cruz, Mexico. Pp. 6 – 13.

Caubel, G., R. Duchesne, and R. Rivoal. 1978. Use of aerial photography in Nematology. Annales de Phytopathologie 10: 273.

Christie, J. R. 1959. Plant Nematodes. Their Bionomics and Control. Jacksonville, Florida: The H. & W. B. Drew Company.

Christie, J. R., and V. G. Perry. 1951. A root disease of plants caused by a nematode of the genus *Trichodorus*. Science 113: 491 – 493.

Clark, S. 1967. The development and life history of the false root-knot nematode, *Nacobbus serendipiticus*. Nematologica 13: 91 – 101.

Cobb, N. A. 1893. Nematodes, mostly Australian and Fijian. Department of Agriculture for New South Wales Miscellaneous Publication 13.

Cohn, E. 1975. Relations between *Xiphinema* and *Longidorus* and their host plants. Pp. 365 – 386. In F. Lamberti, C. E. Taylor, and J. W. Seinhorst, eds. Nematode Vectors of Plant Viruses. UK: Plenum Press.

Cohn, E., and L. W. Duncan. 1990. Nematode parasites of subtropical and tropical fruit trees. Pp. 347 – 362. In M. Luc, R. A. Sikora, and J. Bridge, eds. Plant Parasitic Nematodes in Subtropical and Tropical Agriculture. Wallingford, UK: CAB International.

Cook, R., and K. Evans. 1987. Resistance and Tolerance. Pp. 179 – 231. In R. H. Brown and B. R. Kerry, eds. Principles and Practice of Nematode Control in Crops. Sydney, Australia: Academic Press.

Corbett, D. C. M., and G. A. Hide. 1971. Interactions between *Heterodera rostochiensis* Woll. and *Verticillium dahliae* Kleb. on potatoes and the effect of CCC on both. Annals of Applied Biology 68: 71 – 80.

Cornu, M. 1879. Études sur le *Phylloxera vastatrix*. Mémoires présentés par divers savants à l' Academie des sciences de l'Institut de France, Paris 26: 164 – 174.

Costilla, M. A. 1985. El falso nematodo del nudo *Nacobbus aberrans* (Thorne, 1935) Thorne y Allen, 1944 y su relacion con el cultivo de la papa en el noroeste argentino. Revista Industrial y Agricola de Tucuman 62: 79 – 97.

Crow, W. T., D. W. Dickson, and D. P. Weingartner. 1997. Stubby-root symptoms on cotton induced by *Belonolaimus longicaudatus*. Journal of Nematology 29: 574 (abstr.).

Crow, W. T., D. W. Dickson, D. P. Weingartner, R. McSorley, and G. L. Miller. 2000a. Yield reduction and root damage to cotton induced by *Belonolaimus longicaudatus*. Journal of Nematology 32: 205 – 209.

Crow, W. T., D. P. Weingartner, R. McSorley, and D. W. Dickson. 2000b. Population dynamics of *Belonolaimus longicaudatus* in a cotton production system. Journal of Nematology 32: 210 – 214.

Davis, E. L., R. S. Hussey, T. J. Baum, J. B. Bakker, A. Schots, M. N. Rosso, and P. Abad. 2000. Nematode parasitism genes. Annual Review of Phytopathology 38: 365 – 396.

Dimalla, G. G., and J. van Staden. 1977. Cytokinins in the root-knot nematode, *Meloidogyne incognita*. Plant Science Letters 10: 25 – 29.

Ding, X., J. Shields, R. Allen, and R. S. Hussey. 1998. A secretory cellulose-binding protein cDNA cloned from the root-knot nematode (*Meloidogyne incognita*). Molecular Plant-Microbe Interactions 11: 952 – 959.

Diomandé, M., and M. K. Beute. 1981. Effects of *Meloidogyne hapla* and *Macroposthonia ornata* populations on *Cylindrocladium* black rot on peanut. Phytopathology 71: 491 – 496.

Dorhout, R., C. Kollöffel, and F. J. Gommers. 1992. Alteration of distribution of regions with high net proton extrusion in tomato roots infected with *Meloidogyne incognita*. Physiological and Molecular Plant Pathology 40: 153 – 162.

Dropkin, V. H. 1969. Cellular responses of plants to nematode infections. Annual Review of Phytopathology 7: 101 – 122.

Duncan, L. W., and E. Cohn. 1990. Nematode parasites of citrus. Pp. 321 – 346. In M. Luc, R. A. Sikora, and J. Bridge, eds. Plant Parasitic Nematodes in Subtropical and Tropical Agriculture. Wallingford, UK: CAB International.

Eisenback, J. D. 1999. Root-Knot Nematode Taxonomic Database (CD). Wallingford, UK: CAB International.

Endo, B. Y. 1987. Histopathology and ultrastructure of crops invaded by certain sedentary endoparasitic nematodes. Pp. 196 – 210. In J. A. Veech and D. W. Dickson, eds. Vistas on Nematology. DeLeon Springs, FL: Society of Nematologists.

Evans, K., and P. P. J. Haydock. 2000. Potato cyst nematode management – present and future. Aspects of Applied Biology 59: 91 – 97.

Evans, K., D. L. Trudgill, and N. J. Brown. 1977. Effects of potato cyst-nematodes on potato plants. V. Root system development in lightly and heavily infected susceptible and resistant varieties, and its importance in nutrient and water uptake. Nematologica 23: 153 – 164.

Evans, K., R. Webster, A. Barker, P. Halford, M. Russell, J. Stafford, and S. Griffin. 2003. Mapping infestations of potato cyst nematodes and the potential for spatially varying application of nematicides. Precision Agriculture, 4: 149 – 162.

Fatemy, F., and K. Evans. 1986a. Growth, water uptake and calcium content of potato cultivars in relation to tolerance of cyst nematodes. Revue de Nématologie 9: 171 – 179.

Fatemy, F., and K. Evans. 1986b. Effects of *Globodera rostochiensis* and water stress on shoot and root growth and nutrient uptake of potatoes. Revue de Nématologie 9: 181 – 184.

Fatemy, F., P. K. E. Trinder, J. N. Wingfield, and K. Evans. 1985. Effects of *Globodera rostochiensis*, water stress and exogenous abscisic acid on stomatal function and water use of Cara and Pentland Dell potato plants. Revue de Nématologie 8: 249 – 255.

Fenwick, D. W. 1969. Red ring disease of the coconut palm. Pp. 89 – 98. In J. E. Peachey, ed. Nematodes of Tropical Crops. Farnham Royal, UK: Commonwealth Agricultural Bureaux.

Flor, H. H. 1955. Host-parasite interaction in flax rust – its genetics and other implications. Phytopathology 45: 680 – 685.

Fortuner, R. 1987. A reappraisal of Tylenchina (Nemata). 8. The family Hoplolaimidae Filip'ev, 1934. Revue de Nématologie 10: 219 – 232.

Fortuner, R., and M. Luc. 1987. A reappraisal of Tylenchina (Nemata). 6. The family Belonolaimidae Whitehead, 1960. Revue de Nématologie 10: 183 – 202.

Fortuner, R., and A. R. Maggenti. 1987. A reappraisal of Tylenchina (Nemata). 4. The family Anguinidae Nicoll, 1935 (1926). Revue de Nématologie 10: 163 – 176.

Franco, J. 1980. Effect of potato-cyst nematode, *Globodera rostochiensis*, on photosynthesis of potato plants. Fitopatologia 15: 1 – 6.

Franco, J., R. Montecinos, and N. Ortuño. 1992. Management strategies of *Nacobbus aberrans*. Pp. 240 – 247. In F. J. Gommers and P. W. Th. Maas, eds. Nematology from Molecule to Ecosystem. Wildervank, The Netherlands: Dekker and Huisman.

Gaunt, R. R. 1995. The relationship between plant disease severity and yield. Annual Review of Phytopathology 33: 119 – 144.

Germani, G., and M. D'Héry. 1973. Observations et experimentations concordant le role des nematodes dans deux affections de l' arachide en Haute Volta: la chlorose et le clump. Oleagineaux 28: 235 – 242.

Godfrey, G. H., J. M. Oliveira, and H. M. Hoshino. 1934. Increased efficiency of chloropicrin for nematode control. Phytopathology 24: 1332 – 1346.

Goeldi, E. A. 1892. Relatorio sobre a molestia do cafeeiro na provincia do Rio de Janeiro. Archivos do Museo Nacional, Rio de Janeiro 8: 7 – 121.

Goellner, M., X. Wang, and E. L. Davis. 2001. Endo-β-1, 4-glucanase expression in compatible plant-nematode interactions. The Plant Cell 13: 2241 – 2255.

Gowen, S., and P. Quénéhervé. 1990. Nematode parasites of bananas, plantains and abaca. Pp. 431 – 460. In M. Luc, R. A. Sikora, and J. Bridge, eds. Plant Parasitic Nematodes in Subtropical and Tropical Agriculture. Wallingford, UK: CAB International.

Graham, T. W., and Q. L. Holdeman. 1953. The sting nematode *Belonolaimus gracilis* Steiner: A parasite on cotton and other crops in South Carolina. Phytopathology 43: 434 – 439.

Griffith, R., and P. K. Koshy. 1990. Nematode parasites of coconut and other palms. Pp. 363 – 386. In M. Luc, R. A. Sikora, and J. Bridge, eds. Plant Parasitic Nematodes in Subtropical and Tropical Agriculture. Wallingford, UK: CAB International.

Griffiths, B. S., and W. M. Robertson. 1984. Morphological and histochemical changes occurring during the life-span of root-tip galls on *Lolium perenne* induced by *Longidorus elongatus*. Journal of Nematology 16: 223 – 229.

Gupta, P., and G. Swarup. 1972. Ear-cockle and yellow ear-rot diseases of wheat. II. Nematode bacterial association. Nematologica 18: 320 – 324.

Harlan, J. R. 1976. Diseases as a factor in plant evolution. Annual Review of Phytopathology 14: 31 – 51.

Hewitt, W. B., D. J. Raski, and A. C. Goheen. 1958. Nematode vector of soilborne fanleaf virus of grapevines. Phytopathology 48: 586 – 595.

Holdeman, Q. L., and T. W. Graham. 1954. Effect of the sting nematode on expression of Fusarium wilt in cotton. Phytopathology 44: 683 – 685.

Hooper, D. J., and J. F. Southey. 1978. *Ditylenchus, Anguina* and related genera. Pp. 78 – 97. In J. F. Southey, ed. Plant Nematology. London: Ministry of Agriculture, Fisheries and Food, Her Majesty's Stationery Office.

Huang, C. S., and A. R. Maggenti. 1969. Mitotic aberrations and nuclear changes in developing giant cells in *Vicia faba* caused by root-knot nematode, *Meloidogyne javanica*. Phytopathology 59: 447 – 455.

Hubert, F. P. 1957. Diseases of some export crops in Indonesia. Plant Disease Reporter 41: 55 – 64.

Huettel, R. N., D. W. Dickson, and D. T. Kaplan. 1984. *Radopholus citrophilus* n. sp. (Nematoda), a sibling species of *Radopholus similis*. Proceedings of the

Helminthological Society of Washington 51: 32 – 35.

Hunger, F. W. T. 1901. Een bacterie-ziekte den tomaat. S Lands Plantentium, Mededelingen 48: 4 – 57.

Hunt, D. J. 1993. Aphelenchida, Longidoridae and Trichodoridae: Their Systematics and Bionomics. Wallingford, UK: CAB International.

Hussey, R. S. 1989. Disease-inducing secretions of plant-parasitic nematodes. Annual Review of Phytopathology 27: 123 – 141.

Hussey, R. S., and K. R. Barker. 1976. Influence of nematodes and light sources on growth and nodulation of soybean. Journal of Nematology 8: 48 – 52.

Inomoto, M. M., C. M. Oliveira, P. Mazzafera, and W. Gonçalves. 1998. Effects of *Pratylenchus brachyurus* and *P. coffeae* on seedlings of *Coffea arabica*. Journal of Nematology 30: 362 – 367.

Inserra, R. N., N. Vovlas, G. D. Griffin, and J. L. Anderson. 1983. Development of the false root-knot nematode, *Nacobbus aberrans*, on sugar beet. Journal of Nematology 15: 288 – 296.

Inserra, R. N., M. Di Vito, and H. Ferris. 1984. Influence of *Nacobbus aberrans* densities on growth of sugar beet and Kochia in pots. Journal of Nematology 16: 393 – 395.

Jatala, P., and J. Bridge. 1990. Nematode parasites of root and tuber crops. Pp. 137 – 180. In M. Luc, R. A. Sikora, and J. Bridge, eds. Plant Parasitic Nematodes in Subtropical and Tropical Agriculture. Wallingford, UK: CAB International.

Jepson, S. B. 1987. Identification of Root-Knot Nematodes (*Meloidogyne* species). Wallingford, UK: CAB International.

Jones, B. L., and D. De Waele. 1990. Histopathology of *Ditylenchus destructor* on peanut. Journal of Nematology 22: 268 – 272.

Jones, M. G. K. 1981. Host cell responses to endoparasitic nematode attack: structure and function of giant cells and syncytia. Annals of Applied Biology 97: 353 – 372.

Jones, M. G. K., and D. H. Northcote. 1972. Multinucleate transfer cells induced in *Coleus* roots by the root-knot nematode, *Meloidogyne arenaria*. Protoplasma 75: 381 – 395.

Jones, M. G. K., and H. L. Payne. 1978. Early stages of nematode-induced giant-cell formation in roots of *Impatiens balsamina*. Journal of Nematology 10: 70 – 84.

Kaplan, D. T. 1985. Influence of the sting nematode, *Belonolaimus longicaudatus*, on young citrus trees. Journal of Nematology 17: 408 – 414.

Kaplan, D. T., W. K. Thomas, L. M. Frisse, J. L. Sarah, J. M. Stanton, P. R. Speijer, D. H. Marin, and C. H. Opperman. 2000. Phylogenetic analysis of geographically diverse *Radopholus similis* via rDNA sequence reveals a monomorphic motif. Journal of Nematology 32: 134 – 142.

Kelman, A. 1995. Contributions of plant pathology to the biological sciences and industry. Annual Review of Phytopathology 33: 1 – 21.

Ko, M. P., K. R. Barker, and J. S. Huang. 1984. Nodulation of soybeans as affected by half-root infection with *Heterodera glycines*. Journal of Nematology 16: 97 – 105.

Koch, R. 1884. Die Aetiologie der Tuberculose. Mitteilungen aus dem Kaiserlichen Gesundheitsamte 2: 1.

Koshy, P. K., and J. Bridge. 1990. Nematode parasites of spices. Pp. 557 – 582. In M. Luc, R. A. Sikora, and J. Bridge, eds. Plant Parasitic Nematodes in Subtropical and Tropical Agriculture. Wallingford, UK: CAB International.

Kotcon, J. B., and R. Loria. 1986. Influence of *Pratylenchus penetrans* on plant growth and water relations in potato. Journal of Nematology 18: 385 – 392.

Lambert, K. N., K. D. Allen, and I. M. Sussex. 1999. Cloning and characterization of an esophageal-gland specific chorismate mutase from the phytoparasitic nematode *Meloidogyne javanica*. Molecular Plant-Microbe Interactions 12: 328 – 336.

Lewis, S. A., and M. A. McClure. 1975. Free amino acids in roots of infected cotton seedlings resistant and susceptible to *Meloidogyne incognita*. Journal of Nematology 7: 10 – 15.

Lili, Z., J. Duchesne, H. Nicolas, and R. Rivoal. 1991. Detection infrarouge thermique des maladies du blé d'hiver. OEPP Bulletin 21: 659 – 672.

Loof, P. A. A. 1978. The genus *Pratylenchus* Filipjev, 1936 (Nematoda: Pratylenchidae). A review of its anatomy, morphology, distribution, systematics and identification. Sveriges Lantbruksuniversitet Västskyddsrapporter 5: 1 – 50.

Loveys, B. R., and A. F. Bird. 1973. The influence of nematodes on photosynthesis in tomato plants. Physiological Plant Pathology 3: 525 – 529.

Luc, M. 1987. A reappraisal of Tylenchina (Nemata). 7. The family Pratylenchidae Thorne, 1949. Revue de Nématologie 10: 203 – 218.

Luc, M., A. R. Maggenti, and R. Fortuner. 1988. A reappraisal of Tylenchina (Nemata). 9. The family Heteroderidae Filip'ev & Schuurmans Stekhoven 1941. Revue de Nématologie 11: 159 – 176.

Malek, R. B., and J. E. Appleby. 1984. Epidemiology of pine wilt in Illinois. Plant Disease 68: 180 – 186.

Mamiya, Y. 1983. Pathology of pine wilt disease caused by *Bursaphelenchus xylophilus*, Annual Review of Phytopathology 21: 201 – 220.

Mankau, R., and S. K. Mankau. 1963. The role of mycophagous nematodes in the soil. I. The relationships of *Aphelenchus avenae* to phytopathogenic soil fungi. Pp. 271 – 280. In J. Doeksen and J. van der Drift, eds. Soil Organisms. The Netherlands: North-Holland Publishing Co.

Marbán-Mendoza, N. 1989. Corchosis del cafeto, una amenza en la agricultura. Boletin Informativo, MIP/CATIE, Turrialba 13: 1 – 13.

Marks, R. J., and B. B. Brodie, eds. 1998. Potato Cyst Nematodes, Biology, Distribution and Control. Wallingford, UK: CAB International.

Masood, A., and S. K. Saxena. 1980. Nucleic acid changes in three tomato varieties infected with *Meloidogyne incognita*. Indian Journal of Nematology 10: 102 – 104.

Mateille, T. 1994. Biologie de la relation plantes-nématodes: perturbations physiologiques et mécanismes de défense des plantes. Nematologica 40: 276 – 311.

May, R. N. 2000. Spiral nematode (*Helicotylenchus*) population density correlated to

soybean fields displaying "Platte Valley yellows". Journal of Nematology 32: 444 – 445.
McCarter, J. P., D. McK. Bird, S. W. Clifton, and R. H. Waterston. 2000. Nematode gene sequences, December 2000 update. Journal of Nematology 32: 331 – 333.
McClure, M. A. 1977. *Meloidogyne incognita*: a metabolic sink. Journal of Nematology 9: 88 – 90.
McDonald, D., W. S. Bos, and M. H. Gumel. 1979. Effects of infections of peanut (groundnut) seed by the testa nematode, *Aphelenchoides arachidis*, on seed infection by fungi and on seedling emergence. Plant Disease Reporter 63: 464 – 467.
Meagher, J. W. 1977. World dissemination of the cereal-cyst nematode (*Heterodera avenae*) and its potential as a pathogen of wheat. Journal of Nematology 9: 9 – 15.
Melakeberhan, H., R. C. Brooke, and J. M. Webster. 1986. Relationship between physiological response of French beans of different age to *Meloidogyne incognita* and subsequent yield loss. Plant Pathology 35: 203 – 213.
Melakeberhan, H., R. C. Brooke, J. M. Webster, and J. M. D'Aurias. 1985. The influence of *Meloidogyne incognita* on the growth, physiology and nutrient content of *Phaseolus vulgaris*. Physiological Plant Pathology 26: 259 – 268.
Melakeberhan, H., J. M. Webster, R. C. Brooke, and J. M. D'Aurias. 1988. Effect of KNO_3 on CO_2 exchange rate, nutrient concentration and yield of *Meloidogyne incognita* infected beans. Revue de Nématologie 11: 391 – 397.
Meon, S., J. M. Fisher, and H. R. Wallace. 1978. Changes in free proline following infection of plants with either *Meloidogyne javanica* or *Agrobacterium tumefasciens*. Physiological Plant Pathology 12: 251 – 256.
Miller, L. I. 1985. Economic importance of cyst Nematodes in North America. Pp. 373 – 386. In F. Lamberti and C. E. Taylor, eds. Cyst Nematodes. New York: Plenum Press.
Milne, D. L. 1982. Nematode pests of litchi. Pp. 38 – 41. In D. P. Keetch, and J. Heyns, eds. Nematology in southern Africa. Bulletin No. 400, Scientific Bulletins, Department of Agriculture and Fisheries, Republic of South Africa.
Minton, N. A., and E. B. Minton. 1966. Effect of root knot and sting nematodes on expression of *Fusarium* wilt of cotton in three soils. Phytopathology 56: 319 – 322.
Mountain, W. B. 1959. Studies on the pathogenicity of *Pratylenchus*. Pp. 414 – 417. In Recent Advances in Botany, Vol. 1. Proceedings of 9th International Botanical Congress, Montreal. Toronto: University of Toronto Press.
Mountain, W. B. 1960a. Theoretical considerations of plant nematode relationships. Pp. 419 – 421. In J. N. Sasser and W. R. Jenkins, eds. Nematology, Fundamentals and Recent Advances with Emphasis on Plant Parasitic and Soil Forms. Durnham, NC: The University of North Carolina Press.
Mountain, W. B. 1960b, Acceptable standards of proof and approaches for evaluating plant-nematode relationships. Pp. 422 – 425. In J. N. Sasser and W. R. Jenkins, eds. Nematology, Fundamentals and Recent Advances with Emphasis on Plant Parasitic and Soil Forms. Durnham, NC: The University of North Carolina Press.
Mullholland, V., L. Carde, K. J. O'Donnell, C. C. Fleming, and T. O. Powers. 1996.

Use of the polymerase chain reaction to discriminate potato cyst nematode at the species level. Diagnostics in Crop Protection, British Crop Protection Council Symposium Proceedings No. 65: 247 – 252.

Needham, T. 1743. A letter concerning certain chalky tubulous concretions called malm; with some microscopical observations on the farina of the red eily, and of worms discovered in smutty corn. Philosophical Transactions of the Royal Society 42: 173, 174, 634 – 641.

O'Bannon, J. H. 1977. Worldwide dissemination of *Radopholus similis* and its importance in crop production. Journal of Nematology 9: 16 – 25.

Oerke, E. C., H. W. Dehne, F. Schonbeck, and A. Weber. 1994. Crop Production and Crop Protection: Estimated Losses in Major Food and Cash Crops. Amsterdam: Elsevier.

Oostenbrink, M. 1969. Plant nematodes - proofs of pathogenicity. Pp. 81 – 83. In J. E. Peachey, ed. Nematodes of Tropical Crops. Farnham Royal, UK: Commonwealth Agricultural Bureaux.

Orion, D. 1973. Studies on plant root-knot nematode interrelationships. OEPP/EPPO Bulletin 9: 67 – 71.

Orion, D., G. D. Loots, and T. Orion. 1987. Cell lysis activity of *Meloidogyne* gelatinous matrix. Revue de Nématologie 13: 105 – 107.

Palmer, H. J., H. J. Atkinson, and R. N. Perry. 1991. The use of probes to identify *Ditylenchus dipsaci*. Revue de Nématologie 14: 625 – 628.

Palmer, H. J., H. J. Atkinson, and R. N. Perry. 1992. Monoclonal antibodies (MAbs) specific to surface expressed antigens of *Ditylenchus dipsaci*. Fundamental and Applied Nematology 15: 511 – 515.

Paulson, R. E., and J. M. Webster. 1970. Ultrastructure of the hypersensitive reaction in roots of tomato, *Lycopersicon esculentum* L., to infection by the root-knot nematode, *Meloidogyne incognita*. Physiological Plant Pathology 2: 224 – 227.

Pitcher, R. S. 1963. Role of plant-parasitic nematodes in bacterial diseases. Phytopathology 53: 35 – 39.

Pitcher, R. S. 1965. Interrelationships of nematode and other pathogens of plants. Helminthological Abstracts 34: 1 – 17.

Powell, N. T. 1963. The role of plant-parasitic nematodes in fungus diseases. Phytopathology 53: 28 – 35.

Powell, N. T. 1971. Interactions between nematodes and fungi in disease complexes. Annual Review of Phytopathology 9: 253 – 274.

Prasad, S. K., and J. M. Webster. 1967. Effect of temperature on the rate of development of *Nacobbus serendipiticus* in excised tomato roots. Nematologica 13: 85 – 90.

Price, N. S., and J. Sanderson. 1984. The translocation of calcium from oat roots infected by the cereal cyst nematode *Heterodera avenae* (Woll.). Revue de Nématologie 7: 239 – 243.

Raja, A., and D. R. Dasgupta. 1986. Enhanced synthesis of messenger RNA in relation to resistance expression in cowpea (*Vigna unguiculata*) infected with the root-knot nematode, *Meloidogyne incognita*. Revue de Nématologie 9: 35 – 48.

Ramos, J., J. Franco, N. Ortuño, R. Oros, and G. Main. 1998. Incidencia y severidad de los nematodos parasitos del cultivo de la papa, *Nacobbus aberrans* y *Globodera* spp. y determinacion de especies de *Globodera* spp. en Bolivia. Cochabamba, Bolivia: IBTA/PROINPA.

Rao, V. R., and G. Swarup. 1975. Pathogenicity of the spiral nematode, *Helicotylenchus dihystera*, to sugarcane. Indian Journal of Nematology 4: 160 – 166.

Reed, B. M., P. E. Richardson, and C. C. Russell. 1979. Stem nematode infection of resistant and susceptible cultivars of alfalfa. Phytopathology 69: 993 – 996.

Riggs, R. D. 1977. Worldwide distribution of soybean-cyst nematode and its economic importance. Journal of Nematology 9: 34 – 39.

Riggs, R. D., and J. A. Wrather, eds. 1992. Biology and Management of the Soybean Cyst Nematode. St Paul, MN: APS Press.

Rivoal, R., and R. Cook. 1993. Nematode Pests of Cereals. Pp. 259 – 303. In K. Evans, D. L. Trudgill, and J. M. Webster, eds. Plant Parasitic Nematodes in Temperate Agriculture. Wallingford, UK: CAB International.

Robinson, A. F., R. Inserra, E. P. Caswell-Chen, N. Vovlas, and A. Troccoli. 1997. *Rotylenchulus* species: Identification, distribution, host ranges, and crop plant resistance. Nematropica 27: 127 – 180.

Rossi, M., F. L. Goggin, S. B. Milligan, I. Kaloshian, D. E. Ullman, and V. Williamson. 1998. The nematode resistance gene Mi of tomato confers resistance against the potato aphid. Proceedings of the National Academy of Sciences, USA 95: 9750 – 9754.

Rosso, M. N., B. Favery, C. Piotte, L. Arthaud, J. M. de Boer, R. S. Hussey, J. Bakker, T. J. Baum, and P. Abad. 1999. Isolation of a cDNA encoding a beta-1, 4-endoglucanase in the root-knot nematode *Meloidogyne incognita* and expression analysis during plant parasitism. Molecular Plant-Microbe Interactions 12: 585 – 591.

Sandstedt, R., and M. L. Schuster. 1963. Nematode-induced callus on carrot discs grown in vitro. Phytopathology 53: 1309 – 1312.

Sandstedt, R., and M. L. Schuster. 1966a. Excised tobacco pith bioassays for root-knot nematode-produced plant growth substances. Physiologia Plantarum 19: 99 – 104.

Sandstedt, R., and M. L. Schuster. 1966b. The role of auxins in root-knot nematode-induced growth on excised tobacco stem segments. Physiologia Plantarum 19: 960 – 967.

Sasser, J. N. 1977. Worldwide dissemination and importance of the root-knot nematodes, *Meloidogyne* spp. Journal of Nematology 9: 26 – 29.

Sasser, J. N., and C. C. Carter, eds. 1985. An Advanced Treatise on *Meloidogyne*, Volume I Biology and Control. Raleigh, NC: North Carolina State University Graphics.

Sasser, J. N., and D. W. Freckman. 1987. A world perspective on Nematology: The role of the Society. Pp. 7 – 14. In J. A. Veech and D. W. Dickson, eds. Vistas on Nematology. Hyattsville, MD: Society of Nematologists.

Schacht, H. 1859. Ueber die Kultur der Zuckerrübe. Zeitschrift Ver. Rübenzucker-industrie, Zollverein 9: 375 – 392.

Schmidt, A. 1871. Ueber den Rüben-Nematoden (*Heterodera schachtii* A. S.). Zeitschrift Ver. Rübenzuckerindustrie, Zollverein 21: 1 – 19.

Seinhorst, J. W. 1981. Water consumption of plants attacked by nematodes and mechanisms of growth reduction. Nematologica 27: 34 – 51.

Sequeira, L. 2000. Legacy for the millenium: A century of progress in plant pathology. Annual Review of Phytopathology 38: 1 – 17.

Shaner, G., E. L. Stromber, G. H. Lacy, K. R. Barker, and T. P. Pirone. 1992. Nomenclature and concepts of pathogenicity and virulence. Annual Review of Phytopathology 30: 47 – 66.

Sharma, R. D. 1971. Studies on the plant parasitic nematode *Tylenchorhynchus dubius*. Mededelingen Landbouwhogeschool Wageningen, Nederland 71(1).

Sharma, S. B., ed. 1998. The Cyst Nematodes. Dordrecht, The Netherlands: Kluwer Academic Publishers.

Sharma, S. B., N. S. Price, and J. Bridge. 1997. The past, present and future of plant nematology in International Agricultural Research Centres. Nematological Abstracts 66: 119 – 142.

Shepherd, A. M. 1986. Extraction and estimation of cyst nematodes. Pp. 31 – 49. In J. F. Southey, ed. Plant Nematology. London: Her Majesty's Stationery Office.

Shurtleff, M. C., and C. W. Averre, III. 2000. Diagnosing Plant Diseases caused by Nematodes. St Paul, MN: APS Press.

Siddiqi, M. R. 2000. Tylenchida Parasites of Plants and Insects. Wallingford, UK: CAB International.

Sijmons, P. C., H. J. Atkinson, and U. Wyss. 1994. Parasitic strategies of root nematodes and associated host cell responses. Annual Review of Phytopathology 32: 235 – 259.

Singh, D. B., P. P. Reddy, and J. Syamusundar. 1984. Histological, histopathological and histochemical investigations on root-knot nematode resistant and susceptible lines of cowpea. Nematologia Mediterranea 12: 213 – 219.

Skupski, M. P., M. Booker, A. Farmer, M. Harpold, W. Huang, J. Inman, D. Kiphart, C. Kodira, S. Root, F. Schilkey, J. Schwertfeger, A. Siepel, D. Stamper, N. Thayer, R. Thompson, J. Woryman, J. J. Zhuang, and C. Harger. 1999. The genome sequence database: towards an integrated functional genomics resource. Nucleic Acids Research 27: 35 – 38.

Smant, G., J. P. W. G. Stokkermans, Y. Yan, J. M. de Boer, T. J. Baum, X. Wang, R. S. Hussey, F. J. Gommers, B. Henrissat, E. L. Davis, J. Helder, A. Schots, and J. Bakker. 1998. Endogenous cellulases in animals: isolation of β-1,4-endoglucanase genes from two species of plant-parasitic cyst nematodes. Proceedings of the National Academy of Sciences USA 95: 4906 – 4911.

Solov'eva, G. I., and T. E. Kovalenko. 1980. Morphological aberrations of grass blades caused by nematodes from the family Anguinidae Nicol, 1935 during gall-formation. Pp. 155 – 167. In V. D. Lopatin, ed. Stoenie I Produktivnost' Prirodnykih i Seyanykh lugov. Petrozavodsk, USSR: Karel'skii Filial Akademii Nauk SSSR.

Southey, J. F., ed. 1986. Laboratory Methods for work with Plant and Soil Nematodes. Reference Book 402, Ministry of Agriculture, Fisheries and Food. London: Her Majesty's Stationery Office.

Souza, R. M., and J. G. Baldwin. 1998. Changes in esophagus gland activity during the life cycle of *Nacobbus aberrans* (Nemata: Pratylenchidae). Journal of Nematology 30: 275 – 290.

Spaull, W., and P. Cadet. 1990. Nematode parasites of sugar cane. Pp. 461 – 492. In M. Luc, R. A. Sikora, and J. Bridge, eds. Plant Parasitic Nematodes in Subtropical and Tropical Agriculture. Wallingford, UK: CAB International.

Standifer, M. S., and V. G. Perry. 1960. Some effects of sting and stubby root nematodes on grapefruit roots. Phytopathology 50: 152 – 156.

Strickland, R. M., D. R. Ess, and S. D. Parsons. 1998. Precision farming and precision pest management: the power of new crop production technologies. Journal of Nematology 30: 431 – 435.

Stumm-Tegethoff, B. F. A. 1986. Plant growth hormones as intermediates in stem fasciation caused by nematodes in *Lilium henryi*. Nematologica 32: 234 – 245.

Swarup, G., and C. Sosa Moss. 1990. Nematode parasites of cereals. Pp. 109 – 136. In M. Luc, R. A. Sikora, and J. Bridge, eds. Plant Parasitic Nematodes in Subtropical and Tropical Agriculture. Wallingford, UK: CAB International.

Tanksley, S. D., and S. R. McCouch. 1997. Seed banks and molecular maps: unlocking genetic potential from the wild. Science 277: 1063 – 1066.

Thorne, G. 1961. Principles of Nematology. New York: McGraw-Hill Book Company.

Townshend, J. L., L. Stobbs, and R. Carter. 1989. Ultrastructural pathology of cells affected by *Pratylenchus penetrans* in alfalfa roots. Journal of Nematology 21: 530 – 539.

Trudgill, D. L. 1980. Effects of *Globodera rostochiensis* and fertilisers on the mineral nutrient content and yield of potato plants. Nematologica 26: 243 – 254.

Tyler, J. 1933. Reproduction without males in aseptic cultures of the root knot nematode. Hilgardia 7: 373 – 415.

Van Staden, J., and G. G. Dimalla. 1977. A comparison of the endogenous cytokinins in the roots and xylem exudate of nematode-resistant and susceptible tomato cultivars. Journal of Experimental Botany 28: 1351 – 1356.

Vecht, J. van der. 1950. Op planten parasiterende aaltjes. Pp. 16 – 41. In L. G. E. Kalshoven, ed. De Plagen van de Cultuurgewassen in Indonesie. ' S-Gravenhage, The Netherlands: W. van Hoeve.

Viglierchio, D. R., and P. K. Yu. 1968. Plant growth substances and plant parasitic nematodes. II. Host influence on auxin content. Experimental Parasitology 23: 88 – 95.

Vovlas, N. 1983. Histopathogenesis of galls induced by *Subanguina radicicola* on *Poa annua* roots. Plant Disease 67: 143 – 144.

Wallace, H. R. 1974. The influence of root-knot nematode, *Meloidogyne javanica*, on photosynthesis and on nutrient demand by roots of tomato plants. Nematologica 20: 27 – 33.

Wang, E. L. H., and G. B. Bergeson. 1974. Biochemical changes in root exudate and xylem sap of tomato plants infected with *Meloidogyne incognita*. Journal of Nematology 6: 194 – 202.

Weischer, B., and D. J. F. Brown. 2000. An Introduction to Nematodes. General Nematology. A student's textbook. Sofia-Moscow: PENSOFT.

Wendt, K. R., A. Swart, T. C. Vrain, and J. M. Webster. 1995. *Ditylenchus africanus* sp. n. from South Africa; a morphological and molecular characterization. Fundamental and Applied Nematology 18: 241 – 250.

Wilcox, D. A., and R. Loria. 1986. Water relations, growth, and yield in two snap bean cultivars infected with root-knot nematode *Meloidogyne hapla* (Chitwood). Journal of the American Society of Horticultural Sciences 111: 34 – 38.

Wilcox-Lee, D. A., and R. Loria. 1987. Effects of nematode parasitism on plant-water relations. Pp. 260 – 266. In J. A. Veech and D. W. Dickson, eds. Vistas on Nematology. DeLeon Springs, FL: Society of Nematologists.

Williams, T. D., and J. Bridge. 1983. Plant-parasitic nematodes. Pp. 225 – 249. In A. Johnston and C. Booth, eds. Plant Pathologist's Pocketbook. Farnham Royal, UK: Commonwealth Agricultural Bureaux.

Williamson, V. M. and R. S. Hussey. 1996. Nematode pathogenesis and resistance in plants. The Plant Cell 8: 1735 – 1745.

Yang, H., N. T. Powell, and K. R. Barker. 1976. Interactions of concomitant species of nematodes and *Fusarium oxysporum* f. sp. *vasinfectum* on cotton. Journal of Nematology 8: 74 – 80.

Yeates, G. W. 1971. Feeding types and feeding groups in plant and soil nematodes. Pedobiologia 11: 173 – 179.

Zacheo, G., and T. Bleve-Zacheo. 1987. Stimulation of respiratory pathways in tomato roots infested by *Meloidogyne incognita*. Physiological and Molecular Plant Pathology 30: 461 – 466.

Zacheo, G., and S. Molinari. 1987. Changes in oxygen uptake of mitochondria from susceptible and resistant tomato roots infested by *Meloidogyne incognita*. Nematologia Mediterranea 15: 315 – 321.

Zuckerman, B. M. 1969. Specialized adaptations to parasitism among ectoparasitic plant nematodes. Pp. 84 – 88. In J. E. Peachey, ed. Nematodes of Tropical Crops. Farnham Royal, UK: Commonwealth Agricultural Bureaux.

Zuckerman, B. M., and M. W. Brzeski. 1965. Methods for the study of plant parasitic nematodes in gnotobiotic root culture. Nematologica 11: 453 – 456.

Zunke, U. 1990. Observations on the invasion and endoparasitic behavior of the root lesion nematode *Pratylenchus penetrans*. Journal of Nematology 22: 309 – 320.

Acknowledgements The authors would like to thank all those who have contributed to the preparation of this chapter but are particularly indebted to Dr. David Hunt for the enormous amount of help that he has given.

14 Virus Vectors

Derek J. F. Brown, Jingwu Zheng and Xueping Zhou

14.1 Introduction

Recognition that plant-parasitic nematodes can act as vectors of plant viruses is a relatively new discovery, with most research in this discipline having been done in Europe and North America. Transmission of plant viruses by nematodes is remarkable as only 17 of approximately 350 species belonging to the genera *Longidorus, Paralongidorus,* and *Xiphinema* (Family Longidoridae, Order Dorylaimida), and 13 of approximately 70 species belonging to the genera *Paratrichodorus* and *Trichodorus* (Family Trichodoridae Order Triplonchida) transmit a third of the described nepoviruses and all three tobraviruses, respectively. Thus, of the approximately 4000 described phytonematode species only 30, i. e., less than 1%, are virus-vectors. These include seven *Longidorus,* one *Paralongidorus* and nine *Xiphinema* species as natural vectors of 13 of the 38 known nepoviruses and nine *Paratrichodorus* and four *Trichodorus* of the three tobraviruses. The vector nematode species and their associated viruses occur worldwide and the diseases they cause in fruit, vegetable, and ornamental crops are of substantial economic importance (Weischer and Brown, 2000). The acquisition and transmission of viruses by these nematodes is inextricably linked to their methods of feeding. Recently, with the development of new techniques for studying nematode transmission of viruses and the introduction of increasingly sensitive molecular biology methods there is renewed and increasing interest worldwide in the study of nematode – virus – plant interactions.

14.2 Historical Background

Plant diseases associated with soil, i. e., soil-borne, were first recognized in the latter half of the 19th century, when healthy grapevines were shown to suffer from degeneration if grown in soil collected from old vineyards. Similarly, tobacco plants became diseased when planted in soil collected from seedbeds of particular farms. The infective agent causing the disease in tobacco was reported as a *contagium vivum fluidum*, and considered quite different from bacteria. This latter report is now accepted as representing the seminal paper in the science of Virology (Behrens, 1899). In the first quarter of the 20th century, several plant diseases were attributed to soil-borne viruses and at that time it was suggested that the phytonematode, *Heterodera humuli*, might be responsible for nettle-head disease in hops in southern England. Shortly thereafter this nematode was exonerated as the vector agent (Brown and Trudgill, 1997).

In the 1940s experimental transmission of swine influenza virus in pigs by juveniles of the swine lungworm, *Metastrongylus* spp., was reported (Shope, 1941). These findings stimulated researchers to investigate the possibility that several soil-borne diseases might be transmitted by soil-inhabiting phytonematodes. However, these diseases subsequently were proven not to be transmitted by nematodes. Similarly, in 1958, tobacco and cucumber mosaic viruses and carnation mottle virus were reported not to be transmitted by phytonematodes belonging to the genera *Helicotylenchus, Meloidogyne*, and *Pratylenchus* (Santos, et al., 1997). Later that same year researchers published a seminal paper providing unequivocal evidence that the soil-inhabiting, ectoparasitic phytonematode *Xiphinema index* was a natural vector of grapevine fanleaf nepovirus, present in vineyards in California (Hewitt, et al., 1958). This initial discovery was quickly followed by additional reports identifying phytonematodes as vectors of plant viruses, occurring mainly in crops from North America and Europe. These reports stimulated researchers to investigate the biology, ecology, life-cycle, distribution, and taxonomy of these nematodes. Recent developments in molecular biology have provided new approaches to study nematode-virus-plant interactions resulting in this research discipline being increasingly used as a model system for investigating the fundamental basis of nematode-pathogen interactions.

14.3 The Vector Nematodes and Their Feeding Processes

The complexity and subtlety of interactions between vector nematodes and their associated viruses is not widely appreciated. Longidorid and trichodorid nematodes are vermiform, soil-inhabiting ectoparasites that feed on an extensive range of plant species. These nematodes have relatively long life cycles (2 to 5 years) and most have slow rates of multiplication (Lamberti, et al., 1975). Comprehensive descriptions have been published of the taxonomy and biology of the most important virus-transmitting nematodes (Decraemer, 1995; Hunt, 1993; Jairajpuri and Ahmad, 1992).

14.3.1 Longidorids

Members of the Longidoridae are relatively long nematodes (2 – 12 mm), each having a long hollow feeding stylet (60 – 250 μm) that they use to feed deeply within root-tips. During the feeding process viruses are acquired and transmitted by the nematodes. The stylet has two parts (Fig. 14.1). The anterior section is referred to as the odontostyle, and is used to penetrate root cells. The posterior section is referred to as the odontophore, and contains nerve processes adjacent to the food canal that probably enable the nematode to discriminate between sites deep within plant roots (Robertson and Taylor, 1975). Muscles attached to the rear of the odontophore are used to protract it, and this action can result in the odontostyle being almost entirely protracted due to the length and rigidity of the odontophore. The esophagus is comprised of a long narrow tube connecting the stylet and a large, muscular, cylindrical bulb. The bulb contains three large gland cells. The dorsal gland cell, and its associated gland nucleus, is connected by a duct to the food canal at the anterior end of the bulb. The bulb provides a pumping action that is used during the feeding process to force secretions from the dorsal gland cell forward into the plant cell, and subsequently to withdraw plant cell contents and force them into the gut. Unlike *Longidorus, Xiphinema* species have an extensive duct system connected to the dorsal gland cell that can facilitate rapid ejection of the secretions into the plant cell. The plant cell contents are pumped through a one-way valve at the junction of the bulb into the nematode's gut, against the nematodes hydrostatic body-pressure. A pair of subventral gland ducts and gland nuclei are situated approximately halfway along the bulb, and occasionally a second pair of ducts, without gland nuclei, are present that open into the posterior of the bulb. The subventral gland ducts open into the bulb chamber and it is presumed that the gland cell contents are moved backwards to be ingested during feeding (Wyss, 1975, 1977, 1981, 1987, 1999).

With the exception of members of the *X. americanum* group (Cohn, 1975), feeding by most longidorids results in plant galls that contain large cells with dense cytoplasm. Longidorid nematodes that induce root-tip galls have two distinct feeding behaviors. In the most frequent feeding behavior the nematode feeds on a column of progressively deeper cells. The content of each cell is removed during short periods of ingestion, each lasting from several seconds to several minutes. This activity is interspersed with very brief pauses lasting 1 to 10 s, and during these brief pauses the nematode probably injects dorsal gland cell secretions into the cell, and when virus particles are being retained by the nematode it is at this time that particles are transmitted into the plant. The second, less common, type of feeding involves deep stylet insertion, followed by 15 to 60 min inactivity, and then 1 to 3 h continuous ingestion (Trudgill and Robertson, 1982).

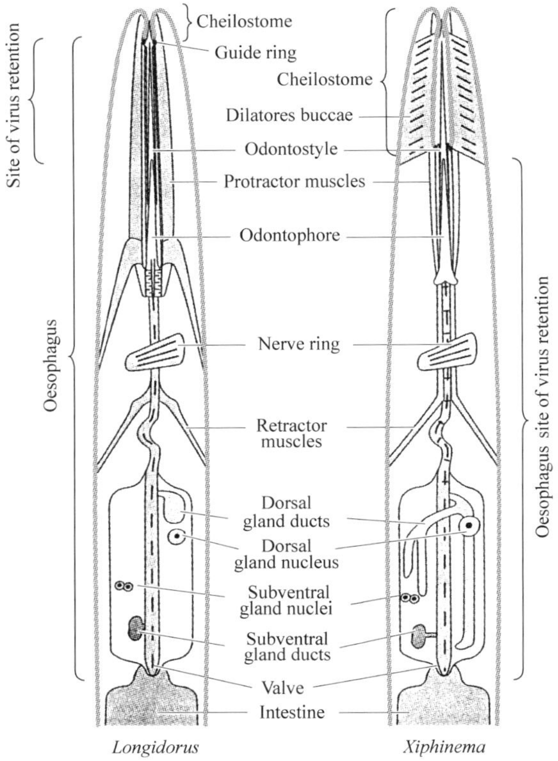

Figure 14.1 Structure of the feeding apparatus of *Longidorus* and *Xiphinema* nematodes showing the sites of retention of nepoviruses, and the glands and ducts in the esophageal bulb (from Brown and MacFarlane, 2001).

Despite *Xiphinema* and *Longidorus* species feeding at root tips, and their feeding resulting in the formation of enlarged root-tip galls, it is relatively common to find mixtures of several species occurring together in the rhizosphere of plant roots (Taylor and Brown, 1976; Weischer and Brown, 2000). Consequently, there appears to be relatively little competition for food between species, even from different genera. Also, a similar situation occurs with trichodorid species.

Root-tip galls caused by *X. index* contain enlarged, multinucleate cells with dense cytoplasm, and those caused by *X. diversicaudatum* contain increased amounts of DNA, RNA and protein. *Longidorus* nematodes induce root-tip galls containing enlarged, amoeboid-shaped cells that contain increased amounts of DNA, but not multinucleate cells. A longidorid nematode feeding for 1 h can remove a volume equivalent to approximately 40 root-tip cells, including the cytoplasm and organelles. Also, in older root-tip galls the interconnecting walls of empty cells have been observed to be holed, believed to have been caused by the secretions from the dorsal gland (Zacheo and Zacheo, 1995).

14.3.2 Trichodorids

Trichodorid nematodes penetrate plant root cells using their onchiostyle, a modified, dorsally convex, mural tooth with a solid tip. The esophagus consists of a long narrow tube that expands posteriorly to form a spatulate bulb, which contains one dorsal, two anterior ventrosublateral, and two posterior ventrosublateral glands (Fig. 14.2). Trichodorids feed on root hairs, and on epidermal cells adjacent to the zone of elongation at root tips. The

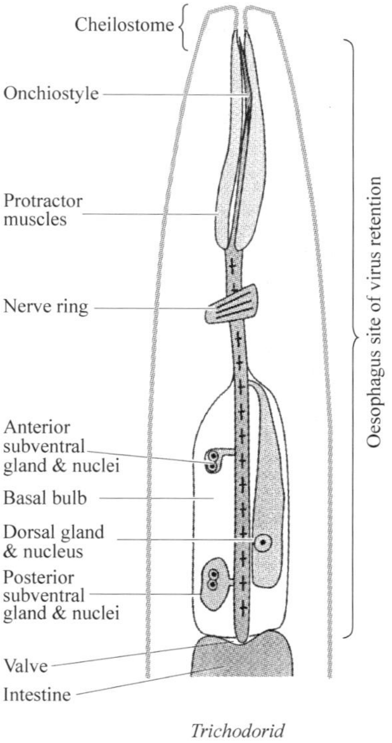

Figure 14.2 Structure of the feeding apparatus of trichodorid nematodes showing the sites of retention of nepoviruses, and the glands and ducts in the esophageal bulb (from Brown and MacFarlane, 2001).

feeding of trichodorids has been intensively studied with the aid of time-lapse cinephotography and computer-enhanced video imaging (Wyss, 1981, 1987). The process consists of five phases: exploration, puncturing of the cell wall, injection of glandular secretions, ingestion, and finally, withdrawal from the cell. To feed, the nematode presses its lips firmly against the cell wall, the stoma is drawn forward, and using rapid thrusting (6/s) of the onchiostyle to perforate the cell wall. This procedure takes approximately 1 minute. The nematode, after penetrating the cell wall, repeatedly thrusts (1/s) its onchiostyle through the puncture hole to a depth of 2 μm to 3 μm. Glandular secretions are injected into the cell and these, in combination with the stimulus of repeated thrusting of the onchiostyle, induce rapid aggregation of cytoplasm at the penetration site. Nuclei lying nearby migrate towards the feeding site and begin to lose their granulated appearance, appearing empty within a short time. Some of the secretions injected into the cell harden rapidly to form a feeding tube through which the aggregated cytoplasm, and usually also the nucleus, are extracted and ingested (Wyss, 1981, 1987).

14.4 The Nematode-transmitted Viruses

Plant viruses transmitted by nematodes belong to two taxonomic groups and have worldwide distributions. The 38 known nepoviruses have isometrical particles, whereas the three tobraviruses have tubular shaped particles. The three tobraviruses are each transmitted naturally by trichodorid nematodes. However, only one third of the nepoviruses are known to have longidorid nematodes as vectors, with the mode of transmission of the remaining two-thirds being through infected seed, pollen, or unknown.

Nepoviruses and tobraviruses have bipartite genomes, with each of the parts containing single-stranded RNA (Figs 14.3 – 14.5). The larger molecule, RNA-1, in nepoviruses carries the genetic determinants for host-range, some types of symptom expression in herbaceous hosts and seed transmissibility. The smaller molecule, RNA-2, carries the determinants for serological specificity (coat protein), nematode transmissibility, and other types of symptom expression in herbaceous hosts. In tobraviruses, each of the two RNAs contains genetic determinants for symptom expression, and the smaller RNA-2 contains determinants for serological specificity and vector transmissibility. Both RNAs of nepoviruses, but only the RNA-1 of tobraviruses, are apparently associated with plant infectivity (Hanada and Harrison, 1977; Harrison and Murant, 1978; Harrison and Robinson, 1981, 1986; Harrison, et al., 1974; MacFarlane, et al., 1996; Ploeg, et al., 1993b).

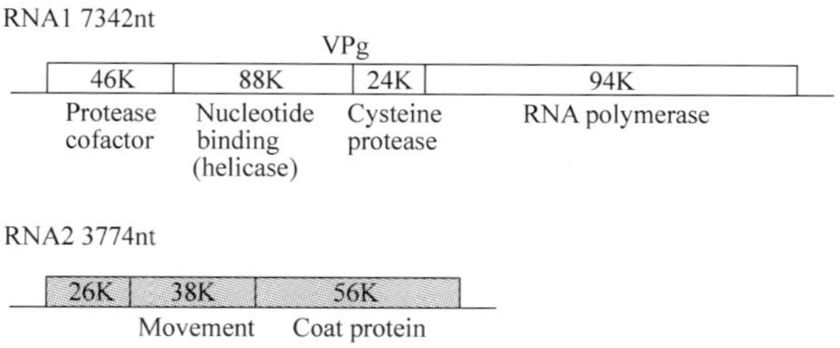

Figure 14.3 The genome structure of grapevine fanleaf nepovirus, detailing the function, where known, and the approximate size in kDa (= K), of each of the proteins produced by cleavage of the two polyproteins. VPg is a peptide covalently bound to the 5' terminus of each of the viral RNAs (from Brown and MacFarlane, 2001).

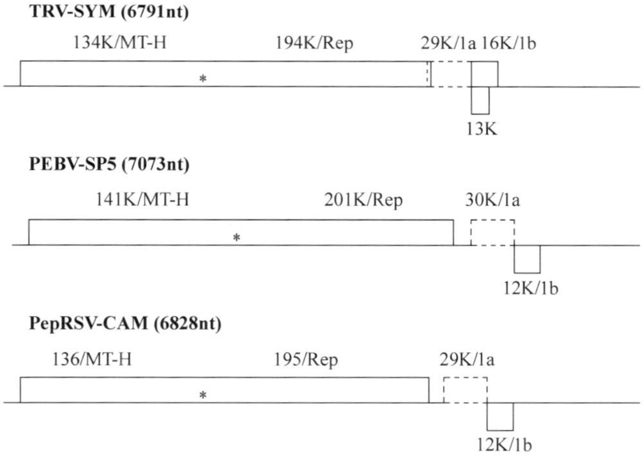

Figure 14.4 The genome structure and number of nucleotides of the RNA-1 of tobacco rattle virus (TRV) isolate SYM, pea early-browning virus (PEBV) isolate SP5, and pepper ringspot virus (PepRSV) isolate CAM. The relative size in kDa (= K) and name of each protein is given: Mt, methyltransferase domain; H, helicase domain; Rep, RNA-dependent RNA polymerase: *, a leaky translation termination codon. The 1a and 1b genes of PEBV and PepRSV overlap, and similarly the 1b and 13K genes of TRV overlap (courtesy of N. Vassilakos, SCRI).

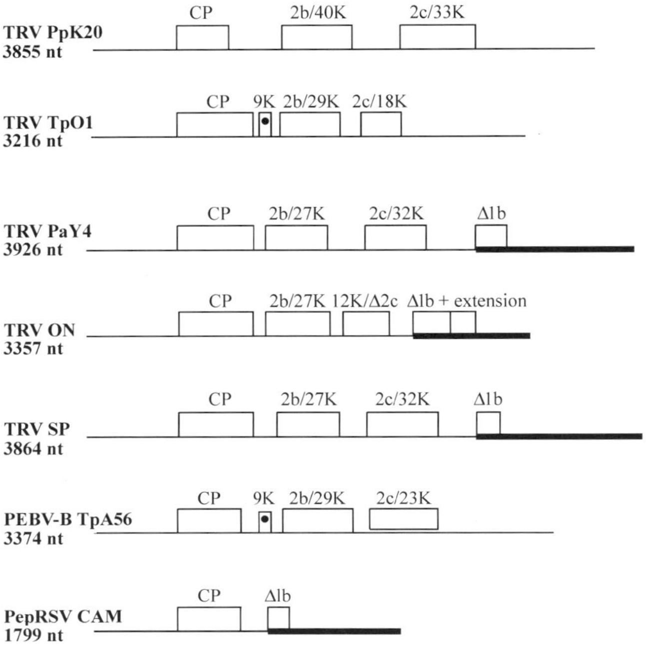

Figure 14.5 The genome structure and number of nucleotides of the RNA-2 of seven tobacco rattle virus (TRV) isolates, two pea early-browning virus (PEBV) isolates, and one pepper ringspot virus (PepRSV) isolate. The relative size in kDa (= K) and name of each protein is given: CP, coat protein; •, a partial or truncated gene. The line in bold denotes sections derived from the RNA-1 from recombination. The five TRV and the PEBV isolates each have their CP and the 2b proteins intact, and therefore may be assumed to be transmissible by vector nematodes. Viruses Tp O1 and TpA56 were each isolated after transmission by individual *Trichodorus primitivus* nematodes, PpK20 after transmission by a single *Paratrichodorus pachydermus,* and PaY4 after transmission by a single *P. anemones*. Isolates On and SP were isolated from an onion and a spinach plant, respectively, however the vector species has not been identified for either virus. The PepRSV isolate does not carry a 2b gene, and therefore is assumed not to be transmissible by vector nematodes (courtesy of N. Vassilakos, SCRI).

Nepoviruses and tobraviruses transmitted by nematodes are primarily pathogens of wild plants, and most are seed transmitted, which is important in their ecology for survival and persistence (Murant, 1983). These viruses have large natural and experimental host ranges, with tobacco rattle virus having the most extensive host range of any plant virus known. Many of the nematode transmitted viruses, particularly tobacco rattle virus, occur naturally as a range of serological and/or symptomatological variants (Harrison and Robinson, 1981, 1986). The major serological variants, especially of tobraviruses, have different nematode species as their natural vectors (Brown, et al., 1989b; Ploeg, et al., 1991, 1992).

The taxonomy of nepoviruses is based principally on serological classification and to a lesser extent on physico-chemical properties. Only 23 of the 38 nepoviruses apparently are serologically unrelated to other members of the genus, and the genus has been subdivided into two, three or four groups based on physico-chemical property differences. Nucleotide sequence data of the coat protein gene of strawberry latent ringspot virus, which is one of two nepoviruses transmitted by *X. diversicaudatum*, has shown that this virus is more distantly related to others than previously thought (Harrison and Murant, 1996). It seems probable that genetic sequence data will increasingly be used in the classification of plant viruses, including nepo- and tobraviruses.

The three tobraviruses, pea early-browning (PEBV), pepper ringspot (PepRV), and particularly tobacco rattle (TRV), occur naturally as numerous serologically distinguishable strains. Serological affinities between the strains have not been fully investigated, but nucleic acid hybridization techniques using cDNA copies of unfractionated viral RNA separated 15 strains of tobraviruses into three distinct groups corresponding to PEBV, PepRV and TRV (Harrison and Robinson, 1986; Robinson and Harrison, 1985).

Pseudo-recombinant nepo- and tobraviruses have been successfully produced in the laboratory with RNA-1 or RNA-2 molecules being exchanged between different strains of the same virus. Such exchanges have not been successful when the RNAs came from different viruses. Several natural tobravirus recombinants have been identified, each of which had the pathogenicity of TRV and the serological properties of PEBV.

Several isolates of PEBV and TRV have been genetically sequenced and full-length cDNA viral copies produced without compromising infectivity and vector transmissibility. The ability to successfully engineer and genetically manipulate tobraviruses has resulted in these viruses and their associated vector nematodes providing a powerful model system with which to investigate the viral genetic determinants of vector transmissibility (see Section 14. 10. 2). Also, plants expressing selected segments of the tobravirus genome have been shown to have quantitative resistance to mechanical infection of tobraviruses, but vector transmission of the virus overcame the transgenic resistance (Ploeg, et al., 1993a; Visser, 2000).

14.5 Transmission Criteria and Procedures

14.5.1 Criteria for Demonstrating Nematode Transmission of Viruses

During the 1960s and 1970s substantial research effort was directed in identifying natural associations between nematodes and viruses. Subsequently, it became apparent that a number of these reports contained insufficient or inadequate evidence to justify the conclusion that the reported nematode species was a virus vector. A set of criteria, based on substantial experience gained by researchers at the SCRI working on virus-vector nematodes and their associated viruses, were established during the 1980s and used for assessing reports of longidorid nematodes transmitting nepoviruses (Trudgill, et al., 1983). Subsequently, these criteria were adapted for application with trichodorids and tobraviruses (Brown, et al., 1989b). The criteria are: (1) infection of the bait plant must be demonstrated; (2) experiments should be done with hand-picked nematodes; (3) appropriate controls should be included to show unequivocally that the nematode is the vector; (4) the nematode should be fully identified; and (5) the virus should be fully characterized. Two thirds of reports of associations between nepo- and tobraviruses and longidorid and trichodorid species, respectively, were found not to fulfill these criteria (Trudgill, et al., 1983; Brown, et al., 1989b).

14.5.2 Virus Transmission Test Procedures

Laboratory and glasshouse experiments to investigate transmission of nepo- and tobraviruses by longidorid and trichodorid nematodes, respectively, are subject to several factors that can influence the final results. Therefore, specialized techniques and methods were established to investigate the interactions between nematodes and their associated viruses. To establish if a nematode is a vector of a virus the nematode must: (1) ingest virus from a virus source plant; (2) retain virus particles that subsequently can be released into the bait plant;

(3) the virus must infect the bait plant, and of particular importance; and (4) all other potential sources of virus infection must be excluded. Specialized techniques for studying virus transmission by longidorid and trichodorid nematodes that fulfil these criteria have been established and provide the means for effectively assessing each stage of the transmission process (Santos, et al., 1997; Taylor and Brown, 1997). The tests involve the use of individual or small numbers of nematodes placed in plastic capsules or small pots, and temperature-controlled boxes or chambers to reduce moisture and temperature fluctuations. Root-tip galls induced by longidorid nematodes feeding on virus source plants and bait plants can be counted to assess the feeding activity of the nematodes. Crushing whole nematodes and examining the resultant suspension by immunosorbent electron microscopy can be used to determine the proportion of nematodes ingesting virus from the source plants. Alternatively, with *Longidorus* spp. the resultant suspension can be rubbed directly onto suitable

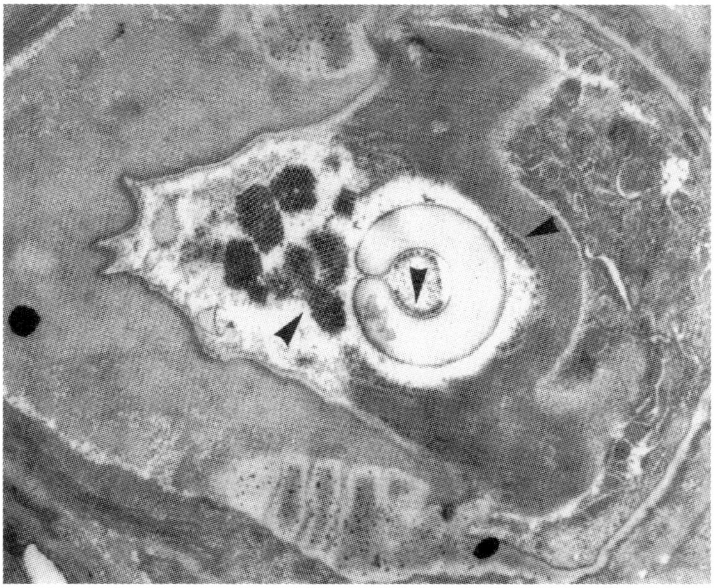

Figure 14.6 A slightly oblique cross-section at the junction of the stoma and the anterior part of the guide sheath in a male *Longidorus elongatus* nematode showing particles (arrowed) of raspberry ringspot nepovirus associated with the odontostyle and the guide sheath, and in crystalline form in the lumen of the stoma (from Brown and Trudgill, 1997).

indicator plants to produce an infection in the plant. Tobacco rattle virus can be detected in suspensions of individual trichodorids using a reverse transcription and polymerase chain reaction (RT-PCR) method (Boutsika, et al., 2000). These methods detect virus present in the nematode gut, but are largely ineffective for detecting virus retained in the nematode head. Therefore, detection of virus by these methods is not indicative that the virus is transmissible by the nematode. Specifically retained virus particles may be detected by electron microscope examination of ultra-thin sections through the specific sites of retention in vector species (Fig. 14.6). Also, virus can be detected in intact heads of nematodes using an indirect immunofluorescent technique (Fig. 14.7) (Wang and Gergerich, 1998). Immunogold labeling techniques with appropriate antisera can be used to confirm the identity of virus-like particles observed in ultrathin sections (Fig. 14.8) (Karanastassi, et al., 2000).

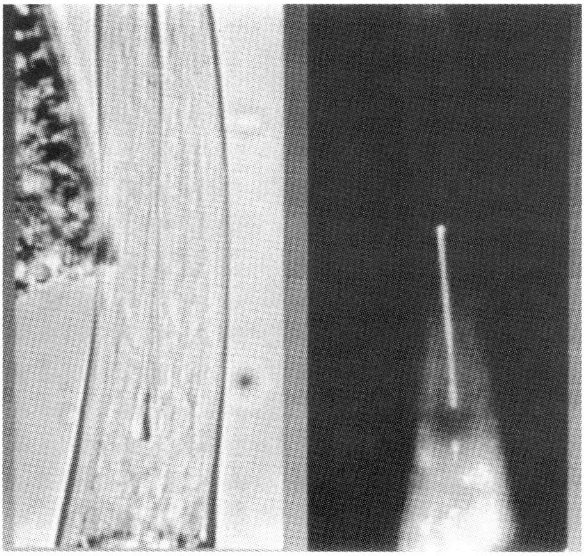

Figure 14.7 Photomicrograph of immunofluorescent labeling of tobacco ringspot nepovirus present at the specific site of retention in the esophageal tract of *Xiphinema americanum* (courtesy of Drs. S. Wang and R. C. Gergerich, Department of Plant Pathology, University of Arkansas, Fayetteville, Arkansas, USA).

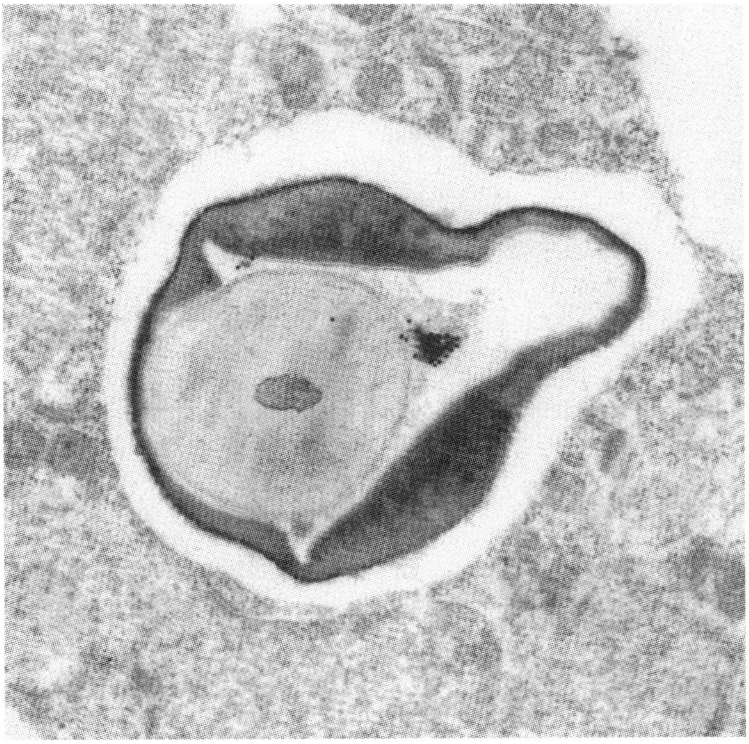

Figure 14.8 A transverse section through the pharyngeal region of a female *Paratrichodorus anemones* nematode showing immunogold labeling of particles of tobacco rattle virus isolate PaY4 at the site of retention in the vector (courtesy of E. Karanastasi, SCRI).

14.6 Vector and Virus Associations

Most vector species are indigenous to Europe and North America, an exception being *L. martini* that is associated with mulberry ringspot virus in Japan. Several *Longidorus* and *Xiphinema* species, and their associated viruses have inadvertently been transported by human activity from their areas of origin to other regions such as Australasia and North America, e.g. *X. diversicaudatum* with arabis mosaic virus (ArMV) and *X. index* with grapevine fanleaf virus (GFLV). Trichodorid nematodes are also widely distributed in Europe and North America with several species associated naturally with tobacco rattle virus (TRV), the most widespread of the tobraviruses. Again, these viruses and vectors have been dispersed through human activity to other countries such as Japan, New Zealand, and Brazil. However, pepper ringspot virus (PepRV) appears to be endemic to Brazil, where it is vectored by *P. minor*, and the third tobravirus, pea early-browning (PEBV), occurs only in localized areas in northern Europe and North Africa (Brown, 1989; Taylor and Brown, 1981).

It is now generally accepted that nepo- and tobraviruses are naturally associated with wild plants and have developed specific relationships with nematode species that function as their vectors. Because of their limited mobility, nematode populations tend to be localized in discrete territorial enclaves in which they become specifically associated with viruses through interdependent ecological factors. Wider distribution occurs when viruses and vectors become associated with cultivated plants, which provide the means for their long distance dissemination in plant material and soil through human activity (Brown, et al., 1994b). In most situations nepovirus and tobravirus infections in the field are apparent only in crop plants in which symptoms are relatively severe and, in some instances, induce plant death. By contrast, infection of wild plants is usually symptomless, indicating an ecologically balanced association. Likewise, many longidorid and trichodorid vector species have a wide host range among wild plants, but apparently their feeding causes little appreciable damage whereas the growth of crop plants is often impaired by nematode feeding (Taylor and Brown, 1997).

14.6.1 Nepoviruses and Their Vector Longidorids

14.6.1.1 Arabis Mosaic (ArMV)

ArMV was described from a single *Arabis* plant growing in a glasshouse in England (Smith and Markham, 1944). It was not until 16 years later that it was shown to cause yellow dwarf disease in red raspberry (Cadman, 1960), the disease subsequently being identified in several European countries. The virus also causes mosaic and yellow crinkle of strawberry, and diseases of varying severity in cucumber, grapevine, narcissus, rose, sugar beet and white clover (Brown and Trudgill, 1989). In Belgium, Britain, and the Czech Republic ArMV causes "bare bine", "nettlehead", and "severe split leaf blotch" in hop resulting in substantial yield loses and, in association with *Prunus* necrotic ringspot and prune dwarf ilarviruses, induces a form of "European rasp leaf" of cherry (Taylor and Brown, 1997).

Several isolates of ArMV differ in their respective biological characteristics, i. e., resistance breaking of red raspberry cultivars, nematode transmissibility (Brown, 1986c; Jones, et al., 1989; Taylor, et al., 1966). Four serologically distinguishable strains of the virus occur: (1) the type strain; (2) a strain from woodland in southern England (Clark, 1976); (3) a strain infecting hop in southern England (Valdez, et al., 1974); and (4) a strain causing a yellowing disease in barley cv. Express in western Switzerland (Ramel, et al., 1995). Each of these strains of ArMV has been found naturally associated with its vector, *X. diversicaudatum*.

Under the regulations of the European Community Plant Health Directive, ArMV is listed as a quarantine organism with an Annex designation of II/A2, and is also entered in the quarantine list of the North American Plant Protection Organization. The virus has been reported infecting plants outside Europe in Australia, Canada, Japan, New Zealand, and South Africa. Most reports refer to the virus having been identified in plants imported from Europe, but in New Zealand the vector is also present at several discrete localities (Brown and Trudgill, 1997; Taylor and Brown, 1997).

14.6.1.2 Artichoke Italian Latent (AILV)

AILV was described from infected artichoke crops (*Cyanara scolymus*) growing near Bari, southeastern Italy, where this crop was first established, and subsequently in the neighboring provinces of Brindisi and Foggia (Majorana and Rana, 1970; Roca, et al., 1975a, 1975b). The vector is *L. apulus*. Interestingly, a serologically distinguishable strain was found to infect artichoke in western Greece, and is transmitted by *L. fasciatus*. In Italy the virus also causes ringspot disease in chicory, however depending on the artichoke cultivar the virus may infect the plant symptomlessly, or produce a generalized yellowing in the foliage, or stunting of the infected plant (Rana and Roca, 1973, 1975; Roca, et al., 1982; Vovlas and Roca, 1975).

14.6.1.3 Cherry Rasp Leaf (CRLV)

CRLV was described by Nyland (1961), but Bodine and Newton (1942) first reported the disease it causes in cherry trees in North America. The leaves of affected cherry trees have enations on their underside, appearing as leafy outgrowths. The disease symptoms usually first appear on the lower leaves from where the disease slowly spreads, sometimes causing death of affected spurs and branches producing an open, bare appearance in the affected tree. The virus also causes flat-apple disease in apple (Parish, 1977), has been recovered from balsam, dandelion, peach, and plantain (Hansen, et al., 1974), and from red raspberry plants imported into Scotland (Jones and Badenoch, 1982), which represents the only report of the occurrence of the virus outside North America. Although CLRV causes "rasp leaf disease" in cherry in North America a similar disease in cherry trees in Europe, referred to as "European rasp leaf" is caused by mixed infections of raspberry ringspot (RRSV), or ArMV, in association with *Prunus* necrotic or prune dwarf ilarviruses (Cropley, 1961).

The vector of CRLV has been reported as *X. americanum* (Nyland, et al., 1969), but with subsequent taxonomic revision of the nematodes belonging to the *X. americanum* group (see Section 14.7.1) the correct identification of the vector species requires to be confirmed.

Under the regulations of the European Community Plant Health Directive CRLV is listed as a quarantine organism with an Annex designation of I/A1. It is also entered in the quarantine list of the Interafrican Phytosanitary Council, and as an A1 quarantine organism with the European and Mediterranean Plant Protection Organization.

14.6.1.4 Cherry Rosette (CRV)

CRV was found affecting cherry trees in the Arth region of Switzerland, and is transmitted by *L. arthensis* (Brown, et al., 1994a). The virus, which has not been fully characterized, causes a steady decline in vigour with accompanying leaf symptoms such as distortion, enations, rosetting, and "oil-flecking" in which the leaves appear to have been contaminated with drops of oil, and eventually the affected tree dies.

A similar disease has been reported affecting cherry trees in the nearby Basleland region of Switzerland, but in this instance the virus is RRSV transmitted by *L. macrosoma* (Buser, 1990; Klingler, et al., 1985).

14.6.1.5 Grapevine Fanleaf (GFLV)

GFLV and its vector nematode, *X. index*, occur in most viticultural regions worldwide. It is believed that the virus, through infected planting material, and the vector in soil in which grapevine was transported, has been spread from their center of origin in the Caucuses and ancient Persia (Hewitt, 1970). Interactions between isolates of GFLV and different cultivars of grapevine rootstocks and fruiting scions result in the occurrence of a wide range of symptoms, e.g., chromogenic "yellow mosaic disease", fanleaf in which the leaf veins are abnormally spread giving the leaf a fan-like appearance, leaf enations, leaf veinbanding, "flat trunk disease", and "pitting disease" of bark and wood.

In most viticultural areas worldwide *X. index* has been recorded associated with the spread of GFLV, however in Central Europe and the Palatinate region of Germany there is circumstantial evidence that *X. vuittenezi* may also be a vector of the virus (Rudel, 1985). Also, in Israel *X. italiae* has been reported as a vector of GFLV (Cohn, et al., 1970).

14.6.1.6 Mulberry Ringspot (MLRV)

MLRV has been reported only from Japan where it is transmitted by *L. martini* and causes a disease in mulberry (Yagita and Komuro, 1972).

14.6.1.7 Peach Rosette Mosiac (PRMV)

PRMV was first identified causing a disease in peach trees in Michigan, USA in 1917, and is restricted in its distribution to the Great Lakes area of USA (Klos, 1976). Early formed leaves on infected peach trees show distortion and chlorotic mottling, internodes are shortened producing a rosette appearance, and defoliation is delayed. The virus also infects grapevines causing a delay in the breaking of over-winter dormancy, late and uneven bloom, leaf deformity and mottling, and small and uneven clusters of fruits. The virus also infects blueberry.

The vector of PRMV has been reported as *X. americanum* (Klos, et al., 1967), but with subsequent taxonomic revision of the nematodes belonging to the *X. americanum* group (see Section 14.7.1), the correct identification of the original vector species needs to be confirmed. In laboratory experiments *X. americanum sensu stricto, X. californicum* and *X. rivesi* transmitted the virus (Brown, et al., 1994c). *Longidorus diadecturus* was reported as the vector of PRMV occurring in a single peach orchard in Ontario, Canada. However, *X. americanum sensu stricto* were also present and transmitting PRMV at the site (Eveleigh and Allen, 1982). This single report of *L. diadecturus* transmitting PRMV remains unsubstantiated; consequently the species is not considered a vector (Taylor and Brown, 1997). In a vineyard in the Niagara peninsula, Ontario, Canada specimens of *X. rivesi* were recovered from soil collected from the rhizosphere of grapevines infected with PRMV (Stobbs and van Schagen, 1996).

14.6.1.8 Raspberry Ringspot (RRSV)

RRSV occurs as two major serotypes, the "English" serotype being transmitted by *L. macrosoma* and the "Scottish" serotype by *L. elongatus. Paralongidorus maximus* is the vector of an atypical strain of the virus that infects grapevines in the Palatinate region of Germany (Jones, et al., 1994). The virus was recognized as causing a lethal disease of raspberry in Scotland in 1922, but was not characterized until the 1950s (Cadman, 1956). In Scotland the virus occurs as several minor serological variants, frequently several occurring together in raspberry plantations in association with their natural vector. The "Scottish" serotype was identified as causing "spoon-leaf" in red currant in The Netherlands (Maat, et al., 1962). In central and northern Europe the virus has been recorded causing diseases, with leaf symptoms such as curling, ringspots, stunting, and yellowing, in blackberry, cherry, gooseberry, grapevine, narcissus, raspberry, strawberry, and numerous wild plants.

Under the regulations of the European Community Plant Health Directive, RRSV is listed as a quarantine organism with an Annex designation of II/A2. Also, it is listed as an A2 quarantine organism by the European and Mediterranean Plant Protection Organization, and is listed as a quarantine organism by the North American Plant Protection Organization.

14.6.1.9 Strawberry Latent Ringspot (SLRSV)

SLRSV was first identified from strawberry plants growing in southern England, and *X. diversicaudatum* was shown to be the natural vector (Lister, 1964). Serologically distinguishable strains of the virus have been found associated with olive, raspberry, and peach in northern Italy. The virus from peach was transmitted consistently, but at a low frequency, by *X. diversicaudatum* from Italy, and only sporadically by nematodes from populations from other European countries, New Zealand, and the USA (Brown, 1985). A report of *X. coxi* transmitting the virus (Putz and Stocky, 1970) is considered to be of doubtful validity, with the nematode possibly having been misidentified.

14.6.1.10 Tobacco Ringspot (TRSV)

TRSV is one of four North American nepoviruses transmitted by nematodes. The virus is widespread in North America, occurring in many wild and cultivated plants. Typically, the virus causes concentric patterns of chlorotic and necrotic tissues in leaves of tobacco plants, bud-blight in soybean, necrotic ringspot in blueberry, ringspot in cucurbits, and other symptoms in these hosts include leaf mottling and malformation, and general stunting of the plants. The virus has also been found causing diseases in American spearmint, anemone, blackberry, crocus, cherry, elderberry, grapevine, geranium, and has been recovered from ash and dogwood trees. The virus occurs as several serologically distinguishable strains and has been shown in laboratory experiments to be transmitted by *X. americanum sensu stricto, X. californicum, X. intermedium, X. rivesi*, and *X. tarjanense* (Brown, et al., 1994c, 1996). Unlike most nepoviruses, TRSV has been reported to be transmitted by vectors such as aphids, grasshoppers, *Tetranychus* spider mites, thrips, and tobacco flea beetles (Dunleavy, 1957; Komuro and Iwaki, 1968; Messieha, 1969; Schuster, 1963; Thomas, 1969), however, these reports require confirmation.

The potato calico strain of TRSV was reported as being transmitted by

X. americanum sensu lato, but unequivocal evidence of its association with vector nematodes was not provided (C. E. Fribourg and P. Jatala, in Jones 1981). Subsequently, it was considered to represent a new Nepovirus, potato black ringspot virus (PBRSV), occurring in several regions in South America (Smith, et al., 1992).

TRSV is listed under the regulations of the European Community Plant Health Directive as a quarantine organism with an Annex designation of I/A1. It is also entered as an A1 quarantine organism with the European and Mediterranean Plant Protection Organization.

14.6.1.11 Tomato Black Ring (TBRV)

TBRV occurs as two major serotypes, the "English", or "German", serotype being transmitted by *L. attenuatus* and the "Scottish" serotype by *L. elongatus*. The "English/German" serotype occurs as several minor variants that differ in their efficiency of transmission by their vector nematode (Brown, et al., 1989a). The virus has a wide host range, is seed-transmitted in 25 species belonging to 15 botanical families (Murant, 1983), and causes economically important diseases in crops such as artichoke, asparagus, bean, cabbage, celery, grapevine, leek, lettuce, lucerne, onion, peach, potato, raspberry, strawberry, sugar beet, swede, turnip and tomato. The virus is widespread in Europe, western Russia, and also recorded from Canada, Brazil, India, Japan, Kenya, and the USA. Many of these records, however refer to detection of the virus in plant material imported from Europe.

The virus is listed as a quarantine organism by the North American Plant Protection Organization, and has an II/A2 designation under the regulations of the European Community Plant Health Directive.

14.6.1.12 Tomato Ringspot (ToRSV)

ToRSV is the most economically important and widespread nematode transmitted virus in North America. The virus occurs as several serologically distinguishable strains and in laboratory experiments was shown to be transmitted by *X. americanum sensu stricto, X. bricolensis, X. californicum, X. intermedium, X. rivesi*, and *X. tarjanense* (Brown, et al., 1994c, 1996). The virus has been reported from several countries outside North America, including several European countries, New Zealand, Russia, but these records mainly refer to detection in imported plant material. In Japan, ToRSV was recovered from narcissus, and shown to be transmitted by a local population of the *X. americanum* group (Iwaki and Komuro, 1974). The virus reputedly occurs locally in former Yugoslavia, and in Chile the virus is spreading in a plum orchard where it is transmitted by *X. americanum* group nematodes (J. C. Magunacelaya, pers. comm.). The virus has a wide host range infecting plants belonging to 35 families (Carusso and Ramsdell, 1995) and causes diseases in numerous important crops such as almond, apple, blueberry, cherry, cucumber, *Gladiolus*, grapevine, *Hydrangea*, nectarine, peach, *Pelargonium*, plum, raspberry, strawberry, and tobacco. Disease symptoms range from "yellow blotch-curl", yellow mosaic, "yellow vein", ringspots, and distortion on leaves, to stem pitting and graft union necrosis in fruit trees.

The virus has an I/A1 designation under the regulations of the European Community Plant Health Directive, and whilst listed as an A2 quarantine organism by the European Plant Protection Organization it has an almost A1 status for fruit in Europe (Smith, et al., 1992). The Interafrican Phytosanitary Council consider ToRSV to be of quarantine significance.

14.6.2 Tobraviruses and Their Vector Trichodorids

14.6.2.1 Pea Early-Browning (PEBV)

PEBV occurs only in northern Europe and North Africa in association with its vector trichodorid nematode species. The virus infects leguminous crops such as broad bean, lucerne, and *Phaseolus* bean, but is only of economic importance in pea crops in which it causes severe stunting and premature death of plants. Seed transmission is an important mode of dispersal of the virus in pea crops, with up to 25% of seed being infected in crops in The Netherlands, whereas in Britain only 1 to 2% infection in seed was recorded (Bos and Van Der Want, 1962; Harrison and Robinson, 1986).

14.6.2.2 Pepper Ringspot (PepRSV)

PepRSV has only been recorded from Brazil where it is transmitted by *P. minor* and was found causing diseases in crops such as artichoke, pepper, and tomato (Chagas and Silberschmidt, 1972; Silberschmidt, 1962).

14.6.2.3 Tobacco Rattle (TRV)

TRV is the most widespread and economically important member of the Tobravirus genus. The virus occurs worldwide as a range of serologically distinguishable strains, with each strain being transmitted by only one, or a few, trichodorid species. The virus has the widest host range of any plant virus, infecting many wild and cultivated plants. Many weed species are symptomlessly infected, whereas TRV disease symptoms in crops range from yellowing and necrosis in leaves, to color break in flowers, and necrotic arcs (the disease being commonly referred to as "spraing" in Europe, and "corky ringspot" in North America) in the tuber flesh in potato (Harrison and Robinson, 1981, 1986).

14.7 Geographical Distribution

Whilst agricultural activities have resulted in the widespread dissemination of several vector nematode species, the geographical distributions of most longidorid and trichodorid nematodes are nevertheless established relative to ecological factors, on a geological timescale. Concurrently, specific associations become established between some of the species and their respective nepo- and tobraviruses.

Nematode taxa that are recognized today are considered to have derived from a widespread taxon, which split as the continents separated to their present geographical position (Ferris, 1983). From a comparison of taxonomic characteristics of longidorid genera, Coomans (1985, 1996) suggested that *Xiphinema* originated in Gondwana, and had spread to Laurasia before the break-up of Pangaea. *Longidorus* and *Paralongidorus* are considered to have originated in Southeast Africa and India, when these two areas were still united, and a later spread to Laurasia was followed by a main speciation of *Longidorus* in the Holarctic region, especially in Europe. Although taxonomic characteristics have been used to identify distinct groupings of trichodorid species (De Waele, et al., 1982; Loof, 1975) these do not indicate evolutionary directions or centers of origin. However, the present distribution of trichodorids is broadly the culmination of the influence of changing geological events. Human influence on the geographical distribution of species is relatively small, probably being mainly manifest in outlier populations, and the dispersal of a species over large distances, such as worldwide dissemination of *X. index* with grapevine from its presumed center of origin in Iran (Persia).

The most recent major geological event that has influenced nematode distribution is the glaciation that occurred approximately 11,000 years ago. During this period large tracts of northern Europe were covered in an extensive ice-sheet with a permafrost area extending southwards from the ice-sheet (Fig. 14.9). Thus, the current distribution of plant parasitic nematode species in northern Europe has been established only during the last approximately 10,000 years. This has resulted in Europe of a marked decrease, south to north, in species richness (Topham and Alphey, 1985), with the present northern limit of several species being clearly demarcated. For example, *L. macrosoma* is uniformly distributed throughout France and the Low Countries but does not extend northwards beyond southern England. This northern boundary has been

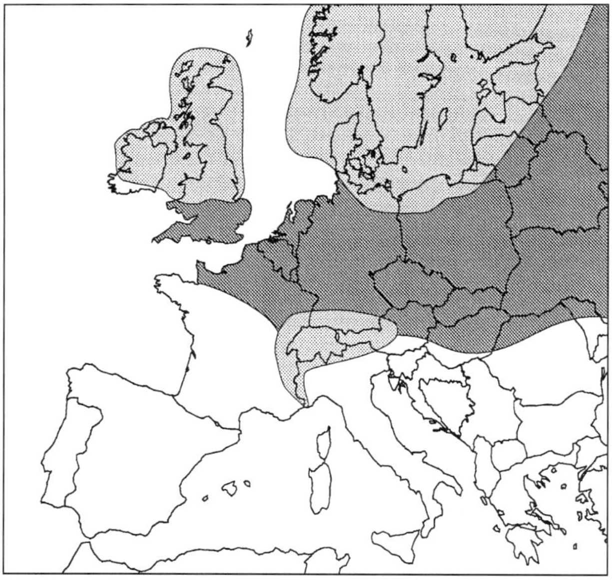

Figure 14.9 The approximate extent of the ice sheet (lighter stippling) and permafrost region (darker stippling) of the last ice-age, c. 11, 000 year before present, superimposed on present day Europe.

correlated with the mean July 15℃ isotherm, implying that the species has certain ecological temperature requirements (Boag, et al., 1991). Also, it has been suggested that *L. macrosoma* populations which survived the period of glaciation were unable to spread northwards because of the destruction of the deciduous forests, which had provided hosts for the species during the bronze and iron ages (Dalmasso, 1970; McNamara and Flegg, 1981). A population of *L. macrosoma* collected from a raspberry plantation in southern England was successfully maintained on raspberry in an outdoor plot at Dundee, eastern Scotland (unpublished data). Therefore, this species would appear to have potential to extend its northern range of distribution in the UK if introduced to appropriate biotopes. *Xiphinema diversicaudatum* is widely distributed in Europe, however the species has a clearly demarcated northern boundary in Scotland for which there are no obvious climatic or edaphic reasons (Topham and Alphey, 1985). The frequency of occurrence of the species suggests that its slow northerly diffusion has not yet been completed. Although these interpretations of geographical distribution are highly conjectural, they nevertheless indicate some of the ecological parameters that influence speciation within the broad ranges of ancestral species. Unfortunately, little is known about the ecological requirements of the majority of longidorid and trichodorid nematodes and elucidation of speciation and species diversity has currently to be approached on the groupings of morphometric characters.

Western Europe appears to be a region of well-established nematode species and with relatively gradual change in distribution and speciation. However, the eastern Mediterranean countries present a complexity of longidorid and trichodorid species that possibly reflects the geological history of the region, with the formation, during the Miocene, of microcontinents movement between the major plates of Africa and Eurasia (McKenzie, 1970, cited in Topham and Alphey, 1985).

The present geographical distribution of *Xiphinema* and *Longidorus* species in Europe and the Mediterranean region (Alphey and Taylor, 1986; Brown and Taylor, 1987) has been quantitatively analyzed by Navas et al. (1990). Two main groups of species, the European-Atlantic and the Mediterranean groups, were recognized by Navas et al. (1990). Furthermore, these authors considered the distribution of *Xiphinema* species to be a relatively recent dispersion process, with the pleisochoric limits of the genus in the southern Mediterranean region and that of *Longidorus*, with a relatively early dispersal, in the northern Mediterranean and southern European countries. In a more detailed appraisal of the distribution of *Longidorus* species in Euromediterranea, Navas et al. (1993) identified the centers of distribution of 32 species within distinctive geographical regions (chorological units). They hypothesized, *inter alia*, that after glaciation the founder species *L. intermedius, L. africanus,* and *L. congoensis* were the origin of dispersive speciation throughout much of Euromediterranea.

Longidorids and trichodorids present in North America are largely distinct from those species present in Europe and other continents. Also, there is an apparent paucity of *Longidorus* species in North America, as compared with more than 50 species reported from Europe. In North America, of the ten species reported only seven are probably indigenous; the three others have been introduced with planting material. This compares with more than 50 species reported from Europe. In contrast, 38 species of *Xiphinema* have been identified in North America, of which more than half have probably been introduced compared with at least 60 species in Europe (Robbins, 1993; Robbins and Brown, 1991). However, the apparent paucity in North America of the long bodied, i. e., greater than 2 mm, *Xiphinema* and *Longidorus* this spece species has been partly attributed to inappropriate nematode extraction procedures having been employed. Thus, it seems probable that many new species will be discovered in this region in the future (Robbins, 1993; Robbins and Brown, 1991).

The apparent effectiveness with which some species have been widely dispersed by human activities or by natural means indicates that the nematodes have genetic and ecological flexibility, e. g., wide host range, adaptability to a range of soil types, and reproduction sustained over a broad temperature range. It may be expected that isolation of populations in diverse biotopes would result in reproductive isolation and hence the establishment of new species. When Brown and Topham (1985) examined *X. diversicaudatum* specimens that were obtained from populations widely distributed throughout the world they found morphological and morphometric differences between populations, but they did not regard these differences as sufficient to distinguish populations as separate species. Subsequently, this conclusion was validated when it was demonstrated that many of these populations could interbreed successfully with a population from Scotland (Brown, 1986a).

Dalmasso and Berge (1983) presented a hypothetical model to explain evolution in the Longidoridae in which populations of ancestral amphimictic forms (species) could be affected by inbreeding resulting in complete homozygosity, subsequent facultative meiotic parthenogenesis with a resulting loss of males, and finally mutations giving rise to "clonal" species. These "clonal" species would be morphologically similar within species complexes. Several such groups are evident today. For example, populations of *X. coxi* from Florida, USA, were identified as being distinct from those in Europe (Dalmasso, 1970; Taylor and Brown, 1976), with Sturhan (1984) recognizing two species, *X. pseudocoxi* and *X. coxi*, the latter being divided into two subspecies (Brown and Taylor, 1987). Morphometric differences between dispersed populations have also been noted in *L. elongatus, L. profundorum* and *L. vineacola*. Such differences also should be considered in relation to their effect on the efficiency of virus transmission, as has been demonstrated with populations of *X. diversicaudatum* (Brown, 1985, 1986c; Brown and Taylor, 1981; Brown and Trudgill, 1983) and in populations of *X. americanum sensu lato* (Griesbach and Maggenti, 1989). However, there is no evidence that a difference in the ability of nematodes to transmit virus provides a basis for distinguishing species or populations of a vector.

14.7.1 The *Xiphinema americanum* Group

During the last century the most widespread longidorid occurring in North America was identified as *X. americanum*. However, a taxonomic reappraisal of populations previously identified as *X. americanum* resulted in 15 new species being described to give a complex of 25 morphologically similar, parthenogenetic species (Lamberti and Bleve-Zacheo, 1979). Currently 20 of the 51 putative species comprising the *X. americanum* group have been described from specimens originating from North America (Robbins, 1993; Robbins and Brown, 1991).

Lamberti et al. (2000) recently provided a comprehensive review of the occurrence and geographical distribution of *X. americanum* group species, and included a series of polytomous keys to facilitate an initial practical means for establishing a preliminary identification of species reported from the major geographical areas. These authors acknowledge that further morphological and molecular research will result in several of the species becoming junior synonyms. Also, they anticipate that nematologists worldwide will contribute to this process, with the final objectives being to establish a list of valid species and a practical morphological identification key that will receive international acceptance.

Halbrendt and Brown (1992, 1993) reported that several *X. americanum* group populations from Europe have four juvenile stages, which is considered typical for Nematoda, but that populations from North America have only three such stages. These North American populations are vectors of three North American nepoviruses (Brown et al., 1993, 1994c). Using Restriction Fragment Length Polymorphism with 16 populations of *X. americanum* group nematodes, Vrain (1993) distinguished four groups of populations representing *X. americanum, X. rivesi, X. pacificum/X. californicum* and *X. bricolense*. Lamberti and Ciancio (1993) studied the morphometrics of 49 populations of 39 species attributed to the *X. americanum* group and by hierarchical cluster analysis placed the populations into five subgroups. The *X. americanum* subgroup they suggested consisted primarily of North American species; *X. pachtaicum* subgroup mainly of European species; *X. lambertii* subgroup mainly of Asian species; *X. brevicolle* subgroup, comprising seven species that have a cosmopolitan distribution, and the *X. taylori* subgroup containing only two species, one each from Europe and North America.

14.8 Persistence of Nematode Populations and Viruses

A characteristic of nematode transmitted viruses is their persistence at sites for long periods, even in the absence of virus infectable crop hosts. This is dependant on both the survival of the vector population, particularly in situations where the soil is cultivated and where crops are rotated, and on the presence of virus-infected weed seeds and/or perennial plants. A study of two populations of *X. diversicaudatum* spanning 30 years provided data on some of the ecological factors involved.

In 1961 the distribution of a population of *X. diversicaudatum* was mapped in an undisturbed, mixed woodland at Geesecroft, southern England (Harrison and Winslow, 1961) and in 1966 and 1967 that of another population in an arable field at Gilliesfaulds, eastern Scotland (Taylor and Thomas, 1968). Resampling at Geesecroft in 1991 revealed the horizontal distribution of the nematode population to have remained virtually unchanged in the intervening years (Taylor, et al., 1994). A reduction in the numbers of nematodes, by two thirds from levels detected in 1961, was believed to have resulted from a serious rainfall deficit in preceding years. This being compounded by the increased density of the woodland canopy that had reduced the weed flora, and hence the hosts for nematodes in the upper soil layers that were sampled. At Gilliesfaulds, the nematode infestation was presumed to have originated in a hedge bordering the eastern perimeter of the field and when sampled in 1966, and again in 1967, had spread about 35 meters into a raspberry crop cv. Malling Jewel (Fig. 14.10(A)). By 1991, the pattern of infestation could still be related to that evident in 1967, but had become scattered into sub-populations with some spread westwards into the remainder of the cultivated area (Fig. 14.10(B)). Continuous cultivation and change in crops during the intervening years would be expected to have some adverse effect on the nematodes as well affecting their horizontal distribution. The largest sub-population recorded in 1991 was in a grass strip at the northern part of the field that, since 1986, had provided a good host for the nematode and an undisturbed habitat.

In 1966 the raspberry crop at Gilliesfaulds was infected with SLRSV but not ArMV, to which it is immune, although the latter virus was detected in several weed species. In 1991, after several crop rotations including the planting of SLRSV-immune raspberry cv. Glen Moy in 1987, only ArMV was detected in soil samples and in some weed species (Taylor, et al., 1994). The virus was present as two strains, differing in the severity of symptoms induced in herbaceous plants, indicating that the natural weed bank is an efficient reservoir for preserving nepoviruses and their variants that occur naturally. Weed seed transmission of nepo- and tobraviruses has been well documented as providing an alternative strategy for virus persistence in nature with up to 100% of seed from an infected mother plant being infected and capable of survival in soil for many years (Murant, 1983).

The existence of two variants of ArMV with a population of *X. diversicaudatum* is an example of a strategy for the survival of nematode transmitted viruses in nature. Other examples of several variants of a virus

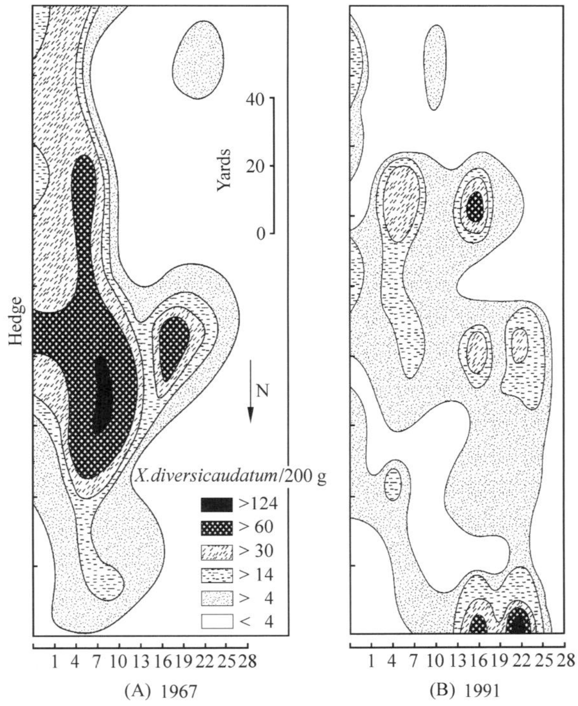

Figure 14.10 The distribution of *Xiphinema diversicaudatum* in a raspberry plantation at Gilliesfaulds, eastern Scotland: (A) as determined during 1967 (data from Taylor and Thomas, 1968); (B) as determined during 1991; (+) occurrence of arabis mosaic nepovirus as detected from soil bait tests (after Taylor, et al., 1994).

occurring naturally together with a vector population have been reported for ArMV and *X. diversicaudatum* from strawberry in Norway; RRSV and *L. elongatus* from raspberry in eastern Scotland (Jones, et al., 1989) and TRV with *T. cylindricus* and *P. pachydermus* in pasture in eastern Scotland (Ploeg, et al., 1992). The appearance of resistance-breaking strains of a virus in new crop cultivars is also an indication of the reservoir of virus variants that exist in nature allowing viruses to adapt to a changing ecological environment, e.g., the MX and Lloyd George strains of RRSV can infect Malling Exploit and Lloyd George raspberry cultivars, respectively, which are immune to the type strain of the virus (Jones, et al., 1989; Murant, et al., 1968).

14.9 Virus and Vector Interactions

14.9.1 Specificity

The soil environment imposes many constraints on the behavior of nematodes. It effectively isolates populations, and as the nematodes are relatively immobile they are able to expand their colonization of a biotope by only a few centimeters annually. Consequently, there is considerable inbreeding that results in increased homozygosity within the population. In such a conserved environment, viruses would tend to develop highly specific associations with their vectors. Specificity of transmission is most evident in the European context, where a particular virus, or a serologically distinct strain, can be transmitted by one species but not by another. For example, ArMV is transmitted by *X. diversicaudatum* but not by any other species tested, Scottish strains of RRSV and TBRV are transmitted by *L. elongatus* but English strains of these viruses by *L. macrosoma* and *L. attenuatus*, respectively, and a strain from grapevine in Germany (Jones, et al., 1994) by *P. maximus,* and apparently not by *L. macrosoma*. Specificity of transmission is also evident with TRV with many of the strains associated with different vector species (Ploeg, et al., 1992).

In North America, *X. americanum* has been implicated as a natural vector of CRLV, PRMV, TRSV, and ToRSV (Breece and Hart, 1959; Fulton, 1962; Klos, et al., 1967; Nyland, et al., 1969) and *X. rivesi* and *X. californicum* as vectors of ToRSV (Table 14.1) (Forer, et al., 1981; Hoy, et al., 1984). This indicates little specific transmission of ToRSV by several members of the *X. americanum* group and that *X. americanum* is unusual in being a vector of each of the four indigenous North American nematode-transmitted nepoviruses. Some of this apparent lack of specificity may be a result of uncertainty in species identification prior to a reappraisal of the taxonomy of the *X. americanum* group (Lamberti and Bleve-Zacheo, 1979) and of the characterization of the specific virus strains transmitted. However, Brown and Halbrendt (1992) and Brown et al. (1993, 1994c) using individual nematodes from three populations of *X. americanum sensu stricto, X. bricolense, X. californicum* and *X. rivesi* showed that *X. americanum, X. californicum* and *X. rivesi* transmitted CRLV, TRSV and two strains of ToRSV but that *X. rivesi* transmitted the viruses more frequently. *Xiphinema bricolensis* transmitted only ToRSV. These data confirm a relative lack of specificity between North American nematode-transmitted nepoviruses and some of their vectors.

Table 14.1 Specific associations between *Longidorus, Paralongidorus*, and *Xiphinema* virus-vector nematode species and Nepoviruses.

Vector species	Virus	Acronym	Reference
L. apulus	artichoke Italian latent (Italian strain)	AILV	Rana and Roca, 1973
L. arthensis	cherry rosette disease	CRDV	Brown et al., 1994a
L. attenuatus	tomato black ring (German/English strain)	TBRV	Harrison, 1964
L. elongatus	raspberry ringspot (Scottish strain)	RRSV	Taylor, 1962
	tomato black ring (Scottish strain)	TBRV	Harrison et al., 1961
L. fasciatus	artichoke Italian latent (Greek strain)	AILV	Roca et al., 1982
L. macrosoma	raspberry ringspot (English strain)	RRSV	Harrison, 1964
L. martini	mulberry ringspot	MRSV	Yagita and Komuro, 1972
P. maximus	raspberry ringspot (German grapevine strain)	RRSV	Jones et al., 1994
X. americanum (*senso lato**)	cherry rasp leaf	CRLV	Nyland et al., 1969
	peach rosette mosaic	PRMV	Klos et al., 1967
	tobacco ringspot	TRSV	Fulton, 1962
	tomato ringspot	ToRSV	Breece and Hart, 1959
X. americanum (*sensu stricto***)	cherry rasp leaf	CRLV	Brown and Halbrendt, 1992
	tobacco ringspot	TRSV	Brown and Halbrendt, 1992
	tomato ringspot	ToRSV	Brown and Halbrendt, 1992
X. bricolense	tomato ringspot	ToRSV	Brown and Halbrendt, 1992
X. californicum	cherry rasp leaf	CRLV	Brown and Halbrendt, 1992
	tobacco ringspot	TRSV	Brown and Halbrendt, 1992
	tomato ringspot	ToRSV	Hoy et al., 1984
X. diversicaudatum	arabis mosaic	ArMV	Jha and Posnette, 1959
	strawberry latent ringspot	SLRSV	Lister, 1964
X. index	grapevine fanleaf	GFLV	Hewitt et al., 1958
X. intermedium	tobacco ringspot	TRSV	Brown et al., 1996
	tomato ringspot	ToRSV	Brown et al., 1996

Continued

Vector species	Virus	Acronym	Reference
X. italiae	grapevine fanleaf	GFLV	Cohn et al., 1970
X. rivesi	cherry rasp leaf	CRLV	Brown and Halbrendt, 1992
	tobacco ringspot	TRSV	Brown and Halbrendt, 1992
	tomato ringspot	ToRSV	Forer et al., 1981
X. tarjanense	tobacco ringspot	TRSV	Brown et al., 1996
	tomato ringspot	ToRSV	Brown et al., 1996

* Unequivocal identification of species not available or prior to the review of the *X. americanum*-group by Lamberti and Bleve-Zacheo (1979).

** Species identification determined by using individual nematodes in virus transmission studies (Brown and Halbrendt, 1992; Brown et al., 1994c).

Martelli and Taylor (1989) speculated that vector populations widely separated geographically might be expected to differ in their ability and efficiency to transmit viruses. This could result from ecological pressure on the virus to adapt to its plant host, whereby if the dominant host were an annual, or short term perennial, selection pressure for frequent (= efficient) transmission would be selected to ensure survival of the virus; whereas with long term perennials, such as fruit trees, such pressure would be much less and relatively infrequent (= inefficient) transmission by vectors might result. Ecologically driven selection pressure can be presumed to be more likely in disturbed (cultivated) than in undisturbed habitats. Viruses are also subject to selection pressure because of the changing flora, as well as biological modifications in the nematodes, and evidence for this may be assumed from the range of virus strains that occur. For example, the type British strain of SLRSV is efficiently transmitted by several geographically disparate populations of *X. diversicaudatum*, i. e., 80 to 100% of nematodes from these populations transmit virus, but only infrequently by populations from France, Italy and Spain, i. e., less than 10% of the nematodes transmit the virus. Conversely, serologically distinguishable strains of SLRSV from Italy were transmitted only, and infrequently by a population of the vector from Italy (Brown 1985; Brown and Taylor, 1981; Brown and Trudgill, 1983).

Recently, specificity of transmission of serologically distinguishable strains of tobraviruses with trichodorid nematodes has been demonstrated (Ploeg, et al., 1992). Tobraviruses occur naturally as a range of serotypes, which in Europe have been shown to be transmitted each by a different nematode species, e. g., isolates of the PRN serotype of TRV from Britain, Sweden and the Netherlands are transmitted only by *P. pachydermus*. In North America, where only *Paratrichodorus* species have been reported as vectors of TRV, it is uncertain if specificity of transmission occurs with TRV isolates and populations/species of (para) trichodorid nematodes.

14.9.2 Sites of Retention

Electron microscopy has provided evidence for the sites of virus retention in nematode vectors. In *Longidorus* species the virus particles are apparently adsorbed in a single layer to the inner surface of the odontostyle, and in *L. elongatus* particles of RRSV or TBRV may also be located between the odontostyle and the guiding sheath (Taylor and Robertson, 1969). In contrast, *Xiphinema* vectors have nepovirus particles specifically associated with the cuticular lining of the odontophore and the esophagus (Taylor and Robertson, 1970a). In trichodorid vectors TRV particles are retained in association with the lining of the food canal from the anterior region of the esophostome to the esophago-intestinal valve, but are not attached to the onchiostyle (Taylor and Robertson, 1970b).

Specificity of the virus/vector association, which probably developed during a long association of the vectors with their associated viruses, implies at its simplest a "gene for gene" type of recognition process. Although in many instances the same part of the coat protein appears to be crucial for both specificity of retention and antibody recognition, this is not so for all nematode and virus associations. For example, a population of *L. attenuatus* from Britain transmitted two serologically distinguishable British isolates of TBRV much more efficiently than two isolates from Germany (Brown, et al., 1989a), implying that efficiency of transmission is correlated more strongly with geographic origin than antigenic relatedness. Nevertheless, more substantive serological differences are usually associated with differences in the vector species. Thus, two serologically distinct strains of artichoke Italian latent nepovirus from Italy and Greece are transmitted by *L. apulus* and *L. fasciatus*, respectively (Table 14.2). Similarly, the German and Scottish serotypes of TBRV are transmitted specifically by *L. attenuatus* and *L. elongatus*, respectively. The Scottish and English serotypes of RRV are transmitted most efficiently by *L. elongatus* and *L. macrosoma*, respectively, but each species can transmit the other virus, albeit less efficiently.

14.9.3 Acquisition and Release of Virus Particles

Brown and Weischer (1998) defined seven distinct processes involved in successful transmission of viruses by nematodes. Three of these involve interactions between the virus and the vector nematode, with the relative efficiency of each determining the final efficiency of transmission of the virus. The three processes are: (1) adsorption of virus particles at the sites of retention

Table 14.2 Specific associations between *Paratrichodorus* and *Trichodorus* virus-vector nematode species and Tobraviruses.

Vector species	Virus	Acronym	Reference
P. allius	tobacco rattle	TRV	Ayala and Allen, 1968
P. anemones	pea early-browning	PEBV	Harrison, 1967
	tobacco rattle	TRV	Hoof, 1968
P. hispanus	tobacco rattle	TRV	Brown and Weischer, 1998
P. minor	pepper ringspot	PepRSV	Salomao, 1975
(syn. *P. christiei*)	tobacco rattle	TRV	Walkinshaw et al., 1961
P. nanus	tobacco rattle	TRV	Cooper and Thomas, 1970
P. pachydermus	pea early-browning	PEBV	Hoof, 1962
	tobacco rattle	TRV	Gibbs and Harrison, 1964b
P. porosus	tobacco rattle	TRV	Ayala and Allen, 1968
P. teres	pea early-browning	PEBV	Hoof, 1962
	tobacco rattle	TRV	Hoof, 1964
P. tunisiensis	tobacco rattle	TRV	Roca and Rana, 1981
T. cylindricus	tobacco rattle	TRV	Hoof, 1968
T. primitivus	pea early-browning	PEBV	Gibbs and Harrison, 1964a
	tobacco rattle	TRV	Sanger, 1961
T. similis	tobacco rattle	TRV	Cremer and Schenk, 1967
T. viruliferus	pea early-browning	PEBV	Gibbs and Harrison, 1964a
	tobacco rattle	TRV	Hoof et al., 1966

in the vector during ingestion of food; (2) retention of the particles until the nematode next feeds; and (3) release of at least some of the particles during egestion of esophageal gland secretions when the nematode next feeds. Also, it is possible that a few as yet undefined plant factors may be involved in some instances as *X. index* can acquire GFLV from mechanically infected *Chenopodium quinoa* plants, but subsequently it is only when the nematodes feed on grapevine that transmission occurs (Trudgill and Brown, 1980). Consequently, it appears that the root cells of *C. quinoa* fed upon by the nematode are unable to support establishment of infection by the virus, or the plant lacks a factor involved in the release of the virus particles from within the vector.

It has been shown with *X. diversicaudatum* that the ability of this species to transmit its associated viruses is genetically inherited, involving the capability of the nematode to specifically retain virus particles. Nematodes from a Scottish population of *X. diversicaudatum*, which were highly efficient as vectors, were crossbred with those from an Italian population, which were inefficient vectors. The Fl progeny were more efficient as vectors than nematodes from the original Italian population, but much less so than those from the Scottish population. The F2 were more efficient than the Fl progeny, but were not as efficient as nematodes from the Scottish population (Brown, 1986b), with the low efficiency of transmission being associated with the nematodes' inability to adsorb virus particles at the site of retention (Brown and Trudgill, 1983).

Two different methods of virus adsorption, retention and release may be utilized by vector nematodes. Virus particles are retained in the esophageal tract of *Xiphinema* nematodes, and in the broadly similar pharyngeal tract of trichodorid nematodes, whereas in *Longidorus* species the particles are retained in the odontostyle region (Figs 14.1, 14.2). There is evidence that specific recognition occurs between vector *Xiphinema* and trichodorid nematodes and their associated viruses, possibly involving an interaction of complimentary molecules at their point of contact. A discontinuous layer of carbohydrate-staining material lines the esophageal tract of *X. diversicaudatum* and *X. index*, and in *X. diversicaudatum* ArMV and SLRSV particles were adsorbed only where this layer occurred. Also, ArMV particles, but not those of SLRSV, were observed to be enveloped by a "cloud" of this material (Robertson and Henry, 1986a). In *P. pachydermus* the entire lining of the pharyngeal tract was observed to be covered with a layer of carbohydrate-staining material (Robertson and Henry, 1986b). Secretory products released from the ducts of the larger esophageal glands possibly are the origin of carbohydrate-staining material observed in these nematodes (Robertson and Henry, 1986a, b). Virus retention in *Xiphinema* and trichodorid nematodes may therefore involve interactions between carbohydrate moieties in the nematodes' esophageal and pharyngeal tracts, respectively, and surface structures of the virus capsids.

Upon completion of feeding *Xiphinema* nematodes retract their odontostyle and appear to use several pulsations of their esophageal bulbs to clean the esophageal tract of residual food (Hunt and Towle, 1979; Trudgill, 1976; Wyss, 1977). These nematodes do not have a mechanism with which to fully ingest or egest such secretions, therefore residual secretions probably remain within the esophageal tract and it is these that account for the carbohydrate-staining layer observed in *Xiphinema* and trichodorid nematodes. Similarly, this may account for the presence of a "mucus-like" layer reported associated with virus retention in several nematode species (McGuire, et al., 1970; Raski, et al., 1973; Taylor and Robertson, 1969, 1970a, b).

Surface charges on virus particles interacting with oppositely charged areas associated with the cuticle lining the feeding apparatus of virus-vector species provides an alternative hypothesis for virus retention by nematodes (Harrison and Roberts, 1968; Raski, et al., 1973; Taylor and Brown, 1981; Taylor and Robertson, 1970a). The cuticle associated with the odontostyle region in nematodes is similar to external cuticle, whereas that associated with the esophageal and pharyngeal tracts represents internal cuticle, with each of these two cuticle types having differential isoelectric points (Inglis, 1966). Cationized ferritin labeling of the odontostyle region in *L. elongatus* revealed a strong negative charge associated with the surface of the odontostyle and the lining of the lumen (Robertson, 1987), and thus with *Longidorus* species surface charges may determine the specific retention of virus particles.

Whereas virus transmission is related to the ability of the vector species to retain virus particles there is one example where an apparent lack of dissociation of virus particles from the site of retention results in the nematode being an inefficient vector. Particles of the Scottish and the English serotypes of RRSV are efficiently adsorbed at the site of retention in *L. macrosoma*, but this species only transmits the English serotype. The nematodes' inability to transmit the Scottish serotype appears to be a consequence of a lack of dissociation of virus particles from the site of retention (Taylor and Robertson, 1975; Trudgill and Brown, 1978).

During feeding, virus-vector nematodes pump esophageal gland secretions forward through the esophageal tract into the root cell. These secretions are considered the principal means by which specifically adsorbed virus particles are injected into the cell. Virus particles may be released from the site of retention in the vector by changes in pH in the esophageal or pharyngeal tracts or the odontostyle region induced by the presence of the secretions. Alternatively, the secretions may contain specific effector molecules, i. e., proteases that induce release.

14.10 Vector-virus Interactions Involved in Vector Transmission

Nepoviruses express their genomes by proteolytic processing of polyproteins, whereas tobraviruses express theirs using subgenomic messenger RNA. Pseudo-recombinant isolates of nepo- and tobraviruses have been produced in the laboratory, with the RNA-1 of one isolate being combined with the RNA-2 of a different isolate of the same virus. Virus transmission studies using pseudo-recombinant isolates have revealed that the RNA-2 segment of the viral genome determines vector transmissibility. For example, *L. elongatus*, the natural vector of isolates of RRSV belonging to the Scottish serotype, readily transmitted a pseudo-recombinant isolate in which the RNA-2 was derived from the Scottish strain but only infrequently transmitted an isolate containing the RNA-2 from an English strain (Harrison, et al., 1974). Similarly, *L. elonagtus* transmitted only a pseudo-recombinant containing the RNA-2 derived from a Scottish strain of TBRV, as the Scottish and German strains of TBRV have *L. elongatus* and *L. attenuatus*, respectively, as their natural vectors (Harrison and Murant, 1978). Similar experiments with pseudo-recombinants of TRV, and of PEBV, and trichodorid nematodes have revealed that the vector transmissibility of tobraviruses is associated with the RNA-2 of these viruses (Brown, et al., 1995; MacFarlane, et al., 1995; Ploeg, et al., 1993b).

The RNA-2 of nepo- and tobraviruses encodes several proteins, including the coat protein (CP) that has an obvious role in nematode transmission as it provides one of the two points of contact between virus and vector, the other being associated with the vectors feeding apparatus. However, a tobravirus isolate prepared by replacing the CP gene of PEBV isolate SP5, with that of TRV isolate PpK20 was not transmitted by *T. primitivus* or by *P. pachydermus*, the natural vectors of PEBV isolate A56 and TRV isolate PpK20, respectively. From this result, it was concluded that vector transmission of tobraviruses was not determined exclusively by the CP, and that other genes encoded by the RNA-2 were probably involved in vector transmissibility of tobraviruses as well (MacFarlane, et al., 1995). This may also occur with nepoviruses.

14.10.1 Virus Particle Structure

Nepoviruses have isometric particles of approximately 28 nm diameter, but the composition of the particles differs between the viruses. Most nepoviruses transmitted by nematodes each has only a single CP of 56 to 60kDa molecular masses, e.g., ArMV, AILV, PRMV, RRSV, TBRV, TRSV, and ToRSV. However, two others that also are transmitted by longidorid nematodes, SLRSV and CRLV, have two and three CPs, respectively (29.4 kDa and 42.6 kDa; and 26 kDa, 23 kDa and 21 kDa) (Mayo and Robinson, 1996). Nematode transmissibility is not correlated with particle composition as *X. diversicaudatum* is the natural vector of ArMV and SLRSV, and CRLV, TRSV and ToRSV are each transmitted by the three species, *X. americanum, X. californicum*, and *X. rivesi*.

Late during infection of a plant, or after purification, the CP contained in particles of TBRV were found to be reduced in size, resulting from a removal of 9 amino acid base pairs on the C-terminus of the CP (Demangeat, et al.,

1992). These results may be indicative of this region being exposed on the surface of the protein where it could interact with nematode surfaces involved in virus retention. An amino acid motif (VQV or VPV) is conserved at the N-terminus of the small capsid protein of SLRSV, and the CP of ArMV, both viruses being naturally transmitted by *X. diversicaudatum* (Kreiah, et al., 1994). However, a similar motif is not present in the CP of GFLV, RRSV, TRSV, and ToRSV, these viruses being transmitted by other longidorid species. This motif may be analogous with a DAG motif present in potyvirus CPs, which is involved in transmission of these viruses by aphids (Blanc, et al., 1997; Harrison and Robinson, 1988). A study of the crystal structure of TRSV revealed several surface-located loops, and two regions of the capsid surface whose sequence is conserved among nine different nepoviruses (Chandrasekar and Johnson, 1998). However, there was no correlation apparent between particular amino acid motifs and the respective vectors of the viruses. Also, alignments of capsid protein sequences revealed that the VQV peptide is not generally conserved, and that a "protruding" C-terminal peptide, as present in TBRV, does not occur in all nepoviruses.

Similarities exist between different nepoviruses in part of the RNA-2 encoded polyprotein that is N-terminal to the CP. The viruses TBRV and RRSV, each transmitted by *Longidorus* species, are similar in this region. Similarly, GFLV and ToRSV, each transmitted by *Xiphinema* species, are similar in this region. However, although each pair of viruses shares similarities in this region, no such similarity exists between the pairs of viruses (Mayo, et al., 1995). This protein may be involved in cell to cell movement, and thus there could be a similarity with potyviruses in which the helper protein (HC-Pro) involved in aphid transmissibility of these viruses also affects virus movement in the plant (Maia, et al., 1996).

Detailed investigation of viral determinants of vector transmission of nepoviruses requires site-specific mutagenesis of infectious cDNA clones of viruses. The only such clones available have been constructed for GFLV, but data have not been published on their transmissibility by nematodes (Viry, et al., 1993). However, several tobravirus isolates have been fully sequenced, and stable, infectious cDNA clones produced that are being used to examine the genetic determinants of vector transmissibility encoded by the viruses.

The particle structure of tobacco mosaic tobamovirus has been comprehensively studied (Namba, et al., 1985), and amino acid sequence alignment of CPs of tobraviruses with that of tobacco mosaic virus has revealed that the CP subunits of all these viruses fold similarly. Tobravirus CP subunits form a tight helical array with their N- and C-termini located on the external surface of the viral particle (Goulden, et al., 1992). The C-terminal domains of tobraviruses are larger than that present in tobacco mosaic virus and in PepRSV it is unstructured, presumably extending outwards from the surface of the virus particle. It was suggested that this "protruding" flexible domain could be involved in the specific attachment of the virus particle to sites of retention in the vector (Mayo, et al., 1995), and subsequently evidence has been obtained confirming the role of these sequences in vector transmissibility of tobraviruses (see Section 14.10.2). However, a gap of 5 to 7 nm between the surface of TRV particles and the cuticle lining the pharyngeal tract in vector trichodorids had been measured using transmission electron microscopy (Robertson and Wyss, 1983). This gap is too wide to be bridged by tobravirus CP C-terminal peptides of only 22 (TRV isolate PpK20) to 38 (PepRSV) amino acids. Consequently, Brown et al. (1995) suggested that other "non-structural" proteins (2b and 2c) known to be encoded by the RNA-2 of nematode transmitted tobravirus isolates might act as a bridge to link the C-terminal peptide to the carbohydrate-staining material lining the nematodes' pharyngeal tract.

14.10.2 Viral Determinants Involved in Vector Transmission of Tobraviruses

Infectious, nematode transmissible, viral cDNA clones of PEBV and TRV have enabled investigation of the genetic determinants of vector transmissibility encoded by tobraviruses. Data from the pseudo-recombinant tobravirus isolates suggest that vector transmissibility is determined by the RNA-2 segment of the bipartite genome. The complete sequence of the RNA-2 of three TRV and one PEBV nematode-transmitted isolates have been determined (Hernandez, et al., 1995; MacFarlane and Brown, 1995; MacFarlane, et al., 1998; N. Vassilakos, pers. comm.), and these RNAs encode the CP and additionally two or three other proteins (2b, 2c and 9 K).

Deletion of, or within, the flexible C-terminal domain of the CP of PEBV isolate TpA56 abolished, or greatly reduced, respectively, transmissibility of the virus by *T. primitivus* (MacFarlane, et al., 1996). Similar results have been obtained with TRV, thus this part of the tobravirus CP is intimately involved in vector transmissibility of the virus. Mutations made to the three other genes encoded by the PEBV RNA-2 also affected vector transmissibility of the virus, with the 2b protein (29 kDa) being found to be essential for vector transmission whereas disruption of either the 2c (23 kDa) or 9 kDa genes reduced the frequency of, but did not necessarily eliminate, vector transmission (Fig. 14.12).

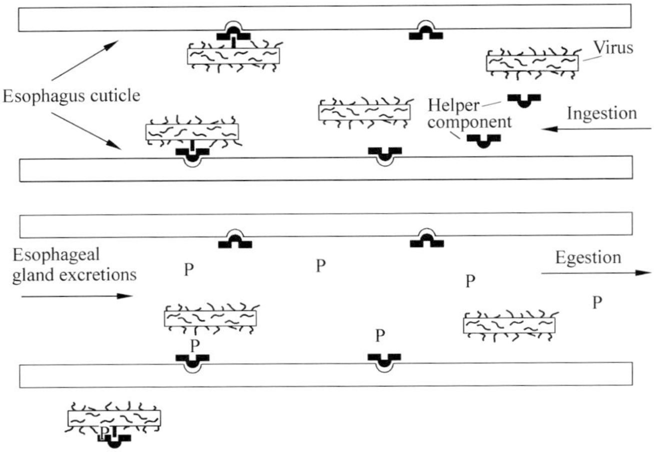

Figure 14.11 Schematic representation of the possible adsorption and release of tobravirus particles at the site of retention in the esophageal tract of trichodorid nematodes. When feeding on a virus infected plant the nematode ingests virus particles, and possibly also viral encoded proteins that interact with the viral capsids and with the site of virus retention in the vector. The helper component forms a "bridge" between the capsids and the esophageal cuticle in the vector, thus specifically anchoring virus particles at the site of retention. When the vector next feeds it pumps esophageal gland secretions into the plant cell and proteases (p) present in these secretions cleave the "bridge", thus releasing the virus particles that are then available to be carried in the secretions into the plant cell to establish infection.

With TRV isolates PpK20 and PaY4, naturally transmitted by *P. pachydermus* and *P. anemones*, respectively, the RNA-2 encodes the CP, 2b and 2c proteins, but not the 9 K protein. Laboratory virus transmission experiments with PpK20 and *P. pachydermus* have revealed that the CP and 2b (29 kDa) proteins are essential for vector transmission, whereas the 2c (32 kDa) protein is not involved in nematode transmission of the viruses (Fig. 14.13) (Hernandez, et al., 1997). Similar results have been obtained with TRV isolate PaY4 and *P. anemones* (N. Vassilakos, pers. comm.). It has yet to be determined if the involvement/non-involvement of the 2c protein in nematode transmissibility of PEBV and TRV is a function of the viruses, or the vector nematode species. *Trichodorus primitivus* transmits TRV isolate TpO1 and PEBV isolate TpA56, and the RNA-2 of each of these viruses encodes the CP, 2b, 2c, and a 9 K protein. Intriguingly, these results suggest

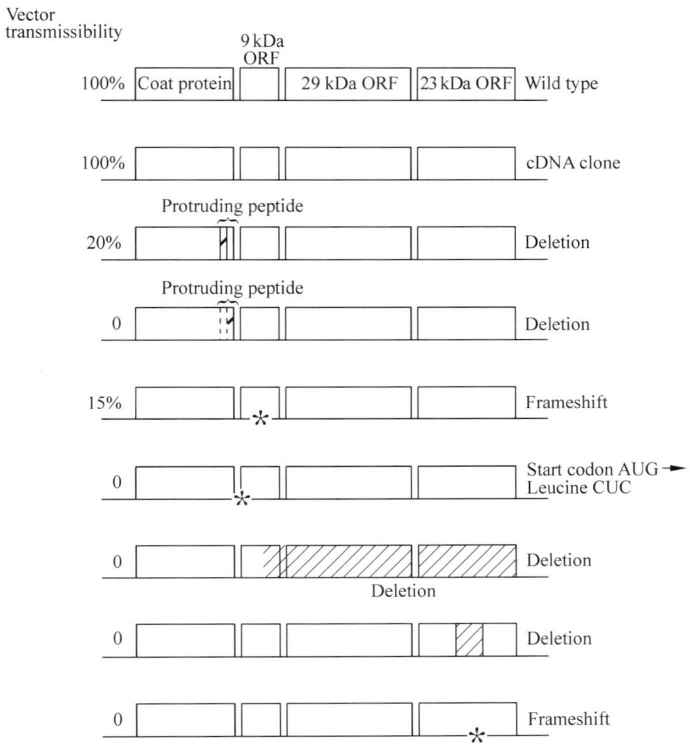

Figure 14.12 Transmissibility by *Trichodorus primitivus* of wild type and a full-length infectious cDNA clone of pea early-browning virus isolate TpA56, and of several mutant viruses derived from the cDNA clone. Removal of the posterior section of the protruding peptide, a flexible C-terminal domain, on the coat protein eliminated vector transmissibility, whereas removal of the anterior section reduced, but did not eliminate, vector transmissibility. A frame-shift in the 9 kDa open reading frame (ORF) reduced vector transmissibility, and a frame-shift 23 kDa and in the start codon of the 9 kDa ORFs, and deletions in, or of, the 9 kDa, 29 kDa and 23 kDa ORFs each eliminated vector transmissibility.

that with these two viruses the nematode, *T. primitivus*, is selecting the number and type of genes encoded by the viruses that are required for successful vector transmission.

Although progress is being made in understanding the genetic determinants of vector transmission encoded by tobraviruses there is as yet no insight of how

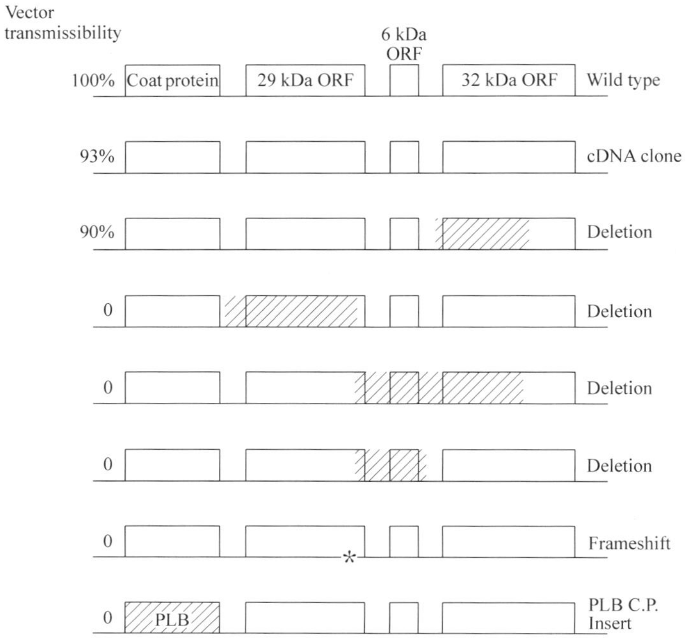

Figure 14.13 Transmissibility by *Paratrichodorus pachydermus* of wild type and a full-length infectious cDNA clone of tobacco rattle virus isolate PpK20, and of several mutant viruses derived from the cDNA clone. A frame-shift in the 29 kDa open reading frame (ORF) and deletions in, or of, the 32 kDa ORF, and a putative 6 kDa ORF each eliminated vector transmissibility. Replacement of the coat protein of TRV isolate PpK20 with that of TRV isolate PLB, a non vector-transmissible isolate, also eliminated vector transmissibility.

these genes facilitate nematode transmission. Current speculation is that the tobravirus 2b and 2c proteins interact physically with the virus particle to facilitate retention of the particles in vector nematodes, as occurs with potyvirus helper protein during transmission by aphids (Gray, 1996). Support for such speculation has recently been proposed by Visser and Bol (1999) who demonstrated, using the yeast 2-hybrid system, that there is an interaction between the CP and 2b proteins of TRV isolate PpK20. Thus, as speculated by Brown et al. (1995), when virus-vector nematodes ingest virus particles they simultaneously ingest viral-encoded protein(s), i. e., the 2b and 2c viral proteins. These proteins may act as helper components by forming a bridge to link the C-terminal peptide to the carbohydrate-staining material lining the nematodes' pharyngeal tract. Subsequently, when the nematode next feeds it pumps esophageal gland secretions into the cell and these secretions induce release of the particles. This may occur through the action of proteases contained in the secretions cleaving the carboxy-terminal domain of the

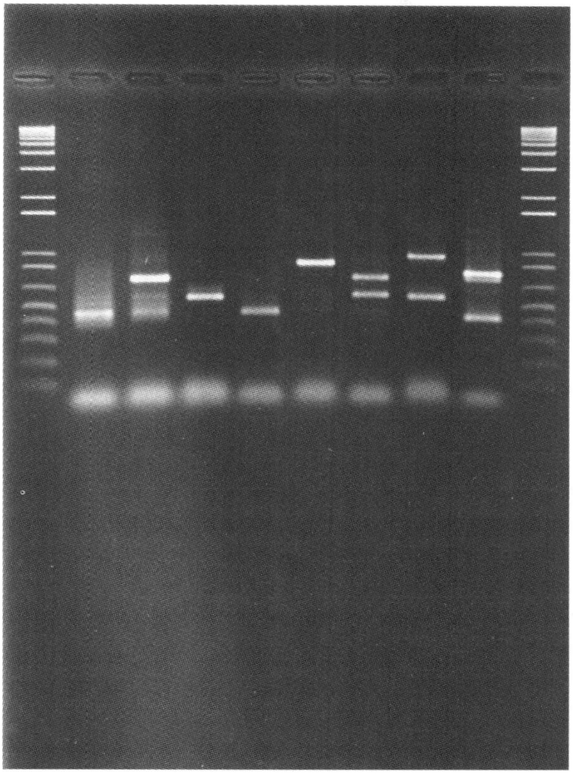

Figure 14.14 Application of a PCR assay with four rDNA primers, derived from ITS-sequences, to reliably distinguish individual *Paratrichodorus anemones, P. pachydermus, Trichodorus primitivus* and *T. similis*. Lanes from left (primers in parenthesis): (1) 1Kb plus DNA ladder; (2) *P. anemones* (664A-ITSB); (3) *P. pachydermus* (664A-ITSB); (4) *P. pachydermus* (ITSA-665A); (5) *T. primitivus* (664A-ITSB); (6) *T. primitivus* (ITSA-665A); (7) *P. anemones* + *P. pachydermus* (664A-ITSB); (8) *P. pachydermus* + *T. primitivus* (ITSA-665A); (9) *P. anemones* + *P. pachydermus* (664A-ITSB); (10) 1Kb plus DNA ladder (courtesy of K. Boutsika, SCRI).

tobravirus particle, which would not adversely affect the virus (Mayo, et al., 1993), and thus release the virus particle from the specific site of retention within the vector (Fig. 14.11).

Increasing interest worldwide of virus transmission by nematodes will undoubtedly result in the identification of new virus and vector associations. Current methods for controlling diseases caused by viruses transmitted by nematodes largely rely on application of highly toxic chemicals used for suppressing vector nematode populations. Societal pressure to reduce, and even to eliminate, the use of such chemicals makes fundamental studies of the mechanisms determining nematode transmission of viruses an essential prerequisite to facilitate identification of novel disease suppression strategies. Also, development of more rapid and sensitive molecular-based methods for detecting and identifying virus-vector nematode species (Fig. 14.14) (Boutsika, et al., 2000), and their associated viruses, will provide potential for more efficiently targeting existing chemical treatments, or alternative resistant crop cultivars.

References

Alphey, T. J. W., and C. E. Taylor, eds. 1986. European Atlas of the Longidoridae and Trichodoridae. Dundee, UK: Scottish Crop Research Institute.

Ayala, A., and M. W. Allen. 1968. Transmission of the Californian tobacco rattle Virus by three species of the nematode genus *Trichodorus*. Journal of the Agricultural University of Puerto Rico 52:101 – 125.

Behrens, J. 1899. Weiter Beitrage zur Kenntnis der Tabakpflanze. XIV. Die Mauche (Mauk) des Tabaks. Landwirtschaftlichen Versuchsstaionen 52:442 – 447.

Blanc, S., J.-J. Lopez-Moya, R. Wang, S. Garcia-Lampasona, D. W. Thornbury, and T. P. Pirone. 1997. A specific interaction between coat protein and helper component correlates with aphid transmission of a potyvirus. Virology 231:141 – 147.

Boag, B., J. W. Crawford, and R. Neilson. 1991. The effect of global warming on the geographical distribution of plant parasitic nematodes. Nematologica 36:312 – 323.

Bodine, E. W., and J. H. Newton. 1942. The rasp leaf of cherry. Phytopathology 32: 333 – 335.

Bos, L., and J. P. H. Van Der Want. 1962. Early-browning of pea, a disease caused by a soil and seed-borne virus. Tijdschrift Plziektenziekten 68:368 – 390.

Boutsika, E., K. R. Wendt, V. C. Blok, S. A. MacFarlane, and D. J. F. Brown. 2000. Molecular detection of trichodorid nematodes and tobacco rattle virus. Russian Journal of Nematology 8:86 (Abstr.).

Breece, J. R., and W. H. Hart. 1959. A possible association of nematodes with the spread of peach yellow bud mosaic virus. Plant Disease Reporter 43:989 – 990.

Brown, D. J. F. 1985. The transmission of two strains of strawberry latent ringspot virus by populations of *Xiphinema diversicaudatum* (Nematoda: Dorylaimida). Nematologia Mediterranea 13: 217 – 223.

Brown, D. J. F. 1986a. Reproduction and interbreeding within and between populations of *Xiphinema diversicaudatum* (Nematoda: Dorylaimoidea). Nematologia Mediterranea 14: 73 – 83.

Brown, D. J. F. 1986b. Transmission of virus by the progeny of crosses between *Xiphinema diversicaudatum* from Italy and Scotland. Revue de Nematologie 9: 71 – 74.

Brown, D. J. F. 1986c. The transmission of two strains of arabis mosaic virus from England by populations of *Xiphinema diversicaudatum* (Nematoda: Dorylaimida) from ten countries. Revue de Nematologie 9: 83 – 87.

Brown, D. J. F. 1989. Viruses transmitted by nematodes. European Plant Protection Organization Bulletin 19: 453 – 461.

Brown, D. J. F., J. Grunder, D. J. Hooper, J. Klingler, and P. Kunz. 1994a. *Longidorus arthensis* sp. n. (Nematoda: Longidoridae) a vector of cherry rosette disease caused by a new nepovirus in cherry trees in Switzerland. Nematologica 40: 133 – 149.

Brown, D. J. F., and J. M. Halbrendt. 1992. The virus vector potential of *Xiphinema americanum* and related species. Journal of Nematology 24: 584 (Abstr.).

Brown, D. J. F., J. M. Halbrendt, A. T. Jones, and C. E. Taylor. 1994b. An appraisal of some aspects of the ecology of nematode vectors of plant viruses. Nematologia Mediterranea 22: 253 – 263.

Brown, D. J. F., J. M. Halbrendt, A. T. Jones, T. C. Vrain, and R. T. Robbins. 1994c. Transmission of three North American nepoviruses by populations of four distinct *Xiphinema americanum*-group species (Nematoda: Dorylaimida). Phytopathology 84: 646 – 649.

Brown, D. J. F., J. M. Halbrendt, R. T. Robbins, and T. C. Vrain. 1993. Transmission of nepoviruses by *Xiphinema americanum*-group nematodes. Journal of Nematology 25: 349 – 354.

Brown, D. J. F., and S. A. MacFarlane. 2001. "Worms" that transmit viruses. Biologist 48: 35 – 40.

Brown, D. J. F., A. F. Murant, and D. L. Trudgill. 1989a. Differences between isolates of the English serotype of tomato black ring virus in their transmissibility by an English population of *Longidorus attenuatus* (Nematoda: Dorylaimoidea). Revue de Nematologie 12: 51 – 56.

Brown, D. J. F., A. T. Ploeg, and D. J. Robinson. 1989b. A review of reported associations between *Tichodorus* and *Paratrichodorus* species (Nematoda: Trichodoridae) and tobraviruses with a description of laboratory methods for examining virus transmission by trichodorids. Revue de Nematologie 12: 235 – 241.

Brown, D. J. F., W. M. Robertson, and D. L. Trudgill. 1995. Transmission of viruses by plant nematodes. Annual Review of Phytopathology 33: 223 – 249.

Brown, D. J. F., R. T. Robbins, D. Zanzinger, and S. Wickizer. 1996. Transmission of tobacco and tomato ringspot nepoviruses by individual *Xiphinema americanum* group nematodes from 29 USA populations. Nematropica 26: 246 (Abstr.).

Brown, D. J. F., and C. E. Taylor. 1981. Variazoni nella trasmissione di virus tra popolazioni di nematodi vettori Longidoridae. Atti Societa Italiano Nematologini,

Giornate Nematologiche, Firenze, 28 – 29 Novembre, 1979. Pp. 191 – 204.

Brown, D. J. F., and C. E. Taylor. 1987. Comments on the occurrence and geographical distribution of longidorid nematodes in Europe and the Mediterranean region. Nematologia Mediterranea 15: 333 – 373.

Brown, D. J. F., and P. B. Topham. 1985. Morphometric variability between populations of *Xiphinema diversicaudatum* (Nematoda: Dorylaimioidea). Revue de Nematologie 8: 15 – 26.

Brown, D. J. F., and D. L. Trudgill. 1983. Differential transmissibility of arabis mosaic and strawberry latent ringspot viruses by three populations of *Xiphinema diversicaudatum* from Scotland, Italy and France. Revue de Nematologie 6: 229 – 238.

Brown, D. J. F., and D. L. Trudgill. 1989. The occurrence and distribution of nepoviruses and their associated vector *Longidorus* and *Xiphinema* nematodes in Europe and the Mediterranean Basin. European Plant Protection Organization Bulletin 19: 479 – 489.

Brown, D. J. F., and D. L. Trudgill. 1997. Longidorid nematodes and their associated viruses. Pp. 1 – 40. In M. S. N. de A. Santos, I. M. de O. Abrantes, D. J. F. Brown, and R. M. Lemos, eds. An Introduction to Virus Vector Nematodes and Their Associated Viruses. Universidade de Coimbra, Portugal: Instituto do Ambiente e Vida.

Brown, D. J. F., and B. Weischer. 1998. Specificity, exclusivity and complementarity in the transmission of plant viruses by plant parasitic nematodes: an annotated terminology. Fundamental and Applied Nematology 21: 1 – 11.

Buser, A. 1990. Untersuchungen uber die feffingerkrankheit der Susskirsche und deren vektor *Longidorus macrosoma* (Nematoda: Longidoridae). Ph. D. dissertation. Eidgenossischen Technischen Hochschule Nr. 9194, Zurich.

Cadman, C. H. 1956. Studies on the etiology and mode of spread of Scottish raspberry leaf curl disease. Journal of Horticultural Research 31: 111 – 118.

Cadman, C. H. 1960. Studies on the relationship between soil-borne viruses of the ringspot type occurring in Britain and Continental Europe. Virology 11: 653 – 664.

Carusso, F. L., and D. C. Ramsdell. 1995. Compendium of Blueberry and Cranberry Diseases. St Paul, MN: The American Phytopathological Society.

Chagas, C. M., and K. Siberschmidt. 1972. Virus de faixa amerela de alcachofra no estado de Sao Paulo. Biologico 38: 35 – 40.

Chandrasekar, V., and J. E. Johnson. 1998. The structure of tobacco ringspot virus: a link in the evolution of icosahedral capsids in the picornavirus superfamily. Structure 6: 157 – 171.

Clark, M. F. 1976. Electrophoretic heterogeneity of the sedimenting components of arabis mosaic virus. Journal of General Virology 32: 331 – 335.

Cohn, E. 1975. Relations between *Xiphinema* and *Longidorus* and their host plants. Pp. 365 – 386. In F. Lamberti, C. E. Taylor, and J. W. Seinhorst, eds. Nematode Vectors of Plant Viruses. London: Plenum Press.

Cohn, E., E. Tanne, and F. E. Nitzany. 1970. *Xiphinema italiae*, a new vector of grapevine fanleaf virus. Phytopathology 60: 181 – 182.

Coomans, A. V. 1985. A phylogenetic approach to the classification of the Longidoridae (Nematoda: Dorylaimida). Agriculture, Ecosystems and Environment 12: 335 – 354.

Coomans, A. 1996. Phylogeny of the Longidoridae. Russian Journal of Nematology 4: 51 – 59.

Cooper, J. I., and P. R. Thomas. 1970. *Trichodorus nanus*, a vector of tobacco rattle virus in Scotland. Plant Pathology 19: 197.

Cremer, M. G., and P. K. Schenk. 1967. Notched leaf in *Gladiolus* spp., caused by viruses of the tobacco rattle virus group. Netherlands Journal of Plant Pathology 73: 33 – 48.

Cropley, R. 1961. Cherry leaf-roll virus. Annals of Applied Biology 49: 524 – 529.

Dalmasso, A. 1970. Influence directe de quelques facteurs ecologiques sur l' activite biologique et la distribution des especes Francaises de la famille des Longidoridae (Nematoda: Dorylaimida). Annales de Zoologie, Ecologie, Animale 2: 163 – 200.

Dalmasso, A., and J. B. Berge. 1983. Enzyme polymorphism and the concept of the genetic species exemplified by *Meloidogyne*. Pp. 187 – 196. In A. R. Stone, H. M. Platt, and L. F. Khalil, eds. Concepts in Nematode Systematics. London: Academic Press.

Decraemer, W. 1995. The family Trichodoridae: stubby root and virus vector nematodes. Dordrecht, The Netherlands: Kluwer Academic Publishers.

Demangeat, G., O. Hemmer, J. Reinbolt, M. A. Mayo, and C. Fitsch. 1992. Virus-specific proteins in cells infected with tomato black ring nepovirus: Evidence for proteolytic processing *in vivo*. Journal of General Virology 73: 1609 – 1614.

De Waele, D., G. Mancini, F. Roca, and F. Lamberti. 1982. *Trichodorus taylori* n. sp. (Nematoda: Dorylaimida) from Italy. Nematologia Mediterranea 10: 27 – 37.

Dunleavy, J. M. 1957. The grasshopper as a vector of tobacco ringspot virus in soyabean. Phytopathology 47: 681 – 682.

Eveleigh, E. S., and W. R. Allen. 1982. Description of *Longidorus diadecturus* n. sp. (Nematoda: Longidoridae), a vector of the peach rosette mosaic virus in peach orchards in southwestern Ontario, Canada. Canadian Journal of Zoology 60: 112 – 115.

Ferris, V. R. 1983. Phylogeny, historical biogeography and the species concept in soil nematodes. Pp. 141 – 161. In A. R. Stone, H. M. Platt, and L. F. Khalil, eds. Concepts in Nematode Systematics. London: Academic Press.

Forer, L. B., N. Hill, and C. A. Powell. 1981. *Xiphinema rivesi*, a new tomato ringspot virus vector. Phytopathology 71: 874 (Abst.).

Fulton, J. P. 1962. Transmission of tobacco ringspot virus by *Xiphinema americanum*. Phytopathology 52: 375 (Abst.).

Gibbs, A. J., and B. D. Harrison. 1964a. A form of pea early-browning virus found in Great Britain. Annals of Applied Biology 54: 1 – 11.

Gibbs, A. J., and B. D. Harrison. 1964b. Nematode-transmitted viruses in sugar beet in East Anglia. Plant Pathology 13: 114 – 150.

Goulden, M. G., J. W. Davies, K. R. Wood, and G. P. Lomonosoff. 1992. Structure of tobraviral particles: A model suggested from sequence conservation in tobraviral and tobamoviral coat proteins. Journal of Molecular Biology 227: 1 – 8.

Gray, S. M. 1996. Plant virus proteins involved in natural vector transmission. Trends in Microbiology 4: 259 – 264.

Griesbach, J. A., and A. R. Maggenti. 1989. Vector capability of *Xiphinema americanum* sensu lato in California. Journal of Nematology 21:517 – 523.

Halbrendt, J. M., and D. J. F. Brown. 1992. Morphometric evidence for three juvenile stages in some species of *Xiphinema americanum* sensu lato. Journal of Nematology 24: 305 – 309.

Halbrendt, J. M., and D. J. F. Brown. 1993. Aspects of the biology and development of *Xiphinema americanum* and related species. Journal of Nematology 25:355 – 360.

Hanada, K., and B. D. Harrison. 1977. Effects of virus genotype and temperature on seed transmission of nepoviruses. Annals of Applied Biology 85:79 – 92.

Hansen, A. J., G. Nyland, F. D. McElroy, and R. Stace-Smith. 1974. Origin, cause, host range and spread of cherry rasp leaf disease in North America. Phytopathology 64: 721 – 727.

Harrison, B. D. 1964. Specific nematode vectors for serologically distinctive forms of raspberry ringspot and tomato black ring viruses. Virology 22:544 – 550.

Harrison, B. D. 1967. Pea early-browning virus (PEBV). Annual Report for 1966 of Rothamsted Experimental Station, England.

Harrison, B. D., W. P. Mowat, and C. E. Taylor. 1961. Transmission of a strain of tomato black ring virus by *Longidorus elongatus* (Nematoda). Virology 14:480 – 485.

Harrison, B. D., and A. F. Murant. 1978. Nematode transmissibility of pseudorecombinant isolates of tomato black ring virus. Annals of Applied Biology 86: 209 – 212.

Harrison, B. D., and A. F. Murant. 1996. The plant viruses. Volume 5. Polyhedral virions and bipartite RNA. London: Plenum Press.

Harrison, B. D., A. F. Murant, M. A. Mayo, and I. M. Roberts. 1974. Distribution of determinants for symptom production, host range and nematode transmissibility between the two RNA components of raspberry ringspot virus. Journal of General Virology 22: 233 – 247.

Harrison, B. D., and I. M. Roberts. 1968. Association of tobacco rattle virus with mitochondria. Journal of General Virology 3:121 – 124.

Harrison, B. D., and D. J. Robinson. 1981. Tobraviruses. Pp. 515 – 50. In E. Kurstak, ed. Handbook of Plant Virus Infection and Comparative Diagnosis. Amsterdam: Elsevier.

Harrison, B. D., and D. J. Robinson. 1986. Tobraviruses. Pp. 339 – 369. In M. H. V. Van Regenmortel and H. Fraenkel-Conrat, eds. The Plant Viruses. The Rod-shaped Plant Viruses. London: Plenum Press.

Harrison, B. D., and D. J. Robinson. 1988. Molecular variation in vector-borne plant viruses: Epidemiological significance. Philosophical Transactions of the Royal Society, London B 321:447 – 462.

Harrison, B. D., and R. D. Winslow. 1961. Laboratory and field studies on the relation of arabis mosaic virus to its nematode vector *Xiphinema diversicaudatum* (Micoletzky). Annals of Applied Biology 49:621 – 633.

Hernandez, C., A. Mathis, D. J. F. Brown, and J. F. Bol. 1995. Sequence of RNA2 of a nematode-transmissible isolate of tobacco rattle virus. Journal of General Virology 76: 2847 – 2851.

Hernandez, C., P. B. Visser, D. J. F. Brown, and J. F. Bol. 1997. Transmission of tobacco rattle virus isolate PpK20 by its nematode vector requires one of the two non-structural genes in the viral RNA 2. Journal of General Virology 78: 465 – 467.

Hewitt, W. B. 1970. Virus and virus-like diseases of the grapevine. Pp. 195 – 196. In N. W. Frazier, ed. Virus Diseases of Small Fruits and Grapevines. Berkley, CA: University of California Division of Agricultural Sciences.

Hewitt, W. B., D. J. Raski, and A. C. Goheen. 1958. Nematode vector of soil-borne virus of grapevines. Phytopathology 48: 586 – 595.

Hoof, H. A. van. 1962. *Trichodorus pachydermus* and *T. teres*, vectors of the early-browning virus of peas. Tijdschrift Plziektenziekten 68: 391 – 396.

Hoof, H. A. van. 1964. *Trichodorus teres* a vector of rattle virus. Netherlands Journal of Plant Pathology 70: 187.

Hoof, H. A. van. 1968. Transmission of tobacco rattle virus by *Trichodorus* species. Nematologica 14: 20 – 24.

Hoof, H. A. van, D. Z. Maat, and J. W. Seinhorst. 1966. Viruses of the tobacco rattle virus group in Northern Italy: the vectors and serological relationships. Netherlands Journal of Plant Pathology 72: 253 – 258.

Hoy, J. W., S. M. Mircetich, and B. F. Lownsbery. 1984. Differential transmission of prunus tomato ringspot virus strains by *Xiphinema californicum*. Phytopathology 74: 332 – 335.

Hunt, D. J. 1993. Aphelenchida, Longidoridae and Trichodoridae: their systematics and bionomics. Wallingford, UK: CAB International.

Hunt, D. J., and A. Towle. 1979. Feeding studies on *Xiphinema vulgare* Tarjan, 1964 (Nematoda: Longidoridae). Revue de Nematologie 2: 7 – 40.

Inglis, W. G. 1966. The origin and function of the cheilostomal complex in the nematode *Falcaustra stewarti*. Proceedings of the Linnean Society, London 177: 55 – 62.

Iwaki, M., and Y. Komuro. 1974. Viruses isolated from narcissus (*Narcissus* spp.) in Japan. V. Arabis mosaic virus. Annals of the Phytopathological Society of Japan 40: 344 – 353.

Jairajpuri, S., and W. Ahmad. 1992. Dorylaimida free-living, predaceous and plant-parasitic nematodes. Leiden, The Netherlands: E. J. Brill.

Jha, A., and A. F. Posnette. 1959. Transmission of a virus to stawberry plants by a nematode (*Xiphinema* sp.). Nature, London 184: 962 – 963.

Jones, A. T., and S. J. Badenoch. 1982. Recent studies on virus and virus-like diseases of *Rubus* in Scotland. Acta Horticulturae 129: 49 – 58.

Jones, A. T., D. J. F. Brown, W. J. McGavin, M. Rudel, and B. Altmayar. 1994. Properties of an unusual isolate of raspberry ringspot virus from grapevine in Germany and evidence for its possible transmission by *Paralongidorus maximus*. Annals of Applied Biology 124: 283 – 300.

Jones, A. T., M. J. Mitchell, and D. J. F. Brown. 1989. Infectibility of some raspberry cultivars with arabis mosaic and raspberry ringspot viruses and further evidence for variation in British isolates of these two viruses. Annals of Applied Biology 115: 57 – 69.

Jones, R. A. C. 1981. The ecology of viruses infecting wild and cultivated potatoes in the Andean region of South America. Pp. 89 – 107. In J. M. Thresh, ed. Pests, Pathogens and Vegetation. London: Pitman.

Karanastasi, E., N. Vassilakos, I. M. Roberts, S. A. MacFarlane, and D. J. F. Brown. 2000. Immunogold localization of tobacco rattle virus particles within *Paratrichodorus anemones*. Journal of Nematology 31: 5 – 12.

Klingler, J., P. Kunz, and H. -P. Havri. 1985. Die Verbreitung des Vektornematoden der Pfeffingerkrankheit in Kerschenanbangebeit der Nordwestschweiz. Schweizerische Zeitschrift fur Obst- und Weinbau 121: 782 – 789.

Klos, E. J. 1976. Rosette mosaic. Pp. 135 – 138. In USDA Agricultural Handbook No. 437. Virus Diseases and Non-infectious Disorders of Stone Fruits in N. America. Washington, D. C., USA.

Klos, E. J., F. Fronek, J. A. Knierim, and D. Cation. 1967. Peach rosette mosaic transmission and control. Quarterly Bulletin of Michigan State University Agricultural Experiment Station 49: 287 – 293.

Komuro, Y., and M. Iwaki. 1968. [Bean yellow mosaic virus and tobacco ringspot virus isolated from crotalaria (*Crotalaria spectabilis*).] Annals of the Phytopathological Society of Japan 34: 7 – 15.

Kreiah, S., G. Strunk, and J. I. Cooper. 1994. Sequence analysis and location of capsid proteins within RNA2 of strawberry latent ringspot virus. Journal of General Virology 75: 2527 – 2532.

Lamberti, F., and T. Bleve-Zacheo. 1979. Studies on *Xiphinema americanum sensu lato* with description of fifteen new species (Nematoda, Longidoridae). Nematologia Mediterranea 7: 51 – 106.

Lamberti, F., and A. Ciancio. 1993. Diversity of *Xiphinema americanum*-group species and hierarchical cluster analysis of morphometrics. Journal of Nematology 25: 332 – 343.

Lamberti, F., S. Molinari, M. Moens, and D. J. F. Brown. 2000. The *Xiphinema americanum* group. I. Putative species, their geographical occurrence and distribution, and regional polytomous identification keys for the group. Russian Journal of Nematology 8: 65 – 84.

Lamberti, F., C. E. Taylor, and J. W. Seinhorst, eds. 1975. Nematode Vectors of Plant Viruses. London: Plenum Press.

Lister, R. M. 1964. Strawberry latent ringspot: a nematode-borne virus. Annals of Applied Biology 54: 167 – 176.

Loof, P. A. A. 1975. Taxonomy of Trichodoridae. Pp. 103 – 126. In F. Lamberti, C. E. Taylor, and J. W. Seinhorst, eds. Nematode Vectors of Plant Viruses. London: Plenum Press.

Maat, D. Z., F. A. ban der Meer, and H. J. Pfaeltzer. 1962. Serological identifications of some soil-borne viruses causing diseases in fruit crops in the Netherlands. Tidschrift Plantenziekten 68: 120 – 122.

MacFarlane, S. A., and D. J. F. Brown. 1995. Sequence comparison of RNA2 of nematode-transmissible and nematode-non-transmissible isolates of pea early browning virus suggests that the gene encoding the 29kDA protein may be involved in nematode transmission. Journal of General Virology 76: 1299 – 1304.

MacFarlane, S. A., D. J. F. Brown, and J. F. Bol. 1995. The transmission by nematodes of tobraviruses is not determined exclusively by the virus coat protein. European Journal of Plant Pathology 101: 535 – 539.

MacFarlane, S. A., N. Vassilakos, and D. J. F. Brown. 1998. Similarities in the genome organization of tobacco rattle virus and pea early-browning virus isolates that are transmitted by the same vector nematode. Journal of General Virology 80: 273 – 276.

MacFarlane, S. A., C. V. Wallis, and D. J. F. Brown. 1996. Multiple virus genes involved in the nematode transmission of pea early browning virus. Virology 219: 417 – 422.

Maia, I. G., A.-L. Haenni, and F. Bernardi. 1996. Potyviral HC-Pro: a multifunctional protein. Journal of General Virology 77: 1335 – 1341.

Marjorana, G., and G. L. Rana. 1970. Un nuovo virus latente isolato da Carciofo in Puglia. Phytopathologia Mediterranea 9: 193 – 196.

Martelli, G. P., and C. E. Taylor. 1989. Distribution of viruses and their nematode vectors. Pp. 151 – 189. In K. F. Harris, ed. Advances in Disease Vector Research. Volume 6. New York: Springer-Verlag.

Mayo, M., K. M. Brierley, and B. A. Goodman. 1993. Developments in the understanding of the particle structure of tobraviruses. Biochemie 75: 639 – 644.

Mayo, M., W. M. Robertson, F. J. Legerburu, and K. M. Brierley. 1995. Molecular approaches to an understanding of the transmission of plant viruses by nematodes. Pp. 277 – 293. In F. Lamberti, C. De Giorgi, and D. McK. Bird, eds. Advances in Molecular Plant Nematology. London: Plenum Press.

Mayo, M., and D. J. Robinson. 1996. Nepoviruses: molecular biology and replication. Pp. 139 – 186. In B. D Harrison and A. F. Murant, eds. The Plant Viruses. Volume 5. Polyhedral Virions and Bipartite RNA. London: Plenum Press.

McGuire, J. M., K. S. Kim, and L. B. Douthit. 1970. Tobacco ringspot virus in the nematode *Xiphinema americanum*. Virology 42: 212 – 216.

McNamara, D. G., and J. J. M. Flegg. 1981. The distribution of virus-vector nematodes in Great Britain in relation to past and present natural vegetation. Pp. 225 – 235. In J. M. Thresh, ed. Pests, Pathogens and Vegetation. London: Pitman.

Messieha, M. 1969. Transmission of tobacco ringspot virus by thrips. Phytopathology 59: 943 – 945.

Murant, A. F. 1983. Seed and pollen transmission of nematode-borne viruses. Seed Science and Technology 11: 973 – 987.

Murant, A. F., C. E. Taylor, and J. Chambers. 1968. Properties, relationships and transmission of a strain of raspberry ringspot virus infecting raspberry cultivars immune to the common Scottish strain. Annals of Applied Biology 61: 175 – 186.

Namba, K., D. L. D. Casper, and G. J. Stubbs. 1985. Computer graphic representation of levels of organization in tobacco mosaic virus structure. Science 227: 773 – 776.

Navas, A., M. Fe. Andres, and M. Arias. 1990. Biogeography of Longidoridae in the Euromediterranea area. Nematologia Mediterranea 18: 103 – 112.

Navas, A., J. G. Baldwin, L. Barrios, and G. Nombela. 1993. Phylogeny and biogeography of *Longidorus* (Nematoda: Longidoridae) in Euromediterranea. Nematologia Mediterranea 21: 71 – 88.

Nyland, G. 1961. Mechanical transmission of the strains of sweet cherry rugose mosaic, mild rusty mottle and rasp leaf viruses. Phytopathology 51: 66 – 67.

Nyland, G., B. F. Lownsbery, B. K. Lowe, and J. F. Mitchell. 1969. The transmission of cherry rasp leaf virus by *Xiphinema americanum*. Phytopathology 59: 1111 – 1112.

Parish, C. L. 1977. A relationship between flat apple disease and cherry rasp leaf disease.

Phyopathology 67: 982 – 984.

Ploeg, A. T., C. J. Asjes, and D. J. F. Brown. 1991. Tobacco rattle virus serotypes and associated nematode vector species of Trichodoridae in the bulb-growing areas in the Netherlands. Netherlands Journal of Plant Pathology 97: 311 – 319.

Ploeg, A. T., D. J. F. Brown, and D. J. Robinson. 1992. The association between species of *Trichodorus* and *Paratrichodorus* vector nematodes and serotypes of tobacco rattle virus. Annals of Applied Biology 121: 619 – 630.

Ploeg, A. T., A. Mathis, J. F. Bol, D. J. F. Brown, and D. J. Robinson. 1993a. Susceptibility of transgenic tobacco plants expressing tobacco rattle virus coat protein to nematode-transmitted and mechanically inoculated tobacco rattle virus. Journal of General Virology 74: 2709 – 2715.

Ploeg, A. T., D. J. Robinson, and D. J. F. Brown. 1993b. RNA-2 of tobacco rattle virus encodes the determinants of transmissibility by trichodorid vector nematodes. Journal of General Virology 74: 1463 – 1466.

Putz, C., and G. Stocky. 1970. Premieres observations su une souche de strawberry latent ringspot virus transmise *Xiphinema* coxi Tarjan et associee a une maladie du framboisier en Alsace. Annales Phytopathologie 2: 329 – 347.

Ramel, M. -E., D. J. F. Brown, R. Vallotton, W. McGavin, and P. Gugerli. 1995. Occurrence of infection by a strain of arabis mosaic nepovirus of barley (cv. Express) and viruliferous *Xiphinema diversicaudatum* in fields near Fribourg in Switzerland. Nematologica 41: 332 – 333 (Abst.).

Rana, G. L., and F. Roca. 1973. Trasmissione con nematodi del virus "latente italiano del carciofo" (AILV). Abstracts of the 2nd Congresso Internazinale Studi Carciofo, Bari, Italy, 21 – 24 novembre. 1973. Pp. 139 – 140 (Abst.).

Rana, G. L., and F. Roca. 1975. Trasmissione con nematodi del virus "latente italiano del carciofo" (AILV). Proceedings of the 2nd Congresso Internazionale Studi Carciofo, Bari, Italy, 21 – 24 novembre 1973. Pp. 855 – 858.

Raski, D. J., A. R. Maggenti, and N. O. Jones. 1973. Location of grapevine fanleaf and yellow mosaic virus particles in *Xiphinema index*. Journal of Nematology 5: 208 – 211.

Robbins, R. T. 1993. Distribution of *Xiphinema americanum* and related species in North America. Journal of Nematology 25: 344 – 348.

Robbins, R. T., and D. J. F. Brown. 1991. Comments on the taxonomy, occurrence and distribution of Longidoridae (Nematoda) in North America. Nematologica 37: 395 – 419.

Robertson, W. M. 1987. Possible mechanisms of virus retention in virus vector nematodes. Pp. 127. In the Annual Report of the Scottish Crop Research Institute for 1986, Invergowrie, Scotland.

Robertson, W. M., and C. E. Henry. 1986a. An association of carbohydrates with particles of arabis mosaic virus retained within *Xiphinema diversicaudatum*. Annals of Applied Biology 109: 299 – 305.

Robertson, W. M., and C. E. Henry. 1986b. A possible role of carbohydrates in the retention of nematode-transmitted viruses. Pp. 113. In the Annual Report of the Scottish Crop Research Institute for 1985, Invergowrie, Scotland.

Robertson, W. M., and C. E. Taylor. 1975. The structure and musculature of the feeding apparatus in *Longidorus* and *Xiphinema*. Pp. 179 – 184. In F. Lamberti, C. E. Taylor,

and J. W. Seinhorst, eds. Nematode Vectors of Plant Viruses. London: Plenum Press.
Robertson W. M., and U. Wyss. 1983. Feeding processes of virus-transmitting nematodes. Pp. 271 – 295. In K. F. Harris, ed. Current Topics in Vector Research. New York: Praeger.
Robinson, D. J., and B. D. Harrison. 1985. Evidence that broad bean yellow band virus is a new serotype of pea early-browning virus. Journal of General Virology 66: 2003 – 2009.
Roca, F., G. P. Martelli, F. Lamberti, and G. L. Rana. 1975a. Distribution of *Longidorus attenuatus* Hooper in Apulian artichoke fields and its relationship with artichoke Italian latent virus. Nematologia Mediterranea 3: 91 – 101.
Roca, F., G. P. Martelli, F. Lamberti, and G. L. Rana. 1975b. Distribution of artichoke Italian latent virus and its nematode vector in Apulia. Pp. 279 – 281. In F. Lamberti, C. E. Taylor, and J. W. Seinhorst, eds. Nematode Vectors of Plant Viruses. London: Plenum Press.
Roca, F., and G. L. Rana. 1981. *Paratrichodorus tunisiensis* (Nematoda, Trichodoridae) a new vector of tobacco rattle virus in Italy. Nematologia Mediterranea 9: 217 – 219.
Roca, F., G. L. Rana, and P. E. Kyriakopoulou. 1982. *Longidorus fasciatus* Roca et Lamberti vector of a serologically distinct strain of artichoke Italian latent virus in Greece. Nematologia Mediterranea 10: 65 – 69.
Rudel, M. 1985. Grapevine damage induced by particular virus vector combinations. Phytopathologia Mediterranea 24: 183 – 185.
Salomao, T. A. 1975. Soil transmission of artichoke yellow band virus. Proceedings of the 2nd Congresso Internazinale Studi Carciofo, Bari, Italy, 21 – 24 novembre 1973. Pp. 831 – 854.
Sanger, H. L. 1961. Untersuchungen uber schwer Ubertragbare Formen des Rattlevirus. Proceedings of the 4th Conference on Potato Virus Diseases, Braunschweig, Germany, 1960. Pp. 22 – 28.
Santos, M. S. N. de A., I. M. de O. Abrantes, D. J. F. Brown, and R. M. Lemos. 1997. An Introduction to Virus Vector Nematodes and Their Associated Viruses. Universidade de Coimbra, Portugal: Instituto do Ambiente e Vida.
Schuster, M. F. 1963. Flea beetle transmission of tobacco ringspot virus in the Lower Rio Grande Valley. Plant Disease Reporter 47: 510 – 511.
Shope, R. E. 1941. The swine lungworm as a reservoir and intermediate host for swine influenza virus. II. The transmission of swine influenza virus by the swine lungworm. Journal of Experimental Medicine 74: 49 – 68.
Silberschmidt, K. 1962. Studies on a Brazilian tomato ringspot disease. Phytopathologisches Zeitschrift 46: 209 – 229.
Smith, I. M., D. G. McNamara, P. R. Scott, and K. M. Harris. 1992. Quarantine Pests for Europe. Wallingford, UK: CAB International.
Smith, K. M., and R. Markham. 1944. Two new viruses affecting tobacco and other plants. Phytopathology 34: 324 – 329.
Stobbs, L. W., and J. G. van Schagen. 1996. Occurrence of peach rosette mosaic virus on grapevine in southern Ontario. Plant Disease 80: 105 (Abstr.).
Sturhan, D. 1984. Untersuchungen über den *Xiphinema* coxi-Komplex (Nematoda: Longidoridae). Nematologica 30: 305 – 323.

Taylor, C. E. 1962. Transmission of raspberry ringspot virus by *Longidorus elongatus* (De Man), (Nematoda: Dorylaimoidea). Virology 17:493 – 494.

Taylor, C. E., and D. J. F. Brown. 1976. The geographical distribution of *Xiphinema* and *Longidorus* in the British Isles and Ireland. Annals of Applied Biology 84: 383 – 402.

Taylor, C. E., and D. J. F. Brown. 1981. Nematode virus interactions. Pp. 281 – 301. In B. M. Zuckerman and R. A. Rohde, eds. Plant Parasitic Nematodes, Volume III. New York: Academic Press.

Taylor, C. E., and D. J. F. Brown. 1997. Nematode Vectors of Plant Viruses. Wallingford, UK: CAB International.

Taylor, C. E., D. J. F. Brown, R. Neilson, and A. T. Jones. 1994. The persistence and spread of *Xiphinema diversicaudatum* (Nematoda: Dorylaimida) in cultivated and uncultivated biotopes. Annals of Applied Biology 124:469 – 477.

Taylor, C. E., and W. M. Robertson. 1969. The location of raspberry ringspot and tomato black ring viruses in the nematode vector, *Longidorus elongatus*. Annals of Applied Biology 64:233 – 237.

Taylor, C. E., and W. M. Robertson. 1970a. Sites of virus retention in the alimentary tract of the nematode vectors, *Xiphinema diversicaudatum* (Micol.) and *X. index* (Thorne and Allen). Annals of Applied Biology 66:375 – 380.

Taylor, C. E., and W. M. Robertson. 1970b. Location of tobacco rattle virus in the nematode vector *Trichodorus pachydermus* Seinhorst. Journal of General Virology 6: 179 – 182.

Taylor, C. E., and W. M. Robertson. 1975. Acquisition, retention and transmission of viruses by nematodes. Pp. 253 – 276. In F. Lamberti, C. E. Taylor, and J. W. Seinhorst, eds. Nematode Vectors of Plant Viruses. London: Plenum Press.

Taylor, C. E., and P. R. Thomas. 1968. The association of *Xiphinema diversicaudatum* (Micoletzky) with strawberry latent ringspot and arabis mosaic viruses in a raspberry plantation. Annals of Applied Biology 62: 147 – 157.

Taylor, C. E., P. R. Thomas, and R. H. Converse. 1966. An outbreak of arabis mosaic virus and *Xiphinema diversicaudatum* (Micoletzky) in Scotland. Plant Pathology 15: 170 – 174.

Thomas, C. E. 1969. Transmission of tobacco ringspot virus by *Tetranychus* sp. Phytopathology 59: 633 – 636.

Topham, P. B., and T. J. W. Alphey. 1985. Faunistic analysis of longidorid nematodes in Europe. Journal of Biogeography 12: 165 – 174.

Trudgill, D. L. 1976. Observations on the feeding of *Xiphinema diversicaudatum*. Nematologica 22: 417 – 423.

Trudgill, D. L., and D. J. F. Brown. 1978. Ingestion, retention and transmission of two strains of raspberry ringspot virus by *Longidorus macrosoma*. Journal of Nematology 10: 85 – 89.

Trudgill, D. L., and D. J. F. Brown. 1980. Effect of the bait plant on transmission of viruses by *Longidorus* and *Xiphinema* spp. Pp. 120. In Annual Report of the Scottish

Horticultural Research Institute for 1979, Invergowrie, Scotland.
Trudgill, D. L., D. J. F. Brown, and D. G. McNamara. 1983. Methods and criteria for assessing the transmission of plant viruses by longidorid nematodes. Revue de Nematologie 6: 133 – 141.
Trudgill, D. L., and W. M. Robertson. 1982. Feeding and salivation behaviour of *Xiphinema diversicaudatum* and *Longidorus elongatus*. Nematologica 28: 177 (Abstr.).
Valdez, R. B., D. G. McNamara, P. J. Ormerod, R. S. Pitcher, and J. M. Thresh. 1974. Transmission of the hop strain of arabis mosaic virus by *Xiphinema diversicaudatum*. Annals of Applied Biology 76: 113 – 122.
Viry, M., M. A. Serghini, F. Hans, C. Ritzenthaler, M. Pinck, and L. Pinck. 1993. Biologically active transcripts from cloned cDNA of genomic grapevine fanleaf nepovirus RNAs. Journal of General Virology 74: 169 – 174.
Visser, P., and J. F. Bol. 1999. Nonstructural proteins of tobacco rattle virus which have a role in nematode transmission: Expression pattern and interaction with viral coat protein. Journal of General Virology 80: 3273 – 3280.
Visser, P. 2000. Role of RNA 2-encoded proteins in nematode-transmission of tobacco rattle virus. Ph. D. thesis, University of Leiden, The Netherlands.
Vovlas, C., and F. Roca. 1975. Trasmissione con nematodi dell' agente della masculatura anulare chorotica della Cicoria, un ceppo del virus latente italiano del Carciofo. Nematologia Mediterranea 3: 83 – 90.
Vrain, T. C. 1993. Restriction fragment length polymorphism separates species of the *Xiphinema americanum* group. Journal of Nematology 25: 361 – 366.
Walkinshaw, C. H., G. D. Griffin, and R. H. Larson. 1961. *Trichodorus christiei* as a vector of potato corky ringspot (tobacco rattle) virus. Phytopathology 51: 806 – 808.
Wang, S., and R. C. Gergerich. 1998. Immunofluorescent localization of tobacco ringspot nepovirus in the vector nematode *Xiphinema americanum*. Phytopathology 88: 885 – 889.
Weischer, B., and D. J. F. Brown. 2000. An Introduction to Nematodes. General Nematology: A Student's Texbook. Sofia, Bulgaria: Pensoft Publishers.
Wyss, U. 1975. Feeding of *Trichodorus, Longidorus,* and *Xiphinema*. Pp. 203 – 221. In F. Lamberti, C. E. Taylor, and J. W. Seinhorst, eds. Nematode Vectors of Plant Viruses. London: Plenum Press.
Wyss, U. 1977. Feeding processes of virus transmitting nematodes. Proceedings of the American Phytopathological Society 4: 30 – 41.
Wyss, U. 1981. Ectoparasitic root nematodes: feeding behaviour and plant cell responses. Pp. 325 – 351. In B. M. Zuckerman and R. A. Rohde, eds. Plant Parasitic Nematodes. Volume III. New York: Academic Press.
Wyss, U. 1987. Video assessment of root cell responses to dorylaimid and tylenchid nematodes. Pp. 211 – 220. In J. A. Veetch and D. W. Dickson, eds. Vistas on Nematology: a Commemoration of the Twenty-fifth Anniversary of the Society of Nematologists. Hyattsville, Maryland, USA: Society of Nematologists.
Wyss, U. 1999. *Xiphinema index*, maintenance and feeding in monoxenic cultures. Pp. 251 – 278. In K. Mamamorosch and F. Mahmood, eds. Maintenance of Animal/human

and Plant Pathogen Vectors. Available from Science Publishers, Enfield, USA. New Delhi: Oxford and IBH Publishing.

Yagita, H., and Y. Komuro. 1972. [Transmission of mulberry ringspot virus by *Longidorus martini* Merny.] Annals of the Phytopathological Society of Japan 38: 275 – 283 (In Japanese).

Zacheo, G., and T. Zacheo. 1995. Plant-nematode interactions: Histological, physiological and biochemical interactions. Pp. 321 – 353. In K. Kohmoto, V. S. Singh, and R. P. Singh, eds. Pathogenesis and Host Specificity in Plant Diseases, Histopathological, Genetic and Molecular Bases. Volume II: Eukaryotes. London: Pergamon Press.

Acknowledgments This chapter was prepared as part of an UK-Chinese collaborative research project sponsored by the Royal Society, London. The financial assistance provided by the Royal Society and Zhejiang University is gratefully acknowledged. Research at the Scottish Crop Research Institute is grant aided by the Scottish Executive Rural Affairs Department.

15 Physiological Interactions between Nematodes and Their Host Plants

Haddish Melakeberhan

15.1 Introduction

As soil-inhabiting microscopic invertebrates, plant-parasitic nematodes parasitize below ground (root) and above ground (shoot) plant tissues in habitats ranging from the driest to the wettest and from the coldest to the hottest climates (Norton, 1978). They parasitize their hosts as ecto- (*anterior body portion embedded*), migratory endo- (*body fully embedded*), or sedentary endoparasites (*body fully or partly embedded or not embedded at all*). Types of cellular damage at the feeding sites include destructive (*host cells killed*), adaptive (*cells modified*), or neoplastic (*cells modified and undergo new growth*) changes. Root-lesion (*Pratylenchus* spp.), cyst (*Heterodera* and *Globodera* spp.), and root-knot (*Meloidogyne* spp.) nematodes represent the three respective feeding behaviors (Dropkin, 1989). Unlike ectoparasites and migratory endoparasites, cyst and root-knot nematodes (sedentary endoparasites) must maintain active feeding sites at all times in order to grow and reproduce (Sijmons, et al., 1994; Williamson and Hussey, 1996). Otherwise, the nematodes die.

Symptoms at the cellular, tissue, and whole plant level have been reviewed extensively (Bleve-Zacheo and Melillo, 1997; Bird, 1974; Dropkin, 1969a; Golinowski, et al., 1997; Hussey, 1989; Hussey and Williamson, 1998; Jones, 1981; Melakeberhan and Webster, 1993; Seinhorst, 1961; Webster, 1967; 1975). Although damage to a plant may differ, depending on the location of the parasite in the plant host and the influence of other organisms and physical factors, plant-parasitic nematodes break down cell structure and consume cell contents. The disruption of the root system by the destructive, adaptive, and neoplastic feeding behaviors interferes with the physiological processes involved in water and nutrient relations and the phytohormones originating in the root (primary factors), thereby creating a cascade effect on chlorophyll synthesis, photosynthesis and respiration in the shoot (secondary factors). The combination of these primary and secondary effects leads to diminished plant productivity and poor growth compared with uninfected plants.

Whether or not plant-parasitic nematodes have destructive, adaptive, or neoplastic feeding behaviors, they cannot be distinguished by the symptoms of leaf chlorosis or stunted growth. Visible symptoms, however, are a function of interactions taking place from the sub-cellular to the ecosystem level. This chapter reviews the physiology of plant-nematode relationships at the organismal and higher levels of interactions, with an emphasis on the basic and applied aspects related to nematode-energy requirements, water-relations, nutrient uptake, CO_2 exchange, and plant growth regulators.

15.2 Effects of Nematodes on Energy Demand

In addition to causing root destruction, all plant-parasitic nematodes consume cell contents (Atkinson, 1985; Williamson and Hussey, 1996). The amount of energy consumed by a nematode depends on individual nematode size, its reproductive potential, and host status (Atkinson, 1985; Melakeberhan and Ferris, 1988). Given their size and sedentary endo-parasitism, cyst and root-knot nematodes are more dependent on the host and likely to have greater energy demands than migratory nematodes (Reversat, 1987; Melakeberhan and Ferris, 1988). Cyst and root-knot nematodes have been shown to divert significant amounts of photosynthate (Bird and Loveys, 1975; McClure, 1977; Melakeberhan, et al., 1990) or nutrients (Bockenhoff and Grundler, 1993; Bockenhoff, et al., 1996; Dorhout, et al., 1983; 1988) to their feeding sites. However, nematode size seems to be influenced by host status. For example, an adult *M. incognita* female grew up to 32 μg and laid (310 ± 126) eggs in a susceptible compared with 29 μg and (195 ± 92) eggs per female in a resistant grape (*Vitis vinifera* L.) cultivar (Melakeberhan and Ferris, 1988). The bigger nematode size associated with the susceptible versus the resistant cultivar has been attributed to a more rapid photosynthate translocation to the nematode feeding site in the former than in the latter cultivar (Melakeberhan, et al., 1990). Similar nematode size differences between susceptible and resistant hosts have been reported on cotton and other crops (Jenkins, et al., 1995; Powers, et al., 1991; Powers and McSorley, 1993).

The cost to the host of supporting a single *M. incognita* from second-stage juvenile to the end of oviposition was 1.176 calories on a susceptible and 0.834 calories on a resistant grape cultivar (Melakeberhan and Ferris, 1988). One calorie is equivalent to approximately 0.213 mg plant dry weight or 0.4 mg CO_2 (Melakeberhan and Ferris, 1988). If the nematode energy demand

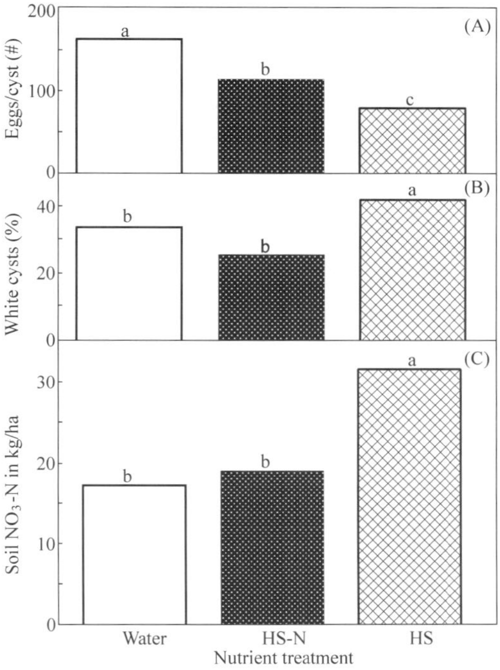

Figure 15.1 The effects of water, complete Hoagland solution (HS) and HS without nitrogen (HS-N) on: (A) the number of *Heterodera glycines* eggs per cysts, (B) percent white cysts per gram fresh root[1), and (C) amount of nitrate in soil from an *H. glycines* susceptible (Tracy M) soybean cultivar 26 days after inoculation with 15,000 eggs per 800 cm^3 of soil. From Melakeberhan (1999b).

Data are means of four replications ((A) and (B)). Because nematode treatments did not affect soil nitrate, data in (C) are means of eight replications per treatment. The nutrient treatments are indicative of soil conditions and show the levels of nutrients that influence the host-parasite interaction in favor of the plant.

Bars followed by the same letters within the same frame and shade are not statistically different according to Tukey's range test ($P > 0.05$).

[1)] As cysts develop, the color changes from white to yellow to dark brown. The presence of more white cysts in the optimum nitrogen treatment shows that *H. glycines* develops more slowly than under nitrogen-deficient conditions.

is proportional to its size (Atkinson, 1985; Melakeberhan and Ferris, 1988), it is fair to assume that cyst nematodes (including *H. glycines*) may consume more energy from their hosts than do root-knot nematodes because of the smaller size of the latter. The slower growth and decreased reproductive potential of *H. glycines* in a susceptible cultivar grown in nutrient rich than in the same cultivar grown in nutrient-deficient conditions (Fig. 15.1) (Melakeberhan, 1999b), suggests that the nematodes may not be getting enough energy to grow.

One of the questions that arises is "can nematode energy demand be alleviated by manipulating nutrient stress?" If nematode energy demand is proportional to size, the answer seems to be "yes" for some nematodes. Recent studies have shown that *H. glycines* grows more slowly and lays fewer eggs per female when plants are grown in nitrogen-rich than nitrogen-deficient conditions (Fig. 15.1 (A), (B)) (Melakeberhan, 1999b). This suggests that energy demand can be decreased and perhaps manipulated in order to alleviate nematode-induced stress.

15.3 Nematodes and Water-relations

15.3.1 Effects of Nematodes on Water Uptake

While the degree to which nematodes affect water-relations may vary with the type of host-parasite interaction, nematodes invariably affect water uptake. For example, decreases in water uptake have been demonstrated for *Globodera rostochiensis* on potato (Evans, et al., 1977; Fatemy and Evans, 1986a; 1986b), *M. incognita* (Alam, et al., 1975), *M. javanica* (Meon, et al., 1978), and *Rotylenchus reniformis* (Islam and Alam, 1975) on tomato, and *Tylenchorhynchus brassicae* (Alam and Saxena, 1975) on cauliflower and cabbage. Depending on the intensity of infection and root size and age, substantial time may elapse before the appearance of effects of root-parasitic nematodes on water uptake, root and stomatal conductance, and wilting (Melakeberhan and Webster, 1993; Wilcox-Lee and Loria, 1987). For example, *Bursaphelenchus xylophilus* girdles *Pinus* spp. trunks (Tamura, et al., 1987) by cavitation (Kuroda, 1989; Kuroda, et al., 1988), which blocks water translocation (Ikeda and Susaki, 1984; Ikeda, et al., 1990) distal to the location of nematode inoculations (Melakeberhan, et al., 1991). Consequently, the plant fails to photosynthesize normally (Melakeberhan and Webster, 1990), loses turgor, wilts and dies (Mamiya and Tamura, 1977; Melakeberhan, et al., 1991). Melakeberhan et al. (1991) showed that the water-splitting complex of photosystem II activity of Scots pine shoots decline within 24 h and wilting 3 days after inoculation of *B. xylophilus*, indicating that water-relations may be one of the primary physiological effects of nematodes.

15.3.2 Gross Physiological Changes and Management Implications

Water shortage is probably the abiotic factor most limiting for plant growth (Boyer, 1985). Decreased water availability resulted in decreased nutrient uptake and translocation of solutes, decreased chloroplast activity, decomposition of proteins and nucleic acids, and increased hydrolytic enzymes (Colhoun, 1973; Schoeneweiss, 1975). Most of these physiological processes result from irreversible changes (Ayres, 1984), and it is logical to expect that nematode infection further accelerates any effects of water shortage. While varying with plant species, the effect of water shortage on the aforementioned physiological processes is influenced by daily fluctuations of temperature and sunlight under field conditions (Boyer, 1985). For example, tomato is very sensitive to water stress levels of below −6 bars (−0.6 MPa) (Atherton and Rudich, 1986). It is entirely possible that tomato plants growing at −0.3 and −0.4 MPa of tissue water potential may appear normal visually, still suffering a level of water stress that is critical to the performance and development of a plant when infected with nematodes. At 20℃ and 65% radiation reaching full canopy, *c*. 585 calories/cm^2 are required to evaporate 1 cm^3 of water (Atherton and Rudich, 1986). It is expected that a plant will experience some level of water stress during its development and that daily fluctuations are accelerated by high temperature (Fiscus, 1975; Hsiao, 1973; Hsiao, et al., 1976). Most nematode water-relations studies to date have focused on cause-and-effect relationships. In order to more completely understand the effect of nematode-induced water shortage on host physiology, however, it is necessary to venture beyond cause-and-effect relationships, to separate abiotic from biotic stresses, and to quantify their additive or interactive effects on plant growth.

15.4 Nematodes and Nutrient-relations

15.4.1 Effects of Nematodes on Nutrient Uptake

Regardless of feeding behavior, plant-parasitic nematodes disrupt nutrient uptake and create an imbalance of macro- and most micro-nutrients (Melakeberhan, 1997a). Yellowing, stunting, and poor growth are manifestations of root destruction, which impedes nutrient uptake and alters the physiology and metabolism of a plant (Baldwin, et al., 1976; Evans, et al., 1977; Fatemy and Evans, 1986a; 1986b; Melakeberhan, et al., 1987; Wilhelm, et al., 1985). Although reports on the actual nutrients and the degree to which nematodes affect host-parasite interactions vary greatly, the normal content of all macro-nutrients (Blevins, et al., 1995; Spiegel, et al., 1982) and most micro-nutrients (Melakeberhan, et al., 1987; Wilhelm, et al., 1985) is disturbed by nematode infection. Depending on the inoculum level and host age at the time of inoculation, significant changes in plant tissue elemental levels can occur within one week of nematode inoculation. Furthermore, a large change in the concentration of any given element may not be necessary in order to significantly affect on host physiology (Melakeberhan, et al., 1987). Regardless of feeding behavior, it is hard to distinguish nematodes by the way they induce leaf chlorosis or stunted growth.

15.4.2 Effects of Nutrition on Alleviating Nematode-induced Stress

Most arable soils need some levels of nutrient amendment (Marschner, 1995). Much of the nutrition work to alleviate the effect of nematodes (Evans, et al., 1977; Melakeberhan, 1997a; Melakeberhan, et al., 1988), microbial pathogens (Engelhard, 1989), and insects (Rajaratnum and Hock, 1975) has focused on changing the soil conditions to benefit the plant. The beneficial effects of nutrition have been demonstrated under field and under controlled conditions in potato (Evans, at al., 1977; Trudgill, 1987), dry bean (Melakeberhan, et al., 1988), soybean (Melakeberhan, 1999b), tomato (Melakeberhan, 1998b), cotton (Oteifa and Elgindi, 1976), and stonefruits (Table 15.1) (Melakeberhan, et al. 1997). Because of the variability in the type of nematodes, nutrient elements, and experimental conditions, comparison of these studies is difficult. However, increased nutrition generally increases plant growth and nutrient accumulation in plant tissue with or without effects on nematode population densities (Melakeberhan, et al., 1987; 1997; Oteifa and Elgindi, 1976; Trudgill, 1987). Furthermore, nutrition seems to be effective against nematodes with destructive (Melakeberhan, et al., 1997), adaptive (Evans, et al., 1977; Melakeberhan, 1997b; 1999b; Trudgill, 1987), and neoplastic (Melakeberhan, et al., 1988; Oteifa and Elgindi, 1976) feeding behaviors. Although the basic underlying physiological mechanisms by which increased nutrition influences the host response to nematodes are largely unknown, it is clear that nutrition does alleviate nematode-induced stress. If nutrition benefits the host in the presence of nematodes, therefore, it is logical to suggest that healthy plant-based nutrient recommendations may not be adequate under nematode infested conditions.

However, long-term economic and ecological reasons may impose limits on nutrient application (Gerloff and Gabelman, 1983). Therefore, there is a need to understand specific nutrient types and their manner of influence on host-nematode interactions, so that the benefits of nutrition can be exploited in the best ways possible.

Table 15.1 Numbers of *Pratylenchus penetrans* (per 100 cc soil plus 1.0 g of root) in four cherry rootstocks grown in nutrient deficient or optimum conditions for 110 days (25℃ ± 2℃) after inoculation with 1500 nematodes per gram of fresh root. From Melakeberhan et al. (1997).

Rootstock	Deficient	Optimum	*t*-test
Mazzard	673 ab	194 a	0.043
Mahaleb	389 b	244 a	0.062
GI-148-1	1418 a	324 a	0.023
GI-148-8	1040 ab	404 a	0.046

Data are means of four replications. Means within each column followed by different letters are significantly different according to Tukey's multiple range test at $P \leq 0.05$. Student *t*-test values are comparisons between optimum and deficient nutrient regime within each rootstock.

Deficient treatments received 40 μg of N, P, and K each at planting whereas optimum treatments received the same treatments twice weekly. The optimum treatment represents current recommendations for the crop.

Difference in *P. penetrans* population densities among rootstocks can be attributed to host genetic differences. However, the response to nutrition across the rootstocks over multiple generations indicates that the effect on the nematode may be host-mediated.

15.4.3 Effects of Nutrition on Nematode and Host Biology

If the effects of nematodes on host physiology are more pronounced under nutrient deficient than under nutrient optimum conditions, it can be hypothesized that increasing nutrient supply may benefit the host-plant with or without affecting the nematode (Melakeberhan, 1997a; 1999a). An increase in plant growth without a negative effect on nematode population density suggests an increase in tolerance of nematode-infected plants (Fatemy and Evans, 1986a; 1986b; Melakeberhan, et al., 1988; Spiegel, et al., 1982). An increase in plant growth and/or physiological efficiency but a decrease in nematode population density suggests that nutrition has a positive effect on host plants and a negative effect on nematodes (Fig. 15.1 (A), (B)). A negative effect on nematodes with or without an increase in plant growth may be an indication that nutrition interferes with infective behavior (Castro, et al., 1991) and/or nematode development in plant roots (Fig. 15.1 (A), (B)) (Melakeberhan, 1999b).

A decrease in plant growth and in nematode population density under nutrient-optimum would suggest that nutrition has a negative (toxic) effect on both organisms. Hence, the types and levels of nutrients that are safe to plants but not to nematodes need to be determined. If an adverse effect on nematode development is documented, it is essential to determine whether or not the effect is direct or indirect, mediated through host factors. Melakeberhan (1999b) showed the presence of greater white cysts and fewer eggs per cyst were present in a susceptible soybean cultivar treated with a complete and balanced nutrient source than in a cultivar treated with water or nitrogen-deficient solution (Fig. 15.1 (A), (B), (C)). These new results suggest that host-mediated factors may be involved, but the mechanisms by which cyst development is slowed down or reproductive potential reduced are unknown. Similar plant-mediated processes have been reported for insects (Orians and Fritz, 1996) and on the suppression of giant cell formation on tomato and soybean (Orion, et al., 1995).

In manipulating soil mineral nutrition to achieve desired physiological changes, it is important to consider nutrient levels, sources, and their ecological implications. As shown in Fig. 15.1(C), certain levels of soil N can result in slow development and reproductive potential of *H. glycines*. At the same time, deciding on mineral nutrition provides us with challenges of balancing the effect of *H. glycines* on nodulation and how to maintain a legume-generated N supply (Hargrove, 1988).

15.4.4 Effects of Nitrogen Source and Temperature on Plant Growth and Resistance

The *Mi*-gene, which confers resistance to warm climate root-knot nematodes (*M incognita, M. javanica* and *M. arenaria*) (Roberts, 1992), represents one of the most significant advances in development of nematode resistance (Williamson, et al., 1992; 1993). While tomato growth extends to 35℃ (Atherton and Rudich, 1986), resistance breaks down around 28℃ and higher temperatures (Dropkin, 1969b). It is not known how this resistance is mediated or other management practices are involved in its breakdown. Depending on the genotype, however, tomato may grow faster at higher than at lower temperatures (Abdalla and Verkerk, 1968) although balancing source/sink relationships may be stressful at high temperatures (Bar-Tsur, et al., 1985; Hanna and Hernendez, 1982). An imbalance between morphometric changes and physiological response is likely to affect the source/sink relationship (Daly, 1976), nutrient uptake and demand (Ganmore-Neumann and Kafkafi, 1980), and internal energy homeostasis (Chen, et al., 1982; Neuman and Stein, 1983).

Tomato is a crop that requires intensive soil management (Geisenberg and Stewart, 1986; Wolcott, et al., 1967) and nutritional amendments.

Generally, tomato and other plants fertilized with nitrate appear to grow bigger than those fertilized with ammonium (Ganmore-Neumann and Kafkafi, 1980). When subjected to high nitrate fertilization, many tomato cultivars grow better at high than at low temperatures (Ganmore-Neumann and Kafkafi, 1980), a fact attributed to accumulation of more balanced organic and inorganic nutrients than those treated with ammonium (Kirby and Mengel, 1967). The adverse effects of ammonium appear to come from it lowering soil pH (Pill and Lambeth, 1977). The effect of level and form of nitrogen is not limited to the effect on plant growth, but also affects plant response to disease development (Huber and Watson, 1974). Thus, there is strong likelihood of management practices affecting the form of nitrogen and causing other nutritional imbalance (Widders and Lorenz, 1979), and thereby introducing factors that affect host physiology as well as the breakdown of the *Mi*-gene.

The effects of water (check), Hoagland solution without nitrogen, or Hoagland solution with NO_3, NH_4, or NH_4NO_3 as nitrogen sources on growth of a resistant ("VFN-8") and a susceptible ("Rutgers") tomato cultivar and their responses to an aggressive and non-aggressive *M. incognita* population were tested at 24℃ and 28℃ (Melakeberhan, 1998a). Both cultivars grew larger at 24℃ than at 28℃ and more so with NH_4 and NH_4NO_3 treatments (Table 15.2). Nematode infection levels were similar in the susceptible cultivar regardless of plant size, whereas larger plants had fewer nematodes in the resistant cultivar. On the resistant cultivar, both NH_4 and NH_4NO_3 sources of nitrogen had fewer nematodes than NO_3 treatments. Both nematode populations infected the resistant cultivar less than the susceptible cultivar, indicating that, even if resistance breaks down, the process may be gradual (Table 15.2).

Although the mechanisms by which the *Mi*-gene mediates recognition and resistance to root-knot nematodes remain unknown (Williamson, et al., 1992; 1993), the inverse relationship between plant size and nematode infection in the resistant cultivar suggests that plant growth regulating (physiological) processes may be altering how the gene mediates resistance. With regard to the role of nitrogen source, there appears to be some relationships with phytohormones. For example, exogenous application of cytokinins, which play an important role in the pathogenesis and gall formation process of root-knot nematodes, has been shown to result in the breakdown of the *Mi*-gene (Dropkin, et al., 1969; Sawhney and Webster, 1975). It is also known that NO_3 increases the concentration of cytokinins more than NH_4 does (Salam and Wareing, 1978), while NH_4 has the advantage in regulating the

Table 15.2 Main effect of Hoagland solution (HS)-based form of nitrogen source, temperature, and the interaction between temperature and nitrogen on shoot dry weights (g) and the numbers of adult female of an aggressive and a non-aggressive *Meloidogyne incognita* population of a resistant (VFN-8) and susceptible (Rutgers) tomato cultivar at 24℃ and 28℃. From Melakeberhan (1998a).

Treatments			Adult nematode females			
N source	Temperature	Shoot weight	Aggressive		Non-aggressive	
28℃	VFN-8	Rutgers	VFN-8	Rutgers	VFN-8	Rutgers
Water (control)	0.34 bc[z]	0.31 ab	234 ab	464	14.8	138 a
HS-N (deficient)	0.31 c	0.26 b	306 a	580	47.3	170 a
HS + NO_3	0.26 c	0.25 b	255 ab	587	49.0	95 b
HS + NH_4	0.46 a	0.51 a	162 b	432	14.0	54 b
HS + NH_4NO_3	0.38 ab	0.33 ab	145 b	229	11.0	47 b
24℃						
Water (control)	0.51 b	0.45 b	99 a	437	0.8	50
HS-N (deficient)	0.57 b	0.52 b	99 a	558	0.3	40
HS + NO_3	0.62 b	0.53 b	136 a	556	0.0	35
HS + NH_4	0.85 a	0.76 a	33 b	502	0.0	22
HS + NH_4NO_3	0.83 ab	0.84 a	92 a	488	0.0	18
Temperature						
28℃	0.33 b	0.29 b	220 a	458	27.2 a	101 a
24℃	0.62 a	0.56 a	92 b	508	0.2 b	33 b
Nitrogen temperature	**	ns	*	ns	***	**

[z] Numbers are means of eight replications. Numbers followed by no letters or the same letters in each column per temperature are not statistically different ($P \leq 0.05$) from each other according to Tukey's test. **/ */ ns = Interactions are significant at $P \leq 0.01$ or $P \leq 0.05$ or not significant, respectively.

photosynthetic carbon flow (Mohammed and Gnanam, 1979). The combination of lower numbers of nematodes in either NH_4 or NH_4NO_3 than in the NO_3 treatments in the resistant cultivar only with greater growth of resistant and susceptible cultivars former nutrient treatments, indicates that the effect of the nitrogen source on the nematode may be direct or indirect through unidentified plant-mediated processes (Melakeberhan, 1998a). Similar plant-mediated processes have been reported on insects (Leru, et al., 1994; Orians and Fritz, 1996), and on the suppression of giant cell formation in tomato and soybean (Orion, et al., 1995).

15.5 Nematodes and the CO_2 Exchange Processes

15.5.1 Effects of Nematodes on Photosynthesis

The threshold levels and the degree to which nematodes affect the carbon assimilation process varies. However, an effect on photosynthesis has been demonstrated with *M. javanica* on tomatoes (Bird and Loveys, 1975; Loveys and Bird, 1973; Meon, et al., 1978; Wallace, 1974), *G. rostochiensis* on potatoes (Evans, et al., 1977; Fatemy and Evans, 1986a, b; Franco, 1980), *B. xylophilus* on Scots pine (Melakeberhan and Webster, 1990, 1992), *M. incognita* on beans (Melakeberhan, et al., 1984, 1985a, b, 1986, 1987, 1988) and on grapes (Melakeberhan and Ferris, 1989; Melakeberhan, et al., 1990), and *H. glycines* on soybeans (Blevins, et al., 1995; Melakeberhan, 1999b). Depending on the age of the plant at the time of inoculation, a significant decline of photosynthesis was observed at 3 – 7 days (Melakeberhan, et al., 1986), chlorophyll beginning at 7 days (Melakeberhan, et al., 1985a, b, 1986, 1988), and growth parameters at two weeks (Melakeberhan, et al., 1985a) after nematode inoculation. Considering the role that water and nutrient relations play in photosynthesis and that nematodes affect both of them, it is likely that the effects of nematodes on photosynthesis are preceded by effect on water and nutrient relations.

Skeptics may think that measuring an effect of nematodes on photosynthesis may not give any new information, as such an effect is expected because of the effects of nematodes on water and nutrient uptake (Table 15.3). Where photosynthesis measurements become especially significant is in understanding the intricate physiological processes by which a plant regulates its reaction to nematode infection and other stresses. For example, tolerant and susceptible cultivars of any plant species do not differ in their nematode carrying capacities, but the former do not suffer the yield loss shown by the latter. This suggests that there has to be a difference in the photo-assimilation process between susceptible and tolerant cultivars. Another example is a plant's response to external nutrient input. An *H. glycines*-infected susceptible soybean cultivar photosynthesizes as much as an uninfected control when grown under optimum or under balanced nutrient conditions, in contrast to much poorer performance under nutrient-deficient conditions (Table 15.3). Resistant and tolerant cultivars did not behave in this way, indicating that the plants mediate their growth differently. Hence, measuring photosynthesis can be helpful in understanding the underlying physiological mechanisms of host-nematode-environment interactions. This, in turn, could lead to an understanding of the genetic bases of the interactions.

Table 15.3 The effect of water, complete Hoagland solution (HS) and HS without N (HS-N) on photosynthetic rates (μmol $CO_2 m^{-2} s^{-1}$) of a *Heterodera glycines* resistant (Bryan), tolerant (G88-20092) and susceptible (Tracy M) soybean cultivars after inoculations with 15,000 eggs of *H. glycines* at 24 days after inoculation. From Melakeberhan (1999b).

Nutrient	Cultivar	Nematode treatment	
		Control	*H. glycines*
Water	Bryan	15.93 a[x]	16.08 a
HS-N		17.60 a	15.43 a
HS		18.60 a	17.80 a
Water	G88-20092	14.70 a	11.05 a*[y]
HS-N		15.68 a	10.82 a**
HS		16.95 a	12.23 a***
Water	Tracy M	15.78 a	8.69 b***
HS-N		18.08 a	9.70 b***
HS		17.13 a	15.55 a

[x] Means followed by the same letters within the same nematode treatment and cultivar are not statistically different according to Tukey's range test ($P \leqslant 0.05$).

[y] Means followed by *, **, *** indicate significant differences at $P \leqslant 0.05$, 0.01, and 0.001, respectively, between nematode treatments within a nutrient treatment and cultivar according to Tukey's range test.

15.5.2 Effects of Nematodes on Photosynthate Translocation

In addition to the effect on photosynthesis, sedentary parasites such as root-knot nematodes divert a significant amount of photosynthate to their feeding sites (Bird and Loveys, 1975; McClure, 1977). This, in turn, affects photosynthate partitioning. Comparative studies of the concentration of sugars in *M. incognita*-induced galls of moderately resistant and susceptible *Vitis vinifera* cultivars showed more of an increase in the concentration of non-reducing sugars with duration of infection and increasing size of nematode inoculum in a susceptible than in a resistant cultivar (Melakeberhan, et al., 1990). This seems to suggest that photosynthate translocation to the nematode feeding site may be more rapid in susceptible than in resistant cultivars (Melakeberhan, et al., 1990). As in photosynthesis, understanding how nematodes influence the translocation of photosynthate material can be helpful in unraveling some of the unknowns of host-nematode interactions. For example, why does a nematode develop slower and lay fewer eggs in a particular host or under certain nutrient conditions than others? Could this be due limited translocation? If so, how?

15.5.3 Effects of Nematodes on Respiration

Little new information has been published on the effects of nematodes on respiration since the last review (Melakeberhan and Webster, 1993). As in most plant-pathogen interactions (Daly, 1976) plant-parasitic nematodes alter the rate of dark respiration. Total respiration (growth and maintenance) decreases as a function of plant biomass accumulation (Melakeberhan, et al. 1990). One of the areas that has been neglected is how plant-parasitic nematodes affect photorespiration, the uptake of O_2 and the evolution of CO_2 in the light that results from glycolate synthesis in the chloroplasts and subsequent glycolate and glycine metabolism in peroxisomes and mitochondria (Nobel, 1983). It is possible that an increase in the amount of energy unaccounted for and in the physiological inefficiency of a nematode-infected plant may be related to an increased activity in the wasteful glycolate pathway (Melakeberhan, et al., 1990). Identifying any links between nematode infection and increase in glycolate pathway activity will be helpful to understanding the physiological bases of host-nematode interactions.

15.6 Effects of Nematodes on Plant-growth Regulators

The effects of plant-parasitic nematodes on plant-growth regulators have been reviewed earlier (Melakeberhan and Webster, 1993). Root-knot nematodes have been the focus of most studies. The concentrations of plant cytokinins vary with the life cycle of the nematode (Bird and Millerd, 1962; Bird, et al., 1980; Van Staden and Dimalla, 1977) and are high in the galls and low in the shoot (Bird, et al., 1980; Dimalla and Van Staden, 1977; Van Staden and Dimalla, 1977). Changes in indoleacetic acid (Viglierchio and Yu; 1968) and increases in abscisic acid (Volmar, 1991) and ethylene (Glazier et al., 1983; 1984) have been reported. The mechanisms by which the gross physiological changes are induced remain largely unknown (Sawhney and Webster, 1975; 1979; Setty and Wheeler, 1968; Webster, 1967).

Gibberellins, required for cell elongation and cytokinins, required for cell division (Kefeli, 1978), translocation of photoassimilates (Thorpe and Lang, 1983), and nutrients (Mauk and Nooden, 1983; Newman and Stein, 1983), and for chlorophyll synthesis (Maunders, et al., 1983) are produced in root tissues (Carmi and Heuer, 1981; Scott and Horgan, 1984). Therefore, it is likely that plant-parasitic nematodes may decrease these root-originating phytohormones as a result of injury to the root system. This, in turn, may explain the common symptoms of stunting and leaf senescence and an uneven distribution and decrease of nutrient concentration in shoots of nematode-infected plants (Melakeberhan, et al., 1987).

15.7 Future Outlook

As we look into the future, it is appropriate to invoke Holt's (1991) analyses of scientific approaches to solving agricultural problems. He suggests two questions: (1) Are we doing it right? and (2) Are we doing the right thing? These are tactical (objective-oriented) and strategic (goal-oriented) questions, respectively.

I would like to suggest five major considerations:

First, the need to increase links between basic and applied nematology. Unraveling the molecular basis of host-nematode interaction undoubtedly holds the keys to future, basic science-driven approaches for solving nematode and agricultural problems. Maintaining basic science-driven advances depends on integration of our knowledge of host-nematode interactions at the subcellular, cellular, organismal, and ecosystem levels (Evans, 1995; Fenoll, et al., 1997; Grundler, et al., 1997; Wallace, 1978, 1987; Yoder, 1993).

Second, the need to extend our understanding of host-nematode interactions beyond cause-and-effect relationships into the physiological mechanisms that drive the interactions. As shown in the preceding sections, most of the host-nematode interaction analyses are based on cause-and-effect relationships. Knowing the physiological mechanisms of host-nematode interactions will be helpful to developing basic and applied approaches to a given problem. Two examples worth noting are resistance mechanisms and nutrition-driven physiological changes. With a few multigenic (e. g., some *H. glycines*) exceptions, most nematode resistance genes identified to date are gene-for-gene (e. g., *Mi*-gene) relationships. Although there are indications that the *Mi*-gene resistance may be related to how the plant regulates its growth (Melakeberhan, 1998a), little is known about the physiological mechanisms by which resistance is mediated in either form of resistance. In this respect, it must borne in mind that no single gene operates on its own within any given plant (Marschner, 1995).

There is lot to be learned from understanding the physiological bases of nutrition-driven changes in the host and in the nematode. Nutrition appears to affect nematodes at both pre-penetration and post-penetration of the host plant stages (Castro, et al., 1991; Ehrlich, et al., 1998; Melakeberhan, 1999a). A decrease in population density of the first generation is likely to be a result of direct effect on nematode hatching and/or infective behavior (Melakeberhan, 1997a, 1999b; Perry, 1997). Slowing down of development and a decrease in reproductive potential are likely to be host-mediated (indirect effect) processes, and provide bases for exploring any genetic links of the physiological processes that influence nematode growth. If nutrition-driven physiological factors are slowing down the development and/or decreasing reproductive potential of a nematode, it has to be considered in light of the biology of the particular nematode. In the case of the host modifiers, translocation of photosynthate material to the feeding sites, forming and maintaining active feeding sites, and nematode growth are among the factors that need to be looked at critically. Furthermore, how the nutrition-driven physiological changes relate to the secretions of the nematode is very important (Davis, et al., 1994).

Third, we need to consider multiple factors and their interactions in solving practical problems. An integrated physiological understanding of host, nematode, other stress-inducing factors, and the environment may provide a better approach to solving many agricultural problems. The interactions range from antagonisms between nematodes and fungi (El-Borai, et al., 2002a, b) to a single factor like soil pH driving the balance of the interacting factors (Melakeberhan, et al., 2000). Another common type of interaction differential response to stresses. For example, the incidence of *Phytophthora sojae* was increased in the presence of high and low *H. glycines* levels and nutrient imbalance stress from Al and Fe in *H. glycines*-susceptible soybean cultivars (Kaitany, et al., 2000). An *H. glycines*-resistant cultivar (Jack), however, absorbed less Al and Fe and suffered less from *P. soja*. This suggests that resistance to *H. glycines* may linked with mechanisms of nutrient uptake (Kaitany, et al., 2000).

Fourth, the need for futuristic applications. Precision agriculture is providing an information revolution utilizing global positioning systems (GPS) and geographic information systems (GIS) and remote and/or ground-based sensing of physiological changes (RS) (Gage, et al., 2000; Melakeberhan, 2002). These technologies enable for better decision making in the management of crop yield-limiting biotic and abiotic factors and their interactions on a site-specific (SSM) basis in a wide range of production systems. Characterizing the nature of the problem(s) and public education are among the challenges that scientists, producers, and industry face when adapting precision agriculture technologies. In order to apply SSM, spatio-temporal characteristics of the problem(s) need to be determined and variations within a field demonstrated (Avendano, et al., 2001). Spatio-temporal characteristics of a given pathogen or pest problem may be known, but it may not be the only or primary cause of the problem. Hence, the exact cause-and-effect relationships need to be established by incorporating GIS, GPS and RS generated data as well as possible interactions from small scale to regional ecosystem levels (Evans, et al., 2001; Gage, et al., 2000; Melakeberhan, 2002).

Fifth, there is a need to increase cross-disciplinary interactions. There have been multiple advances in the genetics of plant nutrition, cellular and subcellular nutrient translocation mechanisms (Graham, et al., 1987; Kochian, 1991), and whole-plant based genomic analysis (Chapin, et al., 1987; Tax and Vernon, 2001). Unfortunately, there is a knowledge gap between the advances in subcellular plant science and nematology as they relate to nematode-induced physiological changes. If the cross-disciplinary gaps are to be bridged, we need to implement Van Gundy's (1980) call to "*Take off our blinders and broaden our horizons*".

References

Abdalla, A. A., and K. Verkerk. 1968. Growth, flowering and fruit-set of the tomato at high temperature. Netherlands Journal of Agricultural Science 16: 71 – 76.

Alam, M. M., and S. K. Saxena. 1975. The effect of *Tylenchorhynchus brassicae* on water absorption capacity of roots. Geobios 2: 161 – 162.

Alam, M. M., Q. Ali, A. Masood, and A. M. Khan. 1975. Studies on the changes induced by the infection of root-knot nematode (*Meloidogyne incognita*) in tomato and egg-plant and the stunt nematode (*Tylenchorhynchus brassicae*) in cabbage and cauliflower roots. Indian Journal of Nematology 14: 715 – 718.

Atherton, J. G., and J. Rudich. 1986. The tomato crop: scientific basis for improvement. London: Chapman and Hall.

Atkinson, H. J. 1985. The energetics of plant parasitic nematodes – a review. Nematologica 31: 62 – 71.

Avendano, F., F. Pierce, O. Schabenberger, and H. Melakeberhan. 2001. Application of geostatistical tools to assess the spatial distribution of *Heterodera glycines*. Journal of Nematology 33: 250 (abstr.).

Ayres, P. G. 1984. The interaction between environmental stress injury and biotic disease physiology. Annual Review of Phytopathology 22: 53 – 75.

Baldwin, J. G., K. R. Barker, and L. A. Nelson. 1976. Effect of *Meloidogyne incognita* on nitrogen fixation in soybean. Journal of Nematology 11: 156 – 161.

Bar-Tsur, A., J. Rudich, and B. Bravo. 1985. High temperature effects on CO_2 gas exchange in heat-tolerant and sensitive tomatoes. Journal of American Society for Horticultural Science 110: 582 – 586.

Bird, A. F. 1974. Plant response to root-knot nematode. Annual Review of Phytophatology

12: 69 – 85.

Bird, A. F., and B. R. Loveys. 1975. The incorporation of photosynthates by *Meloidogyne incognita*. Journal of Nematology 7: 111 – 113.

Bird, A. F., and A. Millerd 1962. Respiration studies on *Meloidogyne*-induced galls in tomato roots. Nematologica 8: 261 – 266.

Bird, A. F., R. N. Sauer, R. N. Chapman, and B. R. Loveys. 1980. The influence of light quality on the growth and fecundity of *Meloidogyne* infecting tomato. Nematologica Mediterania 8: 59 – 66.

Bleve-Zacheo, T., and M. T. Melillo. 1997. The biology of giant cells. Pp. 65 – 79. In C. Fenoll, F. M. W. Grundler, and S. A. Ohl, eds. Cellular and Molecular Aspects of Plant-nematode Interactions. Boston: Kluwer Academic Publishers.

Blevins, D. G., V. H. Dropkin, and V. A. Luedders. 1995. Micronutrients uptake, translocation, and tissue concentration of soybeans infected with cyst nematode and elemental composition of cysts isolated from roots. Journal of Plant Nutrition 18: 579 – 591.

Bockenhoff A., and F. M. W. Grundler. 1993. Studies on the nutrient uptake by the beet cyst nematode *Heterodera schachtii* by in situ microinjection of fluorescent probes into the feeding structures in *Arabidopsis thaliana*. Parasitology 109: 249 – 254.

Bockenhoff, A., D. A. M. Prior, F. M. W. Grundler, and K. J. Oparka. 1996. Induction of phloem unloading in *Arabidopsis thaliana* roots by the parasitic nematode *Heterodera schachtii*. Plant Physiology 112: 1421 – 1427.

Boyer, J. S. 1985. Water transport. Annual Review of Plant Physiology 36: 473 – 516.

Carmi, A., and B. Heuer. 1981. The role of roots in control of bean shoot growth. Annals of Botany 48: 519 – 527.

Castro, C. E., H. E. McKinney, and S. Lux. 1991. Plant protection with inorganic ions. Journal of Nematology 23: 409 – 413.

Chapin, F. S., A. J. Bloom, C. B. Field, and R. H. Waring. 1987. Plant response to multiple environmental factors. Bioscience 37: 49 – 57.

Chen, H., Z. Shen, and P. H. Li. 1982. Adaptability of crop plants to high temperature stress. Crop Science 22: 719 – 725.

Colhoun, J. 1973. Effects of environmental factors on plant disease. Annual Review of Phytopathology 11: 343 – 364.

Daly, J. M. 1976. The carbon balance of diseased plants: changes in respiration, photosynthesis and translocation. Encyclopedia of Plant Physiology 4: 450 – 479.

Davis, E. L., R. Allen, and R. S. Hussey. 1994. Developmental expression of esophageal gland antigens and their detection in stylet secretions of *Meloidogyne incognita*. Fundamental and Applied Nematology 17: 255 – 262.

Dimalla, G. G., and J. Van Staden. 1977. Cytokinins in the root-knot nematode, *Meloidogyne incognita*. Plant Science Letters 10: 25 – 29.

Dorhout, R., F. J. Gommers, and C. Kolloffel. 1983. Phloem transport of carboxyfluorescein through tomato roots infected with *Meloidogyne incognita*. Physiological and Molecular Plant Pathology 43: 1 – 10.

Dorhout, R., C. Kolloffel, and F. J. Gommers. 1988. Transport of an apoplastic fluorescent dye to feeding sites induced in tomato roots by *Meloidogyne incognita*. Phytopathology 78: 1421 – 1424.

Dropkin, V. H. 1969a. Cellular changes of plants to nematode infections. Annual Review of Phytopathology 7: 101 – 122.

Dropkin, V. H. 1969b. The necrotic reaction of tomatoes and other hosts resistant to *Meloidogyne*: reversal by temperature. Phytopathology 59: 1632 – 1637.

Dropkin, V. H. 1989. Introduction to Plant Nematology. New York: John Wiley and Sons.

Dropkin, V. H., J. P. Helgeson, and C. D. Upper. 1969. The hypersensitivity reaction of tomatoes resistant to *Meloidogyne incognita*: Reversal by cytokinins. Journal of Nematology 1: 55 – 61.

Ehrlich, S. M., T. C. Smith, and H. Melakeberhan. 1998. Effect of nutrient solution and pH on embryogenesis of *Meloidogyne incognita*. Journal of Nematology 30: 493 – 494.

El-Borai, F. E., L. W. Duncan, and J. H. Graham. 2002a. Eggs of *Tylenchulus semipenetrans* inhibit growth of *Phytophthora nicotianae* and *Fusarium solani in vitro*. Journal of Nematology 34: in press.

El-Borai, F. E., L. W. Duncan, and J. H. Graham. 2002b. Infection of citrus roots by *Tylenchulus semipenetrans* reduces root infection by *Phytophthora nicotianae*. Journal of Nematology 34: in press.

Engelhard, A. W. 1989. Soilborne plant pathogens: Management of diseases with macro- and micronutrients. St. Paul, MN: The American Phytopathological Society Press.

Evans, K. 1995. Closing the gap between molecular biologists and traditional nematologists. Nematologica 41: 385 – 394.

Evans, K., D. L. Trudgill, and N. J. Brown. 1977. Effects of potato cyst-nematodes on potato plants. V: Root system development in light and heavy-infected susceptible and resistant varieties, and its importance in nutrient and water uptake. Nematologica 23: 153 – 164.

Evans, K., R. M. Webster, P. D. Halford, A. D. Baker, and M. D. Russell. 2001. Site-specific management of nematodes: Pitfalls and practicalities. Journal of Nematology 33: 256 (abstr.).

Fatemy, F., and K. Evans. 1986a. Growth, water uptake and calcium content of potato cultivars in relation to tolerance of cyst nematodes. Revue de Nematologie 9: 171 – 179.

Fatemy, F., and K. Evans. 1986b. Effects of *Globodera rostochiensis* and water stress on shoot and root growth and nutrient uptake of potatoes. Revue de Nematologie 9: 181 – 184.

Fenoll, C., F. A. Aristizabal, S. Sanz-Alferez, and F. F. del Campo. 1997. Regulation of gene expression in feeding sites. Pp. 133 – 149. In C. Fenoll, F. M. W. Grundler, and S. A. Ohl, eds. Cellular and Molecular Aspects of Plant-nematode Interactions. Boston: Kluwer Academic Publishers.

Fiscus, E. L. 1975. The interaction between osmotic and pressure-induced water flow in plant roots. Plant Physiology 55: 917 – 922.

Franco, J. 1980. Effect of potato cyst-nematode, *Globodera rostochiensis*, on photosynthesis of potato plants. Fitopatologia 15: 1 – 6.

Gage, S. H., M. Colunga, G. R. Safir, and A. Momin. 2000. A structural design for managementand visualization of information from regional agricultural models. Computers and Electronics in Agriculture 33: 77 – 84.

Ganmore-Neumann, R., and U. Kafkafi. 1980. Root temperature and percentage NO_3^-/NH_4^+ effect on tomato plant development: Morphology and growth. Agronomy Journal

72: 758 – 761.

Geisenberg, C., and K. Stewart. 1986. Field crop management. Pp. 510 – 555. In J. G. Atherton and J. Rudich, eds. The Tomato Crop: a Scientific Basis for Improvement. London: Chapman and Hall.

Gerloff, G. C., and W. H. Gabelman. 1983. Genetic basis of inorganic plant nutrition. Pp. 453 – 480. In A. Lauchli and R. L. Bieleski, eds. Inorganic Plant Nutrition. New York: Springer-Verlag.

Glazier, I., A. Apelbaum, and D. Orion. 1984. Reversal of nematode-induced growth retardation in tomato plants, *Lycopersicon esculentum*, by inhibition of ethylene action. Journal of American Society of Horticultural Science 109: 539 – 544.

Glazier, I., D. Orion, and A. Apelbaum. 1983. Interrelationships between ethylene production, gall formation, and root-knot nematode development in tomato plants infected with *Meloidogyne javanica*. Journal of Nematology 15: 539 – 544.

Golinowski, W., M. Sobzak, W. Kurek, and G. Grymaszewska. 1997. The structure of syncytia. Pp. 80 – 97. In C. Fenoll, F. M. W. Grundler, and S. A. Ohl, eds. Cellular and Molecular Aspects of Plant-nematode Interactions. Boston: Kluwer Academic Publishers.

Graham, R. D., J. S. Ascher, P. A. E. Ellis, and K. W. Shepherd. 1987. Transfer to wheat of the copper efficiency factor carried on rye chromosome arm 5RL. Plant and Soil 99: 107 – 114.

Grundler, F. M. W., and A. Böckenhoff. 1997. Physiology of nematode feeding and feding sites. Pp. 107 – 119. In C. Fenoll, F. M. W. Grundler, and S. A. Ohl, eds. Cellular and Molecular Aspects of Plant-nematode Interactions. Boston: Kluwer Academic Publishers.

Hanna, H. Y., and T. P. Hernendez. 1982. Response of six tomato genotypes under summer and spring weather conditions in Louisiana. HortScience 17: 758 – 759.

Hargrove, W. L. 1988. Cropping strategies for efficient use of water and nutrients. American Society of Agronomy Special Publication No. 51, Madison, WI, USA.

Holt, D. 1991. Organizational paradigms of agricultural research and development. Agricultural Research Institute Proceedings and Minutes, No. 39, Pp 8 – 20.

Hsiao, T. C. 1973. Plant response to water stress. Annual Review of Plant Physiology 24: 519 – 570.

Hsiao, T. C., E. F. Acevedo, E. Fereres, and D. W. Henderson. 1976. Stress metabolism. Philosophical Transactions of the Royal Society of London 273: 479 – 500.

Huber, D. M., and R. D. Watson. 1974. Nitrogen form and plant disease. Annual Review of Phytopathology 12: 139 – 165.

Hussey, R. S. 1989. Disease-inducing secretions of plant-parasitic nematodes. Annual Review of Phytopathology 27: 123 – 141.

Hussey, R. S., and V. M. Williamson. 1998. Physiological and molecular aspects of nematode parasitism. Pp. 87 – 108. In K. R. Barker, G. A. Pederson, and G. L. Windham, eds. Plant and Nematode Interactions. Madison, WI: ASA/CSSA/SSSA Publishers.

Ikeda, T., and T. Susaki. 1984. Influence of pine wood nematodes on hydraulic conductance and water relations in Pinus thunbergii. Japanese Journal of Forestry Society 66: 412 – 420.

Ikeda, T., T. Kiyohara, and M. Kusunoki. 1990. Changes in water status of *Pinus thunbergii* Parl. inoculated with species of *Bursaphelenchus*. Journal of Nematology 22: 132 – 135.

Islam, W., and M. M. Alam. 1975. Inhibition in water absorption capacity of castor and tomato roots by the infection of the reniform nematode, *Rotylenchulus reniformis*. Geobios 2: 23 – 24.

Jenkins, J. N., R. G. Creech, B. Tang, G. W. Lawrence, and J. C. McCarty. 1995. Cotton Resistance to Root-Knot Nematode: II. Post-Penetration Development. Crop Science 35: 369 – 373.

Jones, M. G. K. 1981. Host response to endoparasitic nematode attack: structure and function of giant cells and syncytia. Annals of Applied Biology 97: 353 – 372.

Kaitany, R., H. Melakeberhan, G. W. Bird, and G. Safir. 2000. Association of *Phytophthora sojae* with *Heterodera glycines* and nutrient stress. Nematropica 30: 193 – 199.

Kefeli, V. I. 1978. Nature of plant growth inhibitors and phytohormones. The Hague: Dr. W. Junk Publications.

Kirby, E. A., and K. Mengel. 1967. Ionic balance in different tissues of the tomato plant in relation to nitrate, urea, or ammonium nutrition. Plant Physiology 42: 6 – 14.

Kochian, L. V. 1991. Mechanisms of micronutrient uptake and translocation in plants. Pp. 229 – 296. In J. J. Mortvedt, ed. Micronutrients in Agriculture. Madison, WI: Soil Science Society of America.

Kuroda, K. 1989. Terpenoids causing tracheid-cavitation in *Pinus thunbergii* infected by the pine wood nematode (*Bursaphelenchus xylophilus*). Annals of the Phytopathological Society of Japan 55: 170 – 178.

Kuroda, K., T. Yamada, K. Minio, and H. Tamura. 1988. Effects of cavitation on the development of pine wilt disease caused by *Bursaphelenchus xylophilus*. Annals of the Phytopathological Society of Japan 54: 606 – 615.

Leru, B., J. P. Diangana, and N. Beringar. 1994. Effects of nitrogen and calcium on the level of resistance of cassava to the mealybug *Planococcus manihoti*. Insect Science and its Application 15: 87 – 96.

Loveys, B. R., and A. F. Bird. 1973. The influence of nematodes on photosynthesis in tomato plants. Physiological Plant Pathology 11: 11 – 21.

Mamiya, Y., and H. Tamura. 1977. Transpiration reduction of pine seedlings inoculated with the pine wood nematode, *Bursaphelenchus lignicolus*. Journal of Japanese Forest Society 59: 59 – 63.

Marschner, H. 1995. Mineral Nutrition of Higher Plants. 2nd edition. New York: Academic Press.

Mauk, C. S., and L. D. Nooden. 1983. Cytokinin control of mineral nutrient distribution between the foliage and seeds in soybean explants. Plant Physiology 72: 167.

Maunders, M. J., S. B. Brown, and H. W. Woolhouse. 1983. The appearance of chlorophyll derivatives in senescing tissue. Phytochemistry 22: 2443 – 2446.

McClure, M. A. 1977. *Meloidogyne incognita*: a metabolic sink. Journal of Nematology 9: 88 – 90.

Melakeberhan, H. 1997a. Plant, nematode and nutrient relations: an overview. Japanese Journal of Nematology 27: 41 – 51.

Melakeberhan, H. 1997b. Role of plant nutrition on alleviating nematode parasitism. Pp. 759 – 760. In T. Ando, K. Fujita, T. Mae, H. Matsumoto, S. Mori, and J. Sekiya, eds. Plant Nutrition for Sustainable Food Production and Environment. London: Kluwer Academic Publishers.

Melakeberhan, H. 1998a. Effects of temperature and nitrogen source on tomato genotypes response to *Meloidogyne incognita*. Fundamental and Applied Nematology 21: 25 – 32.

Melakeberhan, H. 1998b. Influence of mineral nutrition on the development of *Heterodera glycines* and plant-nutrient dynamics. Nematologica 44: 535.

Melakeberhan, H. 1999a. Is there a potential for nutrient-based *Heterodera glycines* management in soybeans. VI World Soybean Research Conference, Chicago, IL, USA. Pp. 613 – 614.

Melakeberhan, H. 1999b. Effect of nutrient source on the physiological mechanisms of *Heterodera glycines* and soybean genotype interactions. Nematology 1: 113 – 120.

Melakeberhan, H. 2002. Embracing the emerging precision agriculture technologies for site-specific management of yield-limiting factors. Journal of Nematology 34: 185 – 188.

Melakeberhan, H., G. W. Bird, and R. Gore. 1997. Impact of nutrients on *Pratylenchus penetrans* infection of *Prunus avium* rootstocks. Journal of Nematology 29: 381 – 388.

Melakeberhan, H., R. C. Brooke, J. M. Webster, and J. M. D' Auria. 1985a. The influence of *Meloidogyne incognita* on growth, nutrient content and physiology of *Phaseolus vulgaris*. Physiological Plant Pathology 26: 259 – 268.

Melakeberhan, H., and H. Ferris. 1988. Growth and energy demand of *Meloidogyne incognita* in susceptible and resistant *Vitis vinifera* cultivars. Journal of Nematology 20: 545 – 554.

Melakeberhan, H., and H. Ferris. 1989. Impact of *Meloidogyne incognita* on physiological efficiency of *Vitis vinifera*. Journal of Nematology 21: 74 – 80.

Melakeberhan, H., H. Ferris, and J. Dias. 1990. Physiological response of resistant and susceptible *Vitis vinifera* cultivars to *Meloidogyne incognita*. Journal of Nematology 22: 224 – 230.

Melakeberhan, H., A. L. Jones, and G. W. Bird. 2000. Effects of soil pH and *Pratylenchus penetrans* on the pathogenesis of *Pseudomonas syringae* pv. *syringae* and Mazzard seedling mortality. Canadian Journal of Plant Pathology 22: 131 – 137.

Melakeberhan, H., P. M. A. Toivonen, W. E. Vidaver, J. M. Webster, and S. Dube. 1991. Effect of *Bursaphelenchus xylophilus* on water potential and the water-splitting complex of photosystem II of *Pinus sylvestris* seedlings. Physiological and Molecular Plant Pathology 38: 83 – 91.

Melakeberhan, H., and J. M. Webster. 1990. Effect of *Bursaphelenchus xylophilus* on the assimilation and translocation of ^{14}C in *Pinus sylvestris*. Journal of Nematology 22: 506 – 512.

Melakeberhan, H., and J. M. Webster. 1992. The insignificance of the energy requirements of *Bursaphelenchus xylophilus* as a causal factor in *Pinus sylvestris* seedling mortality. Fundamentals and Applied Nematology 15: 178 – 182.

Melakeberhan, H., and J. M. Webster. 1993. The phenology of plant-nematode interaction and yield loss. Pp. 26 – 41. In M. M. Khan, ed. Nematode Interactions. London: Chapman and Hall.

Melakeberhan, H., J. M. Webster, and R. C. Brooke. 1984. Improved techniques for measuring the CO_2 exchange rate of *Meloidogyne*-infected beans. Nematologica 30: 213 – 221.

Melakeberhan, H., J. M. Webster, and R. C. Brooke. 1985b. Response of *Phaseolus vulgaris* to a single generation of *Meloidogyne incognita*. Nematologica 31: 190 – 202.

Melakeberhan, H., J. M. Webster, and R. C. Brooke. 1986. Relationship between physiological response and yield loss of different age French bean to *Meloidogyne incognita*. Plant Pathology 35: 203 – 213.

Melakeberhan, H., J. M. Webster, R. C. Brooke, and J. M. D'Auria. 1988. Effect of KNO_3 on CO_2 exchange rate, nutrient concentration and yield of *Meloidogyne incognita*-infected beans. Revue de Nematologie 11: 391 – 397.

Melakeberhan, H., J. M. Webster, R. C. Brooke, J. M. D'Auria, and M. Cackette. 1987. Effect of *Meloidogyne incognita* on plant nutrient concentration and its influence on the physiology of beans. Journal of Nematology 19: 324 – 330.

Meon, S., J. M. Fisher, and H. R. Wallace. 1978. Changes in free proline following infection of plants with either *Meloidogyne incognita* or *Agrobacterium tumefaciens*. Physiological Plant Pathology 12: 251 – 256.

Mohammed, A. H., and A. Gnanam. 1979. A possible mechanism of ammonium ion regulation of photosynthetic carbon flow in higher plants. Plant Physiology 64: 263 – 268.

Neumann, P. M., and Z. Stein. 1983. Xylem transport and the regulation of leaf metabolism. What's New in Plant Physiology 14: 33 – 36.

Nobel, P. S. 1983. Biophysical Plant Physiology and Ecology. New York: W. H. Freeman.

Norton, D. C. 1978. Ecology of Plant-parasitic Nematodes. New York: John Wiley and Sons.

Orians, C. M., and R. M. Fritz. 1996. Genetic and soil-nutrient effects on the abundance of herbivores on willow. Oecologia 105: 388 – 396.

Orion, D., W. P. Wergin, and D. J. Chitwood. 1995. Root cortical cell spherical bodies associated with an induced resistance in monoxenic culture of *Meloidogyne incognita*. Journal of Nematology 27: 320 – 327.

Oteifa, B. A., and A. Y. Elgindi. 1976. Potassium nutrition of cotton, *Gossypium barbadense*, in relation to nematode infection by *Meloidogyne incognita* and *Rotylenchulus reniformis*. XIIth Colloquium of International Potash Institute Proceedings, Pp. 301 – 306.

Perry, R. N. 1997. Plant signals in nematode hatching and attraction. Pp. 38 – 50. In C. Fenoll, F. M. W. Grundler, and S. A. Ohl, eds. Cellular and Molecular Aspects of Plant-nematode Interactions. Boston: Kluwer Academic Publishers.

Pill, W. G., and V. N. Lambeth. 1977. Effects of NH_4 and NO_3 nutrition with and without pH adjustment on tomato growth, ion composition and water relations. Journal of the American Society for Horticultural Science 102: 78 – 81.

Powers, L. E., R. A. Dunn, and R. McSorley. 1991. Size differences among root-knot nematodes on resistant and susceptible Alyceclover genotypes. Journal of Nematology 23: 243 – 248.

Powers, L. E., and R. McSorley. 1993. Energetics of *Meloidogyne incognita* on resistant and susceptible Alyceclover genotypes. Journal of Nematology 25: 257 – 264.

Rajaratnam, J. A., and L. I. Hock. 1975. Effect of boron nutrition on intensity of red spider mite attack on oil-palm seedlings. Experimental Agriculture 11: 59 – 63.

Reversat, G. 1987. Increase in chemical oxygen demand during the growth in *Heterodera sacchari*. Revue de Nematologie 10: 115 – 117.

Roberts, P. A. 1992. Current status of the availability, development, and use of host plant resistance to nematodes. Journal of Nematology 24: 213 – 227.

Salam, A. M. S., and P. F. Wareing. 1978. Effects of mineral nutrition on endogenous cytokinins in plants of sunflower (*Helianthus annuus* L.). Journal of Experimental Botany 30: 971 – 981.

Sawhney, R., and J. M. Webster. 1975. The role of plant-growth hormones in determining the resistance of tomato plants to the root-knot nematode, *Meloidogyne incognita*. Nematologica 21: 95 – 103.

Sawhney, R., and J. M. Webster. 1979. The influence of some metabolic inhibitors on the response of susceptible/resistant cultivars of tomato to *Meloidogyne incognita*. Nematologica 25: 86 – 93.

Schoeneweiss, D. F. 1975. Predisposition, stress, and plant disease. Annual Review of Phytopathology 13: 193 – 211.

Scott, I. M., and R. Horgan. 1984. Root cytokinins of *Phaseolus vulgaris* L. Plant Science letters 34: 81 – 87.

Seinhorst, J. W. 1961. Plant-nematode inter-relationships. Annual Review of Microbiology 15: 177 – 196.

Setty, K. G. H., and A. W. Wheeler. 1968. Growth substances in roots of tomato infected with root-knot nematodes (*Meloidogyne* species). Annals of Applied Biology 61: 495 – 501.

Sijmons, P. C., H. J. Atkinson, and U. Wyss. 1994. Parasitic strategies of root nematodes and associated host cell responses. Annual Review of Phytopathology 32: 235 – 259.

Spiegel, Y, E. Cohen, U. Kafkafi, and M. Sulami. 1982. Influence of potassium and nitrogen fertilization on parasitism by the root-knot nematode *Meloidogyne javanica*. Journal of Nematology 14: 530 – 535.

Tamura, H., K. Mineo, and T. Yamada. 1987. Blockage of water conduction in *Pinus thunbergii* inoculated with *Bursaphelenchus xylophilus*. Japanese Journal of Nematology 17: 23 – 30.

Tax, F. E., and D. M. Vernon. 2001. T-DNA-associated duplication/translocations in *Arabidopsis*. Implications for mutant analysis and functional genomics. Plant Physiology 126: 1527 – 1538

Thorpe, M. R., and A. Lang. 1983. Control of import export of photosynthate in leaves. Journal of Experimental Botany 34: 231 – 239.

Trudgill, D. L. 1987. Effects of rates of a nematicide and of fertilizer on the growth and yield of cultivars of potato which differ in their tolerance of damage by potato cyst nematodes (*Globodera rostochiensis* and *G. pallida*). Plant and Soil 104: 235 – 243.

van Gundy, S. D. 1980. Nematology-status and prospects: let's take off our blinders and broaden our horizons. Journal of Nematology 12: 158 – 163.

Van Staden, J., and G. G. Dimalla. 1977. A comparison of the endogenous cytokinins in the roots and xylem exudates of nematode resistant and susceptible tomato cultivars.

Journal of Experimental Botany 28: 1351 – 1356.

Vigliechio, D. R., and P. K. Yu. 1968. Plant growth substances and plant parasitic nematodes. II: host influence on auxin content. Experimental Parasitology 23: 88 – 95.

Volmar, K. M. 1991. Abscisic acid and ethylene increase in *Heterodera avenae*-infected tolerant or intolerant oat cultivars. Journal of Nematology 23: 425 – 431.

Wallace, H. R. 1974. The influence of root-knot nematode, *Meloidogyne javanica*, on photosynthesis and on nutrient demand by roots of tomato plants. Nematologica 2: 27 – 33.

Wallace, H. R. 1978. The diagnosis of plant diseases of complex etiology. Annual Review of Phytopathology 16: 379 – 402.

Wallace, H. R. 1987. Effect of Nematode parasites on photosynthesis. Pp. 253 – 259. In J. A. Veech and D. W. Dickson, eds. Vistas on Nematology. Hyattsville, MD: Society of Nematologists.

Webster, J. M. 1967. The influence of plant-growth substances and their inhibition on the host-parasite relations of *Aphelenchoides ritzemabosi* in culture. Nematologica 13: 256 – 262.

Webster, J. M. 1975. Aspects of the host-parasite relationships of plant parasitic nematodes. Advances in Parasitology 13: 225 – 250.

Widders, I. E., and O. A. Lorenz. 1979. Tomato root development as related to potassium nutrition. Journal of Americian Society of Horticultural Science 104: 216 – 220.

Wilcox-Lee, D., and R. Loria. 1987. Effects of nematode parasitism on plant-water relations. Pp. 260 – 266. In J. A. Veech and D. W. Dickson, eds. Vistas on Nematology. Hyattsville, MD: Society of Nematologists.

Wilhelm, M. J., J. M. Fisher, and R. D. Graham. 1985. The effect of manganese deficiency and cereal cyst nematode infection on the growth of barley. Plant and Soil 85: 23 – 32.

Williamson, V. M., J. Y. Ho, and H. M. Ma. 1992. Molecular transfer of nematode resistance genes. Journal of Nematology 24: 234 – 241.

Williamson, V. M., and R. S. Hussey. 1996. Nematode pathogenesis and resistance in plants. The Plant Cell 8: 1735 – 1745.

Williamson, V. M., K. N. Lambert, and J. Y. Ho. 1993. Root-knot nematode resistance in tomato. Pp. 275 – 284. In J. I. Yoder, ed. Molecular Biology of Tomato: Fundamental Advances and Crop Improvement. Lancaster, PA: Technomic Publishing.

Wolcott, A. R., F. H. Liao, and J. I. Kirkwood. 1967. Effects of fumigation, temperature, and level of nitrate on microbial numbers, CO_2 production, and transformations in an organic soil. Soil Science 103: 131 – 138.

Yoder, J. I., ed. 1993. Molecular biology of tomato: Fundamental advances and crop improvement. Lancaster, PA: Technomic Publishing.

Acknowledgements The author would like to thank Drs. Ken Evans (Rothamsted, England) and David Chitwood (USDA, Beltsville) for critical reviews of the manuscript. Part of the work reported here was funded by a competitive grant from USDA/NRI/CGP and Hatch Funds through the Michigan Agricultural Experiment Station.

16 Taxonomy of Insect Parasitic Nematodes

Khuong B. Nguyen and Grover C. Smart, Jr.

16.1 Introduction

Insect-parasitic nematodes have been known since the 17th century and perhaps earlier, but mostly as a curiosity, and by sometimes giving them names. The reports of von Siebold, Charvet, Berthold, Dujardin, Hope, Diesing, and von Linstow on filariids, mermithids, and a few others were beginning studies in insect nematology (Nickle and Welch, 1984). Extensive studies were carried out in the 19th and 20th centuries. At the beginning of the 20th century, some of these insect parasites were considered as possible biological control agents. Studies on the economic importance and life histories of two mermithid parasites of grasshoppers, *Agamermis decaudata* Cobb et al., 1923 and *Mermis subnigrescens* Cobb, 1926, were excellent contributions (Christie, 1936; Cobb, 1929; Steiner, 1925). The paper by Cobb et al. (1923) about *Agamermis decaudata* stands as a classic in insect nematology.

Nematodes referred to currently as entomopathogenic, although entomophilic might be a better descriptor, were first described in 1923 (Steiner, 1923) and one species, *Steinernema glaseri* Steiner, was studied seriously a decade later by Glaser and his colleagues to control the Japanese beetle, *Popillia japonica* Newman (Glaser and Farrell, 1935; McCoy and Glaser, 1936). After extensive research, including many successful field trials, interest in the use of entomophilic nematodes as biological control agents waned until the 1970s and 1980s. In the interim, increasing environmental concerns about the use of chemical pesticides, and the availability of fewer of them, reawakened interest in entomophilic nematodes for insect control.

From the late 1960s to the early 1980s, Peterson and his colleagues in Louisiana (Peterson, 1984; Peterson, et al., 1968), conducted extensive research on the mermithid nematode, *Romanomermis culicivorax* Ross and Smith, and showed that it could provide good to excellent control of mosquitoes. In Australia, Bedding successfully controlled siricid woodwasps in coniferous forests with a tylenchid nematode, *Deladenus siricidicola* Bedding,

1968 (Bedding, 1984). These examples demonstrated that entomophilic nematodes could be effective biological control agents. Nickle and Welch (1984) presented an excellent and extensive account of the history and development of insect nematology.

In this chapter, we present an overview of the taxonomy of entomophilic nematodes in the orders Rhabditida (Families Rhabditidae, Steinernematidae, Heterorhabditidae, Diploscapteridae), Diplogasterida (Family Diplogasteridae), Tylenchida (Families Allantonematidae, Sphaerulariidae, Parasitylenchidae, Iotonchidae, Fergusobiidae, Entaphelenchidae), Mermithida (Families Mermithidae, Tetradonematidae), and Oxyurida (Families Thelastomatidae, Hystrignathidae, Travassosinematidae). Not all of the nematodes that we include here are proven parasites of insects, but we include them because they have been reported to be insect parasites.

16.2 Taxonomy

16.2.1 Order Rhabditida (Oerley, 1880) Chitwood, 1933

Diagnosis: Number of lips variable, from none to six. Stoma tubular, stoma walls lined with rhabdions. Esophagus with procorpus, metacorpus (sometimes lacking), isthmus, and terminated by a valvate basal bulb. Excretory system with lateral canals, terminal duct cuticularized. Females with one or two ovaries; if one, vulva located posteriorly. Bursa with ribs when present. Suborders: Rhabditina and Cephalobina.

Four families in the suborder Rhabditina will be discussed. Since all species of the Steinernematidae and Heterorhabditidae are pathogenic, each species will be described, and a key to species for the genera *Steinernema* and *Heterorhabditis* will be included. Three genera in the Rhabditidae, and one genus of the Diploscapteridae are included because there is good evidence that they are parasites of insects, or in one case, to snails.

16.2.1.1 Family Steinernematidae Chitwood and Chitwood, 1937 (syn.: Neoaplectanidae Sobolev, 1953)

Taxonomic status: In 1923 Steiner described the genus and species *Aplectana kraussei*, but since the generic name was preoccupied, Travassos (1927) renamed the genus *Steinernema*. Two years later Steiner described *Neoaplectana glaseri*. In 1934, Filipjev placed *Steinernema* and *Neoaplectana* in the new subfamily Steinernematinae, and stated that *Neoaplectana* was probably congeneric with *Steinernema*. Chitwood and Chitwood (1937) raised the subfamily Steinernematinae to the family Steinernematidae. Wouts et al.

(1982) concluded that the two genera were identical and that *Neoaplectana* was a junior synonym of *Steinernema*, leaving the family with a single genus. In 1994, Nguyen and Smart described a new genus, *Neosteinernema*, adding it to the family and emending the family description to accommodate the new genus. Currently, the family Steinernematidae contains only the two genera, *Steinernema* with 33 species and *Neosteinernema* with one species. An on-line document that is updated regularly as new taxonomic research is published is available at: http://kbn.ifas.ufl.edu/kbnstein.htm

Diagnosis: (After Nguyen and Smart, 1994 as emended to include the new genus *Neosteinernema*): ***Females***: (Fig. 16.1) Large, size variable. Cuticle smooth or annulated. Lateral fields absent. Excretory pore distinct, anterior to nerve ring. Head rounded or truncate, rarely offset. Six lips present, partly or

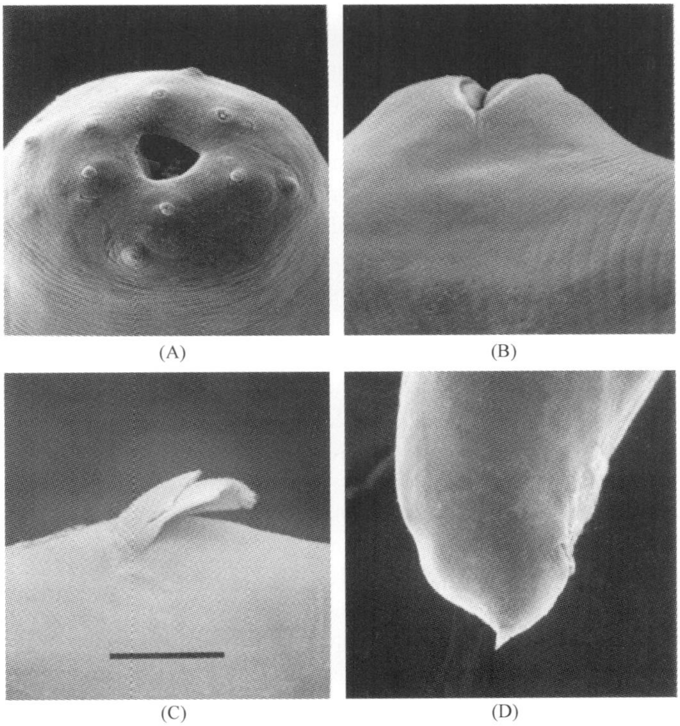

Figure 16.1 SEM of *Steinernema* female. (A) Face view. (B) Vulval region. (C) Vulval region showing epiptygma. (D) Tail. (Modified from Nguyen and Smart, 1996.)

completely fused, each lip with one labial papilla (Fig. 16.1(A)), sometimes, additional papillae-like structures present near labial papillae. Four cephalic papillae. Amphids present, small. Stoma collapsed; cheilorhabdions pronounced, forming a ring resembling two large sclerotized dots in lateral view. Other parts of stoma forming an asymmetrical funnel with thick anterior end. Esophagus rhabditoid with metacorpus slightly swollen, narrow isthmus surrounded by nerve ring, and large basal bulb with reduced valve. Esophago-intestinal valve usually pronounced. Reproductive system didelphic, amphidelphic, reflexed. Vulva at mid-body, sometimes on a protuberance (Fig. 16.1(B)), with (Fig. 16.1(C)) or without (Fig. 16.1(B)) epiptygma. Females oviparous or ovoviviparous with juveniles developing up to the infective stage (IJ) before emerging from the body of the female. Tail longer or shorter than anal body width, with or without prominent phasmids.

Males: (Fig. 16.2) Smaller than female. Anterior end usually with six labial papillae, four large cephalic papillae and usually with perioral disc. Esophagus similar to that of the female. Testis single, reflexed. Spicules paired. Gubernaculum long, sometimes as long as spicule. Bursa absent. Tail tip rounded, digitate or mucronate. One single and 10 to 14 pairs of genital papillae present with 7 to 10 pairs precloacal.

Infective juveniles (= third-stage infective juvenile): (Fig. 16.3) Stoma collapsed. Body slender, with or without a sheath (cuticle of second-stage juvenile). Cuticle annulated. Lateral fields present with four to nine incisures and three to eight smooth ridges (Fig. 16.3(B)). Esophagus and intestine appearing reduced. Excretory pore distinct, anterior to nerve ring. Tail conoid or filiform. Phasmids, located about mid-tail, prominent, inconspicuous, or not observed.

Type genus: *Steinernema* Travassos, 1927 (syn.: *Neoaplectana* Steiner, 1929; *Steineria* Travassos, 1927 nec *Steineria* Micoletzki, 1922). Other genus: *Neosteinernema* Nguyen and Smart, 1994.

Steinernema Travassos, 1927 (Figs 16.1 – 16.3)

Diagnosis: Female without phasmids, tail (T) shorter than anal body width (ABW) (T/ABW = 0.52 – 0.81); oviparous but some eggs often retained and hatch in the body. Male smaller than female, posterior part usually with one single and 11 pairs of genital papillae; phasmids not observed, tail terminus rounded or with mucron. Infective juvenile with phasmids small or inconspicuous, tail conoid and much shorter than esophagus (at most 65% of esophagus length), ratio c (= body length/tail length) greater than or equal to 10.

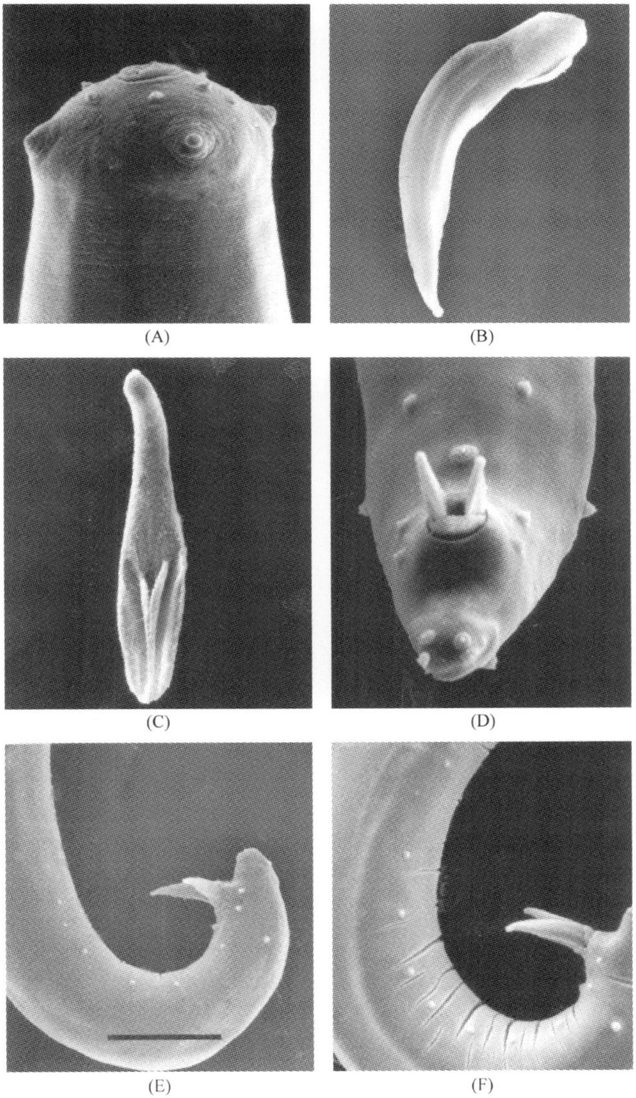

Figure 16.2 SEM of *Steinernema* male. (A) Head showing labial and cephalic papillae. (B) Spicule. (C) Gubernaculum showing Y-shaped cuneus. (D) Ventral view of male tail showing papillae and spicule tips. (E), (F) Male tails with seven and ten preanal papillae. (Modified from Nguyen and Smart, 1996.)

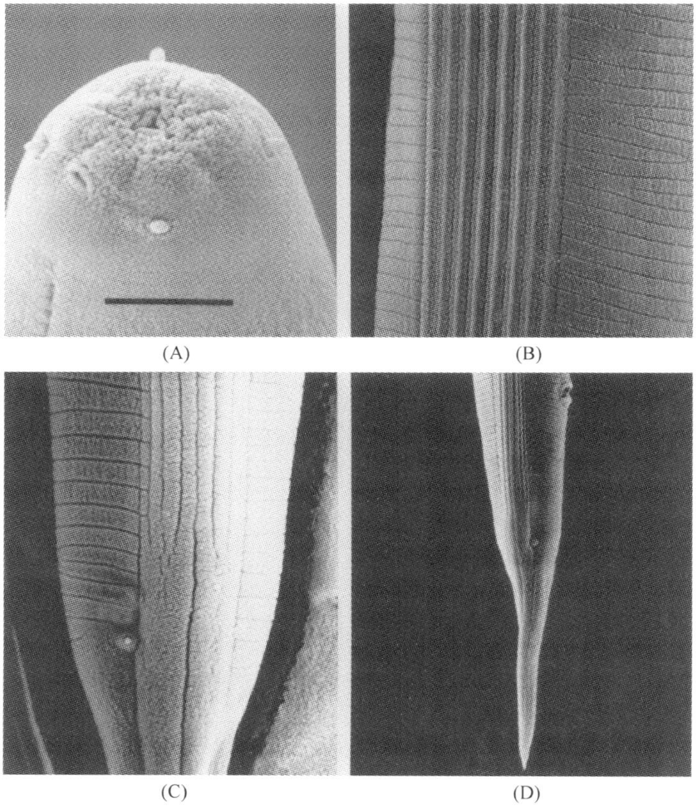

Figure 16. 3 SEM of *Steinernema* infective juvenile. (A) Anterior region showing almost closed oral opening, amphid, cephalic papillae and an incisure in lateral field. (B) Lateral field showing eight ridges and nine incisures. (C) Tail showing lateral field at phasmid level; note that the lateral field with many ridges becomes two larger ridges. (D) Tail with anus, lateral field, phasmid and tail terminus. (Modified from Nguyen and Smart, 1996.)

Steinernema kraussei (Steiner, 1923) Travassos, 1927 (syn.: *Aplectana kraussei* Steiner, 1923). Type species

Diagnosis: **Important notes**: Ratios and abbreviations used in the diagnoses and keys are: D% = EP/ES × 100 (EP = distance from anterior end to excretory pore; ES = esophagus length); E% = EP/T × 100, (T = tail length); IJ = infective juvenile; SW = spicule length divided by anal body width; ratio GS = gubernaculum length/spicule length.

Males: (This diagnosis is a combination of the original description by Steiner, 1923 and the redescription by Mracek, 1994). Anterior end truncate, slightly offset with six labial and four cephalic papillae. Excretory pore well anterior to nerve ring, D% about 53. Tail with rounded tip bearing a fine mucron. Spicule head (manubrium) length as long as wide (Nguyen and Smart, 1997). Spicule shaft (calomus) present, sometimes indistinct. Lamina curvature variable. Velum present, extending about two-thirds lamina length. Spicule length 52 – 57 μm (along the curve) or 42 – 53 μm (straight line from spicule head to tip), spicule length/spicule width averaging 5.1 (4.3 – 6.0). SW = 1.10 and GS = 0.68. In ventral view, gubernaculum tapering anteriorly, cuneus small, Y-shaped.

Infective juveniles: Body length averaging 951 (797 – 1102) μm. Distance from anterior end to excretory pore 63 (53 – 67) μm, *E%* about 80. *T* = 79 (69 – 86) μm. This nematode species was first collected from the web spruce sawfly, *Cephaleia abietis* (L), in Germany and described by Steiner in 1923 as *Aplectana kraussei.* In 1927, Travassos renamed it *Steinernema kraussei* because the generic name, *Aplectana,* was preoccupied. The nematode was found in Czechoslovakia and was redescribed by Mracek in 1977. Later, Mracek (1994) found the nematode at the type locality in Germany and redescribed it again.

Steinernema abbasi Elawad, Ahmad and Reid, 1997

Diagnosis: ***Males***: Anterior end truncate, continuous with body with six labial and four cephalic papillae. Excretory pore anterior to nerve ring, *D%* about 60 (51 – 68). Tail rounded without mucron. Spicule-head length equal to or shorter than width, shaft indistinct, lamina gradually tapering, rostrum and velum present, and tip bluntly pointed. Spicule length 65 (57 – 74) μm. SW = 1.56 (1.1 – 1.9), GS = 0.7 (0.6 – 0.9). Gubernaculum boat-shaped.

Infective juveniles: Body length averaging 541 (496 – 579) μm long. Distance from anterior end to excretory pore 48 (46 – 51) μm, *E%* about 86 (79 – 94). Tail length 56 (52 – 61) μm. This nematode species was collected from soil in the Sultanate of Oman. The nematode develops well up to 35℃.

Steinernema affine (Bovien, 1937) Wouts, Mracek, Gerdin and Bedding, 1982 (syn.: *Neoaplectana affinis* Bovien, 1937)

Diagnosis: ***Males***: Anterior end rounded, continuous with the body. Labial and cephalic papillae as in other species. Excretory pore anterior to nerve ring, *D%* about 61 (60 – 66). Tail with a minute mucron at the end, easily confused with one of the subterminal papillae but the latter are always paired. Spicule head slightly longer than width. Shaft present, sometimes indistinct. Spicule lamina moderately curved. Velum present. Spicule length 70 (67 –

86) μm, spicule length/spicule width =5.2 (4.9 – 5.6). SW = 1.17; GS = 0.66. In ventral view, gubernaculum tapering gradually anteriorly to a ventrally curved end. Cuneus long, needle-shaped.

Infective juveniles: Body length averaging 693 (608 – 880) μm. EP = 62 (51 – 69) μm, *E%* about 94 (74 – 108). *T* = 66 (64 – 74) μm, tail tip with a spine internally. This species and *S. feltiae* (= *bibionis*) were described by Bovien in 1937 from a bibionid fly larva in Denmark. In 1979 Poinar discussed the possibility that *S. affine* and *S. feltiae* represented an example of morphological variation within a single species but, in 1988 Poinar redescribed *S. affine* as a valid species.

Steinernema arenarium (Artyukhovsky, 1967) Wouts, Mracek, Gerdin and Bedding, 1982 (syn.: *Neoaplectana arenaria* Artyukhovsky, 1967; *N. anomali* Kozodoi, 1984; *S. anomalae* (Curran, 1989; Kozodoi, 1984)

Diagnosis: ***Males***: Anterior end rounded with six labial and four cephalic papillae. Excretory pore 2 μm wide, located posterior to nerve ring. *D%* about 93 (88 – 102). Tail without mucron. Spicule head slightly longer than wide (Fig. 16.4(C)). Spicule shaft indistinct. Lamina moderately curved. A small velum sometimes present. Spicules moderately curved, length 84 (81 – 91) μm (Kozodoi, 1984). Spicule tip enlarged with prominent dorsal aperture, making the spicule tip appear constricted. SW = 2.10; GS = 0.65. In ventral view, gubernaculum short and wide, cuneus V-shaped (Fig. 16.5(I)).

Infective juveniles: Body length averaging 1217 (930 – 1580) μm. EP = 83 (76 – 86) μm, E% = 119. *T* = 65 (64 – 84) μm. This species was collected in Central Russia and first described by Artyukhovsky in 1967. In 1997, Artyukhovsky et al. redescribed this species adding molecular data to compare *S. arenarium* and *S. anomalae* and made *S. anomalae* a junior synonym of *S. arenarium*.

Steinernema bicornutum Tallosi, Peters and Ehlers, 1995

Diagnosis: ***Males***: Anterior end rounded, continuous with the body bearing 6 labial and 4 cephalic papillae. Excretory pore anterior to nerve ring, *D%* about 52 (50 – 60). Tail tip rounded without a mucron (second-generation males with mucron). Spicule head with ventral projection, width greater than length (Nguyen and Smart, 1997). Shaft prominent. Lamina well curved, posterior part almost straight. Velum present. Spicule length 65 (53 – 70) μm, spicule length/spicule width = 5.4 (5.3 – 5.5). SW = 2.22 (2.18 – 2.26); GS = 0.71 (0.59 – 0.88). In ventral view, gubernaculum tapering anteriorly to a ventrally curved end. Cuneus arrowhead-shaped (Nguyen and Smart, 1997).

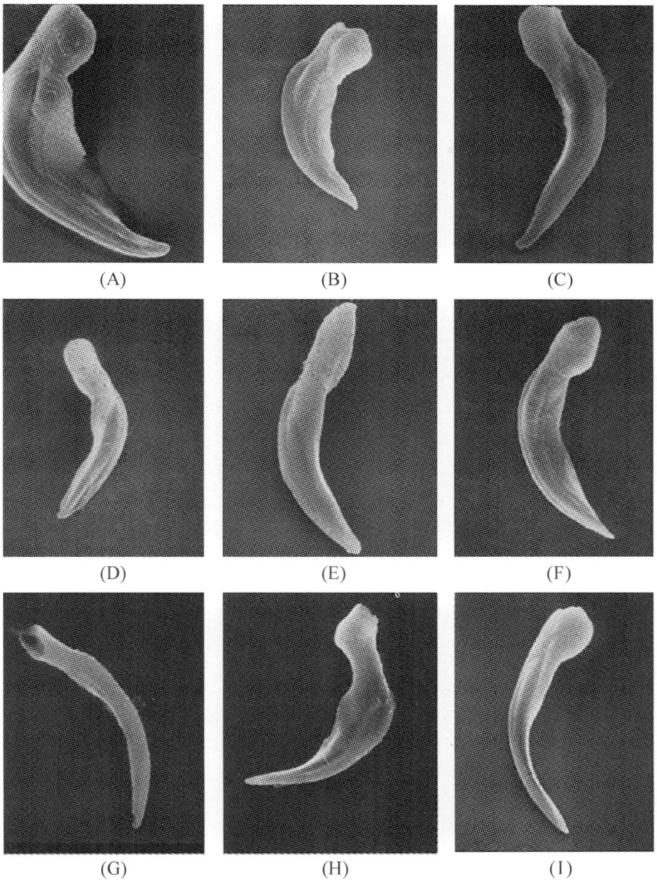

Figure 16.4 SEM of spicules of nine species of *Steinernema*. (A) *S. intermedium*. (B) *S. kushidai*. (C) *S. arenarium*. (D) *S. neocurtillae*. (E) *S. feltiae*. (F) *S. riobrave*. (G) *S. glaseri*. (H) *S. scapterisci*. (I) *S. carpocapsae*.

Infective juveniles: Body length averaging 769 (648 – 873) μm. Lip region with two horn-like papillae. EP = 61 (53 – 65) μm, *E%* about 84 (80 – 100). *T* = 72 (63 – 78) μm. This species was collected in Vojvodina, Yugoslavia.

Steinernema carpocapsae (Weiser, 1955) Wouts, Mracek, Gerdin and Bedding, 1982 (syn. *Neoaplectana carpocapsae* Weiser, 1955; *N. feltiae*

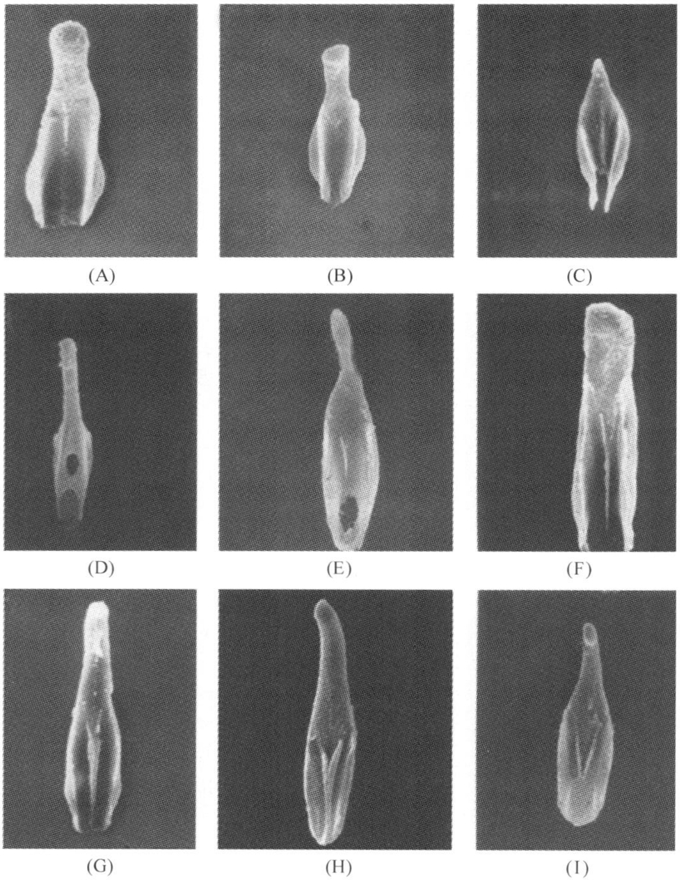

Figure 16.5 SEM of gubernacula of nine species of *Steinernema*. (A) *S. intermedium*. (B) *S. kushidai*. (C) *S. carpocapsae*. (D) *S. neocurtillae*. (E) *S. feltiae*. (F) *S. riobrave*. (G) *S. scapterisci*. (H) *S. glaseri*. (I) *S. arenarium*.

sensu Stanuszek, 1974, nec Filipjev, 1934)

Diagnosis: ***Males***: Anterior end truncate or rounded with six labial and four cephalic papillae. Excretory pore usually anterior to nerve ring. *D%* about 41 (27 – 55). Tail with mucron. Spicule head wider than long, ventrally projected (Fig. 16.4(I)). Shaft present. Lamina moderately curved, tapering to a pointed, or bluntly pointed tip. Velum present. Spicule length 66 (58 – 77) μm, spicule length/spicule width about 5.2 (4.7 – 6.0). SW =

1.72 (1.40 – 2.00); GS = 0.71 (0.59 – 0.88). In ventral view, gubernaculum tapering anteriorly to become short and narrow (Fig. 16.5 (C)). Cuneus arrowhead-shaped or Y-shaped (Nguyen and Smart, 1997).

Infective juvenile: Body length averaging 558 (438 – 650) μm. EP = 38 (30 – 60) μm, *E%* about 60 (54 – 66). *T* = 53 (46 – 61) μm. This species was first described from Czechoslovakia, but now has been found worldwide. There are many recognized strains of the species.

Steinernema caudatum Xu, Wang and Li, 1991

Diagnosis: ***Males***: Anterior end truncate with six labial and four cephalic papillae. Excretory pore anterior to nerve ring. *D%* about 71 (no range available). Tail rounded without mucron (second-generation males with mucron). Spicule head slightly longer than wide, anterior end bluntly rounded. Spicule shaft short or indistinct. Lamina smoothly curved with bluntly rounded tip without a notch. Velum present. Spicule length 75 μm. SW = 2.22, GS = 0.69. Gubernaculum boat-shape in lateral view, anterior end curved. In ventral view, bifurcate posteriorly.

Infective juveniles: Body length averaging 1106 (933 – 1296) μm. EP = 82 (76 – 89) μm, *E%* = 94 (87 – 100). *T* = 88 (80 – 100) μm. This species resembles *S. glaseri* except that the EP is greater, the tail is longer, and the spicule tip does not have a notch. The species was found in China.

Steinernema ceratophorum Jian, Reid and Hunt, 1997

Diagnosis: ***Males***: Anterior end truncate or slightly rounded with six labial and four cephalic papillae. Excretory pore usually anterior to nerve ring. *D%* about 51 (33 – 65). Tail terminus rounded without mucron in both generations. Spicule head length elongate, length/width ratio about 1.5. Spicule shaft present followed by a small rostrum on ventral side. Lamina curved with velum and bluntly pointed tip. Spicule length 71 (54 – 90) μm, spicule length/spicule width about 6.5. SW = 1.4 (1.0 – 2.0), GS = 0.6 (0.4 – 0.8). Gubernaculum boat-shaped in lateral view, and bifurcate in dorsal or ventral view.

Infective juveniles: Body length averaging 706 (591 – 800) μm. Lip region with two horn-like papillae. EP = 55 (47 – 70) μm, *E%* = 84 (74 – 96). *T* = 66 (56 – 74) μm. This species resembles *S. bicornutum* but has a shorter body. The nematode was found in a survey in Jilin and Liaoning provinces, Northeastern China.

Steinernema cubanum Mracek, Hernandez and Boemare, 1994

Diagnosis: ***Males***: Anterior end rounded, head slightly offset with six labial and four cephalic papillae. Excretory pore either anterior or posterior to nerve ring. *D%* about 70. Tail bluntly conoid without mucron. Spicule head elongate, ratio length/width about 1.5 or less (Nguyen and Smart, 1997). Shaft present, sometimes indistinct. Lamina slightly curved, terminus blunt. Velum narrow. Spicule length 58.1 (50 – 67) μm, spicule length/spicule width averaging 7 (6.4 – 8.0). SW = 1.4; GS = 0.67. In lateral view, gubernaculum boat-shaped, tapering to hooked end anteriorly. In ventral view, cuneus Y-shaped (Nguyen and Smart, 1997).

Infective juveniles: Body length averaging 1283 (1149 – 1508) μm. EP = 106 (101 – 114) μm, *E%* = 160. *T* = 67 (61 – 77) μm. This species resembles *S. glaseri* and *S. arenarium* but differs from them by the spicule shape and tail length. The species was collected from a citrus grove in Western Cuba.

Steinernema feltiae (Filipjev, 1934) Wouts, Mracek, Gerdin and Bedding, 1982 (syn.: *Neoaplectana feltiae* Filipjev, 1934; *N. bibionis* Bovien, 1937; *S. bibionis* (Bovien, 1937; Wouts, Mracek, Gerdin and Bedding, 1982)

Diagnosis: ***Males***: Anterior end truncate or rounded with six labial and four cephalic papillae. Excretory pore well anterior to nerve ring. *D%* about 60 (51 – 64). Tail with long mucron. Spicule head elongate, ratio length/width about 1.5 – 2.0 (Fig. 16.4(E)). Spicule shaft present, rostrum absent. Lamina slightly curved without a velum. Spicule length 70 (65 – 77) μm, spicule length/spicule width about 6.0 (5.8 – 6.2). SW = 1.1 (1.0 – 1.3); GS = 0.59 (0.52 – 0.61). Gubernaculum boat-shaped in lateral view. In ventral view, it tapers anteriorly, a narrow neck present. Cuneus short, Y-shaped (Fig. 16.5(E)).

Infective juveniles: Body length averaging 849 (736 – 950) μm. EP = 62 (53 – 67) μm, *E%* = 78 (69 – 86). *T* = 81 (70 – 92) μm. This species has been found in many places in America, Australia, Europe, and New Zealand. *Steinernema glaseri* (Steiner, 1929) Wouts, Mracek, Gerdin and Bedding, 1982 (syn.: *Neoaplectana glaseri* Steiner, 1929)

Diagnosis: ***Males***: Anterior end truncate or rounded, somewhat offset with six labial and four larger cephalic papillae. Excretory pore usually posterior to nerve ring. *D%* about 70 (60 – 78). Tail terminus rounded without mucron. Spicule head longer than wide (Fig. 16.4(G)). Shaft present, prominent. Lamina long, narrow, its tip bearing a ventral aperture which may make it appear hook-like. Velum absent. Rostrum sometimes present, but not prominent. Spicule length 77 (64 – 90) μm, spicule length/spicule width about 8 (Nguyen and Smart, 1995b). SW = 2.05 (1.64 – 2.43); GS = 0.71 (0.64 – 0.85). Gubernaculum boat-shaped in lateral view with anterior end curved ventrally. In ventral view, gubernaculum tapering gradually anteriorly, cuneus Y-shaped with posterior end reaching gubernaculum terminal end (Fig. 16.5(H)).

Infective juveniles: Body length averaging 1130 (864 – 1448) μm. EP = 102 (87 – 110) μm, *E%* = 131 (122 – 138). *T* = 78 (62 – 87) μm. This species was first found in New Jersey, USA, but is widespread in the USA, and has been found in Europe and Asia.

Steinernema intermedium (Poinar, 1985) Mamiya, 1988 (syn.: *Neoaplectana intermedia* Poinar, 1985)

Diagnosis: ***Males***: Anterior end rounded with six labial and four cephalic papillae. Excretory pore usually near nerve ring level. *D%* about 67 (58 – 76). Tail conoid without mucron. Spicule head short, truncate anteriorly (Fig. 16.4(A)). Shaft present but short. Lamina with anterior part well curved, posterior part slightly curved or almost straight, velum large, rostrum present. Spicule tip bluntly rounded. Spicule length 91 (84 – 100) μm. SW = 1.24 (1.03 – 1.39); GS = 0.69 (0.62 – 0.77) μm. In ventral view, gubernaculum tapering anteriorly to a slightly enlarged head. Cuneus short, needle-shaped pointed posteriorly (Nguyen and Smart, 1997).

Infective juveniles: Body length averaging 671 (608 – 800) μm. EP = 65 (59 – 69) μm, *E%* = 96 (89 – 108). *T* = 66 (53 – 74) μm with a depression on dorsal side. In water, infective juveniles are usually immobile and ventrally curved. This species resembles *S. affine* but can be distinguished from it by the absence of a spine-like structure inside the tail tip. The species was found in Charleston, South Carolina, USA.

Steinernema karii Waturu, Hunt and Reid, 1997

Diagnosis: ***Males***: Anterior end truncate or rounded with six labial and four cephalic papillae. Excretory pore anterior to or near nerve ring. *D%* about 66 (53 – 78). Tail conoid with blunt, rounded terminus, mucron absent. Spicule head elongate, about twice as long as wide. Shaft short. Lamina thick, moderately curved with a thin velum and bluntly pointed tip. Spicule length 83 (73 – 91) μm. SW = 1.5; GS = 0.7. Gubernaculum boat-shaped in lateral view. In ventral view, cuneus present and posterior end of gubernaculum bifurcate.

Infective juveniles: Body length averaging 932 (876 – 982) μm. EP = 74 (68 – 80) μm, *E%* = 100. *T* = 74 (64 – 80) μm. This nematode species was isolated in the Central Province of Kenya.

Steinernema kushidai Mamiya, 1988

Diagnosis: ***Males***: Anterior end truncate or slightly rounded with six labial and four cephalic papillae. Excretory pore well anterior to nerve ring. *D%*

about 51 (42 – 59). Tail conical with rounded terminus, without mucron (mucron present on second generation males). Spicule head truncate or rounded anteriorly, almost as long as wide (Fig. 16.4(B)). Shaft very short or indistinct. Lamina wide anteriorly, tapering posteriorly to a blunt tip. Spicule tip usually slightly curved or with a notch dorsally. Velum prominent but short. Spicule length 63 (48 – 72) μm, well curved, spicule length/spicule width about 3.9 (3.6 – 4.3). SW = 1.50, GS = 0.70. In ventral view, gubernaculum tapering anteriorly to form a neck then enlarging slightly to a ventrally curved end. Cuneus present but not obvious (Fig. 16.5(B)).

Infective juveniles: Body length averaging 589 (424 – 662) μm. EP = 46 (42 – 50) μm, *E%* = 92 (84 – 95). *T* = 50 (44 – 59) μm. This species was first isolated from larvae of *Anomala cuprea* (Hope), which were reared in soil collected from Hamakita, Japan.

Steinernema longicaudum Shen and Wang, 1991

Diagnosis: ***Males***: Anterior end truncate with six labial and four cephalic papillae. Excretory pore mostly anterior to nerve ring. *D%* about 62. Tail conoid with rounded terminus, without mucron. Spicule head longer than wide, narrowing posteriorly to form a distinct shaft. Lamina well curved. Near posterior end, lamina reduced in width suddenly to form a tip with flattened terminus (Nguyen and Smart, 1997). Velum present, terminates a distance from spicule tip. Spicule length 77 μm, spicule length/spicule width about 5.1 (4.3 – 5.5). SW = 1.60; GS = 0.62. Gubernaculum mostly boat-shaped, curved ventrally tapering to both ends. In ventral view, cuneus Y-shaped, not reaching posterior end of corpus. Corpus separated posteriorly (Nguyen and Smart, 1997).

Infective juveniles: Body length averaging 1064 μm. EP = 81 μm, *E%* = 85, *T* = 95 μm. This species resembles *S. glaseri* and *S. puertoricense* but can be distinguished from them by its long tail and spicule shape. The nematode was found in a survey for entomopathogenic nematodes in Guangdong and Shandong Provinces, China in 1985.

Steinernema monticolum Stock, Choo and Kaya, 1997

Diagnosis: ***Males***: Anterior end truncate to slightly rounded with six labial and four cephalic papillae. Excretory pore anterior to nerve ring. *D%* about 55 (49 – 61). Tail with mucron. Spicule head as long as wide. Shaft short. Lamina arcuate tapering to a blunt tip. Velum large, making spicule appear very thick. Spicule length 70 (61 – 80) μm, spicule length/spicule width (based on SEM micrographs) about 4.0 (3.8 – 4.2). SW = 1.4 (1.2 – 1.5); GS = 0.6 (0.5 – 0.7). Gubernaculum arcuate, posterior end bifurcate.

Cuneus present but not prominent.

Infective juveniles: Body length averaging 706 (612 – 821) μm. EP = 58 (54 – 62) μm, *E*% = 76 (63 – 86), *T* = 77 (71 – 95) μm. This species was found in a survey for entomopathogenic nematodes in the Republic of Korea.

Steinernema neocurtillae Nguyen and Smart, 1992 (syn.: *S. neocurtillis* Nguyen and Smart, 1992)

Diagnosis: ***Males***: Anterior end truncate or rounded with six labial and four cephalic papillae. Excretory pore close to head region. *D*% about 19 (13 – 26). Tail terminus conoid with mucron (mucron absent in second-generation males). Spicule head elongate, in some, twice as long as wide and about one third of spicule length (Fig. 16.4(D)). Shaft short or absent. Lamina thick, tapering slightly posteriorly, terminus blunt with a depression on ventral side. Velum absent. Spicule length 59 (52 – 64) μm. SW = 1.43; GS = 0.89 (largest in *Steinernema* species). Gubernaculum boat-shaped in lateral view, about three-fourths spicule length.

Infective juveniles: Body length averaging 885 (741 – 988) μm. EP extremely short, 18 (14 – 22) μm, *E*% = 23 (18 – 30). *T* = 80 (64 – 97) μm.

The body length is close to that of *S. feltiae* but it can be distinguished from that nematode by the extremely short distance from anterior end to excretory pore, very low value of *D*% in male, *E*% in infective juvenile, and large GS (= 0.89 compared to 0.79 or less in other species). The nematode was isolated from the northern mole cricket, *Neocurtilla hexadactilla* (Perty), collected in LaCrosse, Alachua County, Florida, USA.

Steinernema oregonense Liu and Berry, 1996

Diagnosis: ***Males***: Anterior end rounded or truncate with six labial papillae and four cephalic papillae. Excretory pore anterior to nerve ring. *D*% about 73 (64 – 75). Tail terminus rounded without mucron (second-generation males with mucron). Spicule head elongate, length/width about 1.5 – 2.0. Shaft present. Lamina slightly curved with blunt tip. Velum very thin. Spicule length 71 (65-73) μm, spicule length/spicule width about 6.1. SW = 1.5; GS = 0.79. Gubernaculum boat-shaped in lateral view. In ventral view, corpus bifurcate posteriorly. Cuneus needle-shaped pointed posteriorly.

Infective juveniles: Body length averaging 980 (820 – 1110) μm. EP = 66 (60 – 70) μm, *E*% = 100 (90 – 110). *T* = 70 (64 – 78) μm. This nematode species resembles *S. feltiae* except the male tail of *S. oregonense* does not have a mucron and the tail of its infective juvenile is shorter. The nematode was found in a survey for entomopathogenic nematodes in Oregon, USA.

Steinernema puertoricense Roman and Figueroa, 1994

Diagnosis: ***Males***: Anterior end truncate to slightly rounded, continuous with body. Excretory pore either anterior or posterior to nerve ring. *D%* about 77. Tail conoid with rounded terminus, mucron absent. Spicule head usually swollen, longer than wide (Nguyen and Smart, 1997). Shaft present. Lamina arcuate with spicule tip curved ventrally. Velum thin, sometimes not observed. Spicule length 78 (71 – 88) μm, spicule length/spicule width about 7.3 (6.6 – 8.6). SW = 1.52; GS = 0.51. In ventral view, gubernaculum tapering anteriorly to a ventrally curved end. Cuneus somewhat Y-shaped reaching end of corpus (Nguyen and Smart, 1997).

Infective juveniles: Body length averaging 1171 (1057 – 1238) μm. EP = 95 (90 – 102) μm, *E%* = 101 (88 – 108). *T* = 94 (88 – 107) μm. The nematode was found in a survey for entomopathogenic nematodes in Puerto Rico.

Steinernema rarum (Doucet, 1986) Mamiya, 1988 (syn.: *Neoaplectana rara* Doucet, 1986)

Diagnosis: ***Males***: Anterior end truncate to slightly rounded with six labial and four cephalic papillae. Excretory pore anterior to nerve ring. *D%* about 50 (44 – 51). Tail conoid with a prominent mucron in both generations. Spicule head short, rounded anteriorly. Shaft present. Lamina wide anteriorly with a short, sometimes large velum, tapering posteriorly to a blunt terminus. Posterior part of spicule occasionally slightly curved dorsally. Spicule length 47 (42 – 52) μm, spicule length/spicule width (from SEM photographs) about 4 (3.5 – 4.5). SW = 0.94 (0.91 – 1.05); GS = 0.71 (0.55 – 0.73). Gubernaculum short, thick, posterior end bifurcate. Cuneus present, short.

Infective juveniles: Body length averaging 511 (443 – 573) μm. EP = 38 (32 – 40) μm, *E%* = 72 (63 – 80). *T* = 51 (44 – 56) μm. This species was found in a survey for entomopathogenic nematodes in Cordoba, Argentina.

Steinernema riobrave Cabanillas, Poinar and Raulston, 1994 (syn.: *S. riobravis* Cabanillas, Poinar and Raulston, 1994)

Diagnosis: ***Males***: Anterior end truncate to rounded with six labial and four cephalic papillae. Excretory pore anterior or posterior to nerve ring. *D%* about 71 (60 – 80). Tail with rounded tip without mucron. Spicule head mostly tapering anteriorly, somewhat longer than wide (Nguyen and Smart, 1997). Shaft distinct but short. Lamina sickle-shaped, moderately curved, tapering gradually to a bluntly pointed tip. Velum present. Spicule length 67 (62.5 – 75) μm, spicule length/spicule width about 5.5 (5.1 – 5.8). SW = 1.14; GS = 0.76. In ventral view, gubernaculum neck has almost same width as posterior corpus (Fig. 16.5(F)), a characteristic typical for this species. Gubernaculum head curved ventrally. Cuneus Y-shaped or needle-shaped, long, pointed posteriorly.

Infective juveniles: Body length averaging 622 (561 – 701) μm. EP = 56 (51 – 64) μm, *E%* = 105 (93 – 111). *T* = 54 (46 – 59) μm. This species was isolated from prepupae and pupae of the corn earworm *Helicoverpa zea* (Boddie) and the fall armyworm, *Spodoptera frugiperda* (J. E. Smith) in the Lower Rio Grande Valley near Weslaco, Texas, USA.

Steinernema ritteri Doucet and Doucet, 1990

Diagnosis: ***Males***: Anterior end truncate or rounded with six labial and four cephalic papillae. Excretory pore anterior to nerve ring. *D%* about 47 (44 – 50). Tail conoid with bluntly rounded terminus without mucron (second generation male with mucron). Spicule head somewhat longer than wide, with rounded anterior end. Shaft present. Lamina with rostrum and thin velum. Spicule tip pointed. Spicule length 69 (58 – 75) μm. SW about 1.56; GS = 0.64 (0.57 – 0.67). Gubernaculum rounded anteriorly in lateral view.

Infective juveniles: Body length averaging 510 (470 – 590) μm. EP = 43 (40 – 46) μm, *E%* = 88 (79 – 97). *T* = 49 (44 – 54) μm. This species was found in a survey of entomopathogenic nematodes in Cordoba, Argentina.

Steinernema scapterisci Nguyen and Smart, 1990 (syn.: *Neoaplectana carpocapsae* "Uruguay strain" of Nguyen and Smart, 1988)

Diagnosis: ***Males***: Anterior end truncate slightly offset with 6 labial papillae and 4 larger cephalic papillae. Excretory pore well anterior to nerve ring. *D%* about 38 (32 – 44). Tail conoid, with mucron (second generation male with mucron also). Spicule head large somewhat angular, anterior end directed dorsally (Fig. 16.4(H)). Shaft prominent. Lamina with a small velum, tapering gradually to pointed tip. Spicule length 83 (72 – 92) μm. SW = 2.52 (2.04 – 2.80); GS = 0.78 (0.69 – 0.84). Gubernaculum boat-shaped in lateral view, in ventral view, it enlarges posteriorly gradually to form the corpus. Cuneus needle-shaped or Y-shaped, pointed posteriorly (Fig. 16.5 (G)).

Females: Excretory duct prominent, forming an elliptically shaped structure, which is so large that it pushes the base of esophagus to one side. Epiptygma well developed.

Infective juveniles: Body length averaging 572 (517 – 609) μm. EP = 39 (36 – 48) μm, *E%* = 73 (60 – 80). *T* = 54 (48 – 60) μm. This nematode species was isolated from the mole cricket *Scapteriscus vicinus* (Scudder) near Rivera, Uruguay in 1985.

Steinernema siamkayai Stock, Somsook and Reid, 1998

Diagnosis: ***Males***: Anterior end truncate to rounded with six labial and four cephalic papillae. Excretory pore anterior to nerve ring. *D%* about 42 (35 – 49). Tail conoid with mucron (second generation male with longer mucron). Spicule head somewhat longer than wide, rounded anteriorly. Shaft present. Lamina moderately curved with rostrum and small velum. Spicule tip blunt. Spicule length 77.5 (75 – 80) μm. SW = 1.7 (1.4 – 2.2); GS = 0.7 (0.6 – 0.8). Gubernaculum slender, bifurcate in ventral or dorsal view.

Females: Epiptygma present in both first and second generations.

Infective juveniles: Body length averaging 446 (398 – 495) μm. EP = 35 (29 – 38) μm, *E%* = 96 (85 – 112). *T* = 35.5 (31 – 41) μm. This species was found in Petchabun Province, Thailand.

Steinernema tami Luc, Nguyen, Reid and Spiridonov, 2000

Diagnosis: ***Males***: Anterior end rounded with six labial and four prominent cephalic papillae. Excretory pore slightly anterior to mid-esophagus, *D%* = 44 (30 – 60). Tail conoid with a mucron. Spicule head 1 – 1.5 times as long as wide. Spicule shaft prominent. Lamina moderately curved. Velum thin. Spicule length 77 (71 – 84) μm. Gubernaculum boat-shaped laterally; in ventral view, tapering gradually anteriorly, cuneus Y-shaped. SW = 2.0 (1.4 – 3.0).

Infective juveniles: Body length averaging 530 (400 – 600) μm. Distance from anterior end to excretory pore 36 (34 – 41) μm, *E%* = 73. Tail length 50 (42 – 57) μm.

This species was found in Cat Tien National Park, Viet Nam.

Key to species of the genus *Steinernema*

1-Infective juvenile (IJ) with a double horn-like structure on labial region 2
IJ without a double horn-like structure on labial region 3
2-Body length of IJ 769 (648 – 873) μm, spicule length 65 (53 – 70) μm, SW = 2.22 (2.18 – 2.26) *S. bicornutum*
Body length of IJ 706 (691 – 800) μm, spicule length 71 (54 – 90) μm SW = 1.4 (1.0 – 2.0) *S. ceratophorum*
3-Body length IJ >800 μm 4
Body length of IJ <800 μm 13
4-Average length of IJ greater than 1000 μm 5
Average length of IJ less than 1000 μm 9
5-Tail of IJ 67 – 88 μm 6
Tail of IJ >90 μm 8
6-*D%* about 93 (88 – 102), spicule length 84 (81 – 91) μm, spicule tip

ball-shaped *S. arenarium*
D% about 70, spicule length about 77 μm, spicule tip not ball-shaped 7

7- EP about 82 (76 – 89) μm, *E%* about 94 (87 – 100) *S. caudatum*
EP about 102 (87 – 110) μm, *E%* about 131 (122 – 138) *S. glaseri*
EP about 106 (101 – 114) μm, *E%* about 160 *S. cubanum*

8- *E%* averaging 101 (88 – 108); in male *D%* about 77 *S. puertoricense*
E% averaging 85 (range not known); in male *D%* about 62 *S. longicaudum*

9- In IJ, EP extremely short, 18 (14 – 22) μm, *E%* = 23; in male, *D%* averaging 19 (13 – 25) *S. neocurtillae*
In IJ, EP = 53 μm or more, *E%* about 69 – 110; in male, *D%* > 50 10

10- Male tail with mucron 11
Male tail without mucron 12

11- Spicule head as long as wide, spicule length 49 (42 – 53) μm, body length averaging 951 (797 – 1102) μm *S. kraussei*
Spicule head 1.5 – 2.0 times as long as wide, spicule length 70 (65 – 77) μm, body length averaging 849 (736 – 950) μm *S. feltiae* (= *bibionis*)

12- Body length averaging 980 (820 – 1110) μm; spicule length 71 (65 – 73) μm *S. oregonense*
Body length averaging 932 (876 – 982) μm; spicule length 83 (73) μm *S. karrii*

13- Average length of IJ > 600 (622 – 693) μm 14
Average length of IJ < 600 (510 – 589) μm 17

14- IJ tail length 77 (71 – 95), *E%* = 76 (63 – 86), *c* = 9.3 (7.6 – 11.1); in male *D%* = 55 (49 – 61) *S. monticolum*
IJ tail length averaging 62 (maximum 74), *E%* > 94, *c* > 10; in male *D%* > 60 15

15- Spine-like structure inside IJ tail tip; mucron present, small *S. affine*
No spine-like structure inside IJ tail tip; mucron absent 16

16- Spicule length about 93 (80 – 106) μm; *E%* about 96 (89 – 108) *S. intermedium*
Spicule length about 67 (63 – 75) μm; *E%* about 105 (93 – 111) *S. riobrave*

17- Average body length of IJ < 540 μm 18
Average body length of IJ > 540 μm 21

18- Average body length 446 (398 – 495) μm, tail length 36 (31 – 41) μm *S. siamkayai*
Average body length about 510 μm or more, tail length 50 μm 19

19- Average body length 530 μm, spicule length 77 μm, SW = 2.0 *S. tami*

Average body length about 510 μm, spicule length shorter, SW smaller 20

20-First generation male without mucron; spicule length 69 (58 – 75) μm, SW = 1.56 (1.44 – 1.57); in IJ, *E%* averaging 88 *S. ritteri*

First generation male with mucron; spicule length 47 (42 – 52) μm, SW = 0.94 (0.91 – 1.05); in IJ, *E%* averaging 72 *S. rarum*

21-Male tail without mucron 22

Male tail with mucron 23

22-Body length of IJ averaging 589 (424 – 662) μm, spicule length/ spicule width about 3.9 (3.6 – 4.3) *S. kushidai*

Body length of IJ averaging 541 (496 – 579) μm, spicule length/spicule width about 5.1 (4.7 – 5.8) *S. abbasi*

23-Spicule length 83 (72 – 92) μm, spicule shaft prominent, SW ratio averaging 2.52 (2.04 – 2.80), female with large, double-flapped epiptygma *S. scapterisci*

Spicule length 66 (58 – 77) μm, spicule shaft very short or indistinct, SW ratio averaging 1.72 (1.40 – 2.00), double-flapped epiptygma rarely present *S. carpocapsae*

Notes: To identify species of *Steinernema* and *Heterorhabditis* the followings should be considered: a) IJ morphometrics usually are insufficient for species identification; male and female characteristics must be considered. b) IJs produced on artificial media (laboratory reared or commercial products) are shorter (rarely longer) than those produced *in vivo*, and usually do not meet the criteria of the original description. Males and females collected 4 or 5 days after the host dies, and IJs collected for one week after they first appear from cadavers, usually meet original species descriptions (Nguyen and Smart, 1995a). c) One should know that keys should be used only to narrow decisions. After using the above key, and all keys in this chapter to identify a nematode, the identity should be verified by comparing its morphology and morphometrics with the data in the original description.

Neosteinernema Nguyen and Smart, 1994 (Fig. 16.6)

Diagnosis: ***Females***: with phasmids prominent, on a protuberance, located in posterior half of tail; tail longer than anal body width (*T*/ABW = 1.10 – 1.68); ovoviviparous, juveniles molting and becoming infective juveniles before exiting the female body. ***Males***: smaller than female, posterior part with one ventral and 13 – 14 pairs of genital papillae, eight of the pairs preanal; phasmids prominent (Fig. 16.6(D)), tail tip digitate; spicule foot-shaped (Fig. 16.6(C)) with a hump on dorsal side. Gubernaculum almost as long as spicule. ***Infective juveniles***: with slightly swollen head (Fig. 16.6(E)); phasmid large, tail elongate or filiform, as long as esophagus, usually curved at end (Fig. 16.6(F)), ratio *c* about 5.5.

Type and only species: *Neosteinernema longicurvicauda* Nguyen and Smart, 1994.

Diagnosis: As for the genus.

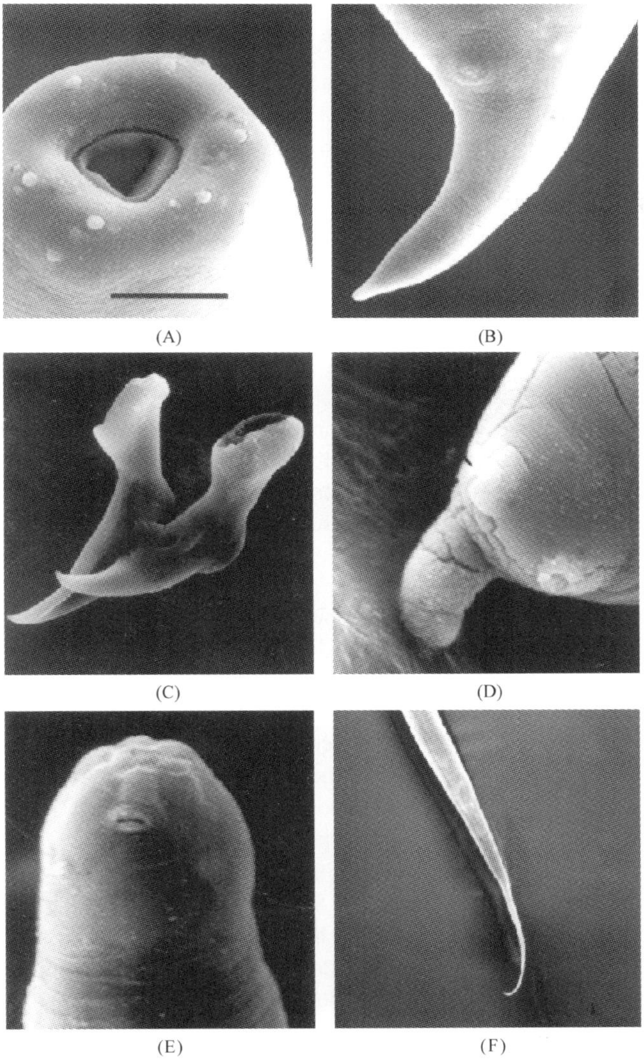

Figure 16.6 SEM of *Neosteinernema longicurvicauda*. (A) Face view showing oral opening with a circular raised structure. (B) Female posterior region showing curved and conical tail and phasmid. (C) Foot-shaped spicules with a hump on dorsal side. (D) Male tail showing phasmid and digitate terminus. (E) Infective juvenile with swollen head. (F) Infective juvenile with long and curved tail. (Modified from Nguyen and Smart, 1996.)

16.2.1.2 Family Heterorhabditidae Poinar, 1976

Diagnosis (emended based on additional information revealed by SEM): Rhabditoidea, Rhabditida. Obligate insect parasites. Infective juveniles carrying symbiotic bacteria. Both hermaphroditic and amphimictic females present.

Hermaphroditic females: After entry into an insect host, infective juveniles develop into hermaphroditic females. Head truncate to slightly rounded, six conical lips well developed, separate, each with a terminal papilla (Fig. 16.7 (A)); one or two small raised structures sometimes visible at the base of each lip; amphidial opening small. Stoma wide but shallow; cheilorhabdions present, forming a ring, which, in lateral view resembles two refractile elongate structures. Other parts of the stoma fused to form a collapsed posterior portion. Posterior part of stoma surrounded by esophagus. Esophagus without metacorpus; isthmus slender; basal bulb swollen; valve in basal bulb reduced. Nerve ring at middle of isthmus. Excretory pore usually posterior to end of esophagus. Vulva median, slit-like, surrounded by elliptical rings; ovotestis amphidelphic, reflexed. Oviparous, later becoming ovoviviparous. Tail pointed, longer than anal body width, postanal swelling usually present.

Amphimictic females: Similar to, but usually smaller than, hermaphroditic female; labial papillae prominent (Fig. 16.7(C)). Reproductive system amphidelphic. Vulva not functional for egg deposition (eggs hatch in female body), but functional for mating.

Males: Testis one, reflexed. Spicules paired, separate, slightly curved ventrally (Fig. 16.7(E)). Spicule head short, offset from lamina by a constriction. Gubernaculum usually about half as long as spicule length. Bursa peloderan with nine pairs of genital papillae.

Infective juveniles: Third-stage infective juvenile usually with sheath (cuticle of second-stage juvenile). Sheath with anterior tessellate pattern and longitudinal ridges throughout the body (Fig. 16.7(G), (H)); cuticle of infective juveniles striated, lateral fields with one smooth band marginated by two ridges. Head with a prominent dorsal tooth anteriorly. Mouth and anus closed. Stoma appearing as a closed chamber with parallel walls. Esophagus and intestine reduced. Excretory pore posterior to nerve ring. Symbiotic bacterial cells found in intestine. Tail pointed.

This family was erected by Poinar in 1976 when he described the genus and species *Heterorhabditis bacteriophora*. The family contains only one genus, *Heterorhabditis*. The genus has 8 species. For new updates in taxonomic research, readers may check website: http://kbn.ifas.ufl.edu/kbnstein.htm
Type and only genus: *Heterorhabditis* Poinar, 1976 (Fig. 16.7) (syn.:

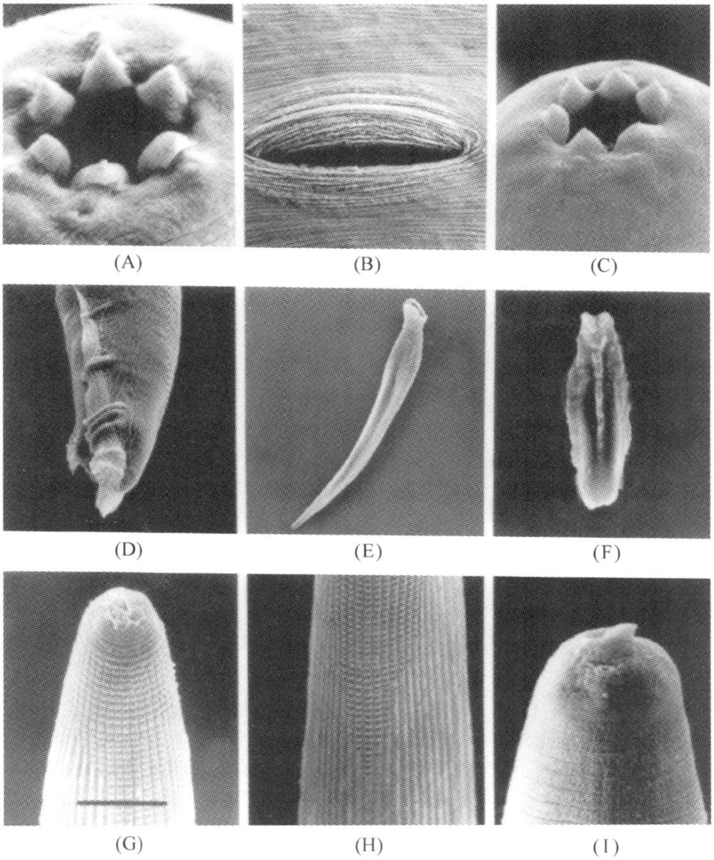

Figure 16.7 SEM of *Heterorhabditis* spp. (A) Face view of hermaphroditic female showing six lips and labial papillae. (B) Vulva. (C) Face view of amphimictic female. (D) Male tail with bursa and ribs (= genital papillae). (E) Spicule. (F) Ventral view of gubernaculum. (G) Anterior region of a third-stage infective juvenile (IJ) in cuticle of second-stage juvenile, showing anterior tessellate pattern. (H) Body of IJ with longitudinal ridges and termination of tessellate pattern. (I) Anterior end of IJ with prominent tooth and amphid. (Modified from Nguyen and Smart, 1996.)

Chromonema Khan, Brooks and Hirschmann, 1976)

Diagnosis: As for family.

Heterorhabditis bacteriophora Poinar, 1976 (syn.: *Chromonema heliothidis* Khan, Brooks and Hirschmann, 1976); *H. heliothidis* (Khan, Brooks and Hirschmann, 1976; Poinar, Thomas and Hess, 1977). Type species

Diagnoses: **Important note**: Ratios and abbreviations used in the diagnoses and keys are: *D%* = EP/ES × 100 (EP = distance from anterior end to excretory pore; ES = esophagus length); *E%* = EP/*T* 100 (*T* = tail length); IJ = infective juvenile; SW = spicule length divided by anal body width; GS = gubernaculum length/spicule length.

Males: Male body width 43 (38 – 46) µm. Distance from anterior end to excretory pore 121 (114 – 130) µm. Distance from anterior end to base of esophagus 103 (99 – 105) µm. *D%* = 117 (115 – 124). Reflexion of testis 79 (59 – 87) µm. Spicule length 40 (36 – 44) µm, rarely with ventral expansion. Gubernaculum length 20 (18 – 25) µm. SW = 1.74 (1.6 – 1.8); GS = 0.50.

Infective juveniles: Body length averaging 588 (512 – 671) µm. Body width 23 (18 – 31) µm. EP = 104 (94 – 109) µm, *E%* = 112 (103 – 130). ES = 125 (100 – 139) µm. *T* = 94 (93 – 99) µm. *c* = 6.2 (5.5 – 7.0). This species was first isolated from a pupa of *Heliothis punctigera* (Hall) in Brecon, South Australia.

Heterorhabditis argentinensis Stock, 1993 (may be synonym of *H. bacteriophora*)

Diagnosis: ***Males***: Body width 56 (42 – 70) µm. EP = 157 (145 – 170) µm. ES = 113 (103 – 120) µm. *D%* = 139 (128 – 141). Reflexion of testis 133 (100 – 194) µm. Spicule length 46 (42 – 49) µm, with ventral expansion. Gubernaculum length 23 (20 – 26) µm. SW = 1.28 (1.25 – 1.37); GS = 0.50.

Infective juveniles: Body length averaging 657 (610 – 710) µm. Body width 31 (24 – 38) µm. Distance from anterior end to excretory pore 107 (68 – 122) µm, *E%* = 127. ES = 132 (101 – 150) µm. *T* = 84 (70 – 105) µm; c = 7.8. This species was first isolated from larvae of the white fringed weevil *Graphognathus* sp. in Rafaela, Argentina.

Heterorhabditis brevicaudis Liu, 1994

Diagnosis: ***Males***: Male body width = 43 (40 – 48) µm. EP = 95 (92 – 100) µm. ES = 108 (104 – 112) µm. *D%* = 88. Reflexion of testis 194 (162 – 240) µm. Spicule length = 47 µm (44 – 48). Gubernaculum length = 22 (18 – 26) µm. SW = 2.38; GS = 0.47.

Infective juveniles: Body length averaging 572 (528 – 632) µm. Body width = 22 μm (20 – 24). EP = 111 (104 – 116) μm, *E%* = 146. ES = 76 (68 – 80) μm. *T* = 76 (68 – 80) μm; *c* = 7.6 (6.6 – 8.6). This species was first collected from Fujian Province, China.

Heterorhabditis hawaiiensis Gardner, Stock and Kaya, 1994 (may be synonym of *H. indica*)

Diagnosis: ***Males***: Body width 63 (49 – 84) μm. EP = 130 (71 – 146) μm. ES = 118 (100 – 149) μm. *D%* = 110. Reflexion of testis 168 (115 – 198) μm. Spicule length = 47 (40 – 51) μm with ventral expansion. Ratios SW = 1.75 (1.65 – 1.85), GS = 0.46 (0.44 – 0.53).

Infective juveniles: Body length = 575 (506 – 631) μm, body width = 25(24 – 28) μm. EP = 114 (95 – 132) μm. ES = 133 (115 – 181) μm, *E%* = 127. *T* = 90 (82 – 108) μm. *c* = 6.4. This species was found in a survey for entomopathogenic nematodes in Hawaii, USA.

Heterorhabditis indica Poinar, Karunaka and David, 1992

Diagnosis: ***Males***: Body width = 42 (35 – 46) μm. EP = 123 (109 – 138) μm. ES = 101 (93 – 109) μm. *D%* = 122. Reflexion of testis 91 (35 – 144) μm. Spicule length = 43 (35 – 48) μm. Gubernaculum = 21 (18 – 23) μm. SW = 1.90 (1.80 – 2.00); GS = 0.49.

Infective juveniles: Body length 528 (479 – 573) μm. Body width 20 (19 – 22) μm. EP = 98 (88 – 107) μm. ES = 117 (109 – 123) μm, *E%* = 94 (83 – 103). *T* = 101 (93 – 109) μm. c = 5.3 (4.5 – 5.6). This species was first isolated from larvae of the sugarcane top borer, *Scirpophaga excerptalis* (Walker) in Coimbatore, India.

Heterorhabditis marelatus Liu and Berry, 1996 (syn.: *H. hepialius* Stock, Strong and Gardner, 1996)

Diagnosis: ***Males***: Body width 51 (48 – 56) μm. EP = 130 (110 – 168) μm. ES = 115 (99 – 123) μm. *D%* = 113. Reflexion of testis 89 (51 – 138) μm. Bursal rays with enlarged tips. Spicules with dorsal expansion. Spicule length = 45 (42 – 50) μm, gubernaculum length = 19 (18 – 22) μm. SW = 1.96; GS = 0.41.

Infective juveniles: Body length averaging 654 (588 – 700) μm. Body width = 28 (24 – 32) μm. EP = 102 (81 – 113) μm. ES = 133 (121 – 139) μm, *E%* = 96. *T* = 107 (99 – 117) μm. *c* = 6.1 (5.5 – 6.6). This species was first found in Seaside, Oregon, USA.

Heterorhabditis megidis Poinar, Jackson and Klein 1987

Diagnosis: ***Males***: Body width = 47 (44 – 50) μm. EP = 156 (139 – 176) μm.

ES = 128 (122 – 134). *D*% = 122. Reflexion of testis 138 (117 – 230) μm. Spicule length 49 (46 – 54) μm; ventral expansion absent. Gubernaculum length 21 (17 – 24) μm. SW = 1.88; GS = 0.43.

Infective juveniles: Membranous ring around oral aperture prominent. Body length = 768 (736 – 800) μm. Body width = 29 (27 – 32) μm. EP = 131 (123 – 142) μm. ES = 155 (147 – 160) μm; *E*% = 110 (103 – 120). *T* = 119 (112 – 128) μm; *c* = 6.5. This species was first isolated from larvae of the Japanese beetle (*Popillia japonica* Newm.) in Jeromesville, Ohio, USA.

Heterorhabditis zealandica Poinar, 1990

Diagnosis: ***Males***: Body width = 41 (36 – 45) μm. EP = 139 (130 – 150) μm. ES = 118 (110 – 128) μm. *D*% = 118. Testis reflexion 115 (97 – 136) μm. Spicule length = 51 (48 – 55) μm, ventral expansion present. Gubernaculum length = 22 (19 – 25) μm. SW = 2.46; GS = 0.43.

Infective juveniles: Body length = 685 (570 – 740) μm. Body width = 27 (22 – 30) μm. EP = 112 (94 – 123) μm. ES = 140 (135 – 147) μm. *E*% = 108 (103 – 109). *T* = 102 (87 – 119) μm. *c* = 6.6 (6.2 – 6.7). This species was first found in Auckland, New Zealand.

Key to species of the genus *Heterorhabditis*

1-Average body length of IJ > 700 (736 – 800) μm — *H. megidis*
 Average body length of IJ < 700 (528 – 685) μm — 2
2-IJ tail short, averaging 76 μm (80 μm or less), *E*% about 147; in male, *D*% about 88 — *H. brevicaudis*
 IJ tail longer, averaging more than 80 (84 – 119) μm, *E*% = 127 or less; in male *D*% > 100 — 3
3-IJ body length averaging > 600 μm — 4
 IJ body length averaging < 600 μm — 6
4-In IJ, tail length about 84 (70 – 105) μm, *E*% > 120, *c* > 7 — *H. argentinensis*
 In IJ, tail length about 100, *E*% < 120, *c* < 7 — 5
5-*E*% about 96, *c* about 6.1; male body width averaging 51 μm, spicule length averaging 45 μm, rostrum mostly absent — *H. marelatus*
 E% about 108, *c* about 6.6; male body width averaging 41 μm, spicule length averaging 51 μm, rostrum prominent — *H. zealandica*
6-IJ body length averaging 528 (479 – 573) μm, *E*% about 94 — *H. indica*
 IJ body length averaging 570 μm, *E*% > 100 — 7
7-*E*% of IJ about 127; spicule averaging 47 μm, lamina with ventral expansion — *H. hawaiiensis*

E% of IJ about 112, spicule averaging 40 μm, lamina without ventral expansion *H. bacteriophora*

16.2.1.3 Family Rhabditidae Oerley, 1880

Diagnosis: Rhabditoidea, Rhabditida. Head usually with six lips. Stoma tubular, its length at least three times as long as wide. Esophagus with or without metacorpus, terminated by a valvated bulb. Gonads one or two, mostly two, reflexed. Spicules separate or fused distally. Bursa mostly well developed, peloderan or leptoderan, occasionally small or rudimentary. There are nine to ten pairs of genital papillae in posterior part of males. Insect-parasitic nematodes in this group were reported by Poinar, 1972.

Rhabditis Dujardin, 1845

Diagnosis: Rhabditinae. As in family. Stoma long with glottoid apparatus. Esophageal collar present. Gonads two. Spicules separate. Bursa leptoderan or sometimes pseudopeloderan mostly with nine pairs of genital papillae.

Rhabditis insectivora Körner *in* Osche, 1952: This is a large nematode, 2.0 – 3.0 mm long, rectum three to four times as long as anal body width. Körner (1954) found 60% of lucanid beetles, *Dorcus paralelopipedus* (L), larvae parasitized by this nematode.

Rhabditis sp. found in silkworms, *Bombyx mori* (L), in Japan (Misuta and Sato, 1965), and the palm weevil *Rhynchophorus palmarum* (L) (Griffith, 1968).

Parasitorhabditis (Fuchs, 1936) Chitwood and Chitwood, 1950

Diagnosis: Protorhabditinae. As in family. Stoma long without glottoid apparatus, usually somewhat divergent posteriorly. Esophageal collar absent. Gonad one, without postvulval sac. Vulva close to anus. Spicule slender, distally fused. Bursa peloderan with ten pairs of papillae. Tail of both sexes short.

Parasitorhabditis spp. associated with different stages of wood boring insects, especially members of the family Scolytidae (Lazarevskaya, 1962; Rühm and Chararas, 1957).

Phasmarhabditis Andrassy, 1976

Diagnosis: Peloderinae. As in family. Large nematode (0.9 – 3.3 mm). Stoma 1.2 – 2 times as long as head width. Cheilorhabdions cuticularized; metarhabdions isoglottoid, each swelling bearing three minute denticles. Esophageal collar present. Gonads paired, vulva at mid-body. Spicules separate. Bursa peloderan, open with nine pairs of genital papillae. Tail of female conoid, spicate, cupola-shaped or sharply pointed. Male tail conoid.

Phasmids prominent, protruding.

Phasmarhabditis papillosa (Schneider, 1866) Andrassy, 1976, and *P. neopapillosa* (Mengert *in* Osche, 1952) Andrassy, 1983 are parasites of snails.

16.2.1.4 Family Diploscapteridae Micoletzky, 1922

Diagnosis: Rhabditoidea. Dorsal and ventral lips transformed into fossors. Stoma rhabditiform without glottoid apparatus and without denticles or warts. Corpus of esophagus cylindrical or slightly swollen. Gonads two, reflexed in females. Spicules separate. Bursa peloderan, narrow with four to nine pairs of papillae. Female tail elongate-conoid. Male tail short.

Diploscapter Cobb, 1913

Diagnosis: As for family.

Diploscapter lycostoma Volk, 1950. A small nematode, 460 – 580 μm long. Corpus cylindrical, longer than isthmus + basal bulb. The nematode was reported to enter the mouth of ants, go into the pharyngeal glands where they remained and fed.

16.2.2 Order Diplogasterida Maggenti, 1982

Several nematodes in this order have been reported as associates or parasites of different insects. These nematodes were reported to kill different kinds of insects in the nature and in the laboratory (Merrill and Ford, 1916; Poinar, 1969a).

Diagnosis: Nematoda. Stoma slender and elongate or spacious, usually with teeth, denticles or bristles. Corpus always muscular with distinctly valved median bulb. Posterior part of esophagus with isthmus and glandular basal bulb. Female reproductive system with one or two ovaries. Males with nine pairs (7 – 12) of genital papillae, three preanal and six caudal, and with or without caudal alae. Spicules separate. Gubernaculum always present.

16.2.2.1 Family Diplogasteridae Mikoletzky, 1922

Diagnosis: As for order. Stoma wide, occasionally oblong, with tooth, teeth, denticles or bristles. Procorpus and metacorpus distinct.

Diplogaster Schultze in Carus, 1857

Diagnosis: Diplogasterinae. Stoma wide. Lip ring with fine grooves. Cheilorhabdion longer than promesorhabdion. Dorsal metarhabdion with a large pyramidal tooth. Subventral rhabdions with a small tooth each. Gonads paired. Tail tapering gradually to thread-like distal section. Bursa absent.

The nematodes in this genus were found associated with the fringed beetle

Pantomorus peregrinus (Buch.) and killed the host (Swain, 1945). Another association was found with white grubs of the genus *Phyllophaga* (Chamberlin, 1944). The dauer stage of the nematode was found in the cephalic region of the grub until it died then the nematode reproduced in the cadaver.

Micoletzkya (Weingartner, 1955) Paesler, 1962

Diagnosis: Neodiplogasterinae. Anterior end wide, flattened. Cheilorhabdion divided into six ribs. Stoma longer than wide. Two large teeth, one dorsal and one right subventral, and one small tooth left subventral. Telostom long. Gonads paired. Tail mostly subulate short or long.

Mikoletzkya aerivora (Cobb *in* Merrill and Ford, 1916) Baker, 1962. This nematode was found in the head of a termite. Heavy infestations killed termites (Merrill and Ford, 1916).

Mikoletzkya labiata (Cobb *in* Merrill and Ford, 1916) Andrassy, 1984. This nematode was found in the gut of cerambicids, where it ruptured the host's gut wall, entered the hemocoel, and killed the insect (Merrill and Ford, 1916).

Pristionchus Kreis, 1932 (syn.: *Mesodiplogaster* Goodey, 1963)

Diagnosis: Neodiplogasterinae. Stomatal dimorphism. In stenostoma form, stoma narrow, lip ring not divided. Metarhabdion variable. In eurystoma form, stoma twice as wide as in stenostoma form and more strongly cuticularized. Lip ring with 12 indentations. Metarhabdion with claw-like tooth and pyramidiform tooth, saw-like structure usually present.

Pristionchus lheritieri (Maupas, 1919) Paramonov, 1952. This nematode was associated with the potato beetle, *Leptinotarsa decemlineata* (Say) in Poland (Sandner and Stanuszek, 1967), and was thought to be a good parasite of the beetle keeping populations in check.

16.2.3 Order Tylenchida – Suborder Tylenchina Chitwood in Chitwood and Chitwood, 1950

Diagnosis: Tylenchida. Cuticle annulated. Amphid on labial region. Stoma with a protrusible, hollow stylet bearing three knobs posteriorly. Esophagus with procorpus, metacorpus, isthmus and glandular portion. Dorsal esophageal gland opening between stylet knobs and median bulb. Female gonads single or double; when single, postvulval sac not very well developed. Males mostly with bursa. Nematodes in this order live in soil or in different parts of plants. Most of them are parasites or associates of plants, but some are parasites or associates of insects.

The following families were reported to contain insect parasitic nematodes:

(in the order of more important first): Allantonematidae, Sphaerulariidae, Tylenchidae, Iotonchidae, Parasitylenchidae, and Fergusobiidae.

16.2.3.1 Family Allantonematidae (Pereira, 1931) Chitwood and Chitwood, 1937

This family contained three subfamilies: Allantonematinae, Contortylenchinae, and Deladinae. Fortuner and Raski (1987) reviewed Neotylenchoidea and suggested that the genus *Deladenus* belong to Allantonematidae. We agree and herein make the transfer.

Diagnosis: (Emended to accommodate Deladinae) Sphaerularioidea. Nematodes in this family have mycetophagous females or have a very short free-living life stage, females penetrate the host after mating.

Females: Mycetophagous females: Lateral field with 6 – 15 incisures (observed with silver deposition). Four lips, each with a single papilla. Stylet 8 – 12 μm long, lumen narrow, basal knobs well developed. Esophagus with fusiform corpus. Isthmus narrow, joining with intestine immediately posterior to nerve ring. Dorsal esophageal gland opening close to stylet base, subventral gland opening near mid-corpus. Excretory pore prominent. Hemizonid present. Vulva with protuberant lips, close to anus. Gonad single without postvulval sac.

Free-living female: Stylet strong, length 15 μm or less. Esophageal glands elongate, lobe – like. Gonad one. Postvulval sac short or absent. Uterus elongate often containing minute sperms. Tail conoid with bluntly pointed terminus.

Parasitic female: Body obese, round, oval, spindle-shaped, or elongate saclike. Body cavity mostly filled with reproductive organs. For *Deladenus,* females very large, 3 – 25 mm long, 0.1 – 0.5 mm wide, usually greenish in color. Esophagus and glands degenerate. Reproductive tube filled with eggs and juveniles. Position of vulva discernible. Tail rounded or tapering.

Males: Not parasitic and have a short life span in the environment. Morphologically similar to females but esophageal glands not enlarged. Esophagus may be degenerate. Testis outstretched. Spicules arcuate, pointed, usually less than 25 μm long. Gubernaculum rarely absent. Tail conoid with or without bursa.

Subfamily Allantonematinae Pereira, 1931

Parasitic female body round, oval, sac – like, never dorsally curved. Vulva indistinct. The type genus is *Allantonema* Leuckart, 1884.

Key to genera of the subfamily Allantonematinae
(Modified after Siddiqi, 1986, and Remillet and Laumond 1991)

1-Free-living females without stylet (Fig. 16.8) *Bradynema* zur Strassen, 1892
Free-living females with stylet 2
2-Males without stylet; esophagus degenerate 3
Males with stylet; esophagus normal or partially degenerate 4
3-Free-living forms generally with clavate tails; parasitic female small, round to oval, with one or two eggs in uterus; parasites of thrips (Fig. 16.9) *Thripinema* Siddiqi, 1986
Free-living forms without clavate tails; parasitic female large, generally tuboid, with a number of eggs or juveniles in uterus; not parasites of thrips (Fig. 16.10) *Howardula* Cobb, 1921
4-Parasitic female round, oval, or bean-shaped; labial region of free-living forms offset (Fig. 16.11) *Allantonema* Leuckart, 1884
Parasitic female elongate, sausage-shaped; labial region of free-living forms not offset 5
5-In free-living female, excretory pore at anterior margin of nerve ring or more anteriorly (Fig. 16.12) *Metaparasitylenchus* Wachek, 1955 (Nickle, 1967)
In free-living female, excretory pore posterior to nerve ring 6
6-Bursa absent (Fig. 16.13) *Protylenchus* Wachek, 1955
Bursa present 7
7-Body surface of parasitic female wavy, with constrictions and swellings (Fig. 16.14) *Sulphuretylenchus* Rühm, 1956 (Nickle, 1967)
Body surface of parasitic female neither wavy nor with constrictions and swellings 8
8-In free-living females, excretory pore about 105 – 125 μm from anterior end; stylet with distinct knobs (Fig. 16.15) *Parasitylenchoides* Wachek, 1955
In free-living females, excretory pore at less than 100 μm from anterior end 9
9-Stylet of free-living forms distinctly knobbed; parasites of staphilinid beetles (Fig. 16. 16) *Proparasitylenchus* Wachek, 1955 (Nickle, 1967)
Stylet of free-living forms without distinct knobs but with basal thickenings, parasites of bark beetles (Fig. 16.17) *Neoparasitylenchus* Nickle, 1967

Subfamily Contortylenchinae Rühm, 1956
Parasitic female body dorsally curved, vulva distinct, in a depression. The type genus is *Contortylenchus* Rühm, 1956.

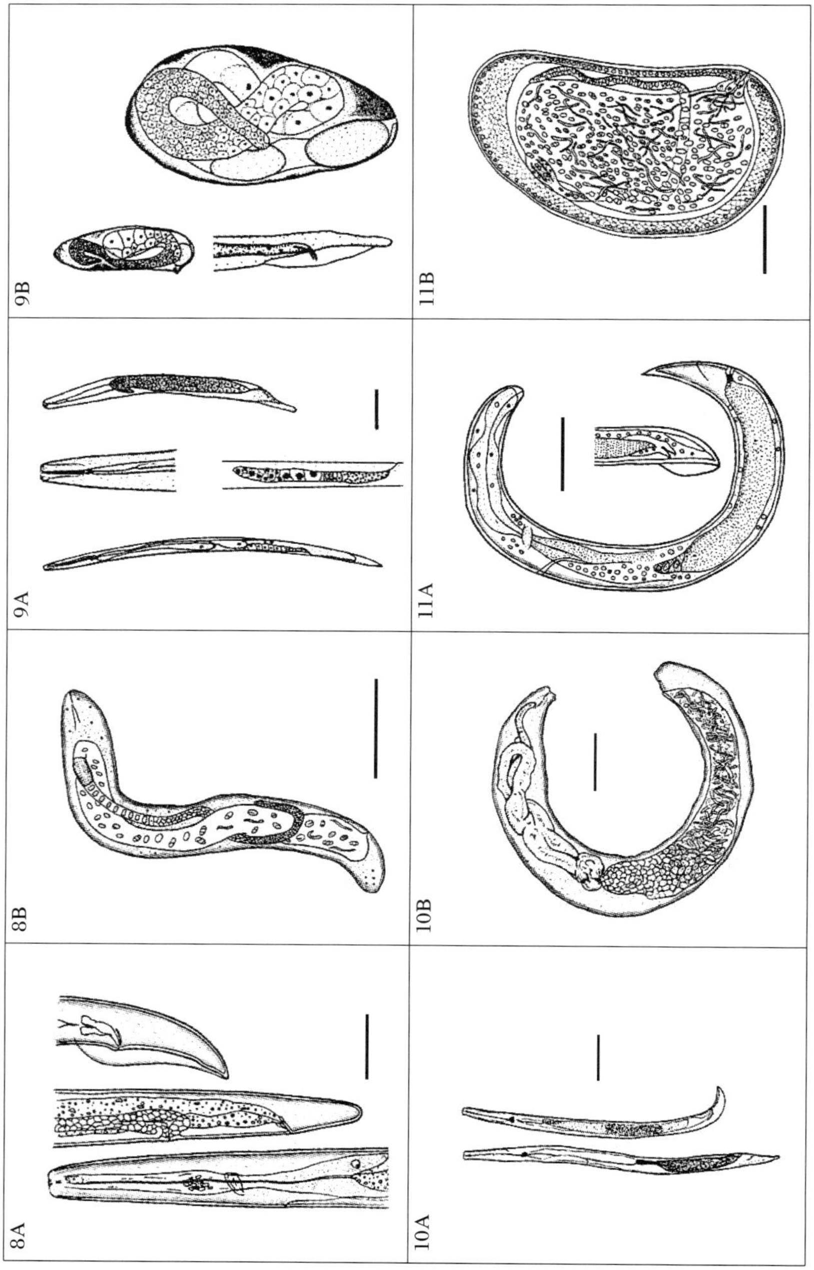
9B
9A
8B
8A
11B
11A
10B
10A

Figures 16.8 – 16.11 Family Allantonematidae. (All illustrations are either redrawn or modified from the cited authors.)

Figure 16.8 *Bradynema.* 8(A) Free-living forms, scale bar = 50 μm. 8(B) Parasitic female, scale bar = 130 μm (Wacheck, 1955).

Figure 16.9 *Thripinema.* 9(A) Infective and developing females. 9(B) Parasitic females and male tail, scale bar (in 9(A)) for Fig. 16.9 = 12 μm for infective female head and male tail, 26 μm for all others (Tipping and Nguyen, 1998).

Figure 16.10 *Howardula.* 10(A) Free living forms, scale bar = 100 μm. 10(B) Parasitic female, scale bar = 20 μm (Poinar et al. 1980).

Figure 16.11 *Allantonema.* 11(A) Preparasitic female after copulation and a male tail, scale bar = 50 μm. 11(B) Full grown parasitic female, scale bar = 400 μm (Wülker, 1923).

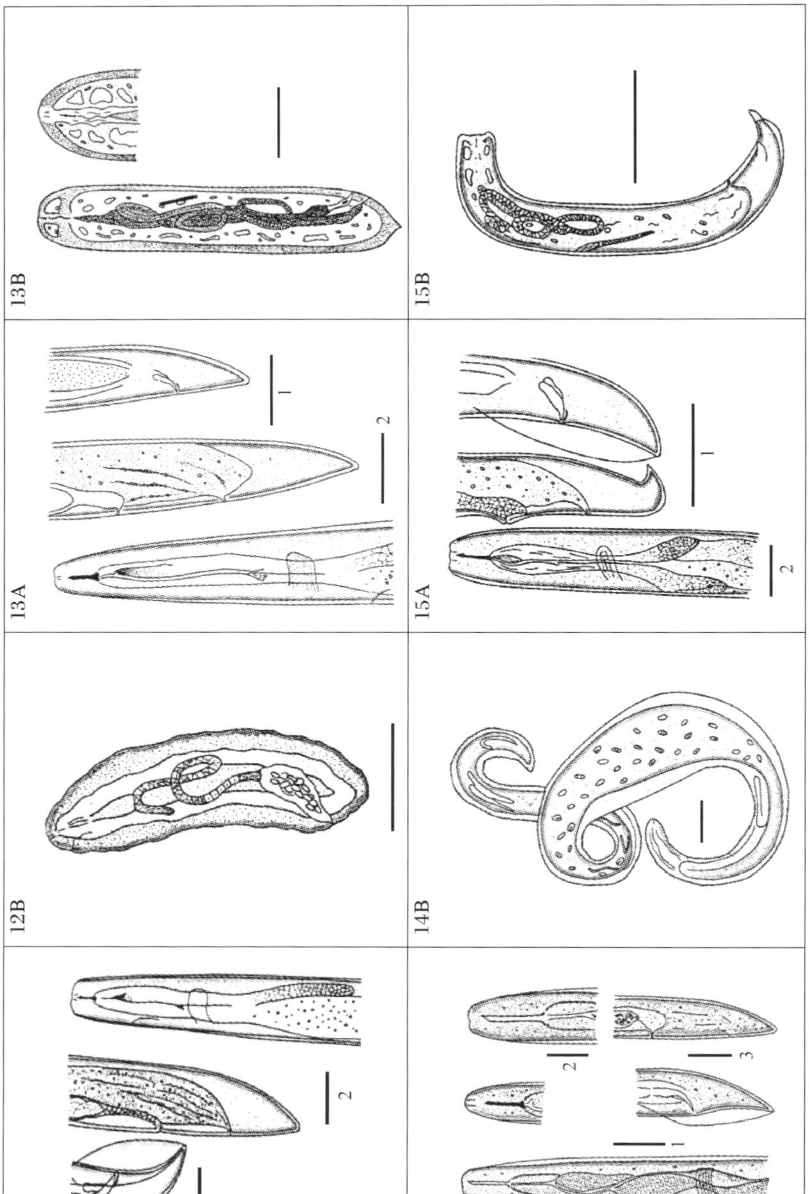
12A
12B
13A
13B
14A
14B
15A
15B

Figures 16.12 – 16.15 Family Allantonematidae. (All illustrations are either redrawn or modified from the cited authors.)

Figure 16.12 *Metaparasitylenchus*. 12(A) Free-living forms, scale bar 1 = 25 μm, scale bar 2 = 20 μm. 12(B) Parasitic female, scale bar = 200 μm (Wacheck, 1955).

Figure 16.13 *Protylenchus*. 13(A) Free-living forms, scale bar 1 = 50 μm, scale bar 2 = 25 μm. 13(B) Parasitic female, scale bar = 200 μm (Wacheck, 1955).

Figure 16.14 *Sulfuretylenchus*. 14(A) Free-living forms, scale bar 1 = 25 μm, scale 2 = 10 μm, and scale bar 3 = 50 μm. 14(B) Parasitic female, scale bar = 100 μm (Rühm, 1956).

Figure 16.15 *Parasitylenchoides*. 15(A) Free-living forms, scale bar 1 = 25 μm, scale bar 2 = 50 μm. 15(B) Parasitic female, scale bar = 500 μm (Wachek, 1955).

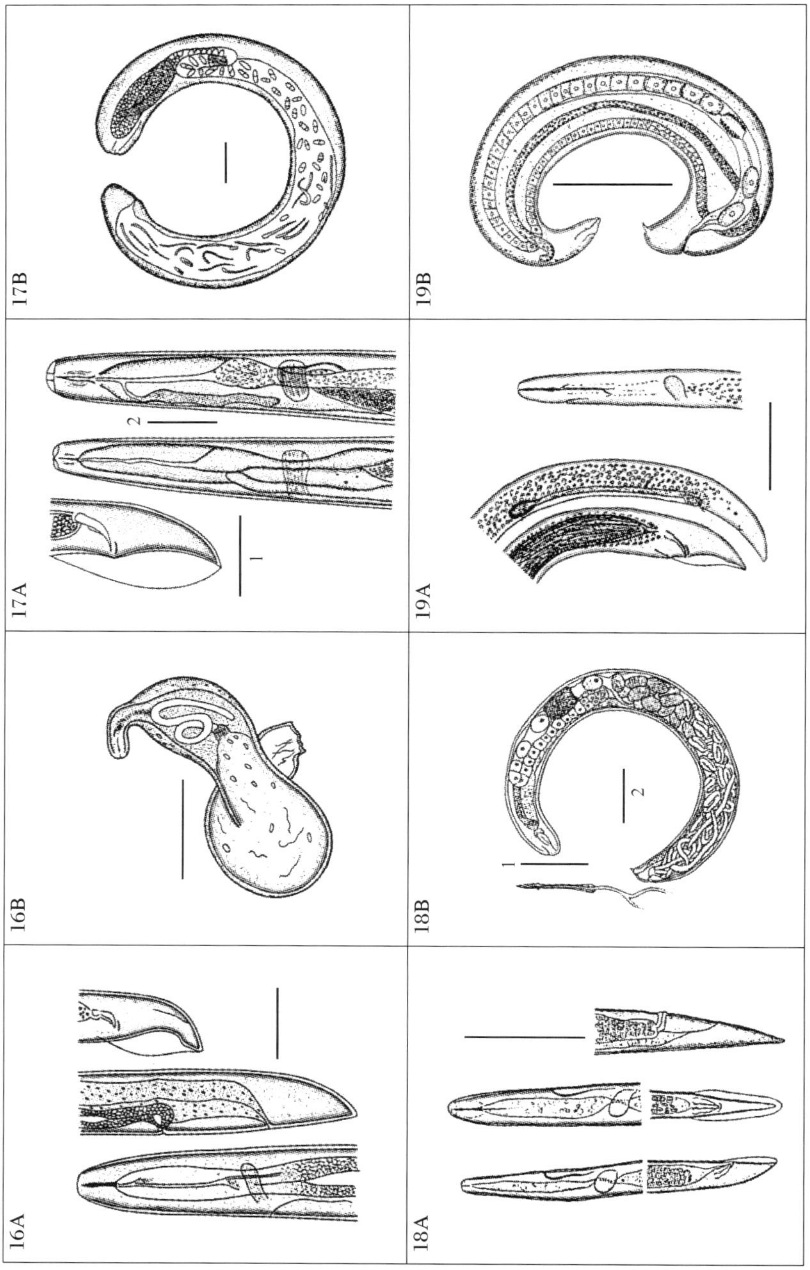
16A
16B
17A
17B
18A
18B
19A
19B

Figures 16.16 – 16.19 Family Allantonematidae. (All illustrations are either redrawn or modified from the cited authors.)

Figure 16.16 *Proparasitylenchus*. 16(A) Free-living forms, scale bar = 25 μm. 16(B) Parasitic female, scale bar = 100 μm (Wachek, 1955).

Figure 16.17 *Neoparasitylenchus*. 17(A) Free-living forms, scale bars 1, 2 = 25 μm (Rühm, 1956). 17(B) Parasitic female, scale = 100 μm (Nickle, 1967).

Figure 16.18 *Spilotylenchus*. 18(A) Free-living forms, scale bar = 50 μm. 18(B) Parasitic female and its stylet, scale bar 1 = 10 μm, scale bar 2 = 100 μm (Deunff, 1984).

Figure 16.19 *Bovienema*. 19(A) Free-living forms, scale bar = 50 μm. 19(B) Parasitic form, scale bar = 100 μm (Bovien, 1937).

Key to genera of the subfamily Contortylenchinae
(Modified after Remillet and Laumond, 1991)

1-Stylet without basal knobs or thickenings; parasites of Siphonaptera (Fig. 16.18) *Spilotylenchus* Launay, Deunff and Bain, 1983
Stylet with basal knobs or thickenings; parasites of Coleoptera 2
2-Parasitic female length < 0.8 mm long; body length/body width < 8; parasites of Scolytidae (Fig. 16.19) *Bovienema* Nickle, 1963
Parasitic female > 0.8 mm long; body length/body width >10 3
3-Parasitic female strongly curved dorsally; parasites of Scolytidae (Fig. 16.20) *Contortylenchus* Rühm, 1956
Parasitic female slightly curved dorsally or straight; parasites of Cerambycidae (Fig. 16.21) *Aphelenchulus* Cobb, 1920

Subfamily Deladinae Siddiqi, 1986

Mycetophagous stage present. Parasitic females long and large. Type and only genus: *Deladenus* Thorne, 1941 (syn.: *Beddingia* Blinova and Korenchenko, 1986).

All species of *Deladenus* described by Bedding (1974, 1968) were transferred to the new genus *Beddingia* by Blinova and Korenchenko (1968). Chitambar (1991) restudied morphologically all species of *Deladenus*, and considered *Beddingia*, a synonym of *Deladenus*. We concur with Chitambar.

Diagnosis: ***Mycetophagous females***: (Fig. 16.26(A)) Lateral fields with 10 – 15 incisures and reduced to seven to nine at vulva (observed with silver deposition). Four lips, each with a single papilla. Stylet 8 – 12 μm long, lumen narrow, basal knobs well developed. Esophagus with fusiform corpus. Isthmus narrow, joining with intestine immediately posterior to nerve ring. Dorsal esophageal gland opening close to stylet base, subventral gland opening near mid-corpus. Excretory pore prominent. Hemizonid present. Vulva with protuberant lips, close to anus. Gonad single without postvulval sac.

Males associated with mycetophagous females: Similar to mycetophagous female. Bursa striated enveloping tail tip. Spicules and gubernaculum present, tylenchoid type. Spicule length variable 13 – 28 μm. Males associated with these mycetophagous females have large amoeboid spermatozoa (10 – 12 μm in diameter).

Infective females (Fig. 16.26(B)): Lateral field with 8 – 13 incisures, three to five on tail. Stylet length variable 14 – 31 μm, very different from mycetophagous female: lumen wide, lacking conspicuous knobs, not tapering anteriorly but extended farther dorsally than ventrally. Esophagus with cylindrical corpus, slightly constricted near stylet base. Isthmus short, broad. Vulva close to anus, barely protruding. Gonad single with a small postvulval sac.

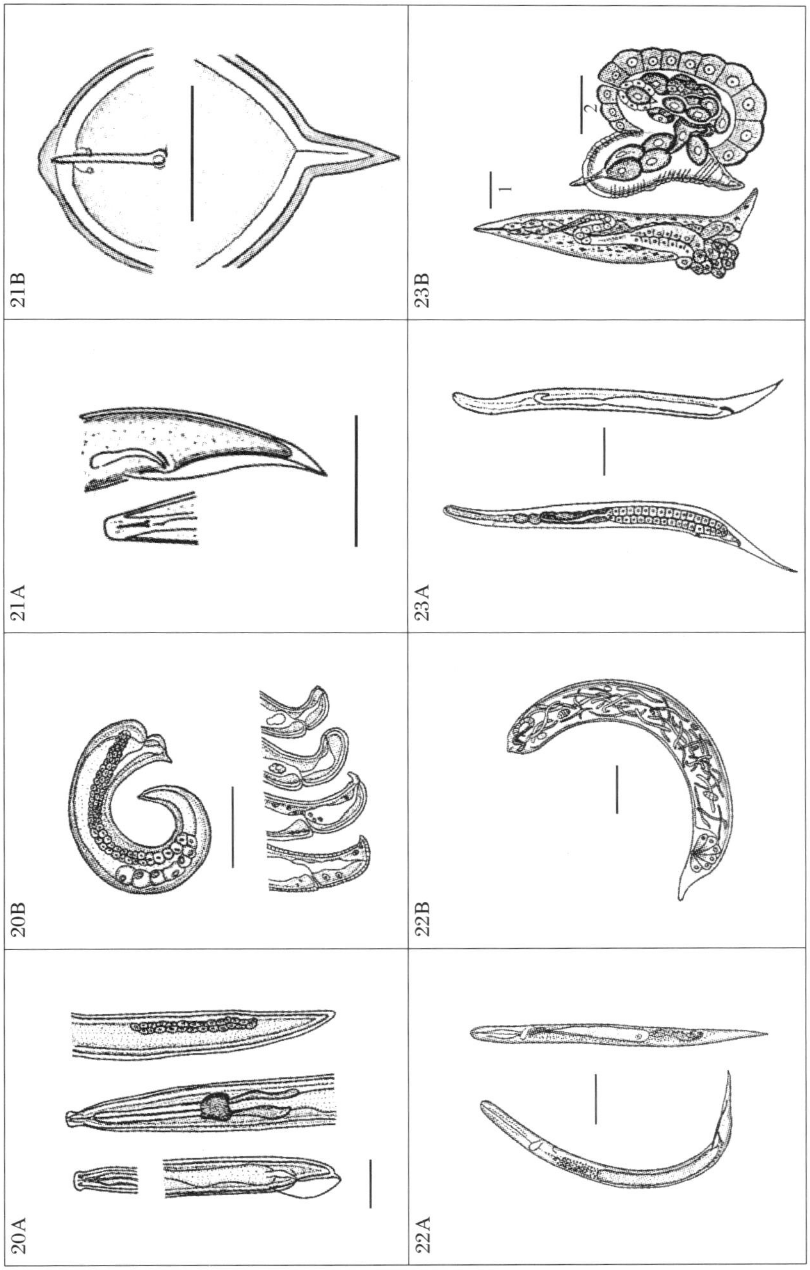
21B
23B
2
1
21A
23A
20B
22B
20A
22A

Figures 16. 20 – 16. 23 Families Allantonematidae and Sphaerulariidae. (All illustrations are either redrawn or modified from the cited authors.)

Figures 16. 20 – 16. 21 Family Allantonematidae.

Figure 16. 20 *Contortylenchus*. 20(A) Free-living forms, scale bar = 20 μm. 20(B) Parasitic female, and variation in tail shapes, scale bar = 100 μm (Massey, 1974).

Figure 16. 21 *Aphelenchulus*. 21(A) Free-living male, scale bar = 42 μm. 21(B) Parasitic female, scale bar = 42 μm (Cobb, 1920).

Figures 16. 22 – 16. 23 Family Sphaerulariidae.

Figure 16. 22 *Scatonema*. 22(A) Free-living forms, scale bar = 50 μm; 22(B) Parasitic female, scale bar = 200 μm (Bovien, 1932).

Figure 16. 23 *Tripius*. 23(A) Free-living forms, scale bar = 50 μm. 23(B) Parasitic females: developing female (left), mature female (right), scale bars 1, 2 = 100 μm (Leuckart, 1887).

Males associated with infective females: Males associated with infective females have very small spermatozoa (1 – 2 μm in diameter) consisting mainly of nucleus.

Mature parasitic females: Very large, 3 – 25 mm long 0.1 – 0.5 mm wide, usually greenish in color. Esophagus and glands degenerate. Reproductive tube filled with eggs and juveniles. Position of vulva discernible. Tail rounded or tapering.

This genus was first found as a parasite of the woodwasp *Sirex noctilio* (F). The type species is *Deladenus durus* (Cobb, 1922) Thorne, 1941. Other insect – parasitic species include *D. canii* Bedding, 1974; *D. imperalis* Bedding, 1974; *D. nevexii* Bedding, 1974; *D. proximus* Bedding, 1974; *D. rudyi* Bedding, 1974; *D. siricidicola* Bedding, 1968; and *D. wilsoni* Bedding, 1968.

16.2.3.2 Family Sphaerulariidae (Skarbilovich, 1947) Lubbock, 1861

Diagnosis: Sphaerularioidea. Nematodes in this family have a very short free-living life span, females penetrate the host after mating.

Females: Free-living stage Stylet length 11 – 19 μm, well developed, conus with distinct lumen, stylet base tripartite or knobbed. Esophagus with glands not elongate. Vulva indistinct. Postvulval sac present or absent. Uterus very long with sperm in fertilized females. Ovary immature. Tail cylindroid with rounded tip.

Parasitic stage Found in the host hemocoel. Uterus large, everted, usually hypertrophies into a sac containing oviduct, ovary, eggs and juveniles.

Males: Not found in host. Similar to females. Testis outstretched, or reflexed. Spicules cephalate. Bursa present, except absent in *Tripius*.

Type genus: *Sphaerularia* Dufour, 1837.

Key to genera of Sphaerulariidae (Modified after Remillet and Laumond, 1991)

1-Parasitic female with uterus hypertrophied but not everted (Fig. 16.22)
Parasitic female with uterus partially or totally everted 2

2-Parasitic female with partially everted uterus not larger than body; male without bursa (Fig. 16.23) *Tripius* Chitwood, 1935
Parasitic female with completely everted uterus much larger than body; male with bursa 3

3-Stylet knobbed; everted uterus surface smooth; bursa distinct, completely enveloping tail; parasites of bark beetles and hymenopterous parasitoids (Fig. 16.24) *Sphaerulariopsis* Wachek, 1955
Stylet not knobbed; everted uterus surface with numerous large rounded

elevations; bursa indistinct; parasites of bumblebee queens and their hymenopterous parasitoids (Fig. 16.25) *Sphaerularia* Dufour, 1837

16.2.3.3 Family Iotonchidae Goodey, 1953 (Skarbilovich, 1959)

Parasitic females found only in *Paraiotonchium*. Type subfamily: Iotonchinae Goodey, 1963

Diagnosis: (Modified after Siddiqi, 1986) (Siddiqi placed this family in his new superfamily Iotonchoidea, 1986, Maggenti placed it in the suborder Sphaerulariina, 1991). Two types of females occur in the host's body cavity: a primary heterosexual female curving ventrally and spiraled and a secondary parthenogenetic female with larger body. ***Free-living females***: Female stylet more than 18 μm long, usually without knobs but thickening present. Excretory pore at the level of, or posterior to, nerve ring. Ovary small. Post vulval sac absent. Tail elongate or conoid. ***Males***: Stylet and esophagus degenerate, stylet sometimes absent. Spicules strong, L – shaped or straight anteriorly and curved posteriorly. Bursa present or absent.

Paraiotonchium Slobodyanyuk, 1975 (Fig. 16.27)

Diagnosis: Iotonchinae. There are two types of females, heterosexual and parthenogenetic. ***Heterosexual females***: Young females (= infective form): (Fig. 16.27(A)), impregnated in the environment; body straight or slightly curved ventrally. Stylet well developed without knobs but usually slightly thickened at base. Esophageal glands lobe – like with three cells extending posteriorly beyond mid – body overlapping intestine dorsally. Esophageal gland openings prominent. Vulva in posterior fourth of body. Gonad prodelphic with large cylindrical uterus, ovary usually with three cells and terminal cap cell. Anus small, obscure. Tail tapering gradually to a point. Parasitic females: (Fig. 16.27(B), inside) much larger than young female. Body strongly curved ventrally, forming C or O shape. Head truncate or rounded. Somatic muscles with prominent nuclei. Stylet as in young female. Esophagus degenerate. Intestine large, dark in color, nuclei of intestinal cells prominent, larger than those of muscles and gonad. Gonad prodelphic outstretched, extending beyond mid – body but never to base of esophagus. Germinal zone three to four oocytes wide. Tail curved ventrally sometimes spicate. ***Males***: (Fig. 16.27(A)) Posterior end of body well curved dorsally. Head continuous with body. Stylet shorter than that of female, very thin, no knobs or thickenings observed. Esophagus partially degenerate. Gonad monorchic outstretched, extending beyond mid – body, with two rows of cells. Spicules with anterior part almost straight, posterior part curved. Gubernaculum absent. Tail tapering to a point. ***Parthenogenetic females***: (Fig. 16.27(B), outside)

Cigar-shaped or long and swollen. Somatic muscles and intestinal cells with prominent nuclei as in parasitic female. Stylet very thin without knobs. Gonad prodelphic wider and longer than that in heterosexual form. In most individuals, ovary extending into head region, becoming reflexed. Germinal zone of ovary large nine to ten oocytes wide. Many eggs in uterus. Tail subcylindrical or tapering with or without mucron.

16.2.3.4 Family Parasitylenchidae Siddiqi, 1986

Diagnosis: (Modified after Siddiqi, 1986). Sphaerularioidea. Two or three types of adults in host's body cavity: primary heterosexual generation, secondary heterosexual generation and/or parthenogenetic generation. Primary heterosexual female not spiral-shaped when relaxed. Female stylet generally under 18 μm long, distinctly knobbed. Orifice of dorsal gland close to or farther behind stylet base. Excretory pore usually anterior to nerve ring. Vulva less than two body widths from anus; no ventral body pore near vulva. Vagina poorly muscular. Postvulval sac absent. Female tail short, conoid. Male may occur in host's body cavity (*Parasitylenchus*). Spicules slender, ventrally curved, about 20 μm long or less (27 – 30 μm long in *Parasitylenchus macrobursatus*). Bursa enveloping tail, absent in *Heterotylenchus.* Gubernaculum present or absent.

Type subfamily Parasitylenchinae Siddiqi, 1986

Diagnosis: Parasitylenchidae. There are two free-living forms – males and females – in the environment, and three forms – primary females, secondary females and males – in host's body cavity. Primary heterosexual female, sausage- or spindle-shape, produces secondary heterosexual forms which produce male and female juveniles. These juveniles leave the host, develop to adults, and copulate in the environment. Nematodes in this subfamily are parasites of Coleoptera and Diptera.

Type and only genus: *Parasitylenchus* Micoletzky, 1922

Diagnosis: (Fig. 16.28): As for the subfamily.

Subfamily Heterotylenchinae Siddiqi, 1986

Diagnosis: Parasitylenchidae. There are two free-living forms, males and secondary heterosexual females in the environment, and two forms, primary heterosexual females, and parthenogenetic females, in insect body cavity. The primary heterosexual females produce only female juveniles that develop to parthenogenetic females. Parthenogenetic females produce male and female juveniles that exit the host, develop to adults, and copulate in the environment. Nematodes in this subfamily are parasites of Coleoptera and Diptera.

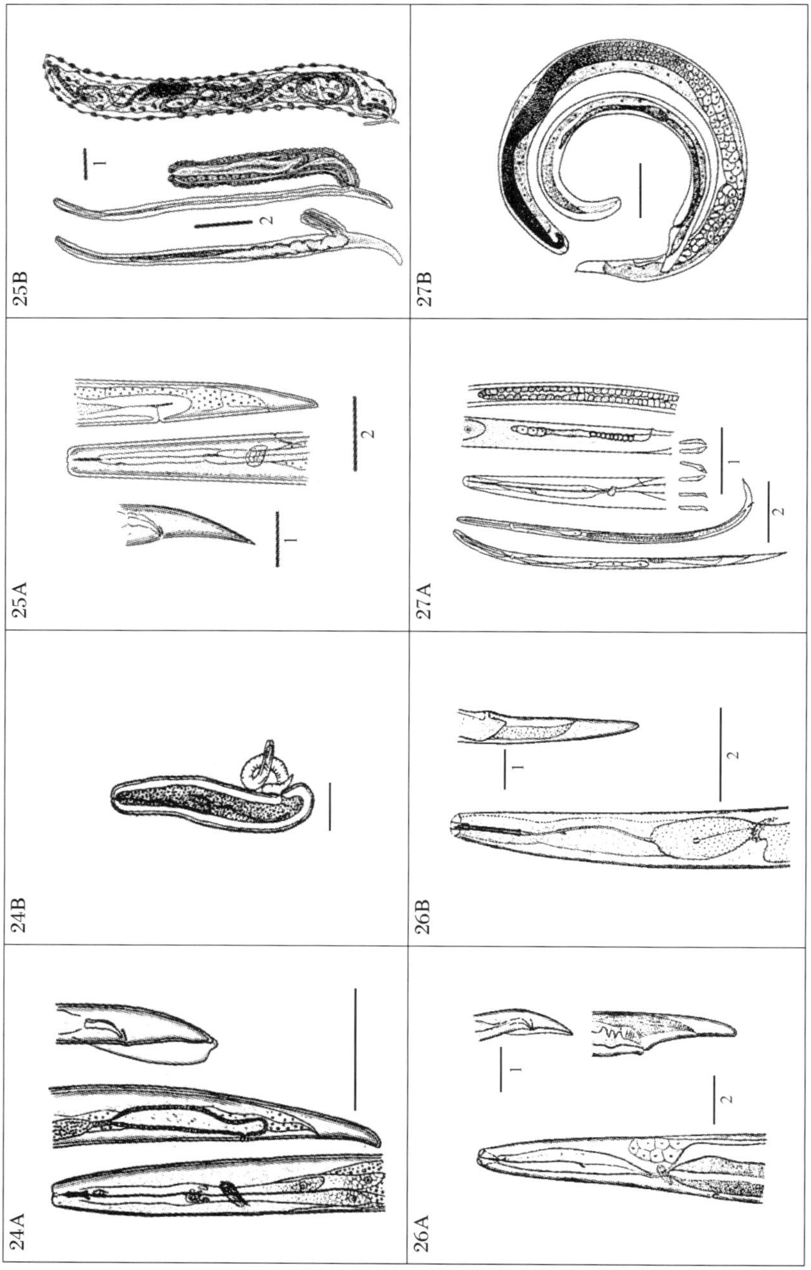
24A
24B
25A
25B
26A
26B
27A
27B

Figures 16.24 – 16.27 Family Sphaerulariidae. (All illustrations are either redrawn or modified from the cited authors.)

Figure 16.24 *Sphaerulaviopsis*. 24(A) Free-living forms, scale bar = 50 μm. 24(B) Parasitic female with everted uterus, scale bar = 50 μm (Wachek, 1955).

Figure 16.25 *Sphaerularia*. 25(A) Free-living forms, scale bar 1 = 20 μm, scale bar 2 = 50 μm. 25(B) Developing and mature females with everted uterus, scale bars 1, 2 = 100 μm (Leuckart, 1887).

Figure 16.26 Family Phaenopsitylenchidae: *Deladenus*. 26(A) Mycetophagous female, and male tail, scale bar 1 = 50 μm, scale bar 2 = 25 μm. 26(B) Infective female, scale bar 1 = 50 μm, scale bar 2 = 25 μm (Bedding, 1968).

Figure 16.27 Family Iotonchidae: *Paraiotonchium*. 27(A) Free-living female and male and enlargements of different structures, scale bar 1 = 20 μm, scale bar 2 = 50 μm. 27(B) Heterosexual (inside) and parthenogenetic (outside) females, scale bar = 250 μm (Coler and Nguyen, 1994).

Type genus: *Heterotylenchus* Bovien, 1937

Key to genera of Heterotylenchinae
(Modified after Remillet and Laumond, 1991)

1-Parasitic females curved dorsally 2
Parasitic females not curved dorsally 3
2-Both parasitic heterosexual and parthenogenetic females curved dorsally, parasites of Siphonaptera (Fig. 16.29)
Psyllotylenchus Poinar and Nelson, 1973
Only parthenogenetic females curved dorsally, parasites of Siphonaptera (Fig. 16.30) *Incurvinema* Deunff, Launay and Beaucournu, 1985
3-Cephalic region and papillae in parasitic females prominent; male with six pairs of caudal papillae (Fig. 16.31)
Paregletylenchus Slobodyanyuk, 1984
Cephalic region and papillae in parasitic females not prominent; male without caudal papillae 4
4-Parasitic heterosexual female with elongate body; male without gubernaculum; parasites of Diptera (Fig. 16.32)
Heterotylenchus Bovien, 1937
Parasitic heterosexual female with sausage-shaped body; male with gubernaculum; parasites of Coleoptera
Wachekitylenchus Slobodyanyuk, 1986

Subfamily Heteromorphotylenchinae Siddiqi, 1986

Diagnosis: Parasitylenchidae. Primary parasitic heterosexual female in insect haemocoel produces only female juveniles that develop to the fourth-stage juveniles before leaving the insect host. These juveniles undergo the last molt and become parthenogenetic females that live on their food reserve, do not feed, and produce a small number of eggs. These eggs develop into males and females, which mate. Fertilized females invade larvae, pupae or nymphs of the insect host. Nematodes in this subfamily are parasites of Coleoptera.

Type and only genus: *Heteromorphotylenchus* Remillet and van Waerebeke, 1978

Diagnosis: (Fig. 16.33): As for the subfamily.

16.2.3.5 Family Fergusobiidae Goodey, 1963 (Siddiqi and Goodey, 1964)

Nematodes in this family are unique, parasitizing both plants and insects. Adult males and females parasitize flower buds, leaf buds and stem tips of various *Eucalyptus* and *Syzigium* spp. in Australia and India. They cause galls to form in flower buds. The fertilized female invades and becomes parasitic in the haemocoel of female larvae of agromyzid flies of the genus *Fergusonina* (see chapter by Giblin-Davis et al.).

Diagnosis: Sphaerularioidea. All forms partially obese. Stylet with knobs. Anterior part of esophagus swollen, cylindrical containing a large valve. Isthmus short, narrow, surrounded by the nerve ring. Posterior bulb glandular, slightly overlapping intestine dorsally. Female with one gonad, without postvulval sac. Male with bursa, spicules strong, gubernaculum absent. Tail of both sexes short, conoid.

Type and only genus: *Fergusobia* Currie 1937.

Diagnosis: As for the family (Fig. 16.34).

16.2.4 Order Tylenchida – Suborder Aphelenchina Geraert, 1966

Diagnosis: Tylenchida. Stylet without knobs but sometimes with basal thickenings. Dorsal esophageal gland duct emptying into esophageal lumen within the large median bulb, anterior to the valve plates. Gonad one. Bursa sometimes with ribs, sometimes short enveloping only the tip of the tail, often absent. Spicules usually rose-thorn-shaped. Nematodes in this order either live free in soil, are predaceous on other nematodes, are mycophagous, are parasitic in leaves, roots, stems, and bulbs, or are insect associates or parasites.

16.2.4.1 Family Entaphelenchidae Nickle, 1970

Nematodes in this family are either endo- or ectoparasites of insects.

Diagnosis: Aphelenchina. Usually with three adult forms, including vermiform males and females, and swollen endoparasitic females. Two ectoparasitic adult forms also present. Stylet with or without basal flanges. Esophagus with large median bulb and overlapping glands. Male usually without bursa. Spicules rose-thorn-shaped. Gubernaculum absent.

Type subfamily Entaphelenchinae Nickle, 1970

Diagnosis: Entaphelenchidae. Nematodes with at least three adult forms. Stylet short, less than 25 μm. Endoparasites of insects.

Type genus: *Entaphelenchus* Wachek, 1955

Entaphelenchus Wachek, 1955

Diagnosis: Entaphelenchinae. (Fig. 16.35), three adult forms. **Small vermiform males and females**: 0.5 – 1.0 mm long, not found in insect body cavity. Labial region offset. Stylet 17 – 22 μm long, without basal knobs. Excretory pore posterior to median bulb. Ovary with few cells, post vulval sac

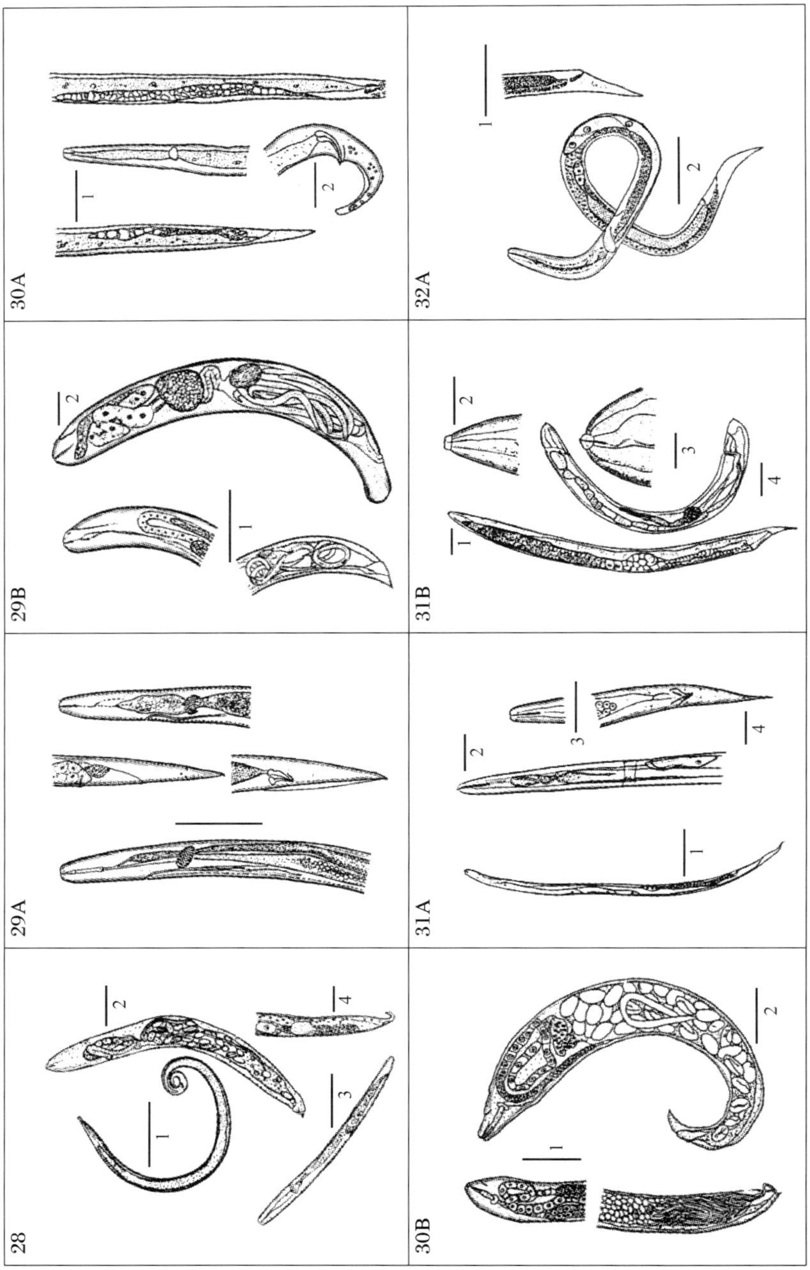
30A
29B
29A
28
32A
31B
31A
30B

Figures 16.28 – 16.32A Family Parasitylenchidae. (All illustrations are either redrawn or modified from the cited authors.)

Figure 16.28 Parasitylenchinae: *Parasitylenchus* male of parasitic generation (top, left) found in host hemocoel, scale bar 1 = 50 μm; parasitic female (top, right), scale bar 2 = 100 μm; second stage juvenile (bottom, left), scale bar 3 = 50 μm; posterior portion of a female (bottom, right) 4 days after penetration of host, scale bar 4 = 50 μm (Iperti and van Waerebeke, 1968).

Figure 16.29 Heterorhabditinae, *Psyllotylenchus*. 29(A) Free-living forms, scale bar = 40 μm. 29(B) Heterosexual female (left), scale bar = 60 μm, and parthenogenetic female, scale bar = 40 μm (Poinar and Nelson, 1973).

Figure 16.30 *Incurvinema*. 30(A) Free-living forms, scale bar 1 = 40 μm, bar 2 (for male tail) = 20 μm. 30(B) Heterosexual (left) and parthenogenetic females, scale bar 1 = 200 μm, bar 2 = 80 μm (Deunff, 1984).

Figure 16.31 *Paregletylenchus*. 31(A) Free-living forms, scale bar 1 = 80 μm, bars 2, 3, 4 = 20 μm. 31(B) Parthenogenetic female (left and top right), and heterosexual female (bottom right), scale bar 1 = 50 μm, bars 2, 3 = 20 μm, bar 4 = 100 μm (Slobodyanyuk, 1984).

Figure 16.32 *Heterotylenchus*. 32(A) Free-living forms, scale bars = 50 μm (Bovien, 1937).

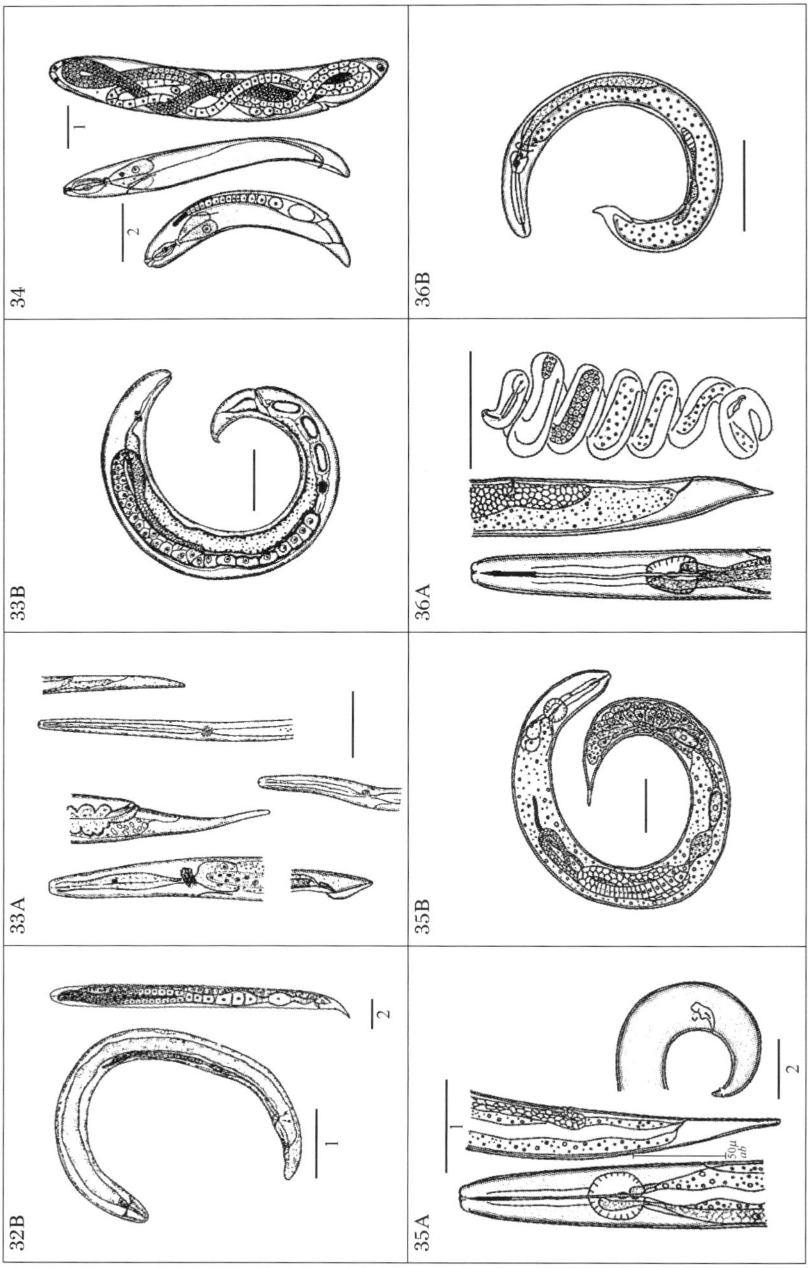
32B
33A
33B
34
35A
35B
36A
36B

Figures 16.32B – 16.36 Families Parasitylenchidae and Fergusobiidae. (All illustrations are either redrawn or modified from the cited authors.)

Figure 16.32(B) *Heterotylenchus*. Heterosexual female (left) and parthenogenetic female (right), scale bar 1 = 500 μm, bar 2 = 100 μm (Bovien, 1937).

Figure 16.33 Heteromorphotylenchinae: *Heteromorphotylenchus*. 33(A) Free-living forms: parthenogenetic female (top left), heterosexual female (top, right), and male (bottom), scale bar = 80 μm. 33(B) Parasitic female, scale bar = 100 μm (Remillet and van Waerebeke, 1978).

Figure 16.34 Family Fergusobiidae: *Fergusobia*: Entomoparasitic female found in body cavity of fly (right), scale bar 1 = 50 μm, and free-living or plant-parasitic forms (left and middle), scale bar 2 = 100 μm.

Figures 16.35 – 16.36 Family Entaphelenchidae.

Figure 16.35 *Entaphelenchus*. 35(A) Non-entomoparasitic forms, scale bars 1, 2 = 50 μm. 35(B) Parasitic female scale bar = 100 μm.

Figure 16.36 *Peraphelenchus*. 36(A) Female and corkscrew-shaped male, scale bar = 50 μm. 36(B) Parasitic female with swollen body, scale bar = 100 μm (Wachek, 1955; Nickle, 1970).

short. Tail conoid with rounded tip. Spicules paired, rostrum prominent. Tail conoid with rounded tip. Bursa and gubernaculum absent. **Parasitic females**: large 1. 3 – 2. 5 mm, found in insect body cavity. Perivaginal glands well developed, filling the body in this region. Body usually convoluted. Tail with spike – like tip.

Type species: *Entaphelenchus oxyteli* Wachek, 1955

Peraphelenchus Wachek, 1955

Diagnosis: (Fig. 16. 36), Entaphelenchinae. Three adult forms (as in family), all found in insect body cavity. Stylet 17 – 18 μm with wide lumen, without knobs. Procorpus cylindrical, lumen wide. Median bulb absent. Basal bulb with valve. Esophageal glands as a long lobe. **Female**: ventrally curved. Ovary with few cells, postvulval sac present. Tail short, conical. **Males**: corkscrew-shaped. Spicules with prominent rostrum. Tail bluntly rounded with two pairs of postanal caudal papillae. **Adult parasitic female**: with larger body, ovary short.

Type species: *Peraphelenchus necrophori* Wachek, 1955

Praecocilenchus Poinar, 1969

Diagnosis: (Fig. 16. 37), Entaphelenchinae. Three adult forms found in insect body cavity. Females and males: found in uterus of adult parasitic female. Labial region not offset. Stylet 8 – 10 μm with wide lumen, without basal knobs. Female: very small, less than 0. 4 mm long. Esophageal gland long, extending to mid-body. Ovary short, postvulval sac short. Tail conical. **Male**: small 0. 4 – 0. 5 mm long, spicules paired, rose-thorn-shaped with prominent rostrum. Tail curved. Bursa and gubernaculum absent. **Parasitic females**: body large, 0. 9 – 2. 3 mm long, and swollen, C-shaped, curved ventrally. Uterus containing sexually mature males and females, postvulval sac absent. Stylet short with wide lumen. Vulval lip protruding. Tail bluntly rounded.

Type species: *Praecocilenchus rhaphidophorus* Poinar, 1969

Roveaphelenchus Nickle, 1970

Diagnosis: Entaphelenchinae. (Fig. 16. 38), three adult forms found in insect body cavity. Stylet slender without basal knobs. Esophageal glands short. **Female**: Ovary one, usually with one large juvenile in uterus, post vulval sac absent. Tail cylindrical, ending in four mucronate points. **Male**: When relaxed, tail well coiled. Testis reflexed. Spicules with prominent rostrum. Tail bluntly rounded. Caudal papillae not seen. Bursa and gubernaculum absent. **Parasitic female**: Body swollen, ovary convoluted, extending to neck region. Uterus containing pre – adult stage juvenile. Tail short, digitate.

Type species: *Roveaphelenchus jonesi* Nickle, 1970

Subfamily Acugutturinae Hunt, 1980

Diagnosis: Entaphelenchidae. Nematodes with two adult forms. Stylet very long, 50 to more than 100 μm. Ectoparasites of insects.

Acugutturus Hunt, 1980

Diagnosis: Acugutturinae. (Fig. 16.39): Females and males: body 0.6 – 0.9 mm long with a single lateral line. Head rounded, offset by constriction. Stylet long 50 – 60 μm, slender with conus three times longer than shaft, basal knobs absent. Procorpus long, reflexed. **Female**: gonad one, without post vulval sac. Tail conical. **Male**: Spicules rose-thorn-shaped with prominent rostrum. Tail without bursa, two pairs of caudal papillae present. Ectoparasites of American cockroach.

Type species: *Acugutturus parasiticus* Hunt, 1980

Noctuidonema Remillet and Silvain, 1988

Diagnosis: Acugutturinae (Fig. 16.40). **Female**: body 0.5 – 0.8 mm, swollen, club - shaped. Stylet very long (over 100 μm) with small basal thickenings. Procorpus long and reflexed. Excretory pore far anterior. Gonad one, outstretched. Tail very short. **Male**: Spicule large, 90 μm long with long ventral arm. Gubernaculum absent. Two pairs of genital papillae, one pair preanal and one pair caudal. A small terminal bursa present. Ectoparasites of noctuid moths.

Type species: *Noctuidonema guyanense* Remillet and Silvain, 1988

16.2.5 Order Mermithida Hyman, 1951

Diagnosis: Nematodes in this order are parasites of invertebrates. The main morphological characteristic of the order is the presence of a stichosome. Early juvenile stages with a protrusible stylet that is absent in the adults. Amphids dorylaimoid.

16.2.5.1 Family Mermithidae Braun, 1883

Diagnosis (Modified after Kaiser, 1991): Mermithoidea. Long, slender nematodes, usually between 10 and 100 mm. Cuticle smooth or containing criss – cross fibers near the outer layers, head containing two, four, or six cephalic papillae and sometimes a pair of lateral labial papillae. Oral opening terminal or shifted ventrally. Amphids tubelike or modified pouch-like. Esophagus, a stichosome usually with four, eight, or 16 stichocytes. Intestine

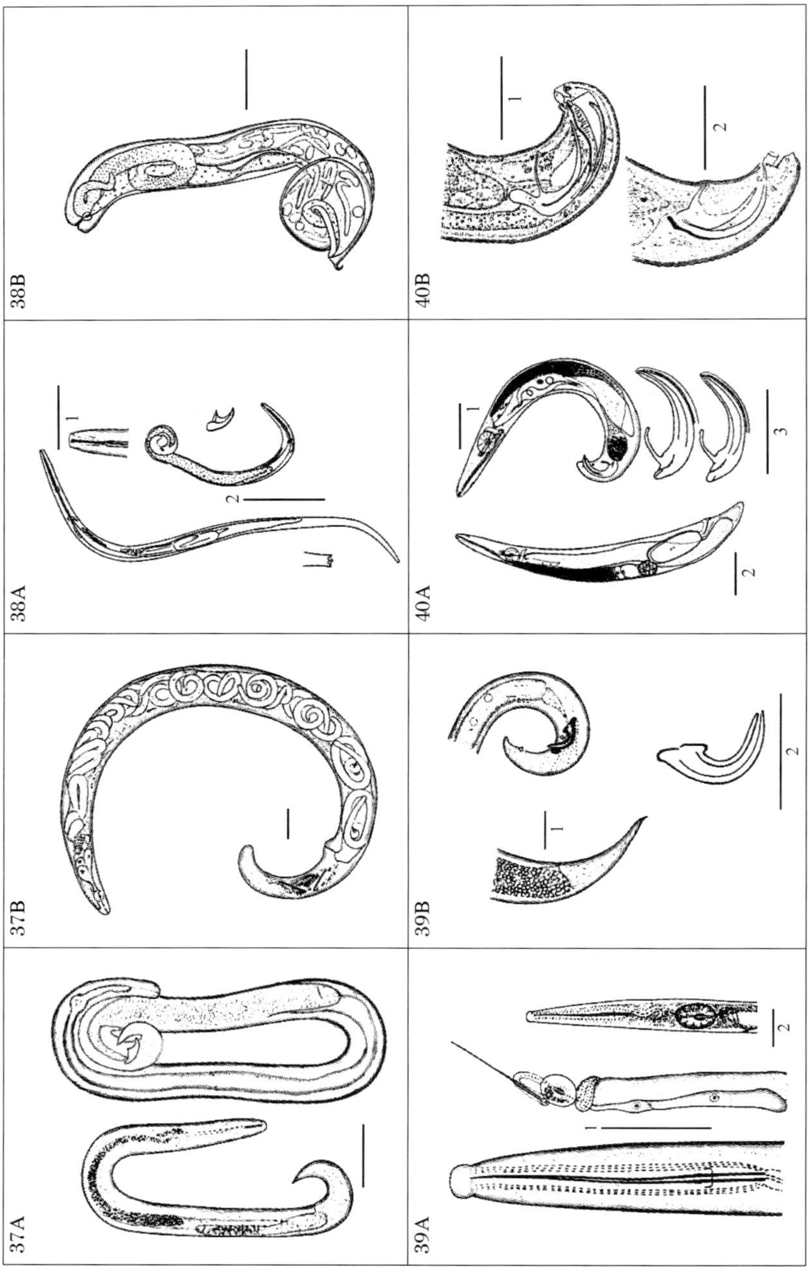
37A
37B
38A
38B
39A
39B
40A
40B

Figures 16.37 – 16.40 Family Entaphelenchidae. (All illustrations are either redrawn or modified from the cited authors.)

Figure 16.37 *Praecocilenchus*. 37(A) Female and male found in uterus of parasitic female, scale bar = 30 μm. 37(B) Parasitic female, scale bar = 70 μm.

Figure 16.38 *Roveaphelenchus*. 38(A) Vermiform adults, for entire female and male, scale bar 1 = 100 μm, for others, bar 2 = 20 μm. 38(B) Swollen parasitic females, scale bar = 20 μm (Nickle, 1970).

Figure 16.39 *Acugutturus*. 39(A) Head and esophagus region, scale bar 1 (head, left), and bar 2 (others) = 20 μm. 39(B) Female tail, male tail and spicule, scale bars = 20 μm (Hunt, 1980).

Figure 16.40 *Noctuidonema*. 40(A) Female, male, and spicules, scale bar 1 = 50 μm, bar 2 = 70 μm, and bar 3 = 25 μm (Marti and Rogers, 1995). 40(B) Posterior region of males, scale bar 1 = 50 μm, bar 2 = 25 μm (Anderson and Laumond, 1990).

is modified as a food storage organ called the trophosome. Functional anus absent. Female with two ovaries; vagina muscular, straight or curved. Males with two testes. Spicules paired, fused, or single. Several rows of genital papillae usually present. Preparasitic juveniles with functional stylet.

Key to genera of terrestrial Mermithidae (Modified after Kaiser, 1991)

1-With four cephalic papillae (Fig. 16.41, top) 2

With six cephalic papillae (Fig. 16.41, bottom) 4

2-With two labial papillae (Fig. 16.41, top); oral opening terminal or slightly shifted to ventral side (Fig. 16.42) *Mermis* Dujardin, 1842

Without labial papillae 3

3-Oral opening shifted ventrally as far posterior as the cephalic papillae; vagina S-shaped (Fig. 16.43, left); postparasitic juveniles without tail appendage *Pheromermis* Poinar, Lane and Thomas, 1976

Oral opening terminal, vagina short with V-like curve (Fig. 16.43, right); postparasitic juveniles with short tail appendage (Fig 16.43) *Melolonthinimermis* Artyukhovsky, 1963

4-Oral opening terminal; amphids small, cup-shaped with small pore-like aperture near the lateral cephalic papillae; vulva without flap, vagina with well-cuticularized lumen, horn-or S-shaped(Fig. 16.44, top right) 5

Oral opening, vagina, and amphids not as above 6

5-Parasitic and postparasitic juveniles with tail appendage *Hexamermis* Steiner, 1924

Parasitic and postparasitic juveniles with a ring on tail end (Fig. 16.44, bottom); preparasitic juveniles lose the tail end during penetration of host *Agamermis* Cobb, Steiner and Christie, 1923

6-Oral opening shifted slightly to ventral side; vagina modified S-shape (Fig. 16.44, top right) reflexed two or three times; spicules paired, one short and thick, the other extremely long and filiform (Fig. 16.45); postparasitic juvenile with short tail appendage *Thaumamermis* Poinar, 1981

Oral opening terminal or slightly shifted to ventral side; vagina and spicules not as above 7

7-Oral opening shifted slightly ventrally, pharyngeal tube very long, can reach nine-tenths of body length; vulva oblique, vagina cylindrical, S-shaped; spicules long, parallel, with arch-like curve, 1.5-2 times of body diameter; parasitic and post parasitic juveniles with small tail appendage *Oesophagomermis* Artyukhovsky, 1969

Oral opening terminal 8

8-Vagina barrel-shaped (Fig. 16.44, right bottom) or cylindrical 9

Vagina S-shaped 10

9-Vagina short, barrel-shaped; spicules long, thin, parallel (Fig. 16.46), 2 – 5.5 times of body diameter; trophosome syncytial *Psammomermis* Polozhentsev, 1941

Vagina short but cylindrical and posteriorly directed; vulva flanked by lateral lips (Fig. 16.48, left); males unknown *Tunicamermis* Schuurmans-Stekhoven and Mawson, 1955

10-Spicules twisted (Fig. 16.47), parallel only in the midpart and on the distal end, fused on tip *Amphimermis* Kaburaki and Imamura, 1932

Spicules parallel throughout their length 11

11-Amphids medium in size; vagina elongate, reflexed two or three times in two different planes (Fig. 16.48, right); spicules paired, long, six or more times as long as body width *Aranimermis* Poinar and Benton, 1986

Amphid very large (Fig. 16.48, bottom), vagina short but S-shaped; spicules short *Amphidomermis* Filipjev, 1934

16.2.5.2 Family Tetradonematidae Cobb, 1919

Diagnosis: (After Poinar, 1975). Mermithoidea. Lips rudimentary or absent; juveniles with a minute stylet which disappears in later stages; head papillae reduced; esophagus consists of a simple hollow tube and associated tetrad (four-celled structure as in Fig. 16.49); anus absent; ovaries and testes paired; vulva mid-body; spicule single, bursa absent. The nematodes mature to the adult stage and mate in the body cavity of the insect host.

Key to the genera of Tetradonematidae (After Poinar, 1975)

1-Life cycle with two parasitic generations; first generation parthenogenetic with female

lacking a vulva; males of second generation producing both sperm and eggs (Fig. 16.50) *Heterogonema* van Waerebeke and Remillet, 1971

Life cycle with single parasitic generation; female with vulva; male producing sperm only 2

2-Pharyngeal region of the female containing a tetrad (four large cells arranged in a row as in Fig. 16.49); mature female with an acute tail; male tail lacking genital papillae *Tetradonema* Cobb, 1919

Pharyngeal region of the female not containing a tetrad; mature female with a round tail; male with or without genital papillae 3

3-Male lacking genital papillae *Aproctonema* Keilin, 1917

Male possessing genital papillae 4

4-Mature forms with three large gland-like cells posterior to the nerve-ring (Fig. 16.51), stichosome and stichocytes lacking; spicule of male with tip

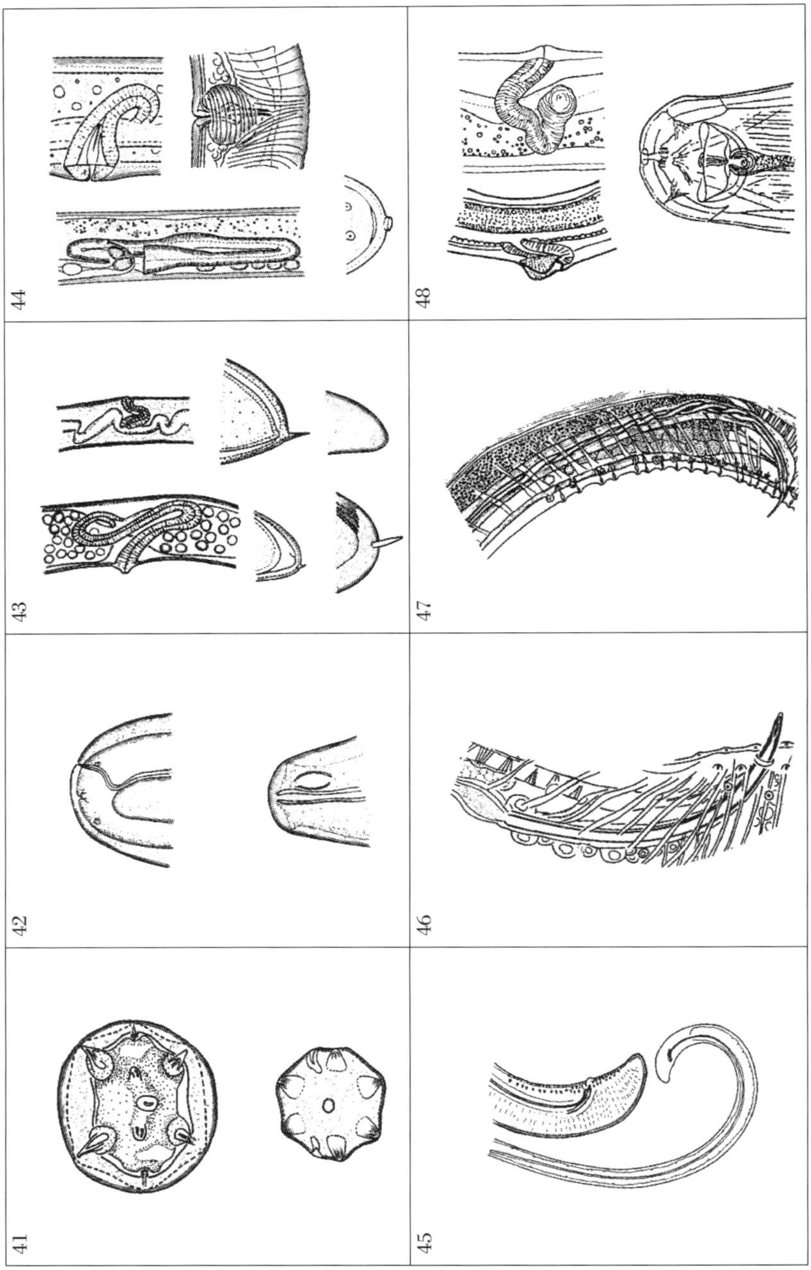
41
42
43
44
45
46
47
48

Figures 16.41 – 16.48 Morphological structures of Mermithidae. (All illustrations are either redrawn or modified from the cited authors.)

Figure 16.41 Face view with four cephalic and two labial (top) and six (bottom) cephalic papillae (Baker and Poinar, 1986; Poinar, 1977).

Figure 16.42 Oral opening shifted ventrally (top) (Poinar et al. 1976), oral opening normal (bottom) (Nickle, 1972).

Figure 16.43 Vagina S-shaped (top, left), vagina V-shaped (top, right), and different shapes of tail tip (bottom).

Figure 16.44 Different shapes of vagina; tail end with a ring.

Figure 16.45 Spicules, one short and thick, other extremely long.

Figure 16.46 Spicules long, parallel.

Figure 16.47 Spicules long and twisted.

Figure 16.48 Vagina long and reflexed two or three times (top, right), vulva flanked by lateral lips (top, left), and amphid large (bottom) (Figures 16.43 – 16.48: Kaiser, 1991; Nickle, 1972).

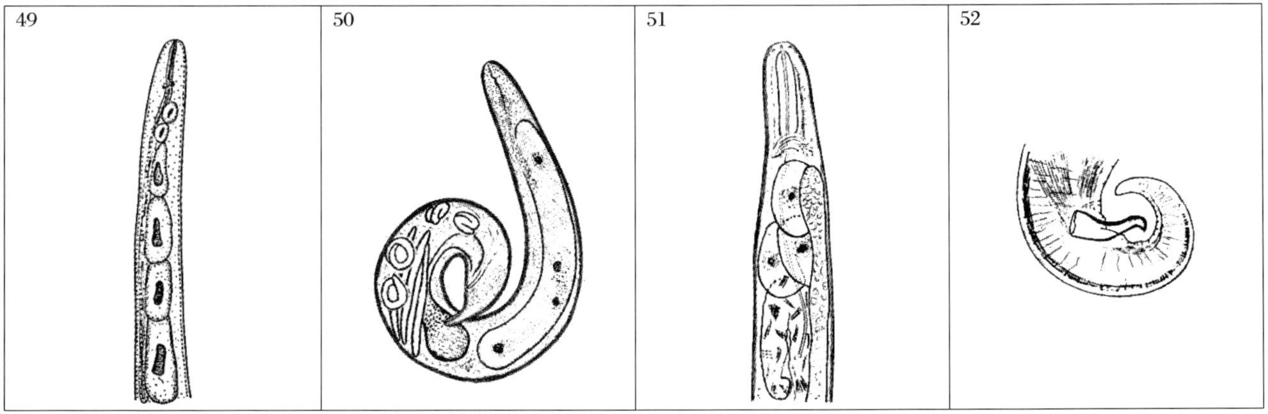

Figures 16.49 – 16.52 Morphological structures of the family Tetradonematidae. (All illustrations are either redrawn or modified from the cited authors.)

Figure 16.49 four-celled structure in esophagus (Poinar, 1975).

Figure 16.50 Male with sperms and eggs (van Waerebeke and Remillet, 1973).

Figure 16.51 Esophagus with three large gland-like cells posterior to nerve ring (Nickle, 1969).

Figure 16.52 Spicule tip curved downward (Nickle, 1969).

curved downward (Fig. 16.52) *Corethrellonema* Nickle, 1969
Mature forms lacking three large cells behind the nerve ring; stichosome with eight stichocytes present in the adults; spicule of male not curved downward at tip *Mermithonema* Goodey, 1941

16.2.6 Order Oxyurida Weinland, 1858 (in Yamaguti, 1961) (syn.: Oxyurata Skrjabin, 1923, in Skryabin et al., 1951)

Diagnosis: Small to medium nematodes, obligatory zooparasites. Eight papillae present in one circle, sometimes they fuse to form four papillae. Amphids small. Stoma cylindrical but short, not cuticularized, not surrounded by esophageal tissue posteriorly. Esophagus with corpus, isthmus and basal bulb; basal bulb valvate. Corpus is either cylindrical, fusiform or composed of procorpus and metacorpus. Female gonads monodelphic or didelphic, terminal portion muscular. Eggs usually with relatively thin shell, and mostly do not hatch until ingested by an appropriate host. Male with one, two, or no spicules. Posterior portion of male with genital papillae; bursa usually present.

This order has two superfamilies: Thelastomatoidea, parasites of invertebrates, and Oxyuroidea, parasites of vertebrates. The superfamily Thelastomatoidea was studied taxonomically by Adamson (1989), and Adamson and Waerebeke (1992a, b, c). They divided the nematodes in this order into five families: Thelastomatidae, Travassosinematidae, Protrelloididae, Pseudonymidae, and Hystrignathidae, and gave the descriptions of genera for each family. Because of space limitations, and lack of information, we will mention only three families, Thelastomatidae, Hystrignathidae, and Travassosinematidae.

16.2.6.1 Family Thelastomatidae Travassos, 1929

Diagnosis: Thelastomatoidea. Anterior end with eight papillae. Amphid rounded or oval. Esophagus with variable corpus, distinct or indistinct isthmus, and a valvular basal bulb. Females with one or two gonads; vulva anterior or posterior to base of esophagus. Male with single spicule or none; tail with one to four genital papillae.

Key to common genera, parasites of Blattaria (Cockroaches)

1-Vulva anterior to posterior end of esophagus (Fig. 16.53) 2
Vulva posterior to posterior end of esophagus 4
2-Esophagus of female long, slender, about 1/4 of body length; male tail digitate *Protrelloides* Chitwood, 1932
Esophagus not long and slender; male tail not digitate 3

3-Lip region bearing eight prominent digitate papillae (Fig. 16.54), eggs without cuticular crest; no spicules in male *Protrelleta* Chitwood, 1932
Lip region not bearing eight prominent digitate papillae, eggs with cuticular crest (Fig. 16.55), one spicule in male *Protrellus* Cobb, 1920
4-Esophagus enlarged subspherically at base of buccal cavity (Fig. 16.56) *Blattophila* Cobb, 1920
Esophagus not so enlarged 5
5-Esophagus with median bulb 6
Esophagus without median bulb 9
6-Vulva in anterior part of the body, 1/4 – 1/3 body length from anterior end *Hammerschmidtiella* Chitwood, 1932
Vulva in mid or posterior part of the body, 2 or more body lengths from anterior end 7
7-Anterior part of esophagus very short, median bulb large (Fig. 16.57), pear-shaped *Aorurus* Leidy, 1849
Anterior part of esophagus long, median bulb not as above 8
8-Median bulb cylindrical (Fig. 16.58) *Leidynema* Schwenk *in* Travassos, 1929
Median bulb spherical (Fig. 16.59) *Leidynemella* Chitwood and Chitwood, 1934
9-Female with one gonad 10
Female with two gonads 12
10-Esophagus long, about 1/3 body length, tail short (Fig. 16.60) *Galebia* Chitwood, 1932
Esophagus not long, about 1/6 or less body length 11
11-Female tail attenuate, vulva located in middle 1/3 of body *Blatellicola* Basir, 1940
Female tail conical, vulva located in posterior 1/3 of body *Blatticola* Schwenk, 1926
12-Eggs fused in pairs along flattened surfaces (Fig. 16.61) *Cameronia* Basir, 1948
Eggs not fused 13
13-Female tail filiform 14
Female tail not filiform (Fig. 16.62, left) 15
14-Excretory pore of female present, male tail filiform or delicately attenuated (Fig. 16.62, right) *Thelastoma* Leidy, 1849
Excretory pore of female not observed, male tail very short, degenerate *Euryconema* Chitwood, 1932
15-Vulva anterior to mid-body *Suifunema* Chitwood, 1932
Vulva near mid-body 16

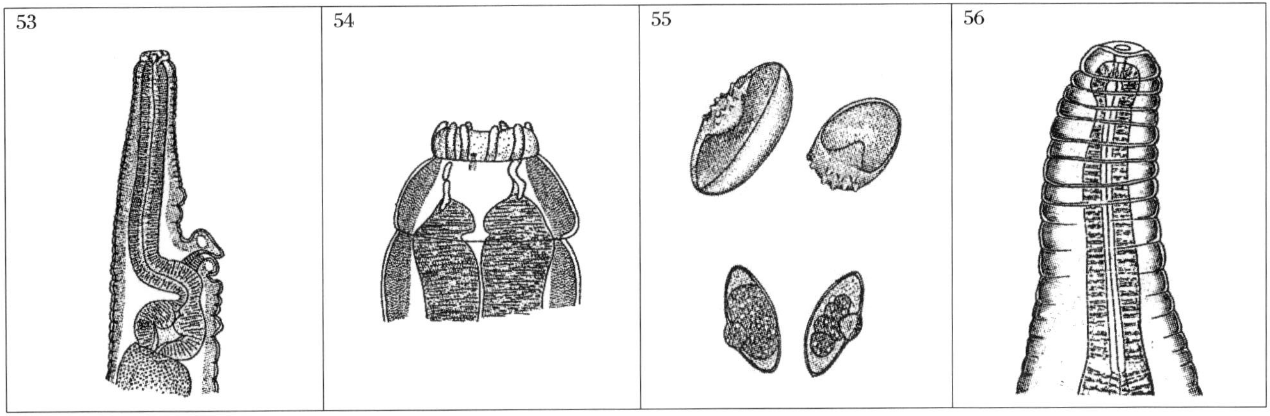

Figures 16.53 – 16.56 Morphological structures of the family Thelastomatidae.
Figure 16.53 Vulva anterior to posterior end of esophagus.
Figure 16.54 Lip region with eight prominent digitate papillae.
Figure 16.55 Eggs with cuticular crest.
Figure 16.56 Anterior end of esophagus subspherical (Chitwood, 1932, 1933).

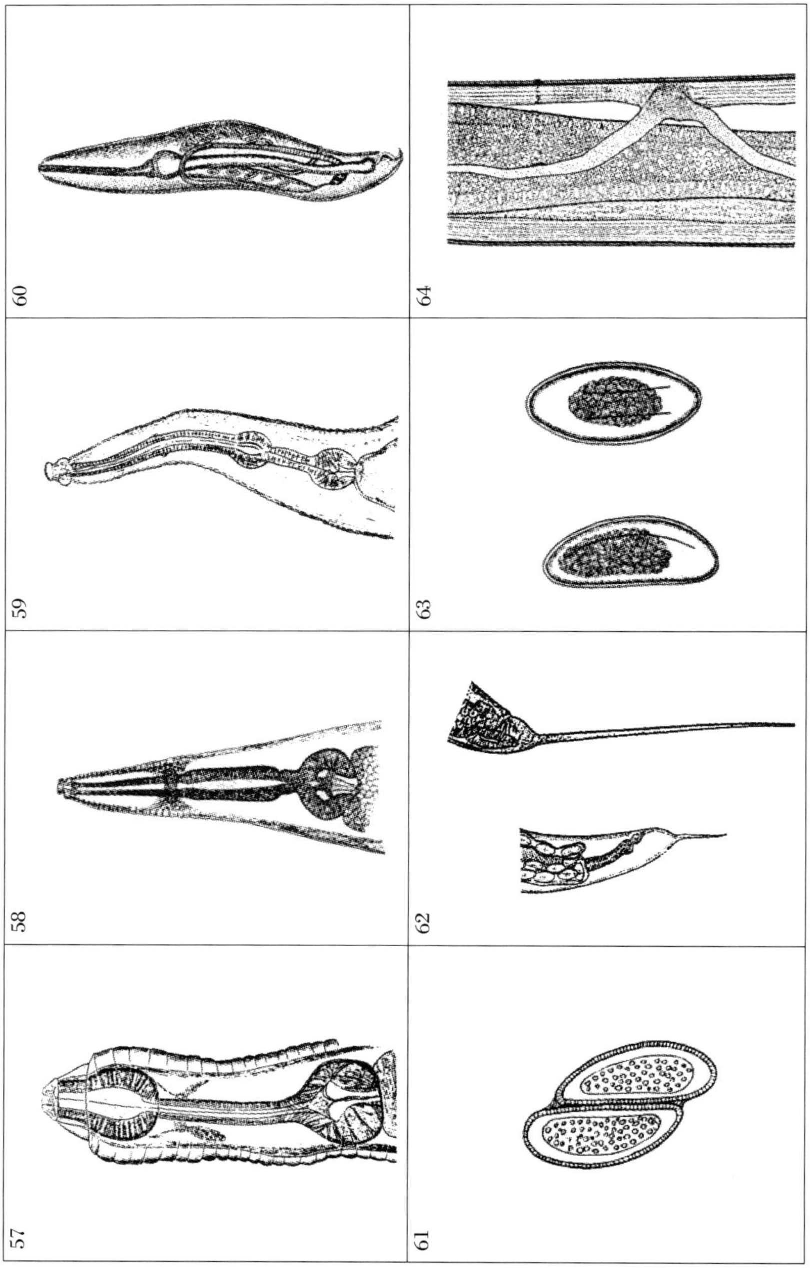
57
58
59
60
61
62
63
64

Figures 16.57 – 16.64 Morphological structures of the family Thelastomatidae. (All illustrations are either redrawn or modified from the cited authors.)

Figure 16.57 Esophagus with short anterior part and large median bulb (Chitwood, 1933).

Figure 16.58 Esophagus with cylindrical median bulb (Chitwood, 1932).

Figure 16.59 Esophagus with spherical median bulb (Chitwood, 1933).

Figure 16.60 A nematode showing long esophagus, about 1/3 of body length, and short tail (Chitwood, 1932).

Figure 16.61 Two eggs fused along flattened surfaces (Parveen and Jairajpuri, 1984).

Figure 16.62 Two types of tail.

Figure 16.63 Eggs showing longitudinal grooves (Chitwood, 1932).

Figure 16.64 Excretory system with large excretory duct (Christie, 1932).

16-Eggs with longitudinal grooves, excretory pore not observed (Fig. 16.63)
Severianoia (Schwenk, 1926) Travassos, 1929
Eggs without longitudinal grooves, excretory pore posterior to basal bulb, excretory duct very large (Fig. 16.64)
Cephalobellus (Syn. *Scarabanema*) Cobb, 1920

Notes: 1 – *Cameronia* is a parasite of mole crickets. 2 – Even though we have included *Blattellicola* in the key, this genus was synonymized with *Blatticola* by Adamson and van Waerebeke (1992a).

16.2.6.2 Family Hystrignathidae Travassos, 1920

Diagnosis: Thelastomatoidea. Cuticle with or without spines or scales. Anterior end with eight papillae. Amphids small. Cephalic region with one or two enlarged annules. Esophageal corpus either cylindrical or divided with posterior part clavate. Isthmus well defined or indistinct. Basal bulb mostly with valves. Females with one or two gonads. Eggs elongate, ornamented with ridges or excrescences. Tail conical, attenuate, sometimes with rounded terminus or with appendage. Male mostly absent, when observed, spicule either single or absent; at least one median single papilla observed. Nematodes in this family are parasites of beetles in family Passalidae.

Key to genera of Hystrignathidae

1-Anterior region with transverse, or longitudinal rows of spines 2
Anterior region without rows of spines 12
2-Female with one gonad 3
Female with two gonads 6
3-Corpus cylindrical anteriorly, clavate posteriorly (Fig. 16.65)
Artigasia Christie, 1934
Corpus cylindrical without clavate portion 4
4-Spines in successive rows alternating
Boraceianema Travassos and Kloss, 1958
Spines in successive rows, aligned 5
5-Oral opening surrounded by pedunculate papillae (Fig. 16.66)
Klossiella Cordeiro, 1981
Oral opening not surrounded by pedunculate papillae
Mentecle Travassos and Kloss, 1958
6-Corpus cylindrical without clavate posterior portion 7
Corpus cylindrical anteriorly, clavate posteriorly 8
7-Cervical cuticle with transverse rows of scale-like projections (Fig. 16.67); male present *Lepidonema* Cobb, 1898
Cervical cuticle with transverse rows of spines, male unknown
Soaresnema Travassos and Kloss, 1958

8-Cervical cuticle with two lateral longitudinal rows of spines (Fig. 16.68)
Carlosia Travassos and Kloss, 1957
Cervical cuticle with transverse rows of spines 9
9-Cervical cuticle with transverse rows of scale-like projections, cephalic region with single circumoral annule (Fig. 16.69)
Salesia Travassos and Kloss, 1958
Cervical cuticle with transverse rows of spines, cephalic region with two annules (Fig. 16.70) 10
10-First row of spines with 16 elements *Hystrignathus* Leidy, 1850
First row of spines with 32 elements 11
11-Buccal cavity divided into anterior spheroid and posterior cylindrical segments (Fig. 16.71) *Urbanonema* Travassos and Kloss, 1958
Buccal cavity without anterior spheroid chamber *Xyo* Cobb, 1898
12-Female with one gonad 13
Female with two gonads 17
13-Anterior end with eight pedunculate papillae (Fig. 16.72)
Coronocephalus Cordeiro, 1981
Anterior end without pedunculate papillae 14
14-Two cephalic annules contiguous *Glaber* Travassos and Kloss, 1958
Two cephalic annules separate 15
15-Body robust, fusiform *Passalidophila* Van Waerebeke, 1973
Body neither robust nor fusiform 16
16-Esophageal corpus short, spindle-shaped (Fig. 16.73)
Christiella Travassos and Kloss, 1957
Esophageal corpus long, cylindrical (Fig. 16.74)
Longior Travassos and Kloss, 1958
17-Esophageal corpus with posterior portion not clavate 18
Esophageal corpus with posterior portion clavate 19
18-Anterior end of esophagus swollen, surrounding base of stoma (Fig. 16.75) *Anomalostoma* Cordeiro, 1981
Anterior end of esophagus not as above
Ventelia Travassos and Kloss, 1958
19-Cephalic end with two annules, tail short, rounded with mucron-like appendage (Fig. 16.76) *Anuronema* Clark, 1978
Cephalic end with one annule, tail not as above 20
20-Circumoral annule in form of truncate cone, isthmus well-defined (Fig. 16.77) *Phalacronema* Clark, 1978
Circumoral annule not as above, isthmus mostly ill-defined 21
21-Vulva in posterior quarter of body *Klossnema* Cordeiro, 1981
Vulva near mid-body 22

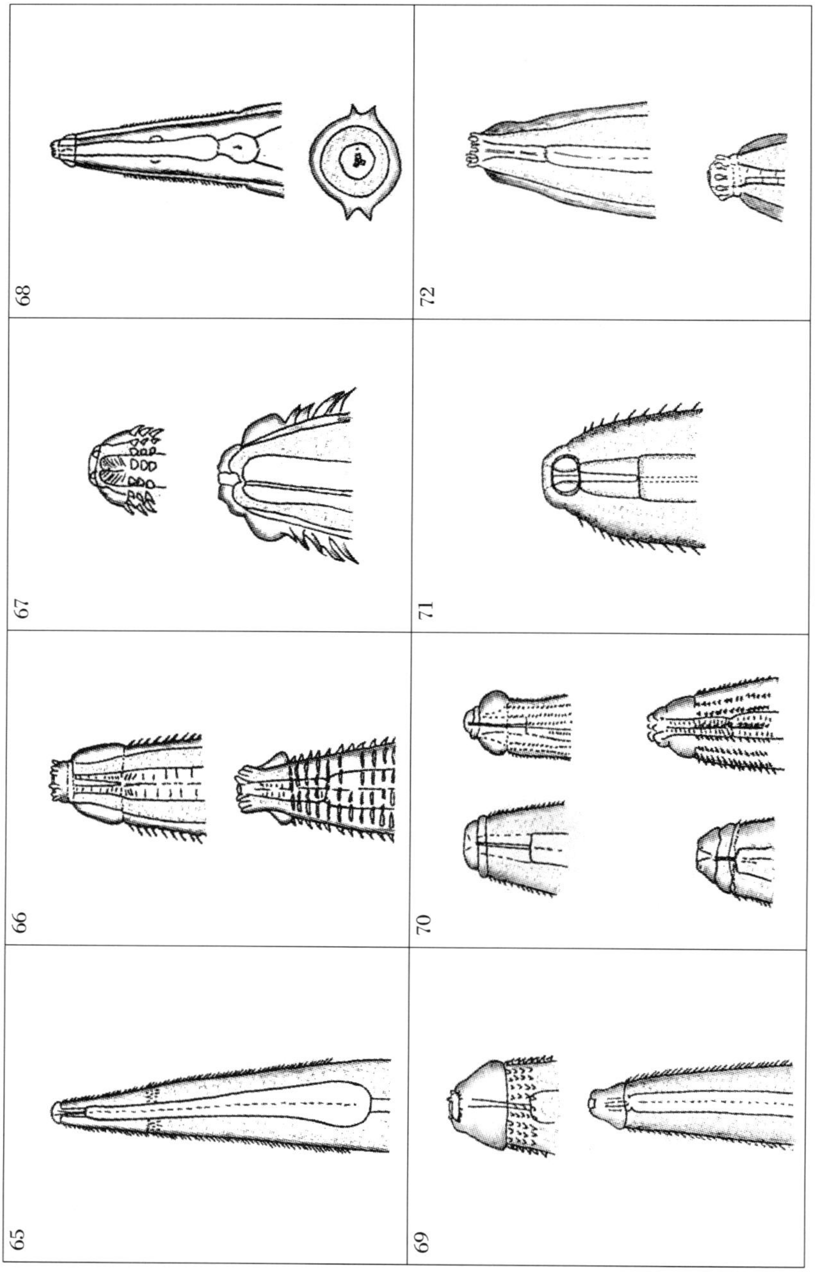
65
66
67
68
69
70
71
72

Figures 16.65 – 16.72 Morphological structures of the family Hystrignathidae. (All illustrations are either redrawn or modified from the cited authors.)

Figure 16.65 Esophagus showing corpus with clavate end (Travassos and Kloss, 1958b).
Figure 16.66 Lip region with pedunculate papillae (Cordeiro, 1981).
Figure 16.67 Cervical cuticle with scale-like structures (Cordeiro, 1981).
Figure 16.68 Cervical region with two lateral longitudinal rows of spines (Travassos and Kloss, 1957).
Figure 16.69 Cephalic region showing single circumoral annule (Travassos and Kloss, 1958b).
Figure 16.70 Cephalic region with two annules (Cordeiro, 1981).
Figure 16.71 Stoma with anterior spheroid and posterior cylindrical segments (Travassos and Kloss, 1958b).
Figure 16.72 Lip region showing eight pedunculate papillae (Cordeiro, 1981).

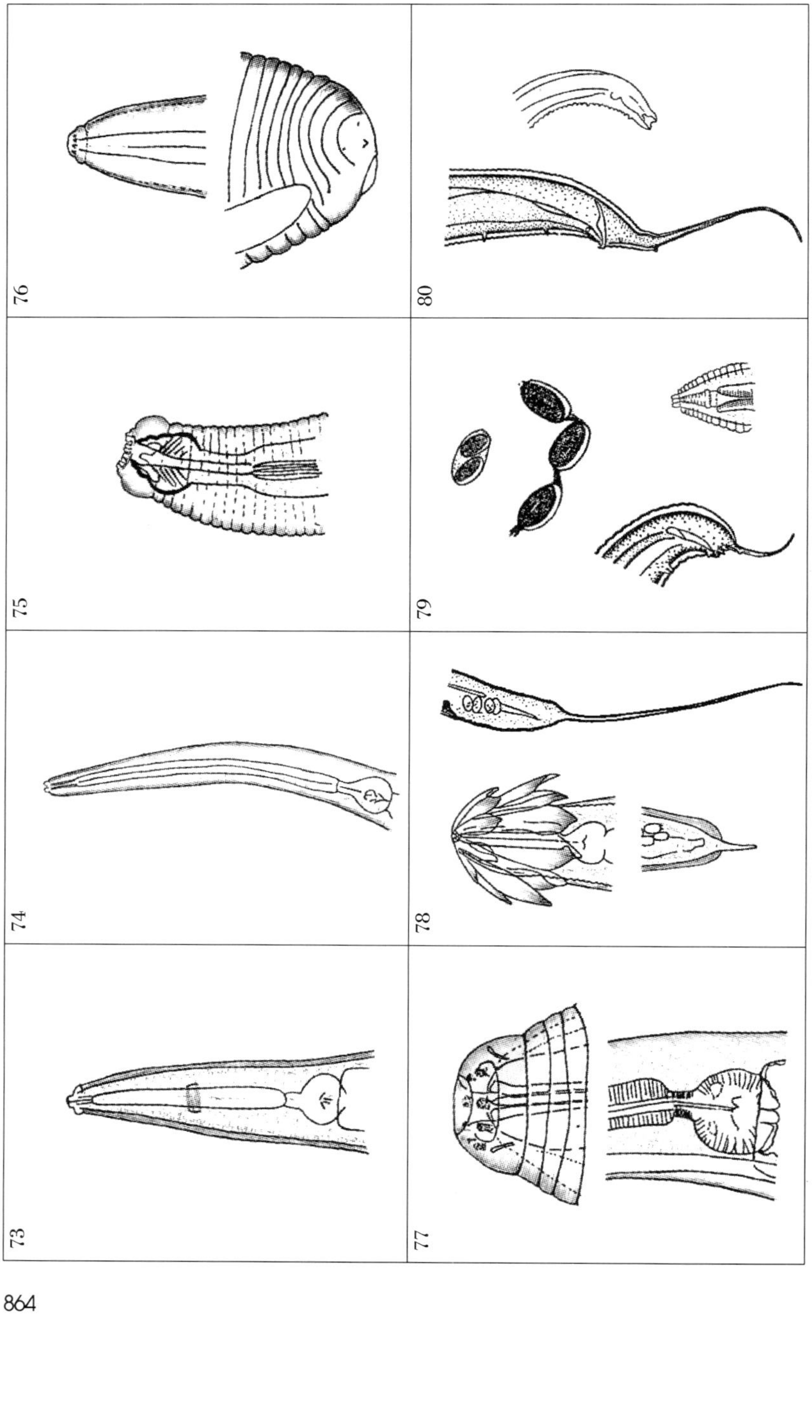
76
75
74
73
80
79
78
77

Figures 16.73 – 16.77 Morphological structures of the family Hystrignathidae.

Figure 16.73 Esophagus with short, spindle-shaped corpus (Travassos and Kloss, 1958c).

Figure 16.74 Esophagus with long, cylindrical corpus (Artigas, 1926).

Figure 16.75 Esophagus showing anterior end swollen surrounding base of stoma (Cordeiro, 1981).

Figure 16.76 Tail terminus short with a mucron-like appendage (Clark, 1978).

Figure 16.77 Labial region and isthmus of *Phalacronema* (Clark, 1978).

Figures 16.78 – 16.80 Morphological structures of the family Travassosinematidae.

Figure 16.78 Head and tail of *Pulchrocephala* (left) (Kloss, 1959), and tail of *Travassosinema* (right) (Rao, 1958).

Figure 16.79 Eggs in a capsule and a chain (top), male tail with small spicule (bottom, left) (Rao, 1958), and long and annulated stoma (bottom right) (Kloss, 1959).

Figure 16.80 Flagellate tail (left) (Rao, 1958), and tail with truncate terminus (right) (Kloss, 1959).

22-Lateral alae broad, bulb without valves *Triumphalisnema* Kloss, 1962
Lateral alae not broad, bulb with poorly developed valves *Sprentia* Clark, 1978

16.2.6.3 Family Travassosinematidae Rao, 1958

Diagnosis: Head with or without leaflike extension. Eggs included in a membrane either connected with each other or separate with one, two or three eggs in a capsule. Male without spicules except in *Binema* and *Isobinema*. Most nematodes in this family are parasites of mole crickets.

Key to genera of Travassosinematidae

1-Female head with leaf-like extensions (Fig. 16.78, top left) 2
Female head without leaf-like extensions 4
2-Body with rows of spines *Indiana* Chakravarty, 1943
Body without rows of spines 3
3-Females with short tail (Fig. 16.78, bottom) *Pulchrocephala* Travassos, 1925
Females with long, filiform tail (Fig. 16.78, right) *Travassosinema* Rao, 1958
4-Eggs in capsules of two or three (Fig. 16.79) 5
Eggs attached together in a chain (Fig. 16.79) 6
5-Male with small spicule (Fig. 16.79, bottom left) *Binema* Travassos, 1925
Male without spicule *Mirzaiella* Basir, 1942
6-Stoma long, annulated (Fig. 16.79, bottom right) *Chitwoodiella* Basir, 1948
Stoma not long, not annulated 7
7-Male tail flagellate (Fig. 16.80, left) *Isobinema* Rao, 1958
Male tail truncate (Fig. 16.80, right) *Singhiella* Rao, 1958

References

Adamson, M. L. 1989. Evolutionary biology of the Oxyurida (Nematoda): biofacies of a haplodiploid taxon. Advances in Parasitology 28: 175 – 228.

Adamson, M. L., and D. van Waerebeke. 1992a. Revision of the Thelastomatoidea, Oxyurida of invertebrate hosts I. Thelastomatidae. Systematic Parasitology 21: 21 – 63.

Adamson, M. L., and D. van Waerebeke. 1992b. Revision of the Thelastomatoidea, Oxyurida of invertebrate hosts 2. Travassosinematidae, Protrelloididae and Pseudonymidae. Systematic Parasitology 21: 169 – 188.

Adamson, M. L., and D. van Waerebeke. 1992c. Revision of the Thelastomatoidea, Oxyurida of invertebrate hosts III. Hystrignathidae. Systemic Parasitology 22: 111 – 130.

Anderson, R. V., and C. Laumond. 1990. *Noctuidonema guyanense* Remillet & Silvain, 1988: morphologie du spicule, redescription du mâle et diagnose amendée du genre *Noctuidonema.* Revue de Nématologie 13: 433: 436.

Andrassy, I. 1976. Evolution as a Basis for the Systematization of Nematodes. London: Pitman.

Andrassy, I. 1983. A Taxonomic Review of the Suborder Rhabditina (Nematoda: Secernentea). Paris: ORSTOM.

Andrassy, I. 1984. Klasse Nematoda (Ordnungen Monhysterida, Desmoscolecida, Araeolaimida, Chromadorida, Rhabditida). Berlin: Akademie-Verlag.

Artigas, P. 1926. Nematoides de invertebrados. Boletim Biologico, São Paulo 1: 1 – 13.

Artyukhovsky, A. K. 1963. A new genus of mermithids, *Melolonthinimermis* gen. nov. (Nematoda: Mermithidae). Materialy Nauchnoi Konferencii Vsesoyuznogo Obshchestva Gel' mintologov, Part 1. Moscow: 22 – 23.

Artyukhovsky, A. K. 1967. *Neoaplectana arenarium* nov. sp. (Steinernematidae: Nematoda) inducing nematode disease in chafers of the Voronezh region. Trudy Voronezhskogo Gosudarstvennogo Zapovednika 15: 94 – 100.

Artyukhovsky, A. K. 1969. New genera of the family Mermithidae (Enoplida: Nematoda). Zoologicheski Zhurnal 48: 1309 – 1319.

Artyukhovsky, A. K., E. M. Kozodoi, A. P. Reid, and S. E. Spiridonov. 1997. Redescription of *Steinernema arenarium* (Artyukhovsky, 1967) topotype from Central Russia and a proposal for *S. anomalae* (Kozodoi, 1984) as a junior synonym. Russian Journal of Nematology 5: 31 – 37.

Baker, A. D. 1962. Check list of the nematode superfamilies Dorylaimoidea, Rhabditoidea, Tylenchoidea and Aphelenchoidea. Leiden, The Netherlands: E. J. Brill.

Baker, G. L., and G. O. Poinar, Jr. 1986. *Mermis quirindiensis* n. sp. (Nematoda: Mermithidae), a parasite of locusts and grasshoppers (Orthoptera: Acrididae) in south-eastern Australia. Revue de Nematologie 9: 125 – 134.

Basir, M. A. 1940. Nematodes parasitic in Indian cockroaches. Proceedings of the Indian Academy of Science 12, Section B: 8 – 16.

Basir, M. A. 1942. Nematodes parasitic in *Grylotalpa.* Records of the Indian Museum 44: 95 – 104.

Basir, M. A. 1948. *Cameronia biovata* gen. et sp. nov. (Thelastomatidae), a nematode parasite of the mole cricket *Gryllotalpa africana* Beauv. Canadian Journal of Research, Section D 26: 201 – 203.

Bedding, R. A. 1968. *Deladenus wilsoni* n. sp. and *D. siricidicola* n. sp. (Neotylenchidae) entomophagus-mycetophagus nematodes parasitic in siricid woodwasps. Nematologica 14: 515 – 525.

Bedding, R. A. 1974. Five new species of *Deladenus* (Neotylenchidae), entomophagous – mycetophagus nematodes parasitic in siricid woodwasps. Nematologica 20: 204 – 225.

Bedding, R. A. 1984. Nematode parasites of Hymenoptera. Pp. 755 – 795. In W. R. Nickle, ed. Plant and Insect Parasitic Nematodes. New York: Marcel Decker.

Blinova, S. L., and E. A. Korenchenko. 1986. *Phaenopsitylenchus laricis* gen. and sp.

n. (Nematoda: Phaenopsitylenchidae n. fam.) parasite of *Phaenops guttulata* and remarks on taxonomy of nematodes of the superfamily Sphaerularioidea. Academia Naukovi SSSR Trudy Gel' mintologicheskoi laboratorii 34: 14 – 23.

Bovien, P. 1932. On a new nematode *Scatonema wülkeri* gen. et spec. n. parasitic in the body cavity of *Scatopse fuscipes* Meig. (Diptera: Nematocera). Videnskabelige Meddelelser Fra Dansk Naturhistorisk Forening Kobenhaven 94: 13 – 32.

Bovien, P. 1937. Some types of association between nematodes and insects. Videnskabelige Meddeleser Fra Dansk Naturhistorisk Forening Kobenhaven 101, 114pp.

Braun, M. 1883. Die thierischer Parasiten des Menchen nebst einer Anleitung zur praktischen Beschä ftigung mit der Helminthologie für Stadierende und Aertze. Wurzburg, 233pp.

Cabanillas, H. E., G. O. Poinar, Jr, and J. R. Raulston. 1994. *Steinernema riobravis* n. sp. (Rhabditida: Steinernematidae) from Texas. Fundamental and Applied Nematology 17: 123 – 131.

Carus, J. V. 1857. Icones zootomicae. Erste Hälfte. Die wirbellosen Thiere. Leipzig: 1 – 32.

Chakravarty, G. K. 1943. On the nematode *Indiana gryllotalpae*, gen. et sp. nov. from *Gryllotalpa* sp. Current Science 12: 257 – 258.

Chamberlin, T. R. 1944. Observations on nematodes associated with white grubs. Journal of Economic Entomology 37: 313 – 314.

Chitambar, J. J. 1991. On the genus *Deladenus* Thorne, 1941 (Nemata: Allantonematidae): Review of mytophagous stage. Revue de Nématologie 14: 427 – 444.

Chitwood, B. G. 1932. A synopsis of nematodes parasitic in insects of the family Blattidae. Zeitschrift für Parasitenkunde 5: 14 – 50.

Chitwood, B. G. 1933. On some nematodes of the superfamily Rhabditoidea and their status as parasites of reptiles and amphibians. Journal of the Washington Academy of Sciences 23: 508 – 520.

Chitwood, B. G. 1935. Nomenclatorial notes. Proceedings of the Helminthological Society of Washington 2: 51 – 54.

Chitwood, B. G., and M. B. Chitwood. 1934. Nematodes parasitic in Philippine cockroaches. Philippine Journal of Science 52: 381 – 393.

Chitwood, B. G., and M. B. Chitwood. 1937. A revised classification of the Nematoda. Pp. 69 – 80. In K. I. Skrjabin, ed. Helminthology, 30 Year Jubileum. Moscow: Izedatel' tsvo vsesoyuznoi Academia Naukovi Lenin.

Chitwood, B. G., and M. B. Chitwood. 1950. An Introduction to Nematology, revised edition. Baltimore, MD: Monumental Printing.

Christie, J. R. 1931. Some nemic parasites (Oxyuridae) of coleopterous larvae. Journal of Agricultural Research 42: 463 – 482.

Christie, J. R. 1934. The nematode genera *Hystrignathus* Leidy, *Lepidonema* Cobb, and *Artigasia* n. g. (Thelastomatidae). Proceedings of the Helminthological Society of Washington 2: 43 – 48.

Christie, J. R. 1936. Life history of *Agamermis decaudata*, a nematode parasite of grasshoppers and other insects. Journal of Agricultural Research 52: 161 – 178.

Clark, W. C. 1978. Nematoda (Oxyuroidea: Thelastomatidae) from Australian passalid beetles (Coleoptera: Passalidae) with descriptions of new species. Australian Journal of Parasitology 26: 603 – 615.

Cobb, N. A. 1898. Extract from a manuscript report on the parasites of stock. Agricultural Gazette of New South Wales 9: 196 – 321, 419 – 454.

Cobb, N. A. 1913. New nematode genera found inhabiting fresh water and non brackish soils. Journal of the Washington Academy of Sciences 3: 432 – 444.

Cobb, N. A. 1919. *Tetradonema plicans* nov. gen. et spec., representing a new family Tetradonematidae as now found parasitic in larvae of the midge-insect *Sciara coprophila* Lintner. Journal of Parasitology 5: 176 – 185.

Cobb, N. A. 1920. One hundred new nemas (Type species of 100 new genera). Contributions to a Science of Nematology 9: 217 – 343.

Cobb, N. A. 1921. *Howardula benigna*, a nemic parasite of the cucumber-beetle (*Diabrotica*). Science 54: 667 – 670.

Cobb N. A. 1922. Two tree-infesting nemas of the genus *Tylenchus*. Annales de Zoologia Aplicada 9: 27 – 35.

Cobb, N. A. 1929. The chromatropism of *Mermis subnigrescens*, a nemic parasite of grass-hoppers. Journal of the Washington Academy of Sciences 19: 159 – 166.

Cobb, N. A., G. Steiner, and J. R. Christie. 1923. *Agamermis decaudata* Cobb, Steiner and Christie; a nema parasite of grasshoppers and other insects. Journal of Agricultural Research 23: 921 – 926.

Coler, R. R., and K. B. Nguyen. 1994. *Paraiotonchium muscadomesticae* n. sp. (Tylenchida: Iotonchidae) a parasite of the housefly (*Musca domestica*) in Brazil and a key to species of the genus *Paraiotonchium*. Journal of Nematology 26: 392 – 401.

Cordeiro, N. S. 1981. Nematoides intestinais de Passalidae; Revisão de sua systematica e catalogacão das especies conhesidas. Doctoral thesis. Institute of biomedical Sciences, University of Sao Paulo, Brazil.

Curran, J. 1989. Chromosome numbers of *Steinernema* and *Heterorhabditis* species. Revue de Nématologie 12: 145 – 148.

Currie, G. A. 1937. Galls on *Eucalyptus* trees: a new type of association between flies and nematodes. Proceedings of the Linnean Society of New South Wales 62: 147 – 174.

Deunff, J. 1984. Parasites de siphonoptères. Doctoral thesis. Université de Rennes U. E. R. Médicales et Pharmaceutiques, France.

Deunff, J., H. Launay, and J. C. Beaucournu. 1985. *Incurvinema helicoides* n. gen. n. sp. Nematodea, Tylenchida: Allantonematidae parasite de *Rhadinopsilla pentacantha* (Rothschild, 1897) (Siphonaptera: Hystrichopsyllidae). Annales de Parasitologie Humaine et Comparée 60: 739 – 746.

Doucet, M. M. A. 1986. A new species of *Neoaplectana* Steiner, 1929 (Nematoda: Steinernematidae) from Cordoba, Argentina. Revue de Nématologie 9: 317 – 323.

Doucet, M. M. A., and M. E. Doucet. 1990. *Steinernema ritteri* n. sp. (Nematoda: Steinernematidae) with a key to the species of the genus. Nematologica 36: 257 – 265.

Dufour, L. 1837. Reserches sur quelques entozoaire et larves parasites des insects Orthoptères. Annales des Sciences Naturelles Zoologie 7: 5 – 20.

Dujardin, F. 1842. Mémoire sur la structure anatomique des *Gordius* et d'un autre helminthe, le *Mermis* qu'on a confondu avec eux. Annales des Sciences Naturelles Zoologie, Second Series 18: 129 – 151.

Dujardin, F. 1845. Histoire naturelle des helminthes ou ver intestinaux. Paris.

Elawad, S., W. Ahmad, and A. Reid. 1997. *Steinernema abbasi* sp. n. (Nematoda: Steinernematidae) from the Sultanate of Oman. Fundamental and Applied Nematology 20: 433 – 442.

Filipjev, I. N. 1934. The classification of the free-living nematodes and their relations to parasitic nematodes. Smithsonian Miscellaneous Collections 89: 1 – 63.

Filipjev, I. N. 1934. Parasites of insects. Chapter 5, Harmful and useful nematodes in rural economy. Moscow: OGIZ-Sel' khozgiez.

Fortuner, R., and D. J. Raski, 1987. A review of Neotylenchoidea Thorne 1941 (Nemata: Tylenchida). Revue de Nematologie 10: 257 – 267.

Fuchs, A. G. 1937. Neue an Borken-und Rüsselkäfern und einige antere Nematoden. I Teil. Die Parasiten der Waldgärtner *Myelophilus piniperda* (L) und *minor* (Hartig) and die Genera *Rhabditis* Dujardin, 1945 and *Aphelenchus* (Bastian, 1865). Zoologiche Jahrbücher Systematik 70: 123 – 190.

Gardner, S. L., S. P. Stock, and H. K. Kaya. 1994. A new species of *Heterorhabditis* from the Hawaiian islands. Journal of Parasitology 80: 100 – 106.

Geraert, E. 1966. The systematic position of the family Tylenchulidae and Criconematidae. Nematologica 12: 362 – 368.

Glaser, R. W., and C. C. Farrel. 1935. Field experiments with Japanese beetle and its nematode parasite. Journal of the New York Entomological Society 43: 345 – 371.

Goodey, T. 1941. On the morphology of *Mermithonema entomophilum* n. g., n. sp., a nematode parasite of the fly *Sepsi cynipsea* L. Journal of Helminthology 19: 105 – 114.

Goodey, T. 1953. On certain eelworms, including Butschli's *Tylenchus fungorum*, obtained from toadstools. Journal of Helminthology 27: 81 – 94.

Goodey, T. 1963. Soil and Fresh Water Nematodes. Second edition, revised by J. B. Goodey. London: Methuen.

Griffith, R. 1968. The mechanism of transmission of the red ring nematode. Journal of the Agricultural Society of Trinidad and Tobago, Paper No. 906.

Hunt, D. J. 1980. *Acugutturus parasiticus* n. g., n. sp., a remarkable ectoparasitic aphelenchoid nematode from *Periplaneta americana* (L.) with proposal of Acugutturinae n. subf. Systematic Parasitology 1: 167 – 170.

Hyman, L. H. 1951. The invertebrates: Acanthocephala, Aschelminthes, and Entoprocta – the Pseudocoelomate Bilateria. Vol. 3. New York: McGraw-Hill.

Iperti, G., and D. van Waerebeke. 1968. Description, biologie et importance d' une nouvelle espèce d' allantonematidae (Nématode) parasite des coccinelles aphidiphage: *Parasitylenchus coccinellinae* n. sp. Entomophaga 13: 107 – 119.

Jian, B., A. P. Reid, and D. J. Hunt. 1997. *Steinernema ceratophorum* n. sp. (Nematoda: Steinernematidae) a new entomopathogenic nematode from northeast China. Systematic Parasititology 37: 115 – 125.

Kaburaki, T., and S. Imamura. 1932. A new mermithid-worm parasitic in the rice borer with notes on its life histories and habits. Proceedings of the Imperial Academy of Japan 8: 109 – 112.

Kaiser, H. 1991. Terrestrial and semiterrestrial Mermithidae. Pp. 899 – 965. In W. R. Nickle, ed. Manual of Agricultural Nematology. New York: Marcel Decker.

Keilin, M. D. 1917. Sur un nematode nouveau, *Aprotonema entomophagum* n. g., n. sp.

Compte Rendu de l' Academie des Sciences, Paris 165: 399 – 401
Khan, A., W. M. Brooks, and H. Hirschmann. 1976. *Chromonema heliothidis* n. gen., n. sp. (Steinernematidae, Nematoda), a parasite of *Heliothidis zeae* (Noctuidae, Lepidoptera), and other insects. Journal of Nematology 8: 159 – 168.
Kloss, G. R. 1959. Nematoides parasitos de Gryllotalpoidea (Orthoptera) do Brasil. Memorias do Instituto Oswaldo Cruz 57: 137 – 170.
Kloss, G. R. 1962. Alguns parasitos do intestino de Coleopteros da familia Passalidae. Papeis Avulsos do Departamento de Zoologia, Sao Paulo 15: 163 – 171.
Körner, H. 1954. Die nematodenfauna desvergehenden holzes und ihre Beziechungen zu den Insecten. Zoologiche Jahrbücher Systematik 82: 157 – 404.
Kozodoi, E. M. 1984. A new entomopathogenic nematode *Neoaplectana anomali* sp. n. (Rhabditida: Steinernematidae) and observations on its biology. Zoological Journal 63: 1605 – 1609.
Kreis, H. A. 1932. Beiträge zur Kenntnis pflanzenparasitischer Nematoden. Zeitschrift Parasitenkunde 5: 184 – 194.
Launay, H., J. Deunff, and O. Bain. 1983. *Spilotylenchus arthuri*, gen. n., sp. n. (Nematodea, Tylenchida: Allantonematidae) parasite of *Spilopsyllus cuniculi* (Dale) (Siphonaptera: Pulicidae). Annales de Parasitologie Humaine et Comparée 58: 141 – 150.
Lazarevskaya, S. 1962. On the nematode fauna of *Orthotomicus laricis* and *Orthomicus proximus* (Coleoptera: Ipidae). Helminthologia 4: 254 – 265.
Leidy, J. 1849. New genera and species of entozoa. Proceedings of the National Academy of Sciences. Philadelphia 4: 225 – 233.
Leidy, J. 1850. Description of some nematoid Entozoa infesting insects. Proceedings of the National Academy of Sciences. Philadelphia 5: 100 – 102.
Leuckart, R. 1884. Über einen neuen heterogenen Nematoden. Tagebl. 57. Verhandlungen Gessellschaft Deutscher Naturforscher und Aerzte 320.
Leuckart, R. 1887. Neue Beiträge zur Kenntnis des Baues und der Lebensgeschichte der Nematoden. Abhandlungen Mathematisch-Physische Klasse der Säechsischen Akademie der Wissenschaften 13: 565 – 704.
Liu, J. 1994. A new species of the genus *Heterorhabditis* from China (Rhabditida: Heterorhabditidae). Acta Zootaxonomica Sinica 19: 268 – 272.
Liu, J., and R. E. Berry. 1996a. *Steinernema oregonensis* n. sp. (Rhabditida: Steinernematidae) from Oregon, USA. Fundamental and Applied Nematology 19: 375 – 380.
Liu, J., and R. E. Berry. 1996b. *Heterorhabditis marelatus* n. sp. (Rhabditida: Heterorhabditidae) from Oregon. Journal of Invertebrate Pathology 67: 48 – 54.
Lubbock, J. 1861. On *Sphaerularia bombi*. Natural History Review 1: 44 – 57.
Luc, P. V., K. B. Nguyen, A. P. Reid, and S. E. Spiridonov. 2000. *Steinernema tami* sp. n. (Rhabditida: Steinernematidae) From Cat Tien Forest, Vietnam. Russian Journal of Nematology 8: 33 – 43.
Maggenti, A. R. 1982. Nemata. Pp. 878 – 923. In S. P. Parker, ed. Synopsis and Classification of Living Organisms. New York: McGraw-Hill.
Maggenti, A. R. 1991. Nemata: Higher classification. Pp. 174 – 187. In W. R. Nickle, ed. Manual of Agricultural Nematology. New York: Marcel Dekker.

Mamiya, Y. 1988. *Steinernema kushidai* n. sp. (Nematoda: Steinernematidae) associated with scarabaeid beetle larvae from Shizuoca, Japan. Applied Entomology and Zoology 23: 313 – 320.

Marti, O. G., Jr., and C. E. Rogers. 1995. *Noctuidonema dibolia* n. sp. (Aphelenchida: Acugutturidae), an ectoparasite of the moth *Micis latipes* (Lepidoptera: Noctuidae). Journal of Nematology 27: 387 – 394.

Massey, C. L. 1974. Biology and taxonomy of nematode parasites and associates of bark beetles in the United States. Agriculture Handbook No. 446. USDA, Washington, D. C.

Maupas, E. F. 1919. Essais d' hybridation chez des nematodes. Bulletin Biologique de la France et de la Belgique 52: 466 – 498.

McCoy, E. E., and R. W. Glaser. 1936. Nematode culture for Japanese beetle control. New Jersey Department of Agriculture Circular No. 265.

Merrill, J. H., and A. L. Ford. 1916. Life history and habits of two new nematode parasites on insects. Journal of Agricultural Research 6: 115 – 127.

Micoletzky, H. 1922. Die freilebenden Erd – Nematoden mit besonderer Berücksichtigung der Steiermark und der Bukowina, zugleich mit einer Revision sämtlicher nichtmariner, freilebender Nematoden in Form von Genus-Beschreibungen und Bestimmungsschlüsseln. Archiv Naturgeschichte, Berlin 87: 1 – 650

Misuta, Y., and M. Sato. 1965. Some observations on a parasitic nematode from the silkworm. Shanshi Kenkyu 57: 22 – 24.

Mracek, Z. 1994. *Steinernema kraussei* (Steiner, 1923) (Nematoda: Rhabditida: Steinernematidae): Redescription of its topotype from Westphalia. Folia Parasitologica 41: 59 – 64.

Mracek, Z., E. A. Hernandez, and N. E. Boemare. 1994. *Steinernema cubana* sp. n. (Nematoda: Rhabditida: Steinernematidae) and the preliminary characterization of its associated bacterium. Journal of Invertebrate Pathology 64: 123 – 129.

Mracek, Z. 1977. *Steinernema kraussei,* a parasite of the body cavity of the sawfly, *Cephaleia abietis* in Czechoslovakia. Journal of Invertebrate Pathology 30: 87 – 94.

Nguyen, K. B., and G. C. Smart, Jr. 1988. A new steinernematid nematode from Uruguay. Journal of Nematology 20: 651.

Nguyen, K. B., and G. C. Smart, Jr. 1990. *Steinernema scapterisci* n. sp. (Steinernematidae: Nematoda). Journal of Nematology 22: 187 – 199.

Nguyen, K. B., and G. C. Smart, Jr. 1992a. *Steinernema neocurtillis* n. sp. (Rhabditida: Steinernematidae) and a key to species of the genus *Steinernema*. Journal of Nematology 24: 463 – 477.

Nguyen, K. B., and G. C. Smart, Jr. 1992b. Addendum to the morphology of *Steinernema scapterisci*. Journal of Nematology 24: 478 – 481.

Nguyen, K. B., and G. C. Smart, Jr. 1993. Scanning electron microscope studies of *Steinernema anomali* Kozodoi, 1984. Journal of Nematology 25: 486 – 492.

Nguyen, K. B., and G. C. Smart, Jr. 1994. *Neosteinernema longicurvicauda* n. gen., n. sp. (Rhabditida: Steinernematidae), a parasite of the termite *Reticulitermes flavipes* (Koller). Journal of Nematology 26: 162 – 174.

Nguyen, K. B., and G. C. Smart, Jr. 1995a. Morphometrics of infective juveniles of *Steinernema* spp. and *Heterorhabditis bacteriophora* (Nemata: Rhabditida). Journal of Nematology 27: 206 – 212.

Nguyen, K. B., and G. C. Smart, Jr. 1995b. Scanning electron microscope studies of *Steinernema glaseri* (Nematoda: Steinernematidae). Nematologica 41: 183 – 190.

Nguyen, K. B., and G. C. Smart, Jr. 1996. Identification of entomopathogenic nematodes in the Steinernematidae and Heterorhabditidae (Nemata, Rhabditida). Journal of Nematology 28: 286 – 300.

Nguyen, K. B., and G. C. Smart, Jr. 1997. Scanning electron microscope studies of spicules and gubernacula of *Steinernema* spp. (Nemata: Steinernematidae). Nematologica 43: 465 – 480.

Nickle, W. R. 1963. *Bovienema* (Nematoda: Allantonematidae), a new genus parasitizing bark beetles of the genus *Pityogenes* Bedel, with notes on other endoparasitic nematodes of scolytids. Proceedings of the Helminthological of Society of Washington 30: 256 – 262.

Nickle, W. R. 1967. On the discussion of the insect parasitic nematodes of the Sphaerulariidae Lubbock, 1861 (Tylenchoidea: Nematoda). Proceedings of the Helminthological Society of Washington 34: 72 – 94.

Nickle, W. R. 1969. *Corethrellonema grandispiculosum* n. gen., n. sp. and *Aproctonema chapmani* n. sp. (Nematoda: Tetradonematidae), parasites of the dipterous insect genera, *Correthrella* and *Culicoides* in Louisiana. Journal of Nematology 1: 49 – 54.

Nickle, W. R. 1970a. Description of Entaphelenchidae n. fam., *Roveaphelenchus jonesi* n. g., n. sp. (Nematoda). Proceedings of the Helminthological Society of Washington 37: 105 – 109.

Nickle, W. R. 1970b. A taxonomic review of the genera of the Aphelenchoidea (Fuchs, 1937) Thorne, 1949 (Nematoda: Tylenchida). Journal of Nematology 2: 375 – 392.

Nickle, W. R. 1972. A contribution to our knowledge of the Mermithidae (Nematoda). Journal of Nematology 4: 113 – 146.

Nickle, W. R., and H. E. Welch. 1984. History, development, and importance of insect nematology. Pp. 627 – 653. In W. R. Nickle, ed. Plant and Insect Nematology. New York: Marcel Dekker.

Oerley, L. 1880. Az anguilluluidák magánrajza. A kir. m. természettudom. térsulat által a bugatdijjal jutalmazott pályamii. Term szetrajzi Füzetek 4: 16 – 150.

Osche, G. 1952. Systematik und Phylogenie der gattung *Rhabditis* (Nematoda). Zoologiche Jahrbücher Systematik 81: 191 – 280.

Paesler, F. 1962. Nematodenin Reisspreu. Beschreibung einiger Arten und Sukzessionserscheinungen. Opuscula Zoologica (Budapest) 4: 105 – 120.

Paramonov, A. A. 1952. Oryt zhologicheskoi klassifikatsii fitonematod. Trudy Gel' mintologicheskoi Laboratorii. 6: 338 – 364.

Parveen, R., and D. S. Jairajpuri. 1984. *Cameronia klossi* n. sp. (Nematoda: Thelastomatidae) from the mole cricket, *Gryllotalpa africana* from Aligarh. Revista Iberica de Parasitologia 44: 153 – 158.

Pereira, C. 1932 *Myenchus botelhoi* n. sp., curiosa nematoide parasite de *Limnobdella brasiliensis* Pinto (Hirudinea). Thesis, Facultad de Medecina , Sao Paulo, Brazil.

Peterson, J. J. 1984. Nematode parasites of mosquitos. Pp. 797 – 820. In W. R. Nickle, ed. Plant and Insect Nematodes. New York: Marcel Decker.

Peterson, J. J., H. C. Chapman, and D. B. Woodard. 1968. Bionomics of a mermithid nematode of larval mosquitos in southwestern Louisiana. Mosquito News 28: 346 – 352.

Poinar, G. O., Jr. 1969a. Diplogasterid nematodes (Diplogasteridae: Rhabditida) and their relationship to insect disease. Journal of Invertebrate Pathology 13: 447 – 454.

Poinar, G. O., Jr. 1969b. *Praecocilenchus rhaphidophorus* n. gen., n. sp. (Nematoda: Aphelenchoidea) parasitizing *Rhynchophorus bilineatus* (Montrouzier) (Coleoptera: Curculionidae) in New Britain. Journal of Nematology 1: 227 – 231.

Poinar, G. O., Jr. 1972. Nematodes as facultative parasites of insects. Annual Review of Entomology 17: 103 – 122.

Poinar, G. O., Jr. 1975. Entomogenous nematodes, a manual and host list of insect – nematode associations. Leiden, The Netherlands: E. J. Brill.

Poinar, G. O., Jr. 1976. Description and biology of a new insect parasitic rhabditoid, *Heterorhabditis bacteriophora* n. gen. n. sp. (Rhabditida: Heterorhabditidae n. fam.) Nematologica 21: 463 – 470.

Poinar, G. O., Jr. 1977. *Empidomermis cozii* n. gen., n. sp. (Mermithidae: Nematoda) a parasite of *Anopheles funestus* (Culicidae: Diptera) in West Africa. Canadian Journal of Zoology 55: 1475 – 1479.

Poinar, G. O., Jr. 1979. Nematodes for biological control of insects. Boca Raton, FL: CRC Press.

Poinar, G. O., Jr. 1981. *Thaumamermis cosgrovei* n. gen., n. sp. (Mermithidae: Nematoda) parasitizing terrestrial isopods (Isopoda: Oniscoidea). Systematic Parasitology 2: 261 – 266.

Poinar, G. O., Jr. 1985. *Neoaplectana intermedia* n. sp. (Steinernematidae: Nematoda) from South Carolina. Revue de Nématologie 8: 321 – 327.

Poinar, G. O., Jr. 1988. Redescription of *Neoaplectana affinis* Bovien (Rhabditida: Steinernematidae). Revue de Nématologie 11: 143 – 147.

Poinar, G. O., Jr. 1990. Taxonomy and biology of Steinernematidae and Heterorhabditidae. Pp. 23 – 60. In R. Gaugler and H. K. Kaya, eds. Entomopathogenic Nematodes in Biological Control. Boca Raton, FL: CRC Press.

Poinar, G. O., Jr., and C. L. B. Benton, Jr. 1986. *Aranimermmis aptispicula* n. g. n. sp. (Mermithidae: Nematoda), a parasite of spiders (Arachnida: Araneida). Systematic Parasitology 8: 33 – 38.

Poinar, G. O., Jr., and B. O. Nelson. 1973. *Psylotylenchus viviparus* n. gen., n. sp. (Nematoda: Tylenchida: Allantonematidae) parasitizing fleas (Siphonaptera) in California. Journal of Medical Entomology 10: 349 – 354.

Poinar, G. O., Jr., G. M. Thomas, and R. Hess. 1977. Characteristics of the specific bacterium associated with *Heterorhabditis bacteriophora* (Heterorhabditidae: Rhabditida). Nematologica 23: 96 – 102.

Poinar, G. O., Jr., T. Jackson, and M. Klein. 1987. *Heterorhabditis megidis* sp. n. (Heterorhabditidae: Rhabditida) parasitic in the Japanese beetle, *Popillia japonica* (Scarabaeidae: Coleoptera), in Ohio. Proceedings of the Helminthological Society of Washington 54: 53 – 59.

Poinar, G. O., Jr., G. K. Karunakar, and H. David. 1992. *Heterorhabditis indicus* n. sp. (Rhabditida: Nematoda) from India: Separation of *Heterorhabditis* spp. by infective juveniles. Fundamental and Applied Nematology 15: 467 – 472.

Poinar, G. O., Jr., R. S. Lane, and G. M. Thomas. 1976. Biology and description of *Pheromermis pachisoma* (V. Linstow) n. gen., n. comb. (Nematoda: Mermithidae), a parasite of yellow jackets (Hymenoptera: Vespidae). Nematologica 22:360–370.

Poinar, G. O., Jr., C. Laumond, and E. Bonifassi. 1980. *Howardula apioni* sp. n. (Allantonematidae: Nematoda), a parasite of *Apion carduorum* Kirby (Curculionidae: Coleoptera) in southern France. Proceedings of the Helminthological Society of Washington 47:218–223.

Polozhentsev, A. P. 1941. The mermithid fauna of *Melolontha hippocastani* Fabr. Trudy Bashkirskogo Nauchno. – Issledovatel Veterinarski Stantsii 3:301–334.

Rao, P. N. 1958. Studies on the nematode parasites of insects and other arthropods. Arquivos do Museu Nacional, Rio de Janeiro 46:33–84.

Remillet, M., and C. Laumond. 1991. Sphaerularoid nematodes of importance in agriculture. Pp. 967–1024. In W. R. Nickle, ed. Manual of Agricultural Nematology. New York: Marcel Dekker.

Remillet, M., and J. F. Silvain. 1988. *Noctuidonema guyanese* n. g., n. sp. (Nematoda: Aphelenchoididae) ectoparasite de noctuelles de genre *Spodoptera* (Lepidoptera: Noctuidae). Revue de Nématologie 11:21–24.

Remillet, M., and D. Van Waerebeke. 1978. Description et cycle biologique de *Heteromorphotylenchus stelidotae* n. g., n. sp. et de *Heteromorphotylenchus carpophilli n.* sp. (Nematoda: Allantonematidae). Nematologica 24:222–238.

Roman, J., and W. Figueroa. 1994. *Steinernema puertoricensis* n. sp. (Rhabditida: Steinernematidae) a new entomopathogenic nematode from Puerto Rico. Journal of Agriculture, University of Puerto Rico 78:167–175.

Rühm, W. 1954. Einige neue epidenspezifische Nematodenarten. Zoologischer Anzeiger 153:221–242.

Rühm, W. 1956. Die Nematoden der Ipiden. Parasitologische Schriftenreihe, 437pp.

Rühm, W., and C. Chararas. 1957. Description, biologie et histologie de quatre especes nouvelles de nématodes parasites de *Dryocoetes hectographus* (Reit). (Col. Scolytidae). Entomophaga 2:253–269.

Sandner, H., and S. Stanuszek. 1967. Limitation of the Colorado beetle by nematodes and prospects for their practical utilization. Proceedings of the International Colloquium on Insect Pathology and Microbial Control. 210–212. Amsterdam: North Holland.

Schneider, A. F. 1866. Monographie der Nematoden. Berlin: Druck und Verlag Von Georg Reimer, 357pp.

Schuurmans-Stekhoven, J. H., Jr., and P. M. Mawson. 1955. Mermithides d' Alsace. Annales de Parasitologie Humaine et Comparée 50:69–82.

Schwenck, J. 1926. Fauna parasitologicados Blattideos do Brasil. Sciencia Medica. Anno IV 9:3–16.

Shen, C. P., and G. H. Wang. 1991. Description and study of an entomopathogenic nematode: *Steinernema longicaudum* sp. nov. Pp. 220–231. In Proceedings of the First National Academy Symposium. Young and Middle Aged Science and Technology Works, Plant Protection, Beijing, China: Chinese Science and Technology Press.

Siddiqi, M. R. 1986. Tylenchida: parasites of plants and insects. Slough, UK: Commonwealth Institute of Parasitology.

Siddiqi, M. R., and J. B. Goodey. 1964. The status of the genera and subfamily of Criconematidae (Nematoda); with a comment on the position of *Fergusobia*. Nematologica 9:363–377.

Skarbilovich, T. S. 1947. Revision of the systematics of the nematode family Anguillulinidae Baylis and Daubney, 1926. Doklady Akademii Nauk, SSR 57: 307 – 308.

Skarbilovich, T. S. 1959. On the structure and systematics of the nematode order Tylenchida Thorne, 1949. Acta Polonica 7: 117 – 132.

Skryabin, K. I., N. P. Shikhobalova, and A. A. Mozgovoi. 1951. Key to Parasitic nematodes, Vol. 2, Oxyurata and Ascaridata. Moscow: Akademiy Nauk SSSR Publishers.

Slobodyanyuk, O. V. 1975. Erection of *Paraiotonchium* n. g. (Nematoda: Sphaerulariidae) and description of the type species, *P. autumnalis* (Nickle, 1967) n. comb. Trudy Gel' mintologicheskoi Laboratorii. 25: 156 – 168.

Slobodyanyuk, O. V. 1984. Entomopathogenic nematodes of Diptera. Order Tylenchida. Moscow: Nauks.

Slobodyanyuk, O. V. 1986. Revision of the genus *Heterotylenchus* Bovien, 1937 and erection of the genus *Wachekitylenchus* gen. n. (Allantonematidae: Tylenchida). Parazitologiya 20: 466 – 475.

Stanuszek, S. 1974. *Neoaplectana feltiae* complex (Nematoda, Steinernematidae) its taxonomic position within the genus *Neoaplectana* and intraspecific structure. Zeszyty Problemowe Postepow Nauk Rolniczych 154: 331 – 360.

Steiner, G. 1923. *Aplectana krassi* n. sp., eine in der Blattwespe *Lyda* sp. parasitierende Nematodenform, nebst Bomerkungen uber das Seitenorgan der parasitischen Nematoden. Zentralblatt fuer Bakteriologie Parasitenkunde Infektionskrankheiten und Hygiene Aeilung I Originale 59: 14 – 18.

Steiner, G. 1924. Beiträge zur Kenntnis der Mermithidae. 2 Teil. Mermithiden aus Paraguay in der Sammlung des Zoology. Museums zu Berlin. Zentralblatt fuer Bakteriologie Parasitenkunde Infektionskrankheiten. 62: 90 – 110.

Steiner, G. 1925. Mermithids parasitic in the tea bug (*Helopeltis antonii* Sign.). Med. Proefstation Three, Batavia 94: 10 – 16.

Steiner, G. 1929. *Neoaplectana glaseri* n. g., n. sp. (Oxyuridae) a new nemic parasite of the Japanese beetle (*Popillia japonica* Newm). Journal of the Washington Academy of Science 19: 436 – 440.

Stock, S. P. 1993. A new species of the genus *Heterorhabditis* Poinar, 1975 (Nematoda: Heterorhabditidae) parasitizing *Graphognathus* sp. larvae (Coleoptera: Curculionidae) from Argentina. Research and Reviews in Parasitology 53: 103 – 107.

Stock, S. P., H. Y. Choo, and H. K. Kaya. 1997. An entomopathogenic nematode, *Steinernema monticolum* sp. n. (Rhabditida: Steinernematidae) from Korea with a key to other species. Nematologica 43: 15 – 29.

Stock, S. P., D. Strong, and S. L. Gardner. 1996. Identification of *Heterorhabditis* (Nematoda: Heterorhabditidae) from California with a new species isolated from larvae of the ghost moth *Hepialius californicus* (Lepidoptera: Hepialidae) from the Bodega Bay Natural Reserve. Fundamental and Applied Nematology 19: 585 – 592.

Stock, S. P., V. Somsook, and A. Reid. 1998. *Steinernema siamkayai* n. sp. (Rhabditida: Steinernematidae), an entomopathogenic nematode from Thailand. Systematic Parasitology 41: 105 – 113.

Strassen, O. K. L., zur. 1892. *Bradynema rigidum* v. Sieb. Zeitschrift Wissenschaftliche

Zoologie 54: 655 – 747.

Swain, R. B. 1945. The association of nematodes of the genus *Diplogaster* with white – fringed beetles. Journal of Economic Entomology 38: 488 – 490.

Tallosi, B., A. Peters, and R. Ehlers. 1995. *Steinernema bicornutum* sp. n. (Rhabditida: Steinernematidae) from Vojvodina, Yugoslavia. Russian Journal of Nematology 3: 71 – 80.

Thorne, G. 1941. Some nematodes of the family Tylenchidae which do not possess a valvular median esophageal bulb. Great Basin Naturalist 2: 37 – 85.

Tipping, C., and K. B. Nguyen. 1998. *Thripinema fuscum* n. sp. (Tylenchida: Allantonematidae), a parasite of the tobacco thrips, *Frankliniella fusca* (Thysanoptera). Journal of Nematology 30: 232 – 236.

Travassos, L. 1920. Esboco de uma chave geral de nematodeos parasitos. Revista Veterinaria y Zootecnia 10: 59 – 70.

Travassos, L. 1925. Quelques nematodes du *Gryllotalpa*. Compte Rendu Societe Biologie, Compte Rendu de l' Academie des Sciences, Paris 93: 140 – 146.

Travassos, L. 1927. Sobre o genera *Oxysomatium*. Boletim Biologico 5: 20 – 21.

Travassos, L. 1929. Contribuica preliminar a systematica dos nematodeos dos arthropodos. Memoriasdo Instituto Oswaldo Cruz 5: 19 – 25.

Travassos, L., and G. R. Kloss. 1957. Nematodeos de invertebrado. 1a. Nota. Revista Brasilieira do Biologia 17: 295 – 302.

Travassos, L., and G. R. Kloss. 1958a. Nematodeos de invertebrado. 14a. Nota. Alas do Sociedade de Rio de Janeiro 2: 27 – 30.

Travassos, L., and G. R. Kloss. 1958b. Sobre a fauna de Nematodeos dos Coleopteros – Passalidae da Estacao biologica de Boraceia. Arquivos de Zooloca, São Paulo 11: 23 – 57.

Travassos, L., and G. R. Kloss. 1958c. Contribuicaoao conhecimentos dos nematodeos de Coleopteros Passalidae. Revista Brasilieira do Biologia 18: 55 – 57.

Van Waerebeke, D. 1973. Quatre nouvelles espèces d' oxyuroides associés aux Oryctes à Madagascar. Bulletin du Muséum National d' Histoire Naturelle, 3ième série; no 123, Zoologie 95: 535 – 553.

Van Waerebeke, D., and M. Remillet. 1971. Existence de mâle hermaphrodites ovigère intervenant dans le cycle htérogonique d'un nématode parasite de Nitidulidae (Coléoptéres). Paris 273: 2275 – 2278.

Van Waerebeke, D., and M. Remillet. 1973. Morphologie et biologie de *Heterogonema ovomasculis* n. sp. (Nematoda: Tetradonematidae) parasite de Nitidulidae (Coleoptera). Nematologica: 19: 80 – 92.

Volk, J. 1950. Die nematoden der Regenwürmer under aasbesuchenden käfer. Zoologiche Jahrbücher Systematik 79: 1 – 70.

Wachek, F. 1955. Die entoparasitischen Tylenchiden. Parasitologische Schriftenreihe 3: 1 – 119.

Waturu, C. N., D. J. Hunt, and A. P. Reid. 1997. *Steinernema karii* sp. n. (Nematoda: Steinernematidae), a new entomopathogenic nematode from Kenya. International Journal of Nematology 7: 68 – 75.

Weingärtner, I. 1955. Versuch einer Neuordnung der Gattung *Diplogaster* Schultze, 1857 (Nematoda). Zoologiche Jahrbücher Systematik 83: 185 – 322.

Weiser, J. 1955. *Neoaplectana carpocapsae* n. sp. (Anguillulata, Steinernematinae), novy cozopasnik housenek obalece jablecnebo, *Carpocapsa pomonella* L. Vestnik Ceskoslovenske Spolecnosti Zoologicke 19: 44 – 52.

Wouts, W. M., Z. Mracek, S. Gerdin, and R. A. Bedding. 1982. *Neoaplectana* Steiner, 1929 a junior synonym of *Steinernema* Travassos, 1927 (Nematoda: Rhabditida). Systematic Parasitology 4: 147 – 154.

Wülker, G. 1923. Über Fortpflanzung und Entwieklung von *Allantonema* und verwandten Nematoden. Ergebnisse und Fortschritte der Zoologie 5: 389 – 507.

Xu, F. Z., G. H. Wang, and X. F. Li. 1991. A new species of the genus *Steinernema* (Rhabditida: Steinernematidae). Zoological Research 12: 17 – 20.

Yamaguti, S. 1961. The nematodes of vertebrates. Systema Helminthum, Vol. 3, Part 1. New York: Interscience publishers.

17 Resistance to Plant-parasitic Nematodes

J. L. Starr and P. A. Roberts

17.1 Introduction

In plant nematology, the most widely used definition of resistance is based on a measurable effect on nematode reproduction: a resistant plant inhibits the reproduction of a nematode species relative to reproduction on a plant lacking such resistance (Cook and Evans, 1987). Resistance of plants to pathogens is usually defined as the ability of the plant to lessen, inhibit or overcome the attack by the pathogen (Wingard, 1953). Entomologists frequently use a broader definition, defining resistance as the amount of heritable characteristics of the plant that influences the damage done by insect pests (Painter, 1951), with non-preference, antibiosis, and tolerance as types of resistance.

Tolerance is used by nematologists to characterize plant response to nematode parasitism and is distinct from the ability of the plant to support nematode reproduction. Thus, a susceptible plant may be intolerant with a relatively large degree of growth suppression due to nematode parasitism or it may be tolerant with limited growth suppression due to parasitism (Cook and Evans, 1987). Likewise, a resistant host can be either tolerant or intolerant of nematode parasitism. Several reports document differences in tolerance of susceptible plant species (Cook, et al., 1997; Hussey and Boerma, 1989), but relatively few document differences in tolerance in species that are also resistant to nematode reproduction, for example, in sweet potato (Roberts and Schueurman, 1984). In most cases, resistant plants are tolerant of nematode attack, even though resistance is typically a post-infection process. The few examples of intolerance of resistant plants relate to root injury resulting from elicitation of a severe hypersensitivity reaction as part of the resistance response during initial root infection, such as is the case for *Globodera tabacum solanacearum* on resistant *Nicotiana* species (Baalaway and Fox, 1971; Johnson, et al., 1989).

These differences in definition among the various disciplines reflect how resistance is expressed in the plant to pathogens, insects and nematodes, the nature of the interaction of the pest or pathogen with the host, and also the methods used to measure resistance. The resistance and tolerance terminology is useful in defining the two important ways in which these traits are of critical value as management strategies for plant-parasitic nematodes. Resistance, because it is usually associated with tolerance, provides protection of crop yield potential and, in most cases, suppresses nematode population densities in soil, thereby easing the pest pressure on the following crop. In nematology, the emphasis on nematode reproduction to some extent reflects the general lack of discrete symptoms upon which assessment of resistance is often based when dealing with microbial plant pathogens. Nematode reproduction can be measured with sufficient ease, accuracy, and precision to be a practical alternative to measurement of symptom development. However, most important is the proven ability of resistance to protect plants from nematode parasitism, again emphasizing that resistant plants are usually tolerant and yield well under nematode pressure. Disease incited by nematodes is strongly influenced by initial population densities (Seinhorst, 1965), especially when compared to many foliar diseases or insect pests for which the rate of increase is of primary importance to the final amount of crop damage. Therefore, the suppressive effect of resistance on nematode population densities becomes an important aspect of the use of resistance in crop management systems.

Numerous recent reviews are available that discuss definitions and availability of resistance (Cook and Evans, 1987; Roberts, 1992; Trudgill, 1991), the genetic basis for resistance (Roberts, et al., 1998), mechanisms of resistance (Williamson, 1998; Williamson and Hussey, 1996), breeding for resistance (Young, 1998), bioengineering resistance (Opperman and Conkling, 1998; Vrain, 1999) and protocols for screening for resistance (Starr, et al., 2002a). In this chapter we will discuss why nematode resistance is not currently utilized to its fullest potential, the need for greater emphasis to be given to the development of nematode-resistant crop genotypes, and also where opportunities exist for increased use of resistance.

17.2 Background of Resistance to Plant-parasitic Nematodes

Among the first reports of resistance to nematodes was that of Webber and Orton (1902), who described resistance of a cowpea variety "Iron" to root-knot nematodes based on reduced root-galling in field plots. They cited reports by Zimmerman (1897) for observations of resistance to root-knot in coffee, and Wilfarth (1900) on selection of sugarbeets with resistance to nematodes.

In these early reports, the authors do not consider the possibility of crossing resistant plants with susceptible genotypes to move the resistance into improved genotypes, but speculated on the possibilities of making selections from within the resistant populations for individuals with good agronomic characteristics and resistance. Ware (1936) reported that Orton made selections in 1905 from the cotton line "Jackson Limbless" that had good resistance to *Fusarium* wilt and noted that it was "somewhat resistant to root-knot but had little else to recommend it." Moore (1960) cited Nilsson-Ehle (1920) as being the first to study heritability of resistance to nematodes and who identified resistance to *H. schachtii* in barley as being due to a single dominant gene. The lack of knowledge or appreciation for the importance of proper identification of the nematode population hampered early efforts to identify and characterize resistance in host species.

Barrons (1939) was one of the first to study the mechanisms of resistance to nematodes. Working with root-knot nematodes on cowpea, he distinguished resistance from tolerance and noted that resistance was not due to inhibition of root penetration. Barrons speculated resistance might be due to chemical inhibitors in the roots and that these inhibitors "may counteract or neutralize the giant cell inducing effect of salivary secretions of the nematode."

A major achievement in resistance to nematodes was the introgression of the *Mi-1* gene for resistance to *M. arenaria*, *M. incognita*, and *M. javanica* from the wild tomato relative *Lycopersicon peruvianum* into *L. esculentum* (Smith, 1944). Today, root-knot resistant tomato cultivars with *Mi-1* are widely grown commercially. In addition, resistance conditioned by the *Mi-1* gene has been a valuable research model and has added greatly to the understanding of resistance in tomato (Williamson, 1998). With the cloning and determination of DNA sequences of *Mi-1* (Milligan, et al., 1998) and also the $Hs1^{pro\text{-}1}$ gene for resistance to *H. schachtii* in wild beet for use in sugarbeet (Cai, et al., 1997), the progress in the understanding of at least these two types of resistance will likely accelerate.

17.3 Resistance as a Component of Nematode Management

Hundreds of reports document crop yield suppression on a worldwide scale due to parasitism by plant-parasitic nematodes (see Evans, et al., 1993; Luc, et al., 1990; Sasser and Freckman, 1987). However, nematodes are often overlooked as crop pests, or they are considered to be pests of minor significance. Several factors contribute to the general underestimation of the impact of nematodes as crop pests. Most nematodes are soilborne pathogens that rarely cause diagnostic symptoms on the foliage. The typical non-specific symptoms of nematode parasitism in the form of general plant stress are often attributed to other causes. Further, although many soil environments that support plant growth are infested with plant-parasitic nematodes, most soils do not harbor sufficient nematodes to cause measurable yield suppression. Local and regional distribution patterns of nematodes also influence the perceived level of crop loss. For example, where more than 30% of the fields planted to a given crop in a region are infested with a potentially damaging nematode species (e. g., root-knot nematodes on cotton [Starr, et al., 1993] or cyst nematodes on soybean [Noel, 1992] in the USA), two thirds of the crop remains free of measurable damage. In addition, nematodes in infested fields are typically distributed in a highly aggregated pattern such that much of the field may be free of damaging population densities of the nematodes (Noe and Barker, 1985). This is not to argue that nematodes are unimportant as plant pathogens, because to lose 10% to 50% of ones yield on 30% of a field is a considerable burden, but rather to illustrate how easy it is to overlook such losses. Nevertheless, with the advent of successful management practices, starting with the soil fumigant nematicides over 50 years ago, and including the introduction of some highly effective resistant cultivars of major crops, the recognition of nematodes as important pathogens continues to increase. This is true both in high input intensive production systems, as well as in low input subsistence farming in developing countries.

Management of plant-parasitic nematodes to avoid crop losses is difficult, as is true for many crop pests. However, if resistance to nematodes was more widely available, crop productivity could be improved with little effort or direct cost to the producer. Resistance is one of several common tactics used to manage nematode pests of crops. Other tactics include use of nematicides, crop rotation and other cultural manipulations, biological control, and regulatory approaches. Use of resistant cultivars and rootstocks, like the other approaches, has both advantages and disadvantages that must be recognized and researched if viable management programs based on resistance are to be implemented. Still, host-plant resistance has been prioritized over chemical, biological, cultural, and regulatory control components as a major goal for pest management (Barker, et al., 1994). Several advantages and benefits can be achieved by breeding crop plants resistant to injurious parasitic nematodes for production on infested land. Resistant crops provide an effective and economical method for managing nematodes in both high and low value cropping systems. Assuming the resistance is coupled with tolerance to nematode infection, the resistant crop is "self-protected" and should yield well on infested land. Furthermore, resistant crops in annual cropping systems can reduce or suppress nematode population densities to levels that are non-damaging to subsequent crops, thereby enabling shorter and more manageable rotations. Additional important benefits of resistant crops are their environmental compatibility, that they do not require specialized applications, and, apart from preference based on agronomic or horticultural desirability, usually they do not require an additional cost input or deficit. Because resistance and tolerance are amenable to integration with other management tactics, management systems can be developed that promote resistance durability or provide additional protection when resistance or tolerance is not expressed at high levels (Roberts, 1993).

In comparison, traditional nematicides such as the fumigant 1, 3-dichloropropene, the carbamates aldicarb and oxamyl, and the organophosphate fenamiphos, when applied correctly will increase crop yield if nematode population densities exceed damage thresholds, and they remain an important aspect of nematode management (see Whitehead, 1998). However, there is no long term suppression of nematode population densities with the use of nematicides. Additionally, the use of nematicides is frequently cost prohibitive, especially in subsistence agriculture, and environmental and human health concerns have resulted in withdrawal of or increased restrictions on the use of these toxic materials for many nematode-crop combinations. No new nematicide that has widespread use has been developed in the past 20 years. Approximately 10 years and tens of millions US$ are required to develop and bring to market any new pesticide, and the nematicide market potential is relatively small (nematicide sales account for less than 1% of pesticide sales in the USA, whereas herbicides and insecticides account for 60% and 21%, respectively, of total pesticides sales for agriculture [Ware, 1994]). Thus, it is unlikely that any new nematicide based on currently available chemistry will be developed in the near future and nematicides are likely to have a diminished role in crop protection.

Crop rotations can decrease the potential for substantial yield losses due to nematodes (see Luc, et al., 1990; Whitehead, 1998) and provide at least short term suppression of nematode population densities. The magnitude of these benefits generally is positively correlated with the number of cropping seasons between the planting of susceptible crops. However, rotation systems are seldom adopted unless there are additional benefits to the producer beyond nematode management. Regardless of whether the producer is involved in intensive production agriculture or subsistence farming, many factors are involved in the decision about the most appropriate cropping system from the producer's standpoint. Overall, profitability and yield stability are the primary concern of the producer. Because many nematode species are polyphagus with wide host ranges and many fields have polyspecific communities of plant-parasitic nematodes, development of cropping systems that meet all of the needs of the producer and suppress nematode population densities is a formidable challenge. Nonetheless, there are numerous examples of effective nematode management with crop rotation (see Luc, et al., 1990; Whitehead, 1998). A further aspect of crop rotation is the role that resistant cultivars of susceptible crops can play. Many important nematodes, such as root-knot nematodes, have very wide host ranges and few non-host rotation crops are available. Thus, use of a resistant crop, with the same overall effect of suppressing nematode population densities, can be as effective as a non-host and may have greater economic appeal to the producer as a cash crop. Examples of resistant cultivars in crop rotation are given further on.

Biological control of nematodes through the use of parasites, predators, or antagonists holds some promise for the future (see Evans, et al., 1993), but with current knowledge it is difficult to promote or establish a microflora or fauna in soils that effectively suppress nematode population densities, especially in a relatively short period of time of a single growing season. In the foreseeable future, reliable and effective biological control systems are likely to be limited to specialized situations (e.g., intensely managed crop systems where the environment can be manipulated to promote biological activity).

17.4 Benefits of Resistance

Resistance is an effective management tool that improves crop yield (Table 17.1) in the presence of nematode population densities that exceed the damage threshold. Because resistance to nematodes is usually developed by selection of plants with reduced rates of nematode reproduction, nematode population densities are typically lower following a resistant cultivar than a susceptible cultivar. The use of the root-knot nematode resistance gene *Mi-1* in tomato on *Meloidogyne incognita* populations is an excellent example of this important effect (Roberts and May, 1986). However, the result may be different if the crop has only partial resistance. Niblack et al. (1986) demonstrated that at moderate to high initial population densities, *M. incognita* populations reach their maximum densities at about 90 days after planting on susceptible soybean cultivars (presumably the limited population development was due to extensive damage to the host), whereas on partially resistant cultivars that were less damaged by the nematodes the population densities were still increasing at 120 days after planting.

Table 17.1 Selected examples of the effect of resistance to plant-parasitic nematodes on crop yield in nematode-infested and non-infested fields.

Crop cultivar	Nematode species	Host status	Yield	
			Infested	Non-infested
soybean[1]	*Heterodera glycines*	S	2146 kg/ha	3183 kg/ha
		R	2919	3190
soybean[1]	*H. glycines*	S	2392	3825
		R	3180	3554
peanut[2]	*Meloidogyne arenaria*	S	914 kg/ha	4678 kg/ha
		R	3771	5155
tobacco[3]	*M. incognita*	S	301 g/plot	504 g/plot
		R	407	477
cotton[4]	*M. incognita*	S	970 kg/ha	–
		R	1075	–

[1]G. L. Tylka, personal communication; [2]Starr, et al., 1998; [3]Barker, et al., 1981; and [4]Ogallo, et al., 1999. S = susceptible, R = resistant.

Resistance not only complements crop rotation for nematode management, but also improves the ease with which effective rotation systems can be developed. Ogallo et al. (1999) demonstrated that resistance to root-knot nematodes in cotton not only increased lint yields in nematode-infested fields compared to susceptible cultivars but also gave yield stability (Fig. 17.1). Additionally, they demonstrated that yield of susceptible lima beans was increased when planted in infested fields following two crops of resistant cotton relative to the yields following two crops of susceptible cotton (Fig. 17.2). This yield increase was due to suppression of population densities of *M. incognita* by the resistant cotton cultivar. Typically, the direct cost to the grower for the use of resistance is minimal, thus resistance fits all agricultural production systems. This low cost to the grower may change, however, with increasing use of transgenic systems by private industry for development of new cultivars. Private companies will expect to recover the high cost of development of resistant cultivars by transgenic methods through increased cost of seed. Beyond these agronomic and economic traits, a critical benefit of resistance is that it provides an ecologically sound approach to nematode management, especially relative to traditional nematicides, though there is concern by some that use of resistance developed by modern genetic engineering will have adverse effects on the environment and public health.

Resistance is not a panacea that will solve all nematode management problems. No resistance to important nematode species (especially migratory ectoparasites such as *Belonolaimus* and *Hoplolaimus*) is known for some crops

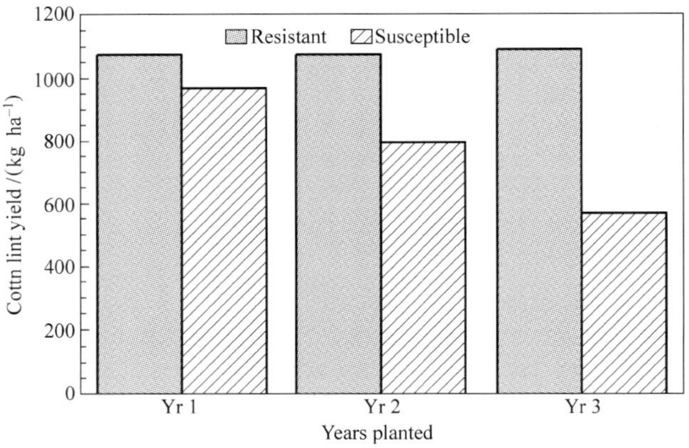

Figure 17.1 Cotton lint yields for three consecutive years for a root-knot resistant cultivar compared to yields of a susceptible cultivar in a field infested with *Meloidogyne incognita* (from Ogallo, et al., 1999).

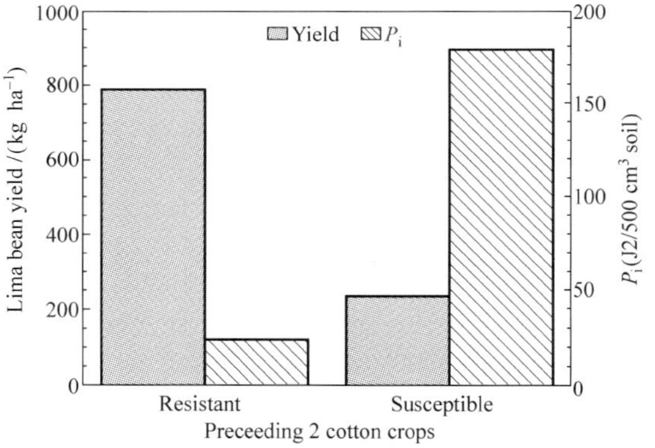

Figure 17.2 Illustration of the value of two years of a resistant host (cotton) on yield of a subsequent susceptible crop (lima bean) in *Meloidogyne incognita*-infested soil, compared to the yield following two years of susceptible cotton. The increased yields of lima bean following the resistant cotton are due to the lower initial nematode population densities (P_i) resulting from lower nematode reproduction on the resistant cultivar (from Ogallo et al., 1999).

or is present only in wild species or undeveloped genotypes. A major effort will be required to develop high yielding crop genotypes with desirable levels of resistance to these and other nematodes. The use of broad-spectrum fumigant nematicides probably will be further restricted. Consequently, the importance of many plant-parasitic nematodes currently considered "non-target" relative to main "target" pest species of root-knot, cyst and some other nematode groups, may take on a greater significance as limiting factors in plant growth. Resistance to the "target" nematode species will not provide a similar broad-based protection against polyspecific nematode communities. Like crop rotation and biological control systems, resistance is typically a highly specific trait and is expected to be effective against only a single nematode species or even a subspecific race or pathotype.

A further consideration in resistance specificity is that after development of a resistant cultivar, the resistance may not be durable if the target nematode species has a high level of genetic variability (Bakker, et al., 1993; Kaloshian, et al., 1996; Roberts, 1995; Young and Hartwig, 1992). In fields where individuals in the nematode population carry virulence to specific resistance, the frequency of those individuals will usually increase in the presence of the resistant cultivar, especially if the resistant cultivar is grown frequently. Thus, a "resistance-breaking" population of nematodes may develop and render the resistance ineffective at that location. However, resistance can be made more durable by pyramiding multiple resistance genes to reduce the intensity of selection. For example, *Meloidogyne* spp. resistance in cowpea based on genes *Rk* plus *rk3* is expressed at a higher level, effective against a broader range of nematode populations and may be more durable than resistance conferred by gene *Rk* alone (Ehlers, et al., 2000). Nematodes with multiple virulence factors matching multiple resistance genes will require multiple mutation and selection events to occur, greatly increasing the probability for long-term durability of the resistance. Specific resistance deployment schemes based on frequency and combination of resistance in cropping systems can be designed that also increase the duration of selection pressure required for development of nematode populations virulent on specific resistance genes. Resistance to multiple nematode species, along with multiple diseases and arthropod pests, can be introgressed into one cultivar.

These limitations aside, resistance in many cases can be readily developed using proven technologies, requiring only that the effort to do so be expended. Transgenic approaches, while useful, are not needed in many cases. There are far more reports of resistance in various crop genotypes than there are resistant cultivars. A bibliography of resistance (Armstrong and Jensen, 1978) contains 1371 citations dealing with resistance in 119 crop species or genera. In the period of 1995 to 1998, Nematological Abstracts contained approximately 300 abstracts annually that dealt with some aspect of resistance. Young (1998) reported that the Crop Science Society of America (CSSA) has registered 143 nematode-resistant cultivars or germplasm lines for 15 field crops. Furthermore, in the texts by Luc et al. (1990) and Evans et al. (1993), resistant cultivars or the potential for their development from known resistant germplasm sources were discussed for nearly every crop. Nearly 90% of all reports involve *Meloidogyne*, *Globodera* or *Heterodera* species. This preponderance of effort on these genera reflects their overall importance as agricultural pests and the relative abundance of resistance to species of these genera.

17.5 Resistance and Crop Yields

Although resistance to plant-parasitic nematodes is usually identified and characterized based on inhibition of nematode reproduction, the primary objective for resistance development is to protect yield potential. As discussed previously, resistance can be an effective tool for management of nematode population densities and thus be of benefit in protecting subsequent susceptible crops, but this must be considered a supplemental benefit. Plant breeders are unlikely to introgress resistance into elite breeding lines or cultivars if the primary benefit will be to another crop through suppression of nematode population densities. It is doubtful if growers would be willing to plant a resistant cultivar if there was no yield benefit to that crop. Thus, when working with host resistance, yield must be the top priority. Ideally, a good yielding resistant cultivar that has the added benefit of suppressing nematode populations in the cropping system will be a strong incentive for crop improvement, particularly as nematicide availability and use decline.

As documented in Table 17.1, when nematode-resistant cultivars are planted in fields where initial nematode population densities exceed the damage threshold, resistant cultivars usually will yield more than susceptible cultivars. A few studies have examined the effects of resistance on the relationship between initial nematode population densities and crop yield. The Seinhorst (1965) model ($Y = m + (1 - m)Z^{P_i - T}$), where Y is the relative yield, m is the minimum yield at the highest possible nematode density, Z a constant, P_i the initial nematode population density, and T the damage threshold density) was used to examine this relationship for *M. incognita* on cotton (Zhou, 1999), pepper (DiVito, et al., 1992), and tomato (DiVito, et al., 1991). In each case, there was little or no change in the damage threshold population density T, but the minimum yield parameter m was increased (Fig. 17.3). Thus, the resistant cultivars were more tolerant (less yield loss) across a range of initial nematode densities. If the resistance of a crop approaches immunity, as with the *Mi-1* gene in tomato or the resistance in peanuts to *M. arenaria*, there may be no effect of initial nematode population density on yield (Fig. 17.4)

(Starr, et al., 2002b). Different types and levels of resistance can be expected to have different effects on the relationship between initial nematode population densities and crop yields.

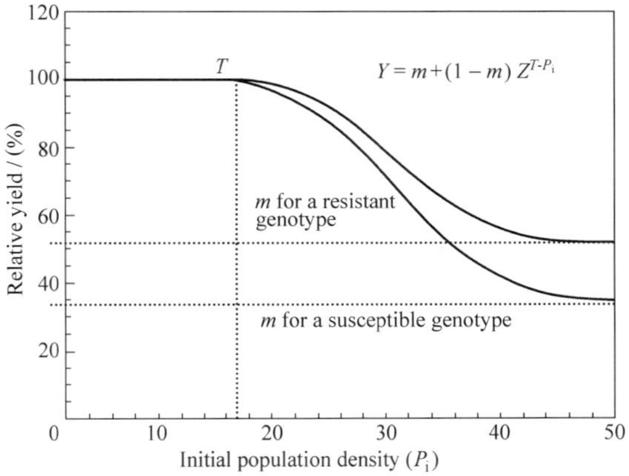

Figure 17.3 Schematic representation of the effect of resistance on the relationship between initial nematode population densities and crop yield based on the Seinhorst model. In some interactions the primary effect of resistance is to increase the value of m, the minimal yield value, with no effect on T, the damage threshold value.

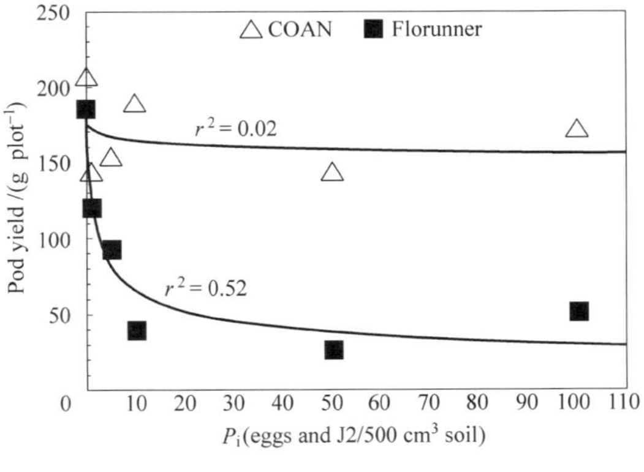

Figure 17.4 Effect of resistance in peanut to *Meloidogyne arenaria* race 1 on the relationship between initial nematode population densities and pod yield. In this case, the slope of response curve for the resistant cultivar "COAN" was not different from zero, indicating that there was no effect of increasing P_i on yield (Starr, et al., 2002b).

Regardless of the specific effect of resistance on the relationship between crop yield and initial nematode population densities, and assuming resistance confers tolerance, one expects greater yield loss at any given nematode population density for a susceptible cultivar than would be observed for a resistant cultivar. The end result is an increase in the economic threshold density (the nematode population density above which the potential economic loss exceeds the cost of implementing a specific management tactic [McSorley and Phillips, 1993]), for resistant cultivars relative to susceptible cultivars. If the direct cost of the resistance to the grower is minimal, then grower profits and (or) yield stability will increase. In many instances, e.g., cotton in the USA, producers are incurring unnecessary losses because nematodes are not recognized as crop pests and therefore are not targeted for management inputs. Improved host resistance, even if it is only partial resistance, will suppress overall crop losses due to nematodes.

Another way to view the benefits that can be derived from the use of resistance is to consider the effects on the portion of the field in which the nematode population density exceeds the damage threshold. Because of the usually clustered distribution of nematode populations, yield losses are seldom uniformly distributed over a field. Indeed, a field with more than 30% of the area exhibiting clear symptoms of nematode damage would be considered severely infested. In most situations were current management relies primarily on nematicides or crop rotation, these practices are applied to the entire field, not just the infested portions. Thus, the increased yield return expected from a portion of the field from these management practices must exceed the cost of treating the entire field. Resistance, which can be used with little or no increase in production costs, can be used to increase yield and profitability in situations where it is uneconomical to use more costly management practices, because the portion of the field infested is limited or the infestation level does not exceed the economic threshold for costly management practices. Even with partial resistance, the increase in yield may result in greater profitability than alternative management practices or no management. Studies that have critically examined the economic benefits of cultivars with partial resistance are lacking and are needed to fully document the benefits of resistance.

17.6 Examples of the Current Use of Resistance

Today, resistance is widely and effectively used in some crops for management of some nematodes. We have chosen a few examples of resistance to root-knot and cyst nematodes to illustrate various considerations and features of current successful resistance development and use. The reader is referred to other reviews for a broader description of examples, including use of resistance in perennial tree and vine crops (Cook and Evans, 1987; Evans, et al., 1993; Luc, et al., 1990; Roberts, 1992; Roberts, et al., 1998; Trudgill, 1991; Young, 1998).

17.6.1 *Meloidogyne*-tobacco

Resistance to root-knot nematodes in tobacco was first reported in the early part of the 20th century (see Clayton, et al., 1958), but resistant cultivars were not grown widely until the 1970s. In North Carolina, 97% of the 84,000 ha tobacco crop is planted with cultivars that are resistant to *M. incognita* (T. Melton, North Carolina State University, pers. comm.). Despite this high percentage use of resistance, more than 70% of the crop also is treated with a nematicide. This reflects the fact that even after more than 20 years of use and a highly effective grower education program, the producers of this high cash value crop are unwilling to put their complete trust in resistance. Factors that contribute to a lack of faith in resistance include the presence of races of *M. incognita* with virulence to the resistance, and the presence in some tobacco fields of *M. arenaria* and *M. javanica* against which the resistance is not effective, and the effectiveness of promotional efforts of the nematicide industry. Because of the value of the crop, growers are willing and able to treat the crop with a nematicide to ensure that root-knot nematodes do not damage the crop. Although the widespread use of this resistance resulted in an increased frequency of *M. arenaria* in North and South Carolina (Fortnum, et al., 1984; Schmitt and Barker, 1988), *M. incognita* remains the most frequently encountered species on tobacco and resistance is still highly effective in most fields.

17.6.2 *Meloidogyne*-tomato

The *Mi-1* gene for resistance to *M. arenaria, M. incognita*, and *M. javanica* has been used for many years in fresh market tomatoes with much success. However, because of *Mi-1* linkage to the undesirable trait of tough fruit attachment that made mechanical harvesting difficult, it was not widely used in commercial processing tomato production in the USA until the 1980s. Currently, many tomato cultivars in the different fruit quality and maturity classes are available with *Mi-1*, and the majority of the tomatoes grown commercially on infested fields in California carry this resistance. Despite the apparent success of the *Mi-1* gene in California, only recently have tomatoes carrying this resistance gene been widely grown in Florida. The recent use in Florida has been the result of the popularity of the cultivar Sanibell that carries the *Mi-1* gene, although its success is because of superior horticultural traits and not because of resistance to *Meloidogyne* spp. Virulence to the *Mi-1* gene can develop in Florida populations of *M. incognita* after as few as five plantings of Sanibell and breakdown of resistance at high soil temperatures also occurs (Noling, 2000). Even so, the *Mi-1* gene can be considered as quite durable, in that it has been overcome by virulence development in few field situations. In California, more than 20 years of intensive use has resulted in only a very few isolated cases of resistance breakdown (Kaloshian, et al., 1996).

Nevertheless, it is fortunate that additional genes for resistance to *M. incognita* and other root-knot species have been identified within the *Lycopersicon peruvianum* germplasm resources and, depending on the gene, several are heat stable or are effective against nematode populations virulent to *Mi-1* (Veremis and Roberts, 1996). Introgression of these genes into tomato cultivars is proving to be a challenging goal, because of problems associated with self- and cross-incompatibility in *L. peruvianum* and between *L. peruvianum* and *L. esculentum*, respectively (Veremis and Roberts, 2000). However, successful introgression of one or more of the novel *Mi* genes along with *Mi-1* may allow development of management systems that overcome problems associated with the use of *Mi-1* as a single gene for resistance.

17.6.3 *Heterodera*-soybean

Resistance to *Heterodera glycines* in soybean was first reported in 1957 (Ross and Brim), with resistant cultivars being widely grown by the mid-1970s. The *H. glycines*-soybean system is representative of cases where the effectiveness of resistance is compromised by virulence in the nematode populations. The race situation in *H. glycines* remains unsettled, with 16 races currently recognized (Riggs and Schmitt, 1988), but debate continues over the appropriate choice of both differential soybean genotypes and the basis of indexing resistant and susceptible reactions. Numerous high yielding soybean cultivars have resistance to races 1 and 3 of *H. glycines*, and a few good cultivars have resistance to races 6 and 14. The cultivar Hartwig has the broadest base of resistance, being resistant to races 1 to 6, 8 and 14, but it has relatively poor yield potential. Hartwig has proved useful as a parent for the development of additional cyst-resistant cultivars. Fortunately, of the 16 described races, eight are rarely encountered. Races 1 and 3 predominate in the northern portion of the USA, whereas races 2 to 6, 9 and 14 predominate in the southern states. In North Carolina, approximately 48% of the soybean crop of 573, 000 ha was planted to cyst nematode-resistant cultivars in 1998

(J. Dunphy, North Carolina State University, pers. comm.), but 60% of the infestations are races against which resistance is not effective.

17.6.4 *Globodera*-potato

Resistance to the potato cyst nematodes was first reported in 1954 (Ellenby). Similar to the genetic variability in *H. glycines*, multiple pathotypes of *G. pallida* and *G. rostochiensis* have been described. However, these remain somewhat controversial, especially those of *G. pallida* for which pathotype differentiation is not clear-cut, due to incomplete data on the genetics of resistance in the host and virulence in the nematodes (Trudgill, 1985). None the less, resistance to *G. rostochiensis* has been widely used in several countries, including The Netherlands and the UK. Currently in The Netherlands about 55% of the ware potatoes (those for human consumption) and 99% of the starch potatoes are resistant to one or more pathotypes of the potato cyst nematodes (F. Gommers, Wageningen Agricultural University, pers. comm.). In the UK, approximately 45% of the potato crop carries the major resistance gene *H1* derived from *Solanum tuberosum* ssp. *andigena* that confers resistance to *G. rostochiensis* pathotype Ro1 and thus is effective against most populations of *G. rostochiensis* in that country (K. Evans, IRAC Rothamsted, pers. comm.). The *H1* gene has remained very effective against the Ro1 pathotype and can be considered a durable resistance, in that few examples exist of *H1* resistance causing emergence of virulent *G. rostochiensis* populations over some 40 years of use. However, a negative consequence of using *H1* resistance is that the frequency of *G. pallida*, which is not controlled by *H1*, is increasing in the UK and only about 1.5% of the potato crop carries effective resistance against prevalent pathotypes of *G. pallida*. Resistance to *G. rostochiensis* was quite effective in managing cyst nematodes until the appearance of *G. pallida* in the late 1970s. Resistance to *G. pallida*, based on resistance genes *Pa2* and *Pa3* from *Solanum vernei*, was then introduced during the 1980s with limited success. The resistance to *G. pallida* from both *S. vernei* and *S. spegazzini* is quantitative in nature based on multiple genes with partial resistance expression. This type of resistance coupled with a high level of genetic variability for virulence in field populations has made it difficult to achieve an adequate level of resistance, and screening for genotypes with tolerance to *G. pallida* has been given priority in some programs (Brodie, et al., 1993). Despite these problems in maintaining effective resistance deployment against such variable pathogens as cyst nematodes, resistance has been useful in alleviating crop losses. Fortunately, the limited host range of these cyst nematodes has made the use of crop rotations that include non-host crops an effective complement to the use of resistance.

17.6.5 *Meloidogyne*-cotton

Cotton is a major crop worldwide and is widely grown in the southern and southwestern regions of the USA. *Meloidogyne incognita* has long been recognized as an important pathogen of cotton (Starr, 1998), because of its damaging effects directly and also because it forms a disease complex with the vascular wilt pathogen *Fusarium oxysporum* f. sp. *vasinfectum* (DeVay, et al., 1997). Resistance to *M. incognita* has been researched since the early part of the 20th century (see Ware, 1936). The first germplasm with a high level of resistance was the transgressive resistance in Auburn 623 RNR, which was selected from the F9 generation of Clevewilt 6 x Wild Mexico Jack Jones (Shepherd, 1974). This early generation material was poorly adapted commercially but Shepherd and his coworkers made substantial improvements in the agronomic traits of this material and eventually released nine germplasm lines with high levels of resistance and good yield potential (Shepherd, et al., 1996). However, this excellent source of resistance has yet to be introgressed into widely grown cotton cultivars.

In the 1990s three cotton cultivars (Acala NemX, Stoneville LA887, and Paymaster 1560) with moderate to good levels of resistance to *M. incognita* were released. Stoneville LA887 and Paymaster 1560 have resistance derived from Clevewilt 6 (Robinson, et al., 1997) whereas the source of resistance in NemX is uncertain. Despite the value of these cultivars in increasing cotton yields in nematode infested fields and in reducing population densities of *M. incognita* (Ogallo, et al., 1999; Zhou, 1999), they accounted for less than 1% of all cotton planted in the USA in 1999 (Anon., 1999). Why these cultivars are not more widely accepted for use in nematode infested fields is unknown, especially given the recent increase in the understanding of nematodes as important pests of cotton. Unfortunately, highly effective cotton IPM programs that have been developed over the past 20 years are focused primarily on insect pests and provide little information to growers on effective strategies for nematode management. With the recent effort in development and sales of the highly profitable transgenic cotton cultivars, the private cotton breeding industry has had little incentive to work with nematode resistance. In California, a reduction in the total cotton acreage in recent years has meant that some of the poorer yielding ground, including that infested with Fusarium wilt and *M. incognita*, has been left out of production, impacting the use of Acala NemX. Another factor limiting current use of nematode resistance is the reluctance of grower education programs to promote or recommend the use of resistance. At least one extension specialist from a cotton producing state expressed reluctance to recommend one of these moderately resistant cultivars because he was concerned that some growers would not understand how to use the resistance, and without nematicide treatment growers may suffer substantial yield losses in severely infested fields. Also, he felt that many growers would continue to use nematicides in infested fields even though resistance was available. Clearly there is a need for improved grower education programs, in addition to expanding the choice of available resistant cultivars.

Rotylenchulus reniformis has become a widespread problem in the USA. In some regions, more than 50% of the cotton fields are infested with either *R. reniformis* or *M. incognita* (Blasingame, 1993). Effective resistance to *R. reniformis* has not been found within *G. hirsutum*, but resistance has been identified in other *Gossypium* species. However, efforts to introgress this resistance into *G. hirsutum* have been complicated by incompatibility barriers. No useful resistance to *R. reniformis* was identified among lines of *G. hirsutum* that carried a single monosomic addition from the highly resistant, but genetically incompatible *G. longicalyx* (Frerich, 1995). Several introgression pathways have been suggested to transfer *R. reniformis* resistance from the diploid species *G. arboreum* and *G. herbaceum* into *G. hirsutum*, including generation of fertile interspecific hexaploids, synthetic allotetraploids, and a 4x

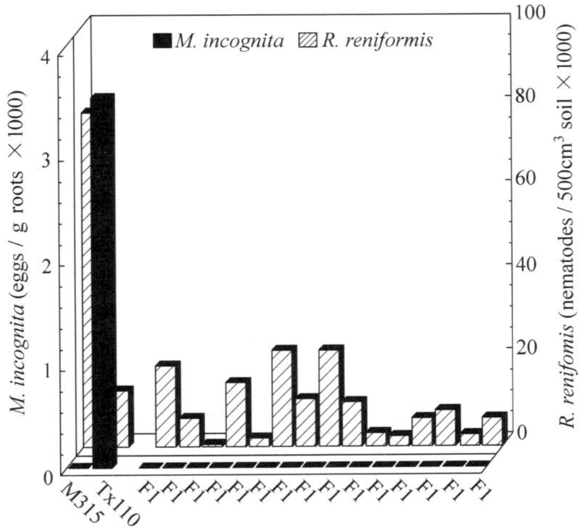

Figure 17.5 Resistance to *Rotylenchulus reniformis* and *Meloidogyne incognita* in F1 individuals from a cross between the reniform resistant *Gossypium barbadense* TX110 and the root-knot nematode resistant *G. hirsutum* M315. Resistance to each nematode species was measured separately on different F1 individuals and was based on nematode reproduction (Starr, unpubl. data).

triple hybrid from a resistant diploid by a 2(ADD) 6x genome cross (Stewart and Robbins, 1996). More recently, the moderately resistant *G. barbadense* line TX110 crossed with the *M. incognita*-resistant *G. hirsutum* genotype M315 resulted in fertile F1 progeny with high levels of resistance to *M. incognita* and moderate resistance to *R. reniformis* (Fig. 17.5) (Starr, unpubl. data). Substantial additional breeding efforts will be required before any resistance to the reniform nematode is available commercially.

17.6.6 *Meloidogyne*-peanut

Recently, success has been achieved in a long term effort to introgress resistance to *M. arenaria* from wild *Arachis* species into the cultivated peanut *A. hypogaea* (Simpson and Starr, 2000). Because *A. hypogaea* is an allotetraploid and the resistant species are diploid with differing genomes, a complex interspecific hybrid was developed first (Table 17.2), resulting in a synthetic tetraploid that was cross-compatible with *A. hypogaea* (Simpson, 1991). A backcross-breeding program was used to introgress the resistance into high yielding genotypes. The first nematode resistant cultivar, COAN, was released in 1999 (Simpson and Starr, 2000) and substantial education programs are in progress to demonstrate the value of the resistance to the growers. Because COAN does not have the yield potential of the best susceptible cultivars, it is only being recommended for nematode infested fields. COAN was a selection from the fifth backcross generation of the breeding program. Selections from the seventh backcross generation appear to have yield potential equal to that of the highest yielding susceptible cultivars (Fig. 17.6) (Church, et al., 2000) and should be more widely acceptable to peanut growers.

Table 17.2 Introgression of resistance to *Meloidogyne arenaria* from wild *Arachis* species into the cultivated peanut *A. hypogaea*.

1. *A. diogoi*, a diploid A genome species with nematode resistance is crossed with *A. cardenasii*, another diploid A genome, nematode-resistant species.
2. The progeny of the *A. diogoi* x *A. cardenasii* is crossed with *A. batizocoi*, which is a diploid B genome and nematode-resistant species.
3. The diploid AB genome progeny of *A. batizocoi* x (*A. diogoi* x *A. cardenasii*) are infertile. Fertility is restored by treatment with colchicine to create a synthetic tetraploid with an AB genome. The synthetic tetraploid (TxAG-6) is highly resistant to *M. arenaria*.
4. The synthetic AB tetraploid TxAG-6 is crossed with the cultivated peanut, *A. hypogaea*, which is also a tetraploid with the AB genome, to transfer nematode resistance to genotypes that can be readily crossed with cultivated peanut.
5. A backcross breeding program, with *A. hypogaea* "Florunner" as the susceptible recurrent parent, with selection for nematode resistance based on reproduction of *M. arenaria* in each generation.
6. The *M. arenaria*-resistant cultivar COAN is a selection from the fifth backcross generation.

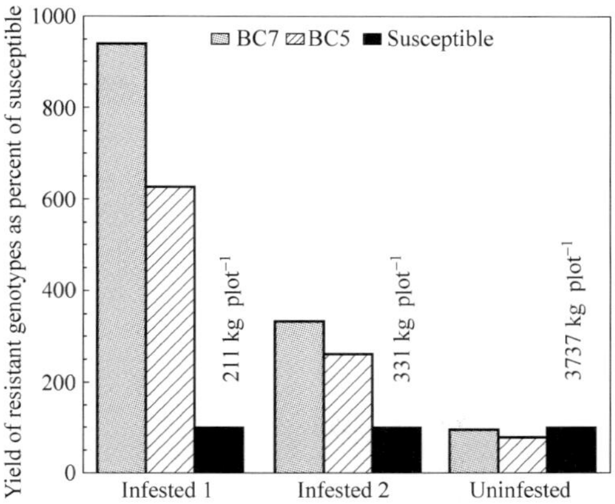

Figure 17.6 Improvement in yield potential of peanut resistant to *Meloidogyne arenaria* race 1 with continued backcrossing and selection for yield. Yield of a selection from the BC7 generation was greater than the selection from the BC5 generation and the susceptible cultivar in the two nematode-infested sites. In the uninfested field, the yield of the BC7 generation selection was equal to that of the susceptible standard, whereas the yield of the selection from the BC5 generation was only 70% of that of the susceptible standard. Note that the yield of the susceptible cultivar was suppressed by more than 85% in the infested fields (from Church, et al., 2000).

The resistance in COAN is due to a single dominant gene derived from *A. cardenasii* (Burow, et al., 1996; Choi, et al., 1999). Restriction fragment length polymorphism (RFLP) markers linked to the resistance loci have been identified and are being used for marker-assisted selection (MAS) in ongoing breeding efforts (Choi, et al., 1999; Church, et al., 2000). MAS is also being used to identify additional nematode-resistance genes within the available germplasm resources, including resistance to *M. javanica* populations parasitic on peanut (Abdel-Momen, et al., 1998). Nematode resistance is being combined with resistance to tomato spotted wilt virus, resistance to Sclerotinia blight, and high oleic to linoleic fatty acid ratio, a desired trait in the peanut industry. Introgression of additional resistance genes will increase the durability of the resistance and promote yield stability.

17.7 Future Projections

Host resistance is a management tactic that has much potential and needs to be utilized more effectively. Many of the problems associated with resistance can be overcome or minimized with additional research, breeding effort, and effective grower education programs. Unfortunately, many nematologists believe or find that their responsibility ends with the research. This attitude may be the greatest impediment to more effective use of resistance.

Many of the available germplasm resources remain to be characterized with respect to resistance to nematodes. Holbrook et al. (2000) advocate the use of core collections, which are subgroupings of the entire germplasm collection representing major geographic regions or other bases for diversity, as a means of more effective screening of large germplasm collections. In this approach, one screens a set number of individuals from each subgroup, with the number being based on the proportion of the entire collection represented by that group. When the trait of interest (i. e., resistance) is identified within a subgroup, one concentrates further screening on that subgroup with corresponding less emphasis on other subgroups.

Even after resistant phenotypes have been identified, further research will be required to determine the number of unique genes for resistance that may be present in different resistant accessions. For example, Robinson and Percival (1997) recently identified accessions of *G. hirsutum* from the Yucatan peninsula of Mexico with resistance to *M. incognita* that is similar to the resistance in Clevewilt 6 and Wild Mexico Jack Jones. However, it is not known whether these accessions carry unique genes for resistance or if their genes for resistance are identical to those already in use in breeding programs. As DNA-based markers (e. g., SSR, RFLP, and RAPDs) tightly linked to resistance loci become more readily available, they can be used to rapidly screen additional resistance phenotypes to identify novel resistance genes. Use of several MAS techniques may allow one to determine more rapidly if the resistance phenotypes are due to unique genes rather than the more time consuming traditional genetic analysis of each candidate.

Recent work on *M. javanica* and *M. incognita* resistance in carrot has demonstrated the utility of MAS (Boiteux, et al., 2000; Simon, et al., 2000; Boiteux, Roberts and Simon, unpubl. data). Analysis of two codominant RAPD-derived STS (sequence-tagged site) markers flanking the *Mj-1* resistance locus have revealed the expression of a weakened level of resistance in heterozygous individuals compared to homozygous individuals due to a gene dosage effect. This finding has important implications for carrot cultivar development, which typically is based on F1 hybrid carrots as the commercial type. Homozygous resistance will be necessary to provide full protection of carrots from *M. javanica*. Not only do these markers allow a rapid screening protocol for selection of homozygous individuals in the resistance breeding program, they also enable direct comparison of resistance conferred by *Mj-1* with other resistance factors in the carrot genome. For example, an unanswered question is whether the more variable and weaker resistance to *M. incognita* in plants with *Mj-1* is due to a different interaction of *M. incognita* with *Mj-1* compared to *M. javanica*, or whether *M. incognita* resistance is conferred or modified by additional gene loci in carrot.

Genetic transformation of plants with cloned resistance genes is an exciting part of the future for resistance to nematodes. The recently cloned genes *Mi-1* from tomato and $Hs1^{pro\text{-}1}$ from sugarbeet represent an essential first step toward that goal. Cloned genes present the opportunity to transfer resistance directly into elite genotypes of the same crop, or into cultivars of different crops. The benefits of this capability are obvious and include avoidance of linkage drag and also the expansion of the resistance utility, particularly in crops where natural genes for resistance have not been identified or are in genotype backgrounds that prohibit simple introgression by crossbreeding. Another opportunity emerging from molecular and genetic analysis of resistance is the common clustering of resistance genes or their effects, either by closely positioned gene loci, or by multiple resistance expression (Michelmore and Meyers, 1998). For example, a novel heat stable resistance gene is located in the *Mi-1* region of tomato (Veremis, et al., 1999), and *Mi-1* has been shown to have a dual resistance function, in that it confers not only nematode resistance but also resistance to potato aphids (Kaloshian, et al., 2000; Rossi, et al., 1998). Thus, quite different pathogens and pests, in this case a root-cell-feeding nematode and a leaf sieve element-feeding insect, are being blocked by at least some portion of the same resistance pathway.

The potential disadvantages of the innovative approaches to resistance usage also must be considered. For example, although *Mi-1* seems quite durable, virulence to this gene in *Meloidogyne* spp. has been documented (Kaloshian, et al., 1996), and an expansion of its use in other crops on the same infested fields would certainly increase the potential for selecting virulence in heterogeneous nematode populations. Other problems exist with cloned genes that must be resolved through research efforts. For example, *Mi-1* has been demonstrated to express resistance when inserted into eggplant, but not in some other solanaceous plants such as tobacco (Frijters, et al., 2000). Clearly, at the fundamental molecular level, there is much we do not understand about resistance gene function and expression in foreign genomes. Interestingly, transgenes for insect resistance (Bt toxin) have been introgressed into the nematode-resistant cotton cultivar Paymaster 1560, but the transgenic plants apparently lost the resistance to *M. incognita* (Colyer, et al., 2000). It remains unclear whether this loss of nematode resistance in transgenic cotton was due to a mutation caused by insertion of the transgene into the nematode-resistance gene. A more likely explanation is that during development of the cultivar subsequent to the insertion of the transgene there was insufficient screening of segregating populations to retain the nematode resistance.

Research has only begun to explore the possibilities for engineered resistance (Opperman and Conkling, 1998; Vrain, 1999) and as yet no cultivar with engineered resistance to nematodes is available for growers. The recent text by Fenoll et al. (1997) lists numerous possibilities for engineered resistance, including anti-nematode genes, anti-feedants, and plantibodies. Many researchers are confident that such sources of resistance will become valuable additions to our arsenal in the near future. It is expected that engineered resistance will help overcome fertility barriers that limit use of some native sources of resistance and will provide sources of resistance to nematodes such as *Belonolaimus* and *Hoplolaimus* spp. for which no resistance is currently known. A major question is whether engineered resistance will be more durable than many currently available resistance genes, especially with respect to *Globodera* and *Heterodera* spp. Based on present knowledge, we can assume that engineered resistance will have some similar and unique constraints with respect to durability compared with natural resistance.

Regardless of the source of resistance, it will be little more than a research tool if nematologists do not form an effective collaboration with plant breeders to move the resistance into appropriate crop genotypes that have the highest yield potentials and other important agronomic and horticultural characteristics. The nematology community has a responsibility to convince the public and private sector plant breeders that introgression of resistance to nematodes into the elite crop germplasm lines or cultivars will be beneficial. Plant breeders need expert input to identify appropriate sources of resistance and to develop or adopt effective screening systems that will permit timely introgression of resistance. The collaborative effort requires an active participation from the initial steps of resistance identification all the way through to the field assessment of the finished cultivar. One challenge is to overcome the common view that resistance frequently comes at the expense of yield. This is true in some cases, including the recently released COAN peanut with resistance to *M. arenaria*. However, there are no data that prove yield must be sacrificed to achieve resistance, and it is often only the first generation of released resistant cultivars that have suboptimal yield potential and later generations overcome this association. As has been recently demonstrated with cotton (Ogallo, et al., 1999), peanut (Church, et al., 2000), and soybean (see Table 17.1), the linkage between lower yield potential and resistance can be broken and resistant genotypes with yield potentials equal to those of the best yielding susceptible genotypes are possible. Similarly, the use of the *Mi-1* gene in tomato was initially limited by problems of linkage drag with undesirable horticultural traits (Williamson, 1998), but this negative linkage has been broken and tomato cultivars carrying the *Mi-1* gene are now widely grown commercially.

Once high yielding cultivars with improved levels of resistance to nematodes are developed, effective grower-based education programs are required to successfully implement their use. Some forms of resistance will lack durability due to variability in the nematode population, or they will express only partial resistance such that some yield loss may occur at high initial nematode population densities. Thus, it is essential that the resistance be deployed in a responsible manner and be integrated with other nematode management tactics to achieve optimal benefits with respect to yield and to enhance durability. Because there are relatively few extension specialists or advisors with nematology as their primary responsibility, we must work in cooperation with extension personnel from a variety of disciplines to develop effective education programs. Often the first task is the education of the extension specialist or crop production advisor as to the value of nematode resistance in the crop and cropping system.

In summary, the identification, development, and deployment of resistance requires a long term and extensive effort, but as has been demonstrated repeatedly, the benefits justify the effort. In one of the few critical studies of the economic benefits of resistance, Brady and Duffy (1982) documented that a $1 million expenditure to develop one soybean cultivar with resistance to *H. glycines* resulted in benefits in the amount of $400 million. Host resistance will not be the solution to all problems caused by plant-parasitic nematodes, but resistance could and should play a bigger role in many nematode management systems. The era of nematicides is approaching an end and we must develop alternative management strategies. Clearly, the use of host plant resistance must be one of these alternatives, and it will play a priority role in many crop production systems, both directly and as a component of an integrated approach to nematode management.

References

Abdel-Momen, S. M., C. E. Simpson, and J. L. Starr. 1998. Resistance of interspecific *Arachis* breeding lines to *Meloidogyne javanica* and an undescribed *Meloidogyne* species. Journal of Nematology 30: 341 – 346.

Anonymous. 1999. Cotton USA buyers guide. Memphis, TN: National Cotton Council of America.

Armstrong, J. M., and H. J. Jensen. 1978. Indexed bibliography of nematode resistance in plants. Bulletin 639, Agriculture Experiment Station, Oregon State University.

Baalaway, H. A., and J. A. Fox. 1971. Resistance to Osborne's cyst nematode in selected *Nicotiana* species. Journal of Nematology 3: 395 – 398.

Bakker, J., R. T. Folkertsma, J. N. A. M. Rouppe van der Voort, J. M. de Boer, and F. J. Gommers. 1993. Changing concepts and molecular approaches in the management of virulence genes in potato cyst nematodes. Annual Review of Phytopathology 31: 169 – 190.

Barker, K. R., F. A. Todd, W. W. Shane, and L. A. Nelson. 1981. Interrelationships of *Meloidogyne* species with flue-cured tobacco. Journal of Nematology 13: 67 – 79.

Barker, K. R., R. S. Hussey, L. R. Krusberg, G. W. Bird, R. A. Dunn, H. Ferris, V. R. Ferris, D. W. Freckman, C. J. Gabriel, P. S. Grewal, A. E. MacGuidwin, D. L. Riddle, P. A. Roberts, and D. P. Schmitt. 1994. Plant and soil nematodes: Societal impact and focus for the future. Journal of Nematology 26: 127 – 137.

Barrons, K. C. 1939. Studies of the nature of root knot resistance. Journal of Agricultural Research 58: 263 – 271.

Blasingame, D. 1993. Nematode distribution and density. Pp. 4 – 6. In Cotton Nematodes - Your Hidden Enemies. Memphis, TN: The Cotton Foundation.

Boiteux, L. S., J. G. Belter, P. A. Roberts, and P. W. Simon. 2000. RAPD linkage map of the genomic region encompassing the root-knot nematode (*Meloidogyne javanica*) resistance locus in carrot. Theoretical and Applied Genetics 100: 439 – 446.

Brady, E. B., and M. Duffy. 1982. The value of plant resistance to soybean cyst nematodes: A case study of Forrest soybean. US Department of Agriculture, Natural Resources Economics Division, Washington, D. C.

Brodie, B. B., K. Evans, and J. Franco. 1993. Nematode parasites of potatoes. Pp. 87 – 132. In K. Evans, D. L. Trudgill, and J. M. Webster, eds. Plant Parasitic Nematodes in Temperate Agriculture. Wallingford, UK: CAB International.

Burow, M. D., C. E. Simpson, A. H. Paterson, and J. L. Starr. 1996. Identification of RAPD markers in peanut (*Arachis hypogaea*) associated with root-knot-nematode (*Meloidogyne arenaria*) resistance, derived from *A. cardenasii*. Molecular Breeding 2: 307 – 319.

Cai, D., M. Kleine, S. Kifle, H.-J. Harloff, N. N. Sandal, K. A. Marcker, R. M. Klein-Lankhorst, E. M. J. Salentijn, W. Lange, W. J. Stiekema, U. Wyss, F. M. W. Grundler, and C. Jung. 1997. Positional cloning of a gene for nematode resistance in sugarbeet. Science 275: 832 – 834.

Clayton, E. E., T. W. Graham, F. A. Todd, J. G. Gates, and F. A. Clark. 1958. Resistance to the root knot disease of tobacco. Tobacco Science 2: 53 – 63.

Choi, K., M. D. Burow, G. Church, G. Burow, A. H. Paterson, C. E. Simpson, and J. L. Starr. 1999. Genetics and mechanism of resistance to *Meloidogyne arenaria* in peanut germplasm. Journal of Nematology 31: 283 – 290.

Church, G. T., C. E. Simpson, M. D. Burow, A. H. Paterson, and J. L. Starr. 2000. Use of RFLP markers for identification of individuals homozygous for resistance to *Meloidogyne arenaria* in peanut. Nematology 2: 575 – 580.

Colyer, P. D., T. L. Kirkpatrick, W. D. Caldwell, and P. R. Vernon. 2000. Root-knot nematode reproduction and root-galling severity on related conventional and transgenic cotton cultivars. Cotton Science 4: 232 – 236.

Cook, C. G., A. F. Robinson, and L. N. Nampken. 1997. Tolerance to *Rotylenchulus reniformis* and resistance to *Meloidogyne incognita* in high-yielding breeding lines of upland cotton. Journal of Nematology 29: 322 – 328.

Cook, R., and K. Evans. 1987. Resistance and tolerance. Pp. 179 – 231. In R. H. Brown and B. R. Kerry, eds. Principles and Practice of Nematode Control in Crops. New York: Academic Press.

DeVay, J. E., A. P. Guiterrez, G. S. Pullman, R. J. Wakeman, R. H. Garber, D. P. Jeffers, S. N. Smith, P. B. Goodell, and P. A. Roberts. 1997. Inoculum densities of *Fusarium oxysporum* f. sp. *vasinfectum* and *Meloidogyne incognita* in relation to the development of Fusarium wilt and the phenology of cotton plants (*Gossypium hirsutum*). Phytopathology 87: 341 – 346.

DiVito, M., V. Cianciotta, and G. Zaccheo. 1991. The effect of population densities of *Meloidogyne incognita* on yield of susceptible and resistant tomato. Nematologia Mediterranea 19: 265 – 268.

DiVito, M., V. M. Cianciotta, and G. Zaccheo. 1992. Yield of susceptible and resistant pepper in microplots infested with *Meloidogyne incognita*. Nematropica 22: 1 – 6.

Ehlers, J. D., W. C. Matthews, A. E. Hall, and P. A. Roberts. 2000. Inheritance of a broad-based form of root-knot nematode resistance in cowpea. Crop Science 40: 611 – 618.

Ellenby, C. 1954. Tuber forming species and varieties of the genus *Solanum* tested for resistance to the potato root eelworm *Heterodera rostochiensis* Wollenweber. Euphytica 3: 195 – 202.

Evans, K., D. L. Trudgill, and J. M. Webster, eds. 1993. Plant parasitic nematodes in temperate agriculture. Wallingford, UK: CAB International.

Fenoll, C., F. M. W. Grundler, and S. A. Ohl. 1997. Cellular and molecular aspects of plant nematode interactions. Dordecht, Germany: Kluwer Academic Publishers.

Fortnum, B. A., J. P. Krausz, and N. G. Conrad. 1984. Increasing incidence of *Meloidogyne arenaria* in flue-cured tobacco in South Carolina. Plant Disease 68: 244 – 245.

Frerich, S. J. 1995. Evaluation of *Gossypium longicalyx* monosomic addition lines to *Gossypium hirsutum* for resistance to *Rotylenchulus reniformis* nematode. M. S. Thesis, Texas A&M University, College Station, TX, USA.

Frijters, A., G. Simons, G. Varga, N. Quaedvlieg, and M. de Both. 2000. The *Mi-1* gene confers resistance to the root-knot nematode *Meloidogyne incognita* in transgenic eggplant. Proceedings Conference, Durable Disease Resistance: Key to Sustainable Agriculture. Ede-Wageningen. Wageningen University, The Netherlands.

Holbrook, C. C., P. A. Timper, and H. Q. Xue. 2000. Evaluation of the core collection approach for identifying resistance to *Meloidogyne arenaria* in peanut. Crop Science 40: 1172 – 1175.

Hussey, R. S., and H. R. Boerma. 1989. Tolerance in maturity groups V-VIII soybean cultivars to *Heterodera glycines*. Supplement to Journal of Nematology 21: 686 – 692.

Johnson, C. S., D. A. Komm, and J. L. Jones. 1989. Control of *Globodera tabacum solanacearum* by alternating host resistance and nematicide. Journal of Nematology 21: 16 – 23.

Kaloshian, I., M. G. Kinsey, V. M. Williamson, and D. E. Ullman. 2000. *Mi*-mediated resistance against the potato aphid *Macrosiphum euphorbiae* (Hemiptera: Aphididae) limits sieve element ingestion. Environmental Entomology 29: 690 – 695.

Kaloshian, I., V. M. Williamson, G. Miyao, D. A. Lawn, and B. B. Westerdahl. 1996. "Resistance-breaking" nematodes identified in California tomatoes. California Agriculture 50: 18 – 19.

Luc, M., R. A. Sikora, and J. Bridge. 1990. Plant parasitic nematodes in subtropical and tropical agriculture. Wallingford, UK: CAB International.

McSorley, R., and M. S. Phillips. 1993. Modelling population dynamics and yield losses and their use in nematode management. Pp. 61 – 85. In K. Evans, D. L. Trudgill, and J. M. Webster, eds. Plant Parasitic Nematodes in Temperate Agriculture. Wallingford, UK: CAB International.

Michelmore, R. W., and B. C. Meyers. 1998. Clusters of resistance genes in plants evolve by divergent selection and a birth-and-death process. Genome Research 8: 1113 – 1130.

Milligan, S. B., J. Bodeau, J. Yaghoobi, I. Kaloshian, P. Zabel, and V. M. Williamson. 1998. The root-knot resistance gene *Mi* from tomato is a member of the leucine zipper, nucleotide binding, leucine rich repeat family of plant genes. Plant Cell 10: 1307 – 1319.

Moore, E. L. 1960. Some problems and progress in the breeding and selection of plants for nematode resistance. Pp. 454 – 460. In J. N. Sasser and W. R. Jenkins, eds. Nematology: Fundamentals and Recent Advances with Emphasis on Plant Parasitic and Soil Forms. Chapel Hill, NC: The University of North Carolina Press.

Niblack, T. L., R. S. Hussey, and H. R. Boerma. 1986. Effects of environments, *Meloidogyne incognita* inoculum levels, and *Glycine max* genotype on root-knot nematode-soybean interactions in field microplots. Journal of Nematology 18: 338 – 346.

Nilsson-Ehle, H. 1920. Ûber resistenz gegen Heterodera schachtii bei gewissen Gersten-Sorten, ihre Vererbungsweise und bedeutung fur die Praxis. Hereditas 1: 1 – 34.

Noe, J. R., and K. R. Barker. 1985. Overestimation of yield loss of tobacco caused by the aggregated spatial pattern of *Meloidogyne incognita*. Journal of Nematology 17: 245 – 251.

Noel, G. R. 1992. History, distribution, and economics. Pp. 1 – 13. In R. D. Riggs and J. A. Wrather, eds. Biology and Management of the Soybean Cyst Nematode. St. Paul, MN: APS Press.

Noling, J. W. 2000. Effects of continuous culture of a resistant tomato cultivar on *Meloiodgyne incognita* soil population densities and pathogenicity. Journal of Nematology 32: 452 (abstr.).

Ogallo, J. L., P. B. Goodell, J. W. Eckert, and P. A. Roberts. 1999. Management of root-knot nematodes with resistant cotton cv. NemX. Crop Science 39: 418 – 421.

Opperman, C. H., and M. A. Conkling. 1998. Bioengineering resistance to plant-parasitic nematodes. Pp. 239 – 250. In G. A. Pederson, G. L. Windham, and K. R. Barker, eds. Plant-nematode Interactions. Madison, WI: Agronomy Society of America.

Painter, R. H. 1951. Insect resistance in plants. New York: MacMillan.

Riggs, R. D., and D. P. Schmitt. 1988. Complete characterization of the race scheme for *Heterodera glycines*. Journal of Nematology 20: 392 – 395.

Roberts, P. A. 1992. Current status of the availability, development, and use of host plant resistance to nematodes. Journal of Nematology 24: 213 – 227.

Roberts, P. A. 1993. The future of nematology: integration of new and improved management strategies. Journal of Nematology 25: 383 – 394.

Roberts, P. A. 1995. Conceptual and practical aspects of variability in root-knot nematodes related to host plant resistance. Annual Review of Phytopathology 33: 199 – 221.

Roberts, P. A., W. C. Matthews, and J. C. Veremis. 1998. Genetic mechanisms of host-plant resistance to nematodes. Pp. 209 – 238. In G. A. Pederson, G. L. Windham, and K. R. Barker, eds. Plant-nematode Interactions. Madison, WI: Agronomy Society of America.

Roberts, P. A., and D. M. May. 1986. *Meloidogyne incognita* resistance characteristics in tomato genotypes developed for processing. Journal of Nematology 18: 353 – 359.

Roberts, P. A., and R. W. Scheuerman. 1984. Field evaluation of sweet potato clones and cultivars for reaction to root-knot and stubby root nematodes in California. Hortscience 19: 270 – 273.

Robinson, A. F., and A. E. Percival. 1997. Resistance to *Meloidogyne incognita* race 3 and *Rotylenchulus reniformis* in wild accessions of *Gossypium hirsutum* and *G. barbadense* from Mexico. Supplement to Journal of Nematology 29: 746 – 755.

Robinson, M. R., J. N. Jenkins, and J. C. McCarty, Jr. 1997. Different sources of root-knot nematode resistance. Pp. 444 In Proceedings of the Beltwide Cotton Conference. Memphis, TN: National Cotton Council of America.

Ross, J. P., and C. A. Brim. 1957. Resistance of soybeans to the soybean cyst nematode as determined by a double row method. Plant Disease Reporter 41: 923 – 924.

Rossi, M., F. L. Goggin, S. B. Milligan, I. Kaloshian, D. E. Ullman, and V. M. Williamson. 1998. The nematode resistance gene *Mi* of tomato confers resistance against the potato aphid. Proceedings National Academy of Sciences, USA 95: 9750 – 9754.

Sasser, J. N., and D. W. Freckman. 1987. A world perspective on nematology: The role of the society. Pp. 7 – 14. In J. A. Veech and D. W. Dickson, eds. Vistas on Nematology. Society of Nematologists.

Schmitt, D. P., and K. R. Barker. 1988. Incidence of plant-parasitic nematodes in the coastal plains of North Carolina. Plant Disease 72: 107 – 110.

Seinhorst, J. W. 1965. The relation between nematode density and damage to plants. Nematologica 11: 137 – 154.

Shepherd, R. L. 1974. Transgressive segregation for root-knot nematode resistance in cotton. Crop Science 14: 872 – 875.

Shepherd, R. L., J. C. McCarty, Jr., J. N. Jenkins, and W. L. Parrot. 1996. Registration of nine cotton germplasm lines resistant to root-knot nematode. Crop Science 36: 820.

Simon, P. W., W. C. Matthews, and P. A. Roberts. 2000. Evidence for a simply inherited dominant resistance to *Meloidogyne javanica* in carrot. Theoretical and Applied Genetics 100: 735 – 742.

Simpson, C. E. 1991. Pathways for introgression of pest resistance into *Arachis hypogaea*. Peanut Science 18: 22 – 26.

Simpson, C. E., and J. L. Starr. 2000. Registration of "COAN" peanut. Crop Science 40: 918.

Smith, P. G. 1944. Embryo culture of a tomato species hybrid. Proceedings of the American Society of Horticultural Science 44:413 –416.

Starr, J. L. 1998. Cotton. Pp. 359 –379. In G. A. Pederson, G. L. Windham, and K. R. Barker, eds. Plant-nematode Interactions. Madison, WI: Agronomy Society of America.

Starr, J. L., C. M. Heald, A. F. Robinson, R. G. Smith, and J. P. Krausz. 1993. *Meloidogyne incognita* and *Rotylenchulus reniformis* and associated soil textures from some cotton production areas of Texas. Supplement to Journal of Nematology 25:895 –899.

Starr, J. L., C. E. Simpson, and T. A. Lee, Jr. 1998. Yield of peanut genotypes resistant to root-kont nematodes. Peanut Science 25:119 –123.

Starr, J. L., J. Bridge, and R. Cook. 2002a. Plant resistance to parasitic nematodes. Wallingford, UK: CAB International.

Starr, J. L., E. R. Morgan, and C. E. Simpson. 2002b. Management of the peanut root-kont nematode, *Meloidogyne arenaria*, with host resistance. Online. Plant Health Progress doi: 10.1094/PHP-2002-1121-01-Hm.

Stewart, J. McD., and R. T. Robbins. 1996. Identification and enhancement of resistance to reniform nematode in cotton germplasm. Proceedings of the Beltwide Cotton Conference 1:255.

Trudgill, D. L. 1985. Potato cyst nematodes: a critical review of the current pathotype scheme. EPPO Bulletin 15:273 –279.

Trudgill, D. L. 1991. Resistance and tolerance of plant parasitic nematodes in plants. Annual Review of Phytopathology 29:167 –192.

Veremis, J. C., and P. A. Roberts. 1996. Relationships between *Meloidogyne incognita* resistance genes in *Lycopersicon peruvianum* differentiated by heat sensitivity and nematode virulence. Theoretical and Applied Genetics 93:950 –959.

Veremis, J. C., and P. A. Roberts. 2000. Diversity of heat-stable resistance to *Meloidogyne* in Maranon races of *Lycopersicon peruvianum* complex. Euphytica 111: 9 –16.

Veremis, J. C., A. W. van Heusden, and P. A. Roberts. 1999. Mapping a novel heat-stable resistance to *Meloidogyne* in *Lycopersicon peruvianum.* Theoretical and Applied Genetics 98:274 –280.

Vrain, T. C. 1999. Engineering natural and synthetic resistance for nematode management. Journal of Nematology 31:424 –436.

Ware, G. W. 1994. The Pesticide Book. Fresno, CA, USA: Thompson Publications.

Ware, J. O. 1936. Plant breeding and the cotton industry. Pp. 657. In Yearbook of Agriculture. US Department of Agriculture: Washington, D. C.

Webber, H. J., and W. A. Orton. 1902. Some diseases of cowpea. II. A cowpea resistant to root knot (*Heterodera radicola*). USDA Bureau of Plant Industry Bulletin no. 17.

Whitehead, A. G. 1998. Plant Nematode Control. Wallingford, UK: CAB International.

Wilfarth, H. 1900. Ein neuer gesichtspunkt zur bek^{3}mpfung der nematoden. Zeitchr. d. ver. d Deut. Zucker-Industrie. Lieferung 529:195 –204.

Williamson, V. M. 1998. Root-knot resistance genes in tomato and their potential for future use. Annual Review of Phytopathology 36:277 –293.

Williamson, V. M., and R. S. Hussey. 1996. Nematode pathogenesis and resistance in plants. The Plant Cell 8: 1735 – 1745.

Wingard, S. A. 1953. The nature of resistance to disease. Pp. 165 – 173. In The Yearbook of Agriculture. Washington, D. C.: US Department of Agriculture.

Young, L. D., and E. E. Hartwig. 1992. Problems and strategies associated with long-term use of nematode resistant cultivars. Journal of Nematology 24: 228 – 233.

Young, L. D. 1998. Breeding for nematode resistance. Pp. 187 – 207. In K. R. Barker, G. A. Pederson, and G. L. Windham, eds. Plant and Nematode Interactions. Madison, WI: American Society of Agronomy.

Zhou, E., 1999. Resistance to *Meloidogyne incognita* in cotton genotypes. Ph. D. Dissertation, Texas A& M University, College Station, TX, USA.

Zimmerman, A. 1897. Het groepsgewijs afsterven der koffie heesters in gesloten plantsoenem. Taysmannia, 23 pp.

18 Crop Rotation and Other Cultural Practices

John M. Halbrendt and James A. LaMondia

18.1 Introduction

Effective nematode control is essential for the profitable production of many commercial agricultural crops. Many techniques have been developed to manage nematodes but for more than 50 years nematicides and soil fumigants have been the primary tools to keep these pests in check. Nematicides are attractive because a wide variety of nematode problems can be easily controlled in a relatively short period of time, thus allowing farmers flexibility in crop production practices. Thurston (1992) noted "It is quite simple to apply a pesticide or utilize a high-yielding resistant variety to manage plant diseases, but one has to know a great deal about the biology of a situation in order to use cultural management."

Despite the advantages and simplicity of chemical control, there are numerous reasons why growers may choose to manage plant-parasitic nematodes without chemicals. Economics is a primary consideration. The cost of nematicides can not be justified for many low value crops. Concern over the environment and human health is another issue. Nematicides and soil fumigants carry the stigma of being dangerous because some have been responsible for ground water contamination, ozone depletion and wildlife kills (Jatala, 1986; Pease, et al., 1995). In the USA and elsewhere, regulatory agencies have banned the most environmentally hazardous compounds and required tougher regulations on the use of the remaining products to make them safer. Nevertheless, nematicides are toxic and their use will always carry some risk.

In some regions, growers are forced to seek alternative nematode controls because repeated chemical use has selected for soil microbes that are capable of enhanced biodegredation and the products are no longer effective (Sexstone, et al., 1992; Smelt, et al., 1996; Verhagen, et al., 1996). In other cropping systems, nematicides provide only a short-term solution because the lack of soil biodiversity allows plant-parasitic nematode populations to rebound soon after treatment. This situation creates the need for repeated applications, sometimes referred to as the "pesticide treadmill" (Jatala, 1986).

There is also growing concern over the sustainability of modern agricultural practices. Over the past 50 years, increased agricultural production and profits have been attributed largely to off-farm inputs and technological advancements including pesticides, fertilizers, irrigation, higher-yielding varieties, pest resistant cultivars, mechanization, and adoption of cost-cutting techniques. While these technological advances have improved the yield and profitability of food and fiber production there has been little work to understand the impact of these practices on soil and soil biology to insure the sustainability of agriculture (Bullock, 1992; Thurston, 1992).

As a consequence of all these concerns, modern agriculture seeks to reduce pesticide use through the development of integrated pest management (IPM) techniques and increased use of biological and biorational pest control practices (Cook, 1991). Undoubtedly, nematicides will continue to be an important tool for nematode management, especially as new chemistries are discovered that are safer for applicators and for the environment. However, the current trend for all pesticides is to minimize their use by developing alternative control methods based on biological and ecological principles.

18.2 Lessons from Traditional Farming

Archeological studies have shown that man has been cultivating crops for thousands of years. Early farmers experienced crop diseases, including nematodes, although they would not have known the causal agents or understood the biology of the problem. Crop failures and yield reductions were no doubt common, but over time and through trial and error, farmers unwittingly developed cultural practices that kept plant-parasitic nematode populations within tolerable levels. Cultural practices that led to sustainable crop production were retained while others that led to crop failure were discarded.

Modern farmers have lost touch with traditional agricultural practices because they are typically labor intensive and low yielding. Economics dictate that we can never return to such basic production methods. Despite this fact, it stands to reason that knowledge of the underlying principles of nematode control in traditional agriculture would be beneficial in the development of acceptable alternative techniques. Many traditional farming practices are still widely used today in regions where pesticides have not been readily available and where hand labor is plentiful and cheap.

Traditional farming practices are often effective because they incorporate aspects of natural ecosystems that help maintain the diversity of soil microfauna and minimize selection pressure for highly virulent pathogens. Crop rotation is one of the oldest cultural practices for maintaining biodiversity in cultivated soil. Rotations can be modified depending upon the needs of the farmer but the end result is essentially the same, i. e., that crops are separated in space and/or time to avoid the build-up of damaging nematode population levels on a particular host (Rodrìguez-Kàbana, 1992).

18.3 Early Cultural Practices

Since ancient times, man has recognized the problem of poor performance when crops are grown in continuous monoculture. Terms such as soil sickness, tired or worn out soil, monoculture injury, monoculture effect, and replant problem have been used to describe this effect. One of the simplest ways to counter this effect is to rotate crops and the practice of crop rotation dates back to antiquity. Early rotation systems were developed from empirical observations leading to sustained production. Traditional rotations were typically long-term, on the order of 10 to 12 years and often included 4 to 6 years in pasture or fallow.

It is now known that crop rotation is a very effective practice to keep nematode populations below damaging levels. However since crop rotation can also help manage other soilborne diseases, weeds, improve fertility and provide other benefits, we can not assume that crop rotation developed solely out of a need to avoid nematode damage. Nevertheless, it seems evident that traditional crop rotations provided sufficient nematode control to permit sustainable crop production.

Numerous references to crop rotation can be found in early writings. Thurston (1992) compiled a number of these to show that farmers throughout history found this to be an extremely valuable practice. For example, crop rotation was considered necessary to maintain production in ancient Chinese records and in Medieval Europe (Anonymous, 1980; Canullo, 1992; Rodrìguez-Kàbana and Yu, 1987). The early Romans developed different rotations for different soil types and they also recognized the value of leaving the land fallow. A translation by Lewis (1941) of advice given to Roman farmers by Virgil (70 – 19 BC) reads in part: "See too that your arable land lies fallow in due rotation, and leave the idle field alone to recoup its strength, ... so too are the fields rested by a rotation of crops and unploughed land promises to pay you." Similarly, farmers in thirteenth century Spain were warned not to plant wheat or millet more than two times in succession because the soil has an aversion to such plantings.

The Incas of the Peruvian Andes enforced a seven-year rotation between potato crops, which is sufficient to control the potato cyst nematodes, *Globodera rostochiensis* and *Globodera pallida* (de la Vega, 1966). The penalty for more frequent potato cultivation was the loss of one's fingers, which provides testimony to the importance of rotation as a control of a devastating nematode pathogen. Modern studies have verified that seven years of fallow or non-host crops were sufficient to reduce potato cyst nematode populations below damaging levels (Brodie, 1984).

The slash and burn system of agriculture has been used since the Neolithic era and may be one of the oldest forms of crop rotation (Conklin, 1961). However instead of rotating crops the practice involves a rotation of fields but provides the same benefits with regard to keeping nematode populations below damaging levels. After burning native vegetation to clear the land, a field is used several times to produce the same crop and then allowed to go fallow again (Thurston, 1992). Based on a study of farming in Nigeria, Wilson and Caveness (1980) reported that this technique was very effective for controlling plant-parasitic nematodes. The practice of slash and burn incorporates burning, rotation, and fallow to control nematodes (and other plant disease problems) and therefore may be considered one of the earliest forms of integrated pest management.

The use of multiple cropping and varietal mixtures is another ancient practice that is still common in many traditional agricultural systems in the tropics today. Reports of multiple cropping systems date back to before the time of Christ in Egypt and India and it was practiced throughout China during the Ming Dynasty (Thurston, 1992). The variety of plants and cultural practices used in multiple cropping systems complicates studies to determine the optimal plant combinations and ratios for maximum nematode control. However, the success of this practice over thousands of years indicates that it is an effective nematode management tool. The effect of multiple cropping and varietal mixtures is to avoid selection pressure and population build-up by maintaining genetic diversity of crops and separating host plants in time and space (Smithson and Lenné, 1996). Research also shows that the diversity of plants in multiple cropping systems fosters the build up of naturally occurring nematode antagonists (Sikora, 1992).

18.4 Modern Crop Rotation Practices

18.4.1 Non-host Crop Rotations

Plant-parasitic nematodes do not move significant distances on their own and their reproductive rate on a susceptible host is slow compared to other plant pathogens such as fungi and bacteria. Therefore simply reducing soil populations of parasitic nematodes below damaging levels may result in dramatic increases in crop growth and yield over a season. As a result, one of the first, most obvious, and important means of controlling nematodes is the use of crop rotation with a non-host plant species. Theoretically, the cultivation of only non-host plants in a particular location would remove food sources and virtually eliminate plant-parasitic nematodes from a field. However numerous biological, practical and economic considerations make cost effective nematode management by rotation more complicated than it would appear. Nevertheless, a good non-host rotation can dramatically reduce nematode populations and increase yields (Rodrìguez-Kàbana and Ivey, 1986).

The efficacy of rotation as a means of reducing nematode populations below damage thresholds depends on many factors including: accurate identification of the nematode species; the host range of the particular nematode; the ability of the nematode to survive in the absence of a host; the presence of alternate hosts in the form of weeds which may also serve as nematode reservoirs; and the economics of the crop rotation. In most instances, the development of rotation practices to manage plant-parasitic nematodes has been done with annual crops. Control of nematodes in perennial crops is extremely difficult as there may be no opportunity to reduce nematode populations after the crop is established. While there has been some success in developing effective crop rotations for annual crops much more research needs to be done.

In the absence of broad-spectrum fumigants or nematicides, the accurate identification of the nematode parasite is the first important step in control. In general, the specialized pathogens such as the potato cyst nematodes *Globodera rostochiensis* and *G. pallida* and the soybean cyst nematode *Heterodera glycines*, are quite host-specific. Other nematodes, such as the migratory endoparasite *Pratylenchus* species, are not host-specific and have a very large host range, making it difficult to find a suitable non-host crop for rotation.

Root-knot nematodes, *Meloidogyne* spp., are specialized parasites but typically have a wider host range than cyst nematodes. Accurate identification of the species or races present may be necessary to determine a suitable non-host or resistant crop. In soils with mixed populations of *Meloidogyne*, different species may predominate depending on the previous crops grown (Johnson, 1985). For example, in mixed race populations of *Meloidogyne arenaria*, races 1 and race 2 each dominated in soil after peanut and tobacco, respectively (Hirunsalee, et al., 1995). Similarly, the sequence of soybean resistant cultivars determined which *Heterodera glycines* race predominated in soil (Young and Hartwig, 1988).

Rotation crops intended to control nematodes may also influence plant yield, soil properties, nutrition and many other factors, including populations of other soil-borne pathogens. The success of the rotation crop against the nematode parasite should be measured by the change in nematode soil population density from before the crop is planted to after the crop is removed or tilled into the soil rather than by plant yield. While a rotation scheme is dependent on nematode control to be successful, the ultimate success of rotation often depends on other factors, such as the economics of the alternative crop, interactions with other pests, the availability of other management tactics, and ultimately, the economics of the rotation management strategy.

A large number of rotation crops and crop sequences have been researched, proposed, or accepted, which take into account factors such as host specificity and nematode biology. A four-year rotation sequence consisting of two years of resistant potato cultivars with one year of a non-host such as oats prior to a single year of susceptible potato has been mandated in potato cyst nematode-infested soils in the USA. This allows potato production in three out of four years without an increase in nematode population over the length of the rotation sequence (Brodie, 1996).

Root-knot nematodes hatch in response to physical conditions such as temperature and moisture rather than in response to signals from a host plant. As a result, root-knot nematode densities may decline more quickly in soils after a single rotation crop. However, root-knot nematodes generally have a wider host range than cyst nematodes, restricting options for a non-host rotation crop. A single year of rotation to cotton greatly reduced the peanut root-knot nematode *Meloidogyne arenaria* race 1 (Rodrìguez-Kàbana, et al., 1987) and a peanut-cotton rotation significantly reduced the level of *Meloidogyne incognita* available to attack cotton (Kirkpatrick and Sasser, 1984). A peanut-cotton rotation effectively manages both nematodes. Other root-knot nematodes such as *Meloidogyne incognita* race 1 may be more effectively controlled by a 1-year rotation to bermudagrass than by fallow or even the use of the nematicide fenamiphos on a susceptible crop (Johnson, et al., 1997). *Meloidogyne incognita* race 3 may be effectively reduced in vegetable production by rotation with nematode-resistant peppers (Thies, et al., 1998). The incorporation of barley in carrot rotations reduces both the northern root-knot nematode *Meloidogyne hapla* and certain weed populations (Leroux, et al., 1996). Rotation with ornamental *Aster* or *Rudbeckia* greatly reduced or eliminated *M. hapla* populations (LaMondia, 1997), perhaps as a result of nematicidal compounds present in roots (Sanchez de Viala, et al., 1998).

The lack of host specificity and wide host range for most migratory endoparasitic and ectoparasitic nematodes makes the identification of effective non-host rotation crops for these nematodes difficult. Still, some effective rotations have been developed. In potato production two-year rotations with legumes such as alfalfa or clover resulted in lower *Pratylenchus penetrans* populations and increased tuber yields in soils infested with *Verticillium dahliae* (Chen, et al., 1995a). *Pratylenchus penetrans* numbers in soil were reduced by rotation with Saia oat (*Avena strigosa*) or sorgho-sudangrass (LaMondia, 1999). Populations of *Belonolaimus* from North Carolina and Georgia reproduced on a wide variety of crops and weeds, but were unable to reproduce on tobacco (Robbins and Barker, 1973). Tobacco also failed to maintain populations of *Criconemella ornata* capable of damaging peanut (Barker, et al., 1982). Rotations of corn, barley, and resistant oat cultivars reduced infection by *Ditylenchus* in subsequent crops of onion and cereals (Rivoal and Cook, 1993).

It is crucial that rotation crops intended to suppress one target nematode be evaluated against other nematode pests of that crop as well. Plant parasitic nematodes often occur in multiple species communities and management of a single species may result in higher populations of another. For example, rotation with sorgho-sudangrass was effective in reducing densities of *Meloidogyne incognita* race 1 and *M. arenaria*, but populations of *Belonolaimus* and *Paratrichodorus minor* increased on that crop and damaged subsequent susceptible crops (McSorley and Dickson, 1995). Similarly, rotation with barley was determined to be a means of reducing *Meloidogyne chitwoodi* populations in potato soils in the Pacific Northwest. Unfortunately, densities of *Pratylenchus neglectus,* another important parasite of potato, increased on the barley crop (Ferris, et al., 1994).

Weed species present in production fields may act as a sanctuary for parasitic nematodes in a non-host rotation. The presence of weeds may support parasitic nematode populations at some level despite the production of an unsuitable non-host or resistant crop (Belair and Benoit, 1996; Bendixen, 1988; Manuel, et al., 1982). Yellow and purple nutsedge were more competitive with chili pepper when *Meloidogyne incognita* race 3 was present and both nutsedge species maintained populations of the nematode throughout the period when non-host commercial crops were grown (Schroeder, et al., 1993). Weeds may even be more competitive to a number of crops in fields infested with sufficient numbers of plant parasitic nematodes to stunt or damage the crop plants (Alston, et al., 1991; Chen, et al., 1995b). Weeds may not only maintain nematodes, but may also serve as reservoirs of nematode-vectored viruses such as tobacco or tomato ringspot (Taylor, et al., 1994). Control of weeds such as dandelion which harbor both dagger nematodes and tomato ringspot virus is an important component in the management of the Xiphinema/Prunus stem pitting complex (Barrat, et al., 1984; Powell and Forer, 1984).

18.4.2 Resistant Crop Rotation

The use of nematode-resistant crops in a rotation scheme may be effective both as a nematode management tactic and also as a means of reducing selection pressure to overcome resistance genes. A rotation utilizing resistant and susceptible soybean cultivars with non-host crops reduced nematode densities, increased soybean yields, and slowed the shift to increased nematode reproduction on resistant cultivars (Young, 1998). Most non-leguminous crops may be used as a non-host rotation for control of the soybean cyst nematode (Riggs and Niblack, 1993). Similarly, most non-cereal crops may be used to reduce densities of the cereal cyst nematode, *Heterodera avenae* (Rivoal and Cook, 1993).

Resistant cultivars are often used in rotations for control of cyst nematodes, as these nematodes are generally difficult to control by rotation with non-hosts alone. Mai and Abawi (1980) determined that the ratio of non-host to host crop production had to be at least 2: 1 or greater (up to 5: 1) to reduce *Heterodera schachtii* densities to below damaging levels over time. These high ratios may be required for nematodes that do not hatch well in the absence of stimulation by a host plant. The decline of *G. rostochiensis* has been demonstrated to be about 30% per year in the absence of a host (Brodie, 1984), which is not particularly efficacious. As a result, *G. rostochiensis* and *H. glycines* may survive at low levels in soil for up to 10 years in the absence of a host (Slack, et al., 1972; Spears, et al., 1968). The slow decline in lipid content of dormant, unhatched potato cyst nematodes may result in reduced juvenile infectivity over a number of years (Atkinson, et al., 2001). However, the utilization of resistant potato cultivars which stimulate hatch without allowing subsequent reproduction, can greatly increase nematode activity and reduce potato cyst nematode populations by 80 – 95% per year (Brodie, 1976). Similar results have been seen for tobacco cyst nematodes. The annual decline of tobacco cyst nematodes may be approximately 40% per year under non-host crops, but under resistant cultivars the nematode mortality may be nearly equivalent to fumigation (LaMondia, 1995). Cyst nematodes with eggs in diapause may respond to a hatching factor not only by juvenile hatch, but also by by increased in-egg mortality (Devine and Jones, 2001).

18.4.3 Fallow Rotations

Many crop rotations include periods of fallow as part of the cropping sequence. The fallow period is commonly interpreted to mean that the land is left uncropped after harvest as in the case of slash and burn agriculture where the land is allowed to revert back to the natural vegetation. Also called "bush" or "weedy" fallow, the invasion of natural vegetation in an uncropped field will increase plant diversity and reduce the plant-parasitic nematode population by limiting the number of suitable hosts or eliminating hosts altogether as in the case of some nematodes with a narrow host range. Such fallow periods may also inhibit plant-parasitic nematodes by increasing the number and diversity of nematode biocontrol agents (Sikora, 1992).

In different cropping systems, variations of fallow have been developed, which may be more effective for nematode suppression over shorter periods of time. The practice of "bare" fallow is intended to starve out the plant-parasitic nematodes by maintaining fields free of all vegetation. This practice can be very successful for nematode control (Duncan, 1986), but it can also have serious detrimental effects on soils such as increased risk of erosion, loss of soil organic matter, and loss of beneficial microflora such as mycorrhizae and *Rhizobium* spp. (Duncan, 1991).

The practice of keeping the land free of vegetation is also known as "clean" or "dry" fallow. It requires input either in the form of repeated herbicide application or cultivation. Cultivation to control weeds can be very effective since nematodes are further reduced by desiccation and heat as roots and soil are exposed to the soil surface (Thurston, 1992).

A "grass" fallow is used for nematode control in some cropping systems. The establishment of a thick stand of grass can effectively reduce populations of certain nematodes that have a narrow host range, such as certain root-knot species. Since the nematodes do not use grass as a host the population will starve just as if there were no crop. However, if certain weed hosts are present the nematodes will survive (Bridge, 1996). The grass fallow system is beneficial in that the soil is protected from erosion and loss of fertility through nutrient cycling. The practice is analogous to use of a non-host rotation except that no marketable crop is produced. Grass fallow is sometimes used as pasture.

Depending upon the crop and nematode to be controlled, a "wet" (intermittent periods of flooding) or "flood" fallow (long periods of water-logged conditions) can also be an effective management tool (Thurston,

1992). Similarly, land that is naturally flooded on a regular cycle also tends to have few plant-parasitic nematodes (Bridge, 1996, 1998; Castillo, et al., 1978). Some nematodes survive only a short time under a wet fallow, probably as a result of anaerobic conditions, pH changes, release of toxic substances or microbial activity (Sasser, 1990; Trivedi and Barker, 1986; Van Gundy, 1985). Thurston (1992) compiled a relatively long list of nematodes that can be controlled by flooding. Some nematodes, such as *Meloidogyne graminicola*, are incapable of entering their host under flooded conditions despite the fact that they can survive extended periods in water-logged soil. Establishing rice seedlings in flooded seedbeds is a useful practice to avoid infection by *M. graminicola* (Bridge and Page, 1982). This practice also protects subsequent vegetable crops that may be grown in the rotation (Thames and Stoner, 1953).

Although flooding can effectively control a number of nematodes, the practice is generally used only in areas where the land is either naturally flooded or where fields are artificially flooded for other purposes such as irrigated rice production. For most farmers, the expense and difficulties of flooding land cannot be justified for the sole purpose of controlling nematodes (Bridge, 1998). In some soils flooding also carries the risk of predisposing plants to other soil-borne diseases such as *Phytophthora, Pythium* and *Aphanomyces* (Barta and Schmitthenner, 1986; Cook and Baker, 1983). There is also a possibility that drainage water from flooded fields may help distribute nematodes into previously uninfested soil.

18.5 Allelopathic Plants

Some plants release nematicidal compounds into the rhizosphere either through root leachates or from decomposing plant residue (Bhatti, 1988; Halbrendt, 1996). When used as a rotation or cover crop, these allelopathic (antagonistic) plants can provide a level of nematode control that is greater than would be achieved by a non-host rotation. Rodrìguez-Kàbana (1992) described nonhosts as providing "passive" nematode control whereas allelopathic plants provide "active" control. For example, in one season of an allelopathic rotation crop such as castor bean or sesame, the control of *Meloidogyne arenaria* was much greater than provided by a non-host (Rodrìguez-Kàbana, et al., 1989). The idea of using allelopathic plants in crop rotation is to suppress the nematode population quickly and thus shorten the rotation cycle necessary to get acceptable levels of control (Lovett, 1986). Allelopathic plants have also been shown to provide nematode control when intercropped with susceptible hosts (Rhode and Jenkins, 1958), although this is generally not an acceptable commercial practice in modern agriculture.

The use of allelopathic rotation crops is limited by practical considerations relating to the farmer's ability to incorporate these plants into his cropping system and economics (Halbrendt, 1996; Sasser, 1990). While the level of nematode control from allelopathic plants is better than that of non-hosts it is still less effective than nematicides. However, realization that the allelopathic potential of plants can be improved through breeding, selection and perhaps genetic manipulation has stimulated interest in the development of more effective allelopathic rotation crops. One company (Fysicon Research, The Netherlands) is now marketing selected marigold varieties that have been bred specifically for enhanced nematicidal activity. In microplot experiments, two-year rotations to Saia oat and Polynema marigold reduced lesion nematode densities, increased potato tuber yields by 40% and reduced potato early dying severity by 25% over potato. These results indicate that rotation crops that reduce nematodes may also aid in management of complex diseases such as potato early dying (LaMondia, 2000).

18.6 Trap Crops

Trap crops offer an additional opportunity to use nematode-host interactions to management advantage. While resistant cultivars may be regarded as trap crops in the broad sense, trap cropping more usually is defined as the destruction of a susceptible crop after nematode infection but before the nematodes mature. This means of reducing nematode densities was postulated as early as 1939 (Carroll and McMahon, 1939) and has been demonstrated as a means of controlling sedentary endoparasitic cyst nematodes that hatch poorly in the absence of host plants (LaMondia, 1996; Whitehead, 1977). Many cyst nematodes are stimulated to hatch by root exudates, and trap crops further reduce populations beyond the decline in numbers due to fallow. Risks associated with trap crops include the possibility of nematode increase in the event of poorly timed trap crop destruction (Whitehead, 1977). The use of resistant plants as short-term trap crops in combination with rotation crops may limit risk and still result in greatly improved nematode control over fallow alone (LaMondia, 1996).

18.7 Green Manure Crops and Soil Amendments

The addition of organic matter as fertilizer has always been an important agricultural practice (Chandler, 1981; Spurr, 1986; von Hagen, 1959). However, research has shown that some types of soil amendments can effectively control plant-parasitic nematodes. This discovery suggests that early reports of improved crop growth resulting from the addition of organic matter may be difficult to interpret if the effect on plant-parasitic nematodes was not considered as plant growth will benefit from both the fertilizer effect and the suppression of soil-borne pathogens (Patrick and Toussoun, 1970). Generally speaking, traditional farmers are only concerned with improved yields and are not concerned with the cause and effect of a particular practice.

The addition of organic matter to soil can essentially be divided into two broad categories, the use of green manure and the use of soil amendments. Green manures are rotation or cover crops that are ploughed back into the soil while still green and allowed to decompose. Soil amendments comprise a much broader category, usually consisting of various waste materials. Often, the waste is a direct byproduct of agricultural production such as pressed seed meal or pomace. In other cases it is waste from other sources such as animal manure, crustacean shells, and even human wastes.

The effects of green manure and other amendments on the agroecosystem are many and varied, and involve changes in soil chemistry, physical properties and microbiology (Lovett, 1986). Furthermore, these changes are dynamic, often changing dramatically within days or even hours as in the case of some chemical changes (Patrick and Toussoun, 1970). The nematicidal effect is usually ephemeral, making research on the mechanism difficult. Often, a particular soil treatment can be shown to be nematicidal but the precise cause and effect relationship is difficult to establish.

When plants are incorporated as green manure there is a rapid influx of organic compounds into the soil as cell membranes loose their integrity and cells begin to leak. The process is not dependent on the slow decomposition of the cell walls. This rapid release of plant compounds may in itself release nematicidal compounds. Many preparations of crude plant homogenates have been shown to kill nematodes (Bhatti, 1988). Similarly, the rupture of cell membranes may release plant defense compounds in high enough concentrations to kill nematodes, as in the case of isothiocyanates released from *Brassica* sp. which form by the myrosinase-mediated hydrolysis of glucosinolate (Brown and Morra, 1997; Halbrendt and Jing, 1996). Other plants accumulate toxic substances such as cyanide, which are released upon decomposition and have been proposed as the possible source of the nematicidal effect (D'Addabbo, 1995; Sikora, 1992; Sterling, 1991; Vawdrey and Sterling, 1997).

Generally, the effect of soil amendments is thought to be indirect by enhancing the activity of naturally occurring antagonists of nematodes. Alternatively, the various soil amendments that have been used to control nematodes may also contain some nematicidal compounds.

18.8 Cultural Practices

A number of cultural practices can effectively reduce nematode population levels. In some situations these can be used as stand alone control tactics but more often they serve as part of an integrated nematode management program. These practices include a variety of exclusion and physical control techniques such as sanitation, solarization, timing of planting or crop destruction, and tillage.

18.8.1 Sanitation

Nematode management by sanitation is a prophylactic measure that precludes nematode problems by preventing nematode introduction into new production areas. Sanitation measures may include the inspection and certification of nematode-free planting material, cleaning of equipment and the use of quarantines to minimize the chance of nematode dispersal (Mai, 1977).

The removal of nematodes from infected plant propagative material may be accomplished by tissue culture, heat treatment, or pruning. Tissue culture has been widely used to eliminate many plant pathogens from plant tissue that can be used for propagation and nematodes are no exception (Delang, et al., 1987; Garg, et al., 1988). Heat treatment has long been used to kill plant parasitic nematodes in living plant tissues (Jenkins, 1960; Towson and Lear, 1982). However, killing all nematodes in seeds, roots, corms, tubers, rhizomes or rootstocks is difficult and may result in plant death. The temperatures required to kill nematodes in plant tissues are often quite similar to the temperatures required to kill plant tissues (Bridge, 1996). Critical temperatures usually range from 45℃ to 55℃, but need to be determined for each plant species almost on a case-by-case basis based on water volume, number and size of plants to be treated, time of treatment, and other factors.

Physical removal of nematodes from plant propagative material may greatly reduce initial populations at planting and also the spread of nematodes to new locations. Root pruning of fibrous roots from bare-root planting stock of several herbaceous perennial ornamentals greatly reduced or eliminated *Meloidogyne hapla* galls and egg production in plants months after propagation (LaMondia, 1997). Paring away diseased or discolored tissues of banana and plantain corms can successfully manage moderate populations of both *Pratylenchus* and *Radopholus* (Bridge, 1996).

18.8.2 Solarization

The use of transparent polyethylene to trap solar heat and disinfest soil is usually considered most effective in hot, arid climates, but may also be effective against soil-borne pathogens, including nematodes, in more humid, temperate regions. Solarization may also control certain weeds, insects, and fungal pathogens as well as influence soil nutrition and ecology (Katan, 1981). Direct effects of temperature may have dramatic effects on nematode populations in soils. Solarization experiments in semi-arid tropical India reduced populations of several genera of nematodes, including *Heterodera*, *Rotylenchulus*, and *Pratylenchus*, comparable to fumigation to a depth of at least 20 cm (Sharma and Nene, 1990).

Under more humid conditions in Florida, control of *Rotylenchulus* and *Meloidogyne* was inconsistent, perhaps due to re-infestation from unaffected areas between rows or from deep in the soil profile. However, the level of control of nematodes such as *Paratrichodorus*, *Criconemella*, and *Helicotylenchus* and diseases such as Fusarium wilt was similar to fumigation (Chellemi, et al., 1997). In South Africa, total nematode control in solarized plots ranged from 37 to 100% at 40 cm and control of the fungus *Phytophthora cinnamomi* was greater than 90% (Barbarcheck and Von-Broembsen, 1986). Free-living nematodes were less affected by high temperatures than plant parasitic nematodes. Thermal inactivation of *Meloidogyne javanica* occurred after exposure to 40℃ for 4 weeks and after exposure to fluctuating conditions up to 45℃ for 2 h per day.

In New York State, soil solarization reduced populations of *G. rostochiensis* by more than 95% to a depth of 10 cm, and significantly reduced nematode survival from 10 to 15 cm deep (LaMondia, 1984). However at greater depths, soil temperatures were not elevated to levels sufficient to kill nematodes. In controlled experiments, *G. rostochiensis* cyst contents were killed after exposure to 40℃ for 7 days or after 2 h exposure to 45℃ (LaMondia, 1990). However, sublethal temperatures may have direct effects on long-term nematode viability, or may affect interactions with other soil microbes, increasing the effects of biological control agents. For example, the attachment of *Pasteuria penetrans* endospores to *Meloidogyne arenaria* juveniles was increased after nematodes were exposed to sublethal temperatures of 23℃ to 30℃ for 4 days (Freitas, et al. 1997). The indirect effects of solarization on nematode survival may be critical for the success of this technique in temperate areas.

18.8.3 Planting Date

Late planting and early harvest or crop destruction can be used to limit nematode reproduction on susceptible crop cultivars. Late planting of soybean, practiced on the 20 to 40% of the soybean production area in a soybean/wheat double crop in any one year, reduces the initial population of *Heterodera glycines* and *Pratylenchus brachyurus* present at planting and results in lower populations at harvest. Escape cropping, planting crops early or late when temperatures are too high or low for nematode infection and development, has been used to reduce nematode damage to vegetable and rice crops in India and Southeast Asia (Bridge, 1996).

18.8.4 Tillage

Tillage can be used to greatly reduce late season population increase and survival of *Meloidogyne* (Barker and Imbriani, 1984) or tobacco cyst nematodes (LaMondia, 1999) by eliminating the roots of host plants that can survive for months after harvest. The impact of tillage in contrast to no or low till or conservation tillage systems on nematode populations has been inconsistent on many crops, perhaps due to the larger effects of weeds, soil structure, nutrition, and other factors on crop growth (Barker and Koenning, 1998).

18.9 Future Outlook

Crop rotation and other cultural tactics have been used to control pathogen and pest complexes in agriculture for centuries, despite the fact that the causal agents of many of these diseases have only been known for a relatively short period of time. In fact, the recognition of nematodes as important plant pathogens coincided with the advent of chemical nematode control within the past 50 – 60 years. In recent years increased awareness of the human and environmental risks associated with pesticide use has tightened restrictions on the availability and use of chemical nematicides. These developments have renewed interest in the use of non-chemical nematode control tactics.

As shown in this chapter, cultural practices to control nematodes are not new but to be useful they must be re-developed to fit with modern agriculture. A seven-year rotation away from potatoes after every potato crop may have worked for the Incas, but would not be appropriate in current economics. Researchers need to develop tactics with increased efficacy, and because no one tactic will likely stand alone as a total nematode control system, integrated systems incorporating several tactics will be needed to achieve adequate and economically feasible nematode management. Successful strategies will likely include crop rotation, chemical control where appropriate, crop resistance, biological control where available, green manuring or soil amendments with antagonistic plants, natural products, and other cultural practices.

Globalization of trade seems already to have increased the spread of pathogens and pests, including nematodes. The interactions of nematodes with other pathogens to increase disease severity will also likely result in increased plant losses due to complex diseases and reduced agricultural sustainability. Sanitation and quarantine will remain important means of restricting global spread of species to new areas.

The use of plant resistance to control nematodes will remain an important and cost-effective nematode control tactic. New sources of plant resistance need to be identified and incorporated into crops by traditional plant breeding or into transgenic crops by genetic engineering biotechnology. The integration of resistance with other tactics will reduce selection pressure and likely extend the availability of effective resistant plants.

New information about allelopathy and the specific chemicals responsible for non-host status need to be identified. Plants may then be selected for, bred for, or transformed to contain higher levels of "active compounds" targeted for control of specific nematodes.

Future research on these topics will no doubt be difficult. Research will need to be long-term, time consuming and adapted to the nematode pathogens in specific locations. Successful nematode control alone does not guarantee success of the overall management strategy. The economics of the rotation or non-host crop must be taken into consideration, as well as the impact of the control tactics on other nematodes, pathogens, insects, weeds, crop growth, and marketable crop yield. All of these factors, including economics, may be specific to certain localities.

The large effort required to develop non-chemical nematode control strategies will be paid back in the benefits of sustained agricultural production.

References

Alston, D. G., R. Bradley, P. Schmitt, and H. D. Coble. 1991. Response of *Helicoverpa zea* (Lepidoptera: Noctuidae) populations to canopy development in soybean as influenced by *Heterodera glycines* (Nematoda: Heteroderidae) and annual weed population densities. Journal of Economic Entomology 84: 267 – 276.

Anonymous. 1980. China: multiple cropping and related crop production technology. Rome, Italy: FAO.

Atkinson, H. J., R. A. Holz, E. Riga, G. Main, R. Oros, and J. Franco. 2001. An algorithm for optimizing rotational control of *Globodera rostochiensis* on potato crops in Bolivia. Journal of Nematology 33: 121 – 125.

Barbarcheck, M. D., and S. L. Von-Broembsen. 1986. Effects of soil solarization on plant – parasitic nematodes and *Phytophthora cinnamomi* in South Africa. Plant Disease 70: 945 – 950.

Barker, K. R., and J. L. Imbriani. 1984. Nematode advisory programs – status and prospects. Plant Disease 68: 735 – 741.

Barker, K. R., and S. R. Koenning. 1998. Developing sustainable systems for nematode management. Annual Review of Phytopathology 36: 165 – 205.

Barker, K. R., D. P. Schmitt, and V. P. Campos. 1982. Response of peanut, corn, tobacco, and soybean to *Criconemella ornata*. Journal of Nematology 14: 576 – 581.

Barrat, J. G., R. Scorza, and B. E. Otto. 1984. Detection of tomato ringspot virus in peach orchards. Plant Disease 68: 198 – 200.

Barta, A. L., and A. F. Schmitthenner. 1986. Interaction between flooding stress and *Phytophthora* root rot. Plant Disease 70: 310 – 313.

Belair, G., and D. L. Benoit. 1996. Host suitability of 32 common weeds to *Meloidogyne hapla* in organic soils of southwestern Quebec. Journal of Nematology 28: 643 – 647.

Bendixen, L. E. 1988. Major weed hosts of nematodes in crop production. Wooster, OH: Ohio State University.

Bhatti, D. S. 1988. Utilization of toxic plants for the control of nematode pests of economic crops. Haryana Agricultural University, Hisar, India. Final Technical Report, Project No. IN-ARS-197.

Bridge, J. 1996. Nematode management in sustainable and subsistence agriculture. Annual Review of Phytopathology 34: 201 – 225.

Bridge, J. 1998. Theory and practice of non-chemical management of nematode pests in tropical farming systems. Brighton Crop Protection Conference: Pests & Diseases. Vol. 3. Brighton, UK: CABI Bioscience. pp. 761 – 768.

Bridge, J., and S. L. J. Page. 1982. The rice root-knot nematode, *Meloidogyne graminicola*, on deep water rice (*Oryza sativa* subsp. *indica*). Revue de Nematologie 5: 225 – 232.

Brodie, B. B. 1976. Managing population densities of *Heterodera rostochiensis*. Journal of Nematology 8: 280 (Abstr.).

Brodie, B. B. 1984. Nematode parasites of potato. Pp. 167 – 212. In W. R. Nickle, ed. Plant and Insect Nematodes. New York: Dekker.

Brodie, B. B. 1996. Effect of initial nematode density on managing *Globodera rostochiensis* with resistant cultivars and nonhosts. Journal of Nematology 28: 510 – 519.

Brown, P. D., and M. J. Morra. 1997. Control of soil-borne plant pests using glucosinolate – containing plants. Advances in Agronomy 61: 167 – 231.

Bullock, D. G. 1992. Crop rotation. Critical Reviews in Plant Sciences 11: 309 – 326.

Carroll, J., and E. McMahon. 1939. Experiments on trap cropping with potatoes as a control measure against potato eelworm (*Heterodera schachtii*). Journal of

Helminthology 17: 101 – 112.
Castillo, M., M. B. Arceo, and J. A. Litsinger. 1978. Effect of geomorphic field position, flooding and cropping pattern on plant-parasitic nematodes of crops following rainfed wetland rice in Iloilo, Philippines. International Rice Research Newsletter 3: 27.
Chandler, R. F. 1981. Land and water resources and management. Pp. 9 – 18. In D. L. Plucknett and H. L. Beemer, Jr., eds. Vegetable Farming Systems in China. Boulder, CO: Westview Press.
Chellemi, D. O., S. M. Olson, D. J. Mitchell, I. Secker, and R. McSorley. 1997. Adaptation of soil solarization to the integrated management of soilborne pests of tomato under humid conditions. Phytopathology 87: 250 – 258.
Chen, J., G. W. Bird, and R. L. Mather. 1995a. Impact of multi-year cropping regimes on *Solanum tuberosum* tuber yields in the presence of *Pratylenchus penetrans* and *Verticillium dahliae*. Supplement to Journal of Nematology 27: 654 – 660.
Chen, J., G. W. Bird, and K. A. Renner. 1995b. Influence of *Heterodera glycines* on interspecific and intraspecific competition associated with *Glycine max* and *Chenopodium album*. Journal of Nematology 27: 63 – 69.
Conklin, H. C. 1961. The study of shifting cultivation. Current Anthropology 2: 27 – 61.
Cook, R. J. 1991. Challenges and rewards of sustainable agricultural research and education. Pp. 32 – 76. In B. J. Rice, ed. Sustainable Agriculture Research and Education in the Field: a Proceedings. Ashington, D. C.: National Academy Press.
Cook, R. J., and K. F. Baker. 1983. The nature and practice of biological control of plant pathogens. St. Paul, MN: American Phytopathological Society.
D'Addabbo, T. 1995. The nematicidal effect of organic amendments: a review of the literature, 1982 – 1994. Nematologia Mediterranea 23: 299 – 305.
De la Vega, G. E. I. 1966. Royal commentaries of the Incas and general history of Peru. Part 1. Austin, TX: University of Texas Press.
Delang, J. H., P. Willers, and M. Nel. 1987. Elimination of nematodes from ginger (*Zingiber officinale* Roscoe) by tissue culture. Journal of Horticultural Science 62: 249 – 252.
Devine, K. J., and P. W. Jones. 2001. Effects of hatching factors on potato cyst nematode hatch and in-egg mortality in soil and *in vitro*. Nematology 3: 65 – 74.
Duncan, L. D. 1991. Current options for nematode management. Annual Review of Phytopathology 29: 469 – 490.
Duncan, L. W. 1986. Effects of bare fallow on plant-parasitic nematodes in the Sahelian zone of Senegal. Revue de Nematologie 9: 75 – 81.
Ferris, H., H. L. Carlson, and B. B. Westerdahl. 1994. Nematode population changes under crop rotation sequences: Consequences for potato production. Agronomy Journal 86: 340 – 348.
Freitas, L. G., D. J. Mitchell, and D. W. Dickson. 1997. Temperature effects on the attachment of *Pasteuria penetrans* endospores to *Meloidogyne arenaria* race 1. Journal of Nematology 29: 547 – 555.
Garg, G. K., U. S. Singh, R. K. Khetrapal, and J. Kumar. 1988. Application of tissue culture in plant pathology. Pp. 83 – 120. In R. S. Singh, U. S. Singh, W. M. Hess, and D. J. Weber, eds. Experimental and Conceptual Plant Pathology. Vol. 1. New Dehli: Oxford & IBH Publishing.

Halbrendt, J. M. 1996. Allelopathy in the management of plant-parasitic nematodes. Journal of Nematology 28: 8 – 14.

Halbrendt, J. M., and G. N. Jing. 1996. Cruciferous green manure as an alternative to nematicide: The effect of glucosinolate content. Pp. 458 – 465. In L. H. Princen and C. Rossi, eds. Nineth International Conference on Jojoba and its Uses and the Third International Conference on New Industrial Crops and Products-1994. Catamarca, Argentina: The Association for the Advancement of Industrial Crops.

Hirunsalee, A. K., R. Barker, and M. K. Beute. 1995. Effects of peanut-tobacco rotations on population dynamics of *Meloidogyne arenaria* in mixed rate populations. Journal of Nematology 27: 178 – 188.

Jatala, P. 1986. Biological control of plant-parasitic nematodes. Annual Review of Phytopathology 24: 453 – 489.

Jenkins, W. R. 1960. Control of nematodes by physical methods. Pp. 443 – 446. In J. N. Sasser and W. R. Jenkins, eds. Nematology - Fundamentals and Recent Advances with Emphasis on Plant Parasitic and Soil Forms. Chapel Hill, NC: North Carolina State University Graphics.

Johnson, A. W. 1985. Specific crop rotation effects combined with cultural practices and nematicides. Pp. 283 – 301. In K. R. Barker, C. C. Carter, and J. N. Sasser, eds. An Advanced Treatise on *Meloidogyne,* Vol. II. Methodology. Raleigh, NC: North Carolina State University Press.

Johnson, A. W., G. W. Burtton, D. R. Sumner, and Z. Handoo. 1997. Coastal bermudagrass rotation and fallow for management of nematodes and soilborne fungi on vegetable crops. Journal of Nematology 29: 710 – 716.

Katan, J. 1981. Solar heating (solarization) of soil for control of soil-borne pests. Annual Review of Phytopathology 19: 211 – 236.

Kirkpatrick, T. L., and J. N. Sasser. 1984. Crop rotation and races of *Meloidogyne incognita* in the coastal plain. Journal of Nematology 16: 323 – 328.

LaMondia, J. A. 1984. Control of *Globodera rostochiensis* by solar heat. Plant Disease 68: 474 – 476.

LaMondia, J. A. 1990. The effects of moisture on the thermosensitivity of *Globodera rostochiensis* (Nematoda). American Potato Journal 67: 349 – 356.

LaMondia, J. A. 1995. Hatch and reproduction of *Globodera tabacum tabacum* in response to tobacco, tomato, or black nightshade. Journal of Nematology 27: 382 – 386.

LaMondia, J. A. 1996. Trap crops and population management of *Globodera tabacum tabacum*. Journal of Nematology 28: 238 – 243.

LaMondia, J. A. 1997. Management of *Meloidogyne hapla* in herbaceous perennial ornamentals by sanitation and resistance. Journal of Nematology 29: 717 – 720.

LaMondia, J. A. 1999. The effects of rotation crops on the strawberry pathogens *Pratylenchus penetrans*, *Meloidogyne hapla*, and *Rhizoctonia fragariae*. Journal of Nematology 31: 650 – 655.

LaMondia, J. A. 2000. The effect of rotation crops on lesion nematodes, *Verticillium dahliae* and potato early dying severity. Phytopathology 90(6S): 44 (Abstr.).

Leroux, G. G., D. L. Benoit, and S. Banville. 1996. Effect of crop rotations in weed control, *Bidens cernua* and *Erieron canaensis* populations, and carrot yields in organic soils. Crop Protection 15: 171 – 178.

Lewis, C. D. 1941. The georgics of virgil: a new translation. Book 1. London: Jonathan Cape.

Lovett, J. V. 1986. Allelopathy: the Australian experience. Pp. 75 – 99. In A. R. Putnam and C. -S. Tang, eds. The Science of Allelopathy. New York: John Wiley and Sons.

Mai, W. F. 1977. Worldwide distribution of potato-cyst nematodes and their importance in crop production. Journal of Nematology 9: 30 – 34.

Mai, W. F., and G. S. Abawi. 1980. Influence of crop rotation on spread and density of *Heterodera schachtii* on a commercial vegetable farm in New York. Plant Disease 64: 302 – 305.

Manuel, J. S., L. E. Bendixen, and R. M. Riedel. 1982. An annotated bibliography of weeds as reservoirs for organisms affecting crops. 1a. Nematodes. Wooster, OH: Ohio State University.

McSorley, R., and D. W. Dickson. 1995. Effect of tropical rotation crops on *Meloidogyne incognita* and other plant-parasitic nematodes. Journal of Nematology 27: 535 – 544.

Patrick, Z. A., and T. A. Toussoun. 1970. Plant residues and organic amendments in relation to biological control. Pp. 440 – 459. In K. F. Baker and W. C. Snyder, eds. Ecology of Soil-borne Plant Pathogens Prelude to Biological Control – an International Symposium on Factors Determining the Behavior of Plant Pathogens in Soil. Berkeley, CA: University of California Press.

Pease, W. S., D. Albright, C. DeRoos, L. Gottsman, A. D. Kyle, R. Morello-Frosch, and J. C. Robinson. 1995. Pesticide contamination of groundwater in California. Berkeley, CA: University of California.

Powell, C. A., and L. B. Forer. 1984. Orchard weeds as hosts of tomato ringspot and tobacco ringspot viruses. Plant Disease 68: 242 – 244.

Rhode, R. A., and W. R. Jenkins. 1958. Basis for resistance of *Asparagus officinalis* var. altilis to the stubby-root nematode *Trichodorus christiei* Allen. Maryland Agricultural Experiment Station Bulletin A-97: 1 – 19.

Riggs, R. D., and T. L. Niblack. 1993. Nematode pests of oilseed crops and grain legumes. Pp. 209 – 258. In K. E. A. Evans, ed. Plant-parasitic Nematodes in Temperature Agriculture. Wallingford, UK: CAB International.

Rivoal, R., and R. Cook. 1993. Nematode pests of cereals. Pp. 259 – 303. In K. E. A. Evans, ed. Plant-parasitic Nematodes in Temperature Agriculture. Wallingford, UK: CAB International.

Robbins, R. T., and K. R. Barker. 1973. Comparisons of host range and reproduction among populations of *Belonolaimus longcaudatus* from North Carolina and Georgia. Plant Disease Reporter 57: 750 – 754.

Rodrìguez-Kàbana, R. 1992. Cropping systems for the management of phytonematodes. Pp. 219 – 233. In F. J. Gommers and P. W. T. Maas, eds. Nematology from Molecule to Ecosystem: Proceedings of Second International Nematology Congress. Invergowrie, Dundee, Scotland: European Society of Nematologists.

Rodrìguez-Kàbana, R., and G. H. Canullo. 1992. Cropping systems for the management of phytonematodes. Phytoparasitica 20: 211 – 224.

Rodrìguez-Kàbana, R., and H. Ivey. 1986. Crop rotation systems for the management of *Meloidogyne arenaria* in peanuts. Nematropica 16: 53 – 64.

Rodrìguez-Kàbana, R., D. G. Robertson, L. Wells, P. S. King, and C. F. Weaver. 1989. Crops uncommon to Alabama for the management of *Meloidogyne arenaria* in peanut. Annals of Applied Nematology 21: 712 – 716.

Rodrìguez-Kàbana, R., H. Ivey, and P. A. Backman. 1987. Peanut-cotton rotations for the management of *Meloidogyne arenaria*. Journal of Nematology 19: 484 – 486.

Sanchez de Viala, S., B. B. Brodie, E. Rodriguez, and D. M. Gibson. 1998. The potential of thiarubrine C as a nematicidal agent against plant-parasitic nematodes. Journal of Nematology 30: 192 – 200.

Sasser, J. N. 1990. Plant-parasitic nematodes: The farmer's hidden enemy. Raleigh, NC: North Carolina State University Graphics.

Schroeder, J., S. H. Thomas, and L. Murray. 1993. Yellow and purple nutsedge and chili peppers host southern root-knot nematodes. *Weed Science* 41: 150 – 156.

Sexstone, A. J., V. G. M. Calabrese, and R. C. Derk. 1992. The role of soil microbiology research in the Northeastern United States. Pp. 27 – 35. In J. T. Sims, ed. Agricultural Research in the Northeastern United States: Critical Review and Future Perspectives. Storrs, CT: American Society of Agronomy.

Sharma, S. B., and Y. L. Nene. 1990. Effects of soil solarization on nematodes parasitic to chickpea and pigeonpea. Journal of Nematology 22: 658 – 664.

Sikora, R. A. 1992. Management of the antagonistic potential in agricultural ecosystems for the biological control of plant parasitic nematodes. Annual Review of Phytopathology 30: 245 – 270.

Slack, D. A., R. D. Riggs, and M. L. Hamblen. 1972. The effect of temperature and moisture on the survival of *Heterodera glycines* in the absence of a host. Journal of Nematology 4: 263 – 266.

Smelt, J. H., A. E. Van De Peppel-Groen, L. J. T. Van Der Pas, and A. Dijksterhuis. 1996. Development and duration of accelerated degradation of nematicides in different soils. Soil Biology and Biochemistry 28: 1757 – 1765.

Smithson, J. B., and J. M. Lenné. 1996. Varietal mixtures: a viable strategy for sustainable productivity in subsistence agriculture. Annals of Applied Biology 128: 127 – 158.

Spears, J. F., W. F. Mai, and D. O. Betz. 1968. A case history of the persistence of *Heterodera rostochiensis* Wollenweber in a treated field. Plant Disease Reporter 40: 632 – 634.

Spurr, M. S. 1986. Arable cultivation in Roman Italy c. 200 B. C. -100 A. D. London: Society for that Promotion of Roman Studies.

Sterling, G. R. 1991. Biological control of plant parasitic nematodes: Progress, problems and prospects. Wallingford, UK: CAB International.

Taylor, C. E., D. J. F. Brown, R. Neilson, and A. T. Jones. 1994. The persistence and spread of *Xiphinema diversicaudatum* in cultivated and uncultivated biotypes. Annals of Applied Biology 124: 469 – 477.

Thames, W. H., and W. N. Stoner. 1953. A preliminary trial of lowland culture rice in rotation with vegetable crops as a means of reducing root-knot nematode infestations in the Everglades. Plant Disease Reporter 37: 187 – 192.

Thies, J. A., J. D. Mueller, and R. L. Fery. 1998. Use of a resistant pepper as a rotational crop to manage southern root-knot nematode. HortScience 33: 716 – 718.

Thurston, H. D. 1992. Sustainable practices for plant disease management in traditional farming systems. 1st ed. Boulder, Colorado: Westview Press.

Towson, A. J., and B. Lear. 1982. Control of nematodes in rose plants of hot-water treatment preceeded by heat-hardening *Meloidogyne hapla* and *Pratylenchus vulnus*. Nematologica 28: 339 – 353.

Trivedi, P. C., and K. R. Barker. 1986. Nematological reviews: management of nematodes by cultural practices. Nematropica 16: 213 – 236.

Van Gundy, S. D. 1985. Ecology of *Meloidogyne* spp. – emphasis on environmental factors affecting survival and pathogenicity. Pp. 177 – 182. In J. N. Sasser and C. C. Carter, eds. An Advanced Treatise on *Meloidogyne*. Vol. 1: Biology and Control. Raleigh, NC: North Carolina State University Graphics.

Vawdrey, L. L., and G. R. Sterling. 1997. Control of root-knot nematode (*Meloidogyne javanica*) on tomato with molasses and other organic amendments. Australasian Plant Pathology 26: 179 – 187.

Verhagen, C., G. Lebbink, and J. Bloem. 1996. Enhanced biodegradation of the nematicides 1, 3-dichloropropene and methyl isothiocyanate in a variety of soils. Soil Biology and Biochemistry 28: 1753 – 1756.

Von Hagen, V. W. 1959. The Incas of Pedro de Cieza de León. Norman, OK: University of Oklahoma Press.

Whitehead, A. G. 1977. Control of potato cyst nematode, *Globodera rostochiensis* Rol, by picrilonic acid and potato trap crops. Annals of Applied Biology 87: 225 – 227.

Wilson, G. F., and F. E. Caveness. 1980. Effect of rotation crops on survival of root-knot, root-lesion and spiral nematodes. Nematropica 10: 56 – 61.

Young, L. D. 1998. Influence of soybean cropping sequences on seed yield and female index of the soybean cyst nematode. Plant Disease 82: 615 – 619.

Young, L. D., and E. E. Hartwig. 1988. Selection pressure on soybean cyst nematode from soybean cropping sequences. Crop Science 28: 845 – 847.

Yu, Y. T. 1987. Agricultural history over seven thousand years. Pp. 19 – 33. In S. Wittwer, Y. T. Yu, H. Sun, and L. Z. Wang, eds. Feeding a Billion: Frontiers of Chinese Agriculture. East Lansing, MI: Michigan State University Press.

19 Use of Antagonistic Plants and Natural Products

Silamar Ferraz and Leandro Grassi de Freitas

19.1 Introduction

The control of plant-parasitic nematodes has been sought, for many years, by the use of nematicides, resistant varieties, and crop rotation. However, pesticides use is dangerous to the environment, human health, wildlife, and beneficial organisms, besides inducing undesirable selection of resistant pests and pathogens. Because of their high toxicity and persistence in the environment, it has become a public concern to face and overcome its dangers (Ajayi, et al. 1993). The high costs of nematicides also make their use prohibitive in many countries, especially in the developing nations where the population survives on subsistence-type agriculture. Such farmers have to rely on non-chemical methods for pest management. Some cultural practices may be used effectively, resulting in increased crop yields and a higher overall income for the farmer. Among those practices, the use of antagonistic plants for the control of nematodes is a very attractive alternative. Some of these plants give the benefit of green manure while controlling the nematodes. Planting legumes (family Poaceae) results in reduction of nematode population density and fixation of nitrogen from the atmosphere into soil, improving soil fertility. Some cover crops, when incorporated, cater a considerable amount of organic material to the soil, and consequently, augment the activity of antagonistic fungi. Another important factor to consider when choosing antagonistic plants is the possibility of isolation of naturally occurring chemicals. Nematicidal substances have been isolated from antagonistic plants and have drawn the attention of researchers and the pesticide industry. According to Quarles (1992), botanical extracts present some advantages over synthetic pesticides, such as: they can provide novel compounds that pests are not yet able to inactivate; they are less concentrated and thus potentially less toxic than pure compounds; they biodegrade rapidly, and may possess multiple modes of action making possible a wide spectrum of use while retaining a selective action within each pest class, and they are derived from

Table 19.1 Nematicidal substances isolated from plants.

Plant Species	Plant Parts	Nematode Species	Active Components	Reference
Acacia auriculiformis	Funicles	*Meloidogyne incognita*	2 Triterpenoid saponins Acaciasides	Babu, et al., 1992 Sukul, et al., 2001
Allium grayi	Plant	*M. incognita*	1-Octanol; methyl 4-hydroxybenzoate; methyl 4-hydroxycinnamate and allygrin	Tada, et al., 1988
Ambrosia artemisiifolia	Roots	*Pratylenchus penetrans*	Dithioacetylenes	Gommers and Voor in't Holt, 1976
Ambrosia psilostachya	Plant	*M. incognita*	Parthenin; coronopilin	Gavilano, et al., 1998
Ambrosia trifida	Roots	*Pratylenchus penetrans*	Dithioacetylenes	Gommers and Voor in't Holt, 1976
Apium graveolens	Seeds	*Panagrellus redivivus* *Caenorhabditis elegans*	3-n-Butyl-4,5-dihydrophthalide	Momin, et al., 2000
Angelica pubescens	Roots	*Aphelenchoides besseyi*	Heptadeca-1,9-diene-4,6-diyne-3,8-diol	Munakata, 1983
Argemone mexicana	Seeds	*M. incognita*	sn-glycerol-1-eicosa-9, 12-dienoate-2-palmitoleate-3-linoleate	Saleh, et al., 1987
Artemisia nilagirica	flowering meristems	*M. incognita*	Santonin	Sukul, et al., 2001
Asclepias sp.	Seeds	*M. chitwoodi*	Cardiac glycosides (cardenolides)	Harry, et al., 1999

Continued

Plant Species	Plant Parts	Nematode Species	Active Components	Reference
Asparagus officinalis	Roots	*Globodera rostochiensis* *Heterodera glycines* *M. hapla* *M. incognita* *Paratrichodorus minor* *Paratylenchus curvitatus* *Pratylenchus penetrans*	Glycoside; asparagusic acid	Rohde and Jenkins, 1958; Takasugi, et al., 1975
Azadirachta indica	Leaves	*M. incognita*	Nimbidine; thionemone	Khan, et al., 1974
Berlinia grandiflora	stem bark	*Caenorhabditis elegans*	betulinic acid	Enwerem, et al., 2001
Bocconia cordata		*Panagrolaimus* sp. *Rhabditis* sp.	Bocconine; chelerytrine; saguinarine	Onda, et al., 1965, 1970
Brassica nigra	Seeds	*Globodera rostochiensis*	Allyl isothiocyanate	Ellenby, 1951
Brassica spp.	Leaves, roots	*Pratylenchus neglectus*	2-phenylethyl glucosinolate	Potter et, al., 1998
Bupleurum salicifolium		*Globodera pallida* *Globodera rostochiensis*	Bursehernin; matairesinol	Gonzalez, et al., 1994, 1995
Canavalia ensiformis	Seeds	*Globodera rostochiensis*	{L-2-amino-4-guanidino-oxybutiric acid} canavanine	Birch, et al., 1993a
Carthamus tinctorius	Flowers	*Aphelenchoides besseyi*	3-cis, 11-trans-Trideca-1,3,11 triene-5,7,9-tryine and 3 trans, 11-trans-trideca-1,3,11 triene-5,7,9-tryine	Kogiso, et al., 1976a, 1976b
Catharanthus roseus	Roots	*M. incognita*	Serpentine (indole alkaloid)	Chandravadana, et al., 1994; Rao, et al., 1996b

Continued

Plant Species	Plant Parts	Nematode Species	Active Components	Reference
Cirsium japonicum	Roots	*Bursaphelenchus xylophilus*	Tridec-1-en-3, 5, 7, 9, 11-pentayne and 9,10-epoxyheptadec-16-en-4,6-diyn-8-ol	Kawazu, et al., 1980
Coreopsis lanceolata		*Bursaphelenchus xylophilus*	1-Phenylhepta-1, 3, 5-triyne and 2-phenyl-5(1-propynyl)-tiophene	Kawazu, et al., 1980
Cornus officinalis	leaves	*Caenorhabditis elegans*	Ellagitannin	Mori, et al., 2000
Crotalaria spectabilis	Leaves, roots	*M. incognita*	Monocrotaline	Fassuliotis and Skucas, 1969
Cymbopogon flexuosus C. martinii C. winterianus	Essential oils	*Anguina tritici* *Heterodera avenae* *M. javanica* *Tylenchulus semipenetrans*	Citral; geraniol; citronellol	Sangwan, et al., 1985
Daphne odora	Roots	*Aphelenchoides besseyi*	Odoracin; odoratrin	Kogiso, et al., 1976c; Munakata, 1983
Derris sp.	Seeds, leaves	*M. incognita* *M. javanica* *Xiphinema diversicaudatum*	[(2R,5R)-dihydroxymethyl-(3R,4R)-dihydroxypyrrolidine] (DMDP)	Birch, et al., 1992, 1993b
Dioscorea deltoidea	Cell cultures	*M. incognita*	Furostanol glycosides	Zinovieva, et al., 1997
Echinops sphaerocephalus		*Meloidogyne* sp.	2-Terthienyl,2-bithienyl and 1-tridecanene-3,5,7,9,11-pentayne	Gommers, 1973
Embelia ribes		*M. incognita*	Embelin (dihydroxiquinone)	Mojumder and Mishra, 1990

Continued

Plant Species	Plant Parts	Nematode Species	Active Components	Reference
Eragrostis curvula	Roots	*Meloidogyne* spp.	Pyrocatechol	Scheffer, et al., 1962
Erigeron philadelphicus	Roots	*Pratylenchus coffeae*	Methyl 2-trans, 8-cis-deca-2, 8-diene-4, 6diynoate (2-trans,8-cis-matricaria ester) and 2-cis, 8-cis-deca-2, 8-diene-4, 6 diynoate (2-cis,8-cis-matricaria ester)	Kimura, et al., 1981
Eriophyllum caespitosum	Roots	*Pratylenchus penetrans*	Dithioacetylenes	Gommers and Voor in't Holt, 1976
Eugenia caryophyllata	Buds	*Anguina tritici* *Heterodera cajani* *M. javanica* *Tylenchulus semipenetrans*	Eugenol; linalool; geraniol	Sangwan, et al., 1990
Gloriosa superba	Seeds	*M. incognita*	Colchicine	Nidiry, et al., 1993
Hannoa undulata	Seeds	*M. javanica*	Chaparrinone, klaineanone; glaucarubolone	Prot and Kornprobst, 1985
Haplophyllum tuberculatum		*M. incognita* *M. javanica*	3 Alkaloids	Stephan, et al., 1989
Helenium hybrid	Roots	*Pratylenchus penetrans*	Pentaynene (2, 3-dihydro-2-hydroxy-3-methylene-6-methybenzo-furan)	Gommers, 1971
Helenium sp.		*Pratylenchus penetrans*	Tridec-1-en-3,5,7,9,11-pentayne	Gommers, 1973
Hyptis suaveolens	Leaves	*M. incognita*	D-limonene; menthol	Babu and Sukul, 1990

Continued

Plant Species	Plant Parts	Nematode Species	Active Components	Reference
Inula helenium		*M. incognita*	Alantolactone	Milman, 1990; Mahajan, et al., 1986
Iris japonica	Roots	*Aphelenchoides besseyi*	Myristic, palmitic and linoleic acids	Munakata, 1983
Iva xanthiifolia	Roots	*Pratylenchus penetrans*	Dithioacetylenes	Gommers and Voor in't Holt, 1976
Knema hookeriana		*Bursaphelenchus xylophilus*	3-undecylphenol 3-(8Z-tridecenyl)-phenol	Alen, et al., 2000
Lantana camara	Aerial parts	*M. incognita*	Lantanoside, linaroside and camarinic acid	Sabira, et al., 2000
Lonchocarpus spp.	Seeds, leaves	*M. incognita* *M. javanica* *Xiphinema diversicaudatum*	[(2R,5R)-dihydroxymethyl-(3R,4R)-dihydroxypyrrolidine] (DMDP)	Birch, et al., 1992, 1993b
Lycopersicum esculentum		*Panagrellus redivivus*	Alpha-tomatine	Allen and Feldmesser, 1970
Mesua ferrea	Heartwood	*M. incognita*	2-Methoxy-xanthone	Singh, et al., 1990
Milleria quinqueflora	Roots	*Pratylenchus penetrans*	Dithioacetylenes	Gommers and Voor in't Holt, 1976
Mucuna aterrima	Stems, roots	*Heterodera glycines*	Fatty acids; allantion; daucosterol + stigmasterol; D-glycoside; L-dopa	Barbosa, et al., 1999
	Stems, roots	*M. incognita*	B-Sitosterol + stigmasterol KNO_3 + $NaNO_3$; unknown alcohol	Barbosa, et al., 1999

Continued

Plant Species	Plant Parts	Nematode Species	Active Components	Reference
	Leaves, roots	*M. incognita*	1-triacontanol; triacontyl-tetracosanate	Nogueira, et al., 1996b
Murraya koenigii	Leaves	*M. incognita*	Koenimbine, koenidin, beta-caryophyllene and beta-gurjunene	Sanjay, et al., 2000
Nicotiana tabacum		*M. incognita* *M. javanica*	Nicotine	Davis and Rich, 1987; Krishnamurthy, et al., 1993
Ocimum basilicum	Leaves	*M. incognita*	Linalool	Chatterjee, et al., 1982
Ocimum basilicum	Essential oil	*Heterodera avenae* *H. cajani* *H. zeae* *M. incognita*	Linalool; metilchavicol	Gokte, et al., 1991
Ocimum gratissimum	Leaves	*Caenorhabditis elegans*	Oleanolic acid	Njoku, et al., 1997b
Ocimum sanctum	Leaves	*M. incognita*	Eugenol	Chatterjee, et al., 1982
Ocimum sanctum	Essential oil	*Heterodera avenae* *H. cajani* *H. zeae* *M. incognita*	Eugenol and "noneugenol"	Gokte, et al., 1991
Paeonia suffruticosa		*Caenorhabditis elegans*	Gallo- and condensed tannins	Mohamed, et al., 2000

Continued

Plant Species	Plant Parts	Nematode Species	Active Components	Reference
Pelargonium graveolens	Leaves, shoot	*M. incognita*	Geraniol; citronellol; linalool	Leela, et al., 1992; Rao, et al., 1966b
Physostigma venenosum	Seeds	*Ditylenchus dipsasi*	Physostigmine	Bijloo, 1966; Welle, 1964
Pimenta dioica	Leaves	*M. incognita*	Eugenol	Leela and Ramana, 2000
Pinus strobus	bark	*Bursaphelenchus xylophilus*	3-O-Methyldihydropinosylvin	Hanawa, et al., 2001
Piper betle		*Caenorhabditis elegans*	Chavicol; chavibetel; allylpyrocatechol; chavibetol acetate; allylpyrocatechol diacetate	Evans, et al., 1984
Polygonum hydropiper	Leaves and stems	*Acrobeles* *Dorylaimus* *Eudorylaimus* *Mononchus* *Tobrilus*	Formic acid; acetic acid; tannic acid	Sukul, 1970
Pycnanthus angolensis	Stem bark	*Caenorhabditis elegans*	Dihydroguaiaretic acid	Njoku, et al., 1997a
Ricinus communis		*M. incognita*	Ricin	Rich, et al., 1989
Rubus idaeus	Canes and roots	*Longidorus elongatus*	Tannins; polyphenols	Taylor and Murant, 1966
Rudbeckia hirta	Roots	*M. incognita* *Pratylenchus penetrans*	Thiarubrine C	Viala, et al., 1998

Continued

Plant Species	Plant Parts	Nematode Species	Active Components	Reference
Schukuhria pinnata	Roots	*Pratylenchus penetrans*	Dithioacetylenes	Gommers and Voor in't Holt, 1976
Sesamum indicum	Roots	*M. incognita*	7 Aminoacids: aspartic acid, glutamic acid, valine, proline, serine, glycine and leucine	Tanda, et al., 1989
Solanum tuberosum	Tuber sprouts	*Panagrellus redivivus*	Alpha-chaconine	Allen and Feldmesser, 1971
Solidago altissima		*Bursaphelenchus xylophilus*	cis-Dehydromatricaria ester	Kawazu, et al., 1980
Sophora alopecuroides	Leaves	*Bursaphelenchus xylophilus*	Aloperine	Zhao and Zhang, 1996
Sophora flavescens		*Bursaphelenchus xylophilus*	N-methylcytisine; anagyrine; (-) - cystisine; (-) - nicotine; sophocarpine; sophoramine	Matsuda, et al., 1989, 1991
Sorghum sudanense		*M. hapla*	Dhurrin	Widmer and Abawi, 2000
Tagetes erecta	Roots	*Anguina tritici* *Ditylenchus dipsaci* *Globodera rostochiensis*	Alpha-terthienyl; 5-butyl-2,2'-bithienyl	Uhlenbroek and Bijloo, 1958, 1959
T. jalisciencis	Roots	*M. incognita*	Thiophene derivatives	Castro and Munoz, 1982
T. patula		*Bursaphelenchus xylophilus*	Hydroxytrematone; Bithienyl derivative	Hatakeda, et al., 1985
Tagetes sp.		*Meloidogyne* spp.	Polythienyls; alpha-terthienyl	Gommers, 1973; Munoz, et al., 1982

renewable resources.

This review will focus on the main species of antagonistic plants, their use and mode of action. Chemically characterized nematicidal substances isolated from plants are listed in Table 19.1. Readers interested on more information on nematicidal compounds can refer to other reviews (Alam, et al., 1990; Chitwood, 1992, 1993, 2002; Gommers, 1973, 1981; Gommers and Bakker, 1988; Khan, et al., 1990; Tsai, et al., 1991).

19.2 *Mucuna* spp.

The genus *Mucuna* (velvet bean) is large, comprising more than 100 species, but its taxonomy remains uncertain. It has not been studied by using modern molecular techniques or phytochemical markers. Duke (1981) recognized six cultivated species, namely, *Mucuna pruriens* (L.) DC., *Mucuna nivea* (syn. *Mucuna lyonii* Merr.) (Lyon velvet bean), *Mucuna hassjoo* (Yokohama velvet bean), *Mucuna aterrima* Holl. (Mauritius or Bourbon velvet bean), *Mucuna utilis* Wall. (Bengal velvet bean), and *Mucuna deeringiana* Merr. (Florida or Georgia velvet bean). *Mucuna pruriens* is extensively cultivated worldwide and is the only species systematically investigated for its chemical and pharmacological properties (Ghosal, et al., 1971). The most commonly encountered varieties of *M. pruriens* are *M. pruriens* var. utilis and *M. pruriens* var. pruriens (Lorenzetti, et al., 1998).

Velvet bean was widely cultivated in the USA during the early part of the 20th century and was included, at that time, in numerous research programs in Africa, Asia, and Latin America (Buckles, 1995). Most *Mucuna* spp. exhibit reasonable tolerance to a number of abiotic stresses, including drought, low soil fertility, and high soil acidity, but they are sensitive to frost and grow poorly in cold, wet soils (Duke, 1981). Species of *Mucuna* grow best under warm, moist conditions, at altitudes below 1500 m, and in areas with plentiful rainfall. Velvet bean is considered to be one of the best plant for green manure because it covers the soil completely and then grows as high as its support allows, yielding around 40 ton/ha of above-ground organic matter and contributing with 150 – 200 kg/ha of nitrogen fixed from the atmosphere. It controls serious weeds such as nutgrass (*Cyperus rotundus*), Bermuda grass (*Cynodon dactilon*) and imperata grass (*Imperata cylindrica*). It is also very efficient to rehabilitate depleted land and to protect soil against erosion, besides being highly palatable to animals.

The most important cultivated species, *M. pruriens*, produces the toxic principle L-dopa and has been reported to contain hallucinogenic compounds related to N, N-dimethyltryptamine (Lorenzetti, et al., 1998). *Mucuna* spp. have also been reported to contain antinutritional factors, in addition to L-dopa, such as phenols and tannins, and to possess trypsin-inhibiting and hemagglutonating activities (Rajaram and Janardhanan, 1991). Duke (1981) also reported tricotine, physostigmine, and serotonin in *Mucuna*. Ghosal et al. (1971) claimed that this last compound, an important neurotransmitter also known as 5-hydroxytryptamine (5-HT), is present in the golden-yellow trichomes of *M. pruriens* pods. In addition, bufotenine, choline, N, N-dimethyltryptamine (DMT), DMT-Nb-oxide, 5-methoxy-DMT, as well as two unidentified 5-oxy-indole-3-alkylamines, an unidentified indole-3-alkylamine, and an unidentified β-carboline, were isolated from a mixture of pods, seeds, leaves, and roots of *M. pruriens* (Ghosal, et al., 1971). L-dopa, used in the treatment of Parkinson's disease, is commercially isolated from velvet bean. L-dopa can also produce a confused state of mind and intestinal disruptions in humans. The toxicity of unprocessed velvet bean may explain why the plant has few problems with insect pests (Duke, 1981; Scott, 1910).

Velvet bean is well known for its nematicidal effects when used in rotation with a number of commercial crops, although it is not itself immune to a number of nematode species (Duke, 1981). It is also susceptible to other soil-borne pathogens, such as *Macrophomina phaseolina* (Berner, et al., 1992), that are detrimental to important food crops.

The mode of action of velvet bean on nematodes is not clear yet. The nematicidal activity of hexane, chloroform, ethyl acetate/acetone, and ethanol/water crude extracts from roots and aerial parts of *M. aterrima* was tested for *Meloidogyne incognita* by Nogueira et al., (1996b). The bioactive substances of these extracts were isolated and identified by chromatography, providing fractions that were purified and subjected to in vitro and in vivo tests with the nematode. Out of the 12 substances isolated, five (HM16, ACM7, ACM63, CHR58, and ACMR17) showed activity towards the nematode. ACM7 and HM16, the alcohol 1-triacontanol ($C_{30}H_{62}O$) and the ester triacontyl tetracosanate ($C_{54}H_{108}O_2$), respectivelly, exhibited in vitro and in vivo (in greenhouse experiment using tomato plants) activity against *M. incognita* race 3 (Nogueira, et al., 1996a).

One interesting aspect of velvet bean and some other antagonistic plants is their effect on soil organisms. Bacterial isolates from rhizosphere (rhizobacteria) of plants that demonstrate an antagonism toward phytopathogenic nematodes, including velvet bean (*M. deeringiana*), castor bean (*Ricinus communis*), sword bean (*Canavalia ensiformis*), and Abruzzi rye (*Secale cereale*), were compared to rhizobacteria isolated from soybean. The results demonstrated that plants with properties antagonistic to phytopathogenic nematodes have a distinct rhizosphere microflora (Kloepper, et al., 1991). Isolates from soybean were predominantly *Bacillus* spp., while those from antagonistic plants included more coryneform and Gram-negative genera. *Pseudomonas cepacia* and *P. gladioli* predominated among Gram-negative bacteria in the rhizosphere of antagonistic plants but were not isolated from soybean. Bacteria from antagonistic plants significantly reduced the incidence of *M. incognita* and *Heterodera glycines* in soybean plants when compared to the bacteria isolated from soybean roots. The results suggest that rhizospheres of antagonistic plants may be useful sources of potential biological control agents for plant-parasitic nematodes (Kloepper, et al., 1992).

The cultivation of velvet bean in potted soil, under greenhouse conditions and without incorporation of the plant, showed that it was a poor host for *M. incognita* (Tenente, et al., 1982) but allowed reproduction of *M. javanica* (Resende, et al., 1987). However, Asmus and Ferraz (1988) observed 65% reduction of *M. javanica* juveniles in a field where *Mucuna aterrima* had been cultivated for 100 days and incorporated into soil whereas a 200% increase in the nematode population occurred after tomato was cultivated instead of *M. aterrima*. Cultivating tomato in the soil previously planted to velvet bean resulted in 14 egg masses per gram of roots compared to 415 in the soil previously planted to tomato. It seems that the incorporation of the aerial part of *Mucuna* is important for an effective control of *M. javanica*.

The effect of velvet bean on pathogenic soil fungi has also been studied. Cotton cultivars IAC 12-2 and IAC RM3, susceptible and resistant, respectively, to *Fusarium* wilt, were planted in both soils naturally infested with *Fusarium oxysporum* f. sp. *vasinfectum* and *M. incognita*, or with *M. incognita* alone, in succession for 6 years or in rotation with groundnut or *M. aterrima* grown in the 1st and 3rd year. In both soils, there was an increase in cotton yield planted after groundnut and a greater increase after *Mucuna*. Rotation of cotton and velvet bean or groundnut was recommended as a complementary measure to control *Fusarium* and nematodes (Ferraz, et al., 1977).

19.3 *Crotalaria* spp.

The name *Crotalaria* means "rattle" refering to the noise made by the seeds when shaken in the mature pods (White and Haun, 1965). It comprises over 350 species located in the tropics and subtropics of both hemispheres (Cook and White, 1996). *Crotalaria juncea* (sunn hemp) is the most important and better-known species of the genus. Its soft, lignified fibers are used for manufacturing pulp and paper, it fixes nitrogen from the air, is antagonistic to nematodes, grows in marginal soils and is quite resistant to drought. *Crotalaria juncea* yields around 30 ton/ha of green manure and is able to fix up to 450 kg of N/ha/year.

The effects of ten *Crotalaria* species on *M. javanica, M. incognita* race 3, and *M. exigua* were studied under greenhouse conditions by Silva et al. (1990a). Sixty-five days after inoculation with 5000 eggs of each nematode species, each *Crotalaria* plant was cut off at soil level and the above-ground part was removed. Soil and root samples were taken to evaluate the nematode population and the remains were thoroughly mixed in the pot. No galls were found in the roots of *C. spectabilis, C. paulina, C. retusa, C. breviflora, C. mucronata, C. striata, C. lanceolata,* and *C. grantiana*. Tomato seedlings were transplanted to the pots and maintained for 30 days when the root galls were counted. Some galls of *M. javanica* and *M. incognita* were found in the roots of the tomatoes planted after *C. pallida* and *C. juncea* but many more were found in the tomato-after-tomato treatment. These data indicate that, in the case of *Crotalaria,* the incorporation of the crop residues to control nematodes is not as essential as for *Mucuna*. This would be an important point in certain situations, such as for the growers that intend to use *C. juncea* for paper production and, at same time, to control nematodes.

Barrons (1939) observed that *Meloidogyne* juveniles were able to penetrate the roots of *C. spectabilis* but they would not develop further. Huang and Mota e Silva (1980) studied four *Crotalaria* species infected by *M. incognita* and came to the same conclusion. The mechanism is still unknown. Silva et al. (1989a) showed that *M. javanica* juveniles were attracted to the roots of *C. spectabilis, C. juncea,* and tomato seedlings in tests conducted in water-agar in petri dishes when egg masses were placed equidistantly (2.5 cm) from tomato and *Crotalaria*. In another assay, these two *Crotalaria* species plus *C. paulina* and tomato were inoculated with 500 *M. javanica* juveniles. No females were seen in the *Crotalaria* roots 45 days after inoculation; in tomato, they were visible at the 24th day. Histopathological studies in which infected tomato roots were compared to those of *C. spectabilis* and *C. juncea* showed that the giant-cells in *Crotalaria* had a granular, dense cytoplasm, with a small number of nuclei. Large vacuoles were frequently absent from the cytoplasm. Giant-cell numbers per female and size were also smaller in the *Crotalaria* species than in the tomato roots (Silva et al., 1990b). Penetration and development of *M. incognita* in the roots of *C. spectabilis* were also studied by Sano et al. (1983) and Sano and Nakasono (1986). Penetration exceeded 30% in the roots of host plants, *L. esculentum* and *Sorghum bicolor*, and in

C. spectabilis, a non-host species. Growth of 2nd stage juveniles was found to occur in the roots of *C. spectabilis*, but they did not develop to the 3rd stage. In *L. esculentum* roots, giant cells started to develop 5 days after inoculation and became well developed after 14 days, showing many enlarged nuclei, a dense cytoplasm, and thickened cell walls, and mature females with egg-masses appeared at the 24th day. In *C. spectabilis*, the infecting juveniles showed small genital primordia and remained sexually undifferentiated despite the formation of developed giant cells, which persisted until 35 days. Necrosis was sometimes observed as well. Some species of *Crotalaria* are also highly resistant to other important nematodes, such as *Pratylenchus brachyurus* (Silva, et al., 1989b), *P. zeae* (Silva, et al., 1989b), and *Rotylenchulus reniformis* (Silva, et al., 1989c).

Even though *Crotalaria* spp. is usually recommended to be included in crop rotation schemes to control nematodes, Villar and Zavaleta-Mejía (1990) found, in two greenhouse experiments, that incorporation of *C. spectabilis* residues into the soil was sufficient to reduce root galling in tomatoes caused by *M. incognita* and *M. arenaria*. Soil incorporation of *C. longirostrata* residues was more effective to control these nematodes than the intercropping of *C. longirostrata* and tomato. These results suggest that the reduction in root galling by *C. longirostrata* was due to toxic compounds in the plant tissues rather than to its role as a trapping crop. Under laboratory conditions root exudates of *C. longirostrata* were found to be nematicidal to *Meloidogyne* spp. juveniles.

The effect of monocrotaline, a pyrrolizidine ester extracted from *C. spectabilis*, on *M. acrita* was investigated by Fassuliotis and Skucas (1969). After a 3-hr exposure to various concentrations of monocrotaline solutions in distilled water, definite alterations in juvenile motion were observed. However, no relationship was found between monocrotaline-containing plants and root-knot nematode resistance. Over 300 plants and plant products are known to contain pyrrolizidine alkaloids. In *Crotalaria* species more than 50 of them have already been found. They cause hepatotoxic effects in livestock and humans and, among them, monocrotaline is one of the most poisonous.

19.4 *Tagetes* spp.

The genus *Tagetes*, family Asteraceae, contains 56 species, of which only six annuals and three perennials are currently cultivated. *Tagetes erecta* L., *T. patula* L., *T. lunulata* Ort., and *T. tenuifolia* Cav. are the four annual species of *Tagetes* commonly cultivated throughout the world as ornamentals.

These four species were brought into cultivation over two millennia ago in Mexico (Soule and Janick, 1996) and were used as ornamental, in rituals, and as medicinal plants (Nuttall, 1920). *Tagetes* are source of essential oils (Lawrence, 1985), are used as condiment (Sweet, 1817), as food colorant (Mejia, et al., 1997; Padma, et al., 1997), for weed control (Pritts, 1992), as insecticide (Macedo, et al., 1997; Perich, et al., 1994), as fungicide (Edwards, et al., 1994; Sadhana and Walia, 1996; Zygadlo, et al., 1994) and source of pigment for poultry feed, to intensify the yellow color of egg yolks (Medina, et al., 1993). Its potential use in Pacific white shrimp diet to enhance pigmentation also has been investigated (Vernon Carter, et al., 1996).

It is a very old practice in India to grow marigolds between beds of solanaceous vegetables. Growers change the directions of the *Tagetes* beds every year, without knowing precisely the significance of this procedure (Khan, et al., 1971). Since the works of Tyler (1938, cited by Gommers, 1981), much research has shown the efficacy of *Tagetes* spp. to control plant-parasitic nematodes (Oostenbrink, et al., 1957; Reynolds, et al., 2000; Slootweg, 1956; Steiner, 1941). Most results indicate that these plants are very efficient in nematode control, especially for *Pratylenchus* and *Meloidogyne* species. However, some *Tagetes* species and varieties have been reported as inefficient to control nematodes such as *Criconemoides mutabile* (Steiner, 1941), *Trichodorus teres* (Kuiper, 1963), *Hemicycliophora similis* (Seinhorst and Klinkenberg, 1963), *Rotylenchus robustus*, *Paratylenchus* spp. (Oostenbrink, 1960), *Longidorus maximus* (Butschli) (Sturhan, 1963), *Tylenchulus semipenetrans* (Cohn, 1971), *Criconemella xenoplax* (Whittington and Zehr, 1992), *Rotylenchulus* sp. (Basu and Roy, 1975), *Belonolaimus longicaudatus, Dolichodorus heterocephalus*, *Paratrichodorus christiei* (Rhoades, 1980) and others. Different species and varietes of *Tagetes* may present distinct reaction to the same nematode. *Tagetes erecta* (seven varieties) and *T. patula* (three varieties) were tested for their reaction against *M. incognita*. All varieties, except *T. erecta* vars. Hawaii and African marigold, showed resistance (Prasad and Haque, 1982). Different nematode populations may also present different behavior when challenging the same *Tagetes* varieties. In an experiment involving 11 different *M. hapla* populations, only four reproduced in *T. erecta* and one of these populations, a race A with a haploid chromosome number of 17, from Virginia, USA, caused extensive galling, lateral root proliferation, and the number of galls and egg masses exceeded 500 per plant (Eisenback, 1987).

Most work on *Tagetes* deal with three species: *T. patula*, *T. erecta,* and *T. minuta*. *Tagetes patula* usually gives better nematode control than the other species. In general they are used in crop rotation but, in many situations, intercropping is very effective. Pepper (*Capsicum annum*)-*T. patula*, pepper-*Crotalaria*, and pepper-maize intercropping reduced the reproduction of *Meloidogyne* spp. and number of galls on pepper roots. Intercroppings normally increased the yield of pepper, except in pepper-cabbage and pepper-tomato plots (Martowo and Rohana, 1987). Populations of *Radopholus similis* and *Pratylenchus coffeae* were reduced on banana roots when intercropped for 4 months with *Tagetes* sp., lucerne (*Medicago sativa*), sunn hemp (*Crotalaria juncea*), or coriander (*Coriandrum sativum*) (Naganathan, et al., 1988). When banana was intercropped with six rows each of *T. erecta* or *T. patula*, the populations of *Meloidogyne*, *Radopholus*, and *Pratylenchus* were suppressed and the nematodes damage to banana roots was reduced (Supratoyo, 1993). The number of root galls due to *M. javanica* in tomato plants grown side by side with *T. erecta* were significantly lower compared to tomato grown alone. Root length, shoot weight, and number and weight of fruits were higher in plants grown with *Tagetes* (Abid and Maqbool, 1990). Intercropping *T. erecta* and eggplant in *M. javanica*-infested soils resulted in better growth of eggplant and reduction of the final nematode population up to 40% over the initial level. Marigold intercultured with eggplant was significantly superior over the carbofuran treatment (Dhangar, et al., 1995). *Tagetes minuta* significantly inhibited the root-knot development caused by *M. incognita* on tomato and eggplant grown in the same pot and reduced the multiplication of *Rotylenchulus reniformis* and *Tylenchorhynchus brassicae* on tomato, eggplant, cabbage, and cauliflower. Growth of all tested plants improved when marigolds were present. The root-exudates of *T. minuta* also exhibited strong nematicidal action (Siddiqui and Alam, 1987a). Root-knot index and soil populations of *M. incognita*, *M. javanica*, and *M. acrita* were reduced and yield of eggplant increased when plants of *T. patula* were planted within or between the rows of eggplant (Varma, et al., 1978).

An interesting experiment was performed by Zavaleta-Mejía and Gomez (1995) to study the effects of planting dates and two spacings in the intercropped tomato-*T. erecta* and their influence on tomato pests and diseases. All intercropped treatments, independently of planting date of *T. erecta* and plant spacing, showed a reduction in infection of tomato roots by the nematode *Nacobbus aberrans* when compared with the treatment with tomatoes only. Tomato foliar damage and incidence of damaged tomato fruit by *Alternaria solani* was significantly reduced in all the intercropped plots. The alate aphid and whitefly populations and incidence of tomato plants with virus symptoms were significantly lower in all intercropped treatments when compared to the control.

The mechanism responsible for the marked reduction of nematode populations in the presence of *Tagetes* is not well understood. A possible explanation was first proposed by Uhlenbroek and Bijloo (1958, 1959) who found the highly nematicidal compounds alpha-terthienyl and 5-(3-buten-1-ynyl)-2,2'-bithienyl in *Tagetes* spp. roots. Seinhorst and Klinkenberg (1963) postulated for *Pratylenchus penetrans* that the low survival was due to the extensive root necrosis they cause in *Tagetes* spp. that impaired the movement of young juveniles. Further studies by Suatmadji (1969) showed that *M. hapla* survives for at least 4 weeks within the roots, but the development beyond the second-stage juvenile (J2) in *T. patula* and *T. minuta* was hardly noticeable, and only a few J2 became adults in *T. erecta*. He also noticed that, unlike tomato, *T. patula* develops few and very small giant cells and galls, the nematode often dies, and giant cells often degenerate. Contrary to the belief that nematodes are damaged or killed outside the roots, the effect appeared centered inside the roots, as it was clearly correlated with the strongly nematicidal tiophenes, present in these plants and rarely elsewhere in nature. His findings were somehow corroborated later by other authors.

Debprasad et al. (2000) studied the chemical composition and nematicidal activity of the volatile and non-volatile fractions of *Tagetes erecta*. In the non-volatile compounds isolated from the hexane extract 4 fractions were obtained, the first two fractions furnished two pure compounds identified as dodecanoic acid and myristic acid. Another fraction, a mixture of two compounds identified as palmitic acid and steric acid, and the last fraction was a mixture of two compounds identified as octaeicosane-8-one and triacontane-1-ol. In the essential oils obtained through hydrodistillation of *Tagetes erecta* flowers five compounds have been identified: the first three compounds were identified as a-sesqui phellandrene, beta-sesquiphellendrene and 2-methyl-6-(4-methyl cyclohexadienyl), and hept-4-en-2-ol ; the remaining two were myristoleic acid, and trieicosane. The methanol extract and essential oils of marigold flowers showed maximum nematicidal activity against *Meloidogyne* incognita juveniles. Among the two purified compounds, the nematicidal activity of myristic acid was more pronounced than dodecanoic acid.

Measurements by Topp et al. (1998) of the size and activity of the microbial population in soils cropped with marigolds, in the field and in the greenhouse, showed that they were not depleted and that it is unlikely that the nematode control by this plant is due to the release into the soil of a biocidal agent. Belcher and Hussey (1977) found that *T. patula* acted as trap crop to *M. incognita*, as it allowed nematode penetration but giant cell formation was not initiated, and juveniles did not develop beyond the early second stage. A hypersensitive necrotic reaction was observed where the juvenile had attempted to feed. Rangaswamy et al. (1993) made similar observations. Motsinger et al. (1977) also noticed that some varieties of *T. patula* behaved as trap crops for the root-knot nematode. Conijn et al. (1996) concluded that *T. patula* proved to be the most satisfactory *Tagetes* species against *Pratylenchus* spp., which are killed by substances produced in the roots invaded by the nematode.

According to Gommers and Bakker (1988), the generation of singlet oxygen by photoactivated alpha-thertienyl is probably the only mechanism in *Tagetes* spp. responsible for the death of the nematodes. Singlet oxygen is toxic to organisms because it oxidizes the amino acids histidine, tryptophan, and methionone, and proteins containing these amino acids, which results in enzyme inactivation. Membranes also may be distorted due the oxidization of the unsaturated fatty acids. It has been shown that alpha-terthienyl and analogues are completely devoid of any nematicidal activity when mixed with soil because photoactivation is necessary. Therefore, to act inside the roots, another mechanism of activation has to be present. Experiments have indicated that the overall activities of peroxidases in roots of *T. patula* increased about six times once *P. penetrans* has penetrated the roots and that these peroxidases could be able to excite alpha-terthienyl. Although these experiments were carried out under artificial conditions, different of those into the roots of marigolds, the authors suggest an explanation on how *P. penetrans* is killed inside the roots of marigold: *Pratylenchus penetrans* accumulates terthienyl and/or related compounds by permeation or ingestion during its stay in the roots, or peroxidases may be ingested from their host, as shown by (Starr, 1979) in the case of *Meloidogyne* spp. The activities of peroxidases in the roots increase after the nematodes invade and may excite terthienyl that is capable of producing the poisonous singlet oxygen in or near the nematode's body. The generation of singlet oxygen may also contribute to the lesion formation in the roots near the nematode.

19.5 *Azadirachta indica*

Neem (*Azadirachta indica* A. Juss.) is a member of the mahogany family, Meliaceae. It has been known by other names in the past (*Melia azadirachta* L., *Melia indica* (A. Juss.) Brand., etc.), which makes the old literature quite confusing. Neem trees are broad-leaved evergreens that can grow up to 30 m tall and 2.5 m in girth. They have straight trunk, moderately thick bark, and round crown. The leaves usually do not fall off, except at extreme drought. The small, white, bisexual flowers are borne in axillary clusters and have a honey like scent. The fruit is a smooth, ellipsoidal drupe, up to almost

2 cm long, and has a sweet pulp enclosing the seed. It is composed of a shell and a kernel, sometimes two or three, each about half of the seed's weight. It is the kernel that is mostly used in pest control, even though pesticidal ingredients are found in other plant parts. The fruits are produced when the tree is 3 – 5 years old and up to 50 kg of them may be produced at 10 years of age and on, when it is fully productive (National Research Council, 1992).

It is native from the Indo-Pakistan subcontinent and can be found in many countries today. Neem grows well from subhumid to semiarid conditions, in warm areas with rainfall of 500 mm/year or less. Seeds are short-lived, losing viability in 2 weeks (Ahmed and Grainge, 1986).

Neem is known to many people as a "wonder tree" due to its many uses in medicine, agriculture, industry, as a shade tree, firewood, etc. It has been found, mainly in the last decade or so, that neem materials can affect more than 200 insect species as well as mites, nematodes, fungi, bacteria, and even a few viruses. Convened during May 1999, Neem 99 – The World Neem Conference, held in Canada, was the 6th international neem meeting and attracted 125 participants from 23 countries.

Neem has been tried in many different ways to control plant-parasitic nematodes: mulching with fresh or dried leaves (Ajith and Sheela, 1996; Grewal, 1988, 1989; Khan, 1992; Khanna and Sharma, 1994; Prasad, et al., 1994; Walia and Gupta, 1995); leaf extracts used as soil amendment (Egunjobi and Afolami, 1975; Qamar, et al., 1989; Salawu, 1992); root exudates (Siddiqui and Alam, 1989b); saw dust (Akhtar, 1998); seed coating with neem extracts or oil (Kathirvel, et al., 1992; Siddiqui and Alam, 1987b; Wani, 1992); seed or kernel powder used as soil treatment or seed coat (Abid, et al., 1995; Gokte and Swarup, 1988; Mojumder and Mishra, 1991); root dipping in neem leaf extracts (Abid and Maqbool, 1991; Akhtar and Mahmood, 1994a, 1994b; Siddiqui and Alam, 1988a, Vats and Nandal, 1993a, b, 1995). This last technique has been tested on several plant species, such as tomato, eggplant, cabbage, cauliflower, and chili, with good results for *M. incognita, M. javanica, Rotylenchulus reniformis, Tylenchorhynchus brassicae,* and other nematodes. The galling index and final population of *M. incognita* decreased significantly in eggplant seedlings given root-dip treatment in neem leaf suspensions mixed with *P. lilacinus* spores. In addition, the colonization of *P. lilacinus* on eggplant roots and egg parasitization of *M. incognita* were significantly increased, an indication of the complementary interaction between these two components for the sustainable management of root-knot nematodes on eggplant (Rao, et al., 1997). Similarly, Nagesh et al. (1997) combined neem and *P. lilacinus* in a field experiment intended for the management of *M. incognita* in *Polianthes tuberosa*. *Paecilomyces lilacinus* was mixed with leaf extracts of neem or castor bean (*Ricinus communis*) and used for bulb treatment and soil drenches. The smallest root-gall index and the highest egg mass and egg percentage infected by *P. lilacinus* were achieved when *P. lilacinus* was mixed with neem leaf extract.

Neem-based pesticidal formulations have been developed in the United States, India, and elsewhere mainly for use as insecticides (Margosan-O, Nimbecidine, Neemgold, Neemazal, Neemax, Fortune Aza, Neemix, Achook, Neemrich, Neemark, Econeem, Rakshak, Repelin, Welgrow, Azatin, Turplex, Align, Bioneem, Benefit, etc.). Some of these products have shown good nematicidal properties. Welgrow reduced *M. incognita* population and increased tomato growth (Dash and Padhi, 1990). The standard aqueous extracts of five neem-based pesticidal formulations (neem seed kernel, neem seed coat, Achook, Neemark, and Nimbecidine) were investigated for their toxicity against *M. incognita* juveniles. All neem products were toxic, with mortality being directly correlated with the extract concentration and the period of exposure. Neem seed kernel was the most effective at all concentrations and exposures, and Achook was the most effective among the commercially available neem pesticides tested (Reshmi and Vijayalakshmi, 1998). Suneem (azadirachtin 80% a. i.) was used to coat tomato seeds for protection against *M. incognita*. The results showed that numbers of juveniles and galls were significantly reduced (Akhtar and Mahmood, 1997).

The nematicidal effect of neem is supposed to be related to the naturally occurring chemicals, e. g., azadirachtin, nimbin, salannin, nimbidin, kaempferol, thionemone, quercetin, and others. Devakumar et al. (1985) made a systematic extraction of neem seed kernels and tested the fractions for their in vitro effects on juvenile emergence and mortality of *M. incognita*. Silica gel column chromatography of crude neem oil enabled the separation of the lipid associated limonoids and the pure oil. The limonoids were highly active against the nematode while the pure oil was inactive. The limonoids are compounds belonging to the beta-furanotriterpenoid group. So far, at least nine neem limonoids have demonstrated an ability to control a range of insects (National Research Council, 1992). New limonoids are still being discovered in neem, but azadirachtin, salannin, meliantrol, and nimbin are the best-known ones.

According to Schmutterer (1997) neem products are safe to spiders, adults of numerous beneficial insect species, and eggs of many predators such as coccinellids. He believes that, due to their relative selectivity, neem products can be recommended for many programs of Integrated Pest Management as it is unlikely that they will cause severe disturbances in ecosystems.

19.6 Grasses

Some grass species (family Poaceae) have shown antagonistic properties against plant-parasitic nematodes. Given a set of circumstances, these plants could be very convenient to use. They fit in rotation schemes for annual crops and, for perennials, they can be interplanted as cover crops. For both, whenever possible, they can be used as pasture. It is well known that root exudates from some grasses may also affect soil fungi such as *Verticillium*, *Fusarium*, and others.

Ten grass species, grown in pots in a soil infested with *M. javanica*, were compared for the control of the nematode. After 60 days from planting, their aerial parts were eliminated and tomato was cultivated for 30 days in the pots. Five species, *Brachiaria decumbens*, *Eragrostis curvula*, *Panicum maximum* cv. Guine, *B. brizantha*, and *digitaria decumbens* cv. Pangola, showed to be very antagonistic to the nematode. The number of galls per tomato roots averaged 10 to 23 in plants of these species, while 2278 galls were observed after *Avena strigosa*. In another experiment, *B. decumbens* and *P. maximum* were inoculated with 5000 eggs of *M. javanica* per pot. After 120 days no galls, egg masses, and J2 were detected on roots and in the soil. Significant reduction on egg hatching was also noted when egg masses were exposed to root exudates of both grasses (Brito and Ferraz, 1987a, 1987b). *Eragrostis curvula* was also efficient against *Pratylenchus loosi* in rotation with tea (Gnanapragasam, 1981). Scheffer et al. (1962) postulated that the effect of *E. curvula* against *M. javanica*, *M. acrita*, *M. thamesi*, and *M. hapla*, was due to the presence of catechol in the roots. This phenolic compound and its derivatives are found in many plants and have shown nematicidal activity against *Hoplolaimus indicus*, *Helicotylenchus indicus*, *Rotylenchulus reniformis*, *Tylenchorhynchus brassicae*, *Tylenchus filiformis* (Alam, et al., 1979), *Caenorhabditis elegans* (Evans, et al., 1984), and *Tylenchorhynchus dubius* (Miller, 1978). In Cameroon, Africa, screening for resistance to *Radopholus similis* identified several *Musa* species and cultivars. The different degrees of nematode infestation in plant tissues was linked to the amounts of flavones and catechol in roots (Sarah, et al., 1997). Catechol has also shown systemic activity. According to Sitaramaiah and Pathak (1979), foliar application of catechol to tomato plants before or after inoculation with *M. javanica*, soil drench or root immersion with this chemical, reduced the number of galls and penetration of juveniles as compared with untreated controls. The total phenol content in roots of susceptible tomato plants sprayed with the chemical was equal to the phenolic content of a resistant variety, Nematex. Exposure of adult female nematodes to the chemical caused a significant reduction in egg liberation and hatch, and in juvenile motility. These effects varied with time of exposure of juveniles and eggs to catechol (Sitaramaiah and Pathak, 1979). This chemical also kept *M. incognita* under control in another experiment (Sitaramaiah and Pathak, 1981).

Pangola grass (*Digitaria decumbens*) was effective against *Helicotylenchus multicinctus*, *Rotylenchulus reniformis,* and *Meloidogyne* sp. in land previously cropped with banana (Stoyanov, 1971, 1973). Haroon and Smart (1983a, b, c) showed that 90 days after the soil infestation with nematodes, the numbers of J2 per root system were nine for *M. javanica*, 11 for *M. hapla*, 12 for *M. incognita,* and 48 for *M. arenaria* in pangola digitgrass roots compared to 475, 307, 429, and 4120 of the same species, respectively, in tomato roots. Roots of tomato seedlings interplanted with the grass were less galled than were roots of tomato seedlings planted alone. Fewer J2 invaded roots of pangola digitgrass than of tomato, and those that entered the grass roots failed to develop beyond the late second stage. Root extracts from older plants killed most of the juveniles within 10 days.

Cymbopogon spp. exudates, extracts, and oils also have shown some positive results. Root exudates of *C. citratus* (lemon grass) reduced *M. incognita* population and egg masses on tomato roots by more than 50% (Sweelam, 1989). Leaf extract prepared from *C. flexuosus* was highly toxic to *M. incognita, Rotylenchulus reniformis*, *Tylenchorhynchus brassicae*, *Hoplolaimus indicus,* and *Helicotylenchus indicus*. The toxicity was related to leaf extract concentration and to the exposure period. *Meloidogyne incognita* suffered 100% mortality after 12 h exposure, being the most affected nematode (Tiyagi, et al., 1986). Water extracts of this plant, used as bare-root dip treatment against *M. incognita*, was effective to suppress the nematode on tomato and aubergine (Tiyagi, et al., 1990). The essential oils of three *Cymbopogon* species (*C. martinii* var. motia, *C. flexuosus,* and *C. winterianus*) and their major constituents, geraniol citral, citronellol, and citronellal were toxic to varying extents against *Anguina tritici*, *Tylenchulus semipenetrans*, *M. javanica,* and *Heterodera avenae* (Sangwan, et al., 1985). Saxena et al. (1987) also tested *C. martinii* var. motia essential oils and found that no *M. incognita* specimen survived the treatment.

In a series of field experiments, Rodríguez-Kábana et al. (1988c, 1989b, 1991a, b, 1994) showed that rotation with *Paspalum notatum* (bahiagrass) improved soybean and peanut yields and was very successful to control *M. arenaria*, *M. incognita,* and *Heterodera glycines*. This last nematode species was also reduced by the pasture grasses *Andropogon gayanus*, *Brachiaria brizantha*, *B. decumbens*, *B. humidicola*, *B. ruziziensis,* and *Panicum maximum* (Valle, et, al., 1996b).

19.7 Other Antagonistic Plant Species

Plant extracts from various *Artemisia* species (family Asteraceae) have substances that show antibacterial, anthelmintic, and pesticidal properties. Toxicity to mosquito larvae and to nematodes has been demonstrated. Allelopathy has also been reported (Sherif, et al., 1987). Mulching with *Artemisia dracunculus* at 2 to 4% plant matter in soil, around phloxes in the field, reduced *Ditylenchus dipsaci* by 90 to 96%. The nematicidal action was considered to be due to substances of the flavonoid group (Timchenko and Maiko, 1989). Extracts of crushed plant material of *Artemisia verlotorum* and *A. absinthium* were effective against *M. incognita* juveniles (Dias, et al., 2000).

In the indian mustard (*Brassica juncea*), Sinigrin is the predominant glucosinolate and it is mainly degraded upon the enzymatic action of myrosinase under normal conditions to give allyl isothiocyanate (AITC) in an aqueous media AITC is considered to be the principal nematicidal ingredient in *B. juncea* (Tsao, et al., 2000). This compound is also active against other soil pathogens, like *Sclerotium rolfsii* (Harvey, et al., 2002). A two-point injection method was used to apply horse radish extract, containig AITC, into the soil and was very effective to control plant parasitic nematodes. The use of polypropylene vinyl sheet to cover the soil after application helped increase the nematicide retention times and nematode control (Mitarai, et al., 1997).

Reports on antagonistic properties of *Cajanus cajan* (pigeon pea) are inconclusive, maybe because of the use of different varieties in various experiments. Good results have been reported for pigeon pea against *M. javanica* (Asmus and Ferraz, 1988; Costa and Ferraz, 1990), *M. incognita* (Haroon and Abadir, 1989; Reddi, 1983), *P. penetrans* (Haroon and Abadir, 1989), *Heterodera glycines* (Valle, et al., 1996a), and others. Rodríguez-Kábana and Ingram (1978) showed that pigeon pea was a good host for a number of ecto- and endoparasitic nematode species. Chavda et al. (1988) screened 82 lines of *C. cajan* against *Rotylenchulus reniformis* and only four were found resistant.

Rodríguez-Kábana et al. (1988a, b, 1989a, 1992) reported that sesame, *Sesamum indicum* (family Pedaliaceae), in rotation with peanut or soybean provided efficient control against *M. arenaria*, *M. incognita*, and

H. glycines. To them, this is an "active" crop in that it generates nematicidal compounds while others, such as maize and sorghum, are simply nonhost species, and is considered "passive". Varma et al. (1978) found that galling index and soil populations of *M. incognita*, *M. javanica*, and *M. acrita* were reduced and yield of eggplant increased when *S. orientale* was planted within or between the rows. However, Starr and Black (1995) suggested that *S. indicum* might be an effective rotation crop for control of *M. arenaria* or *M. incognita* but not against *M. javanica*. Tanda et al. (1988) grew okra (*Hibiscus esculentus*) and *S. orientale* excised roots and callus cultures on synthetic media. Sesame tissues cultured alone or with okra suppressed egg hatch and penetration of roots by J2, delayed adult development, and encouraged development of males in *M. incognita*. Gall formation was inhibited on excised roots of okra by co-culturing with sesame.

Extracts of leaf, shoot and root, dried root powder, and essential oils of *Ocimum* spp. (family Lamiaceae), mainly *O. sanctum*, *O. basilicum*, and *O. americanum*, have been linked to strong nematicidal properties (Gokte, et al., 1991; Haseeb and Butool, 1990; Hussaini, et al., 1996; Kurundkar and Jadhav, 1993; Sharma and Trivedi, 1992a, b; Singh, et al., 1991; Sundarababu, et al., 1990; Widhi and Trivedi, 1995). Bioactivity-guided fractionation of a leaf extract of *O. gratissimum* led to the isolation of oleanolic acid. This compound showed activity against *Caenorhabditis elegans* (Njoku, et al., 1997b). The essential oil of *Ocimum sanctum* and eugenol also showed strong activity against *C. elegans*. Eugenol being the predominant component of the essential oil, is suggested as the putative anthelmintic principle (Asha, et al., 2001).

Extracts of *Ruta graveolens* (family Rutaceae) are effective against *Meloidogyne* spp. (Mareggiani, et al., 1997; Sasanelli, 1995; Sasanelli and D'Addabbo, 1993), *Ditylenchus dipsaci* (Insunza and Valenzuela, 1995), *Xiphinema index* (Sasanelli 1992), and other nematode species. Rutin, a constituent of the aqueous leaf extracts of *R. graveolens*, indicated no nematicidal or nematostatic activity against *Heterodera schachttii* (Sasanelli and D'Addabbo, 1995). Further tests on other constituents of the *R. graveolens* extracts may lead to the discovery of the nematicidal components.

Datura metel and *D. stramonium* (family Solanaceae) have been studied as nematicides mainly as leaf extracts and soil amendment (Ahmad, et al., 1991, 1993; Alam, 1986; Awan, et al., 1992; Firoza and Maqbool, 1996; Goswami and Vijayalakshmi, 1986; Imran and Saxena, 1993; Mani and Chitra, 1989; Mojumder and Mishra, 1991; Nandal and Bhatti, 1986a, 1986b, 1990; Sellami and Mouffarrah, 1994;). Samples of total alkaloids content and of hyoscine extracted from *D. metel* leaves have been assayed for nematicidal action on *Hoplolaimus indicus*, *Helicotylenchus multicinctus*, and *M. incognita*. The alkaloids sample killed 90 to 100% of all nematode species, whereas hyoscine was effective only against *H. indicus* with 90% mortality (Qamar, et al., 1995).

Chenopodium spp. (family Chenopodiaceae) is an ancient worldwide genus that contains substances that are antiviral, antifungal, antibacterial, nematicidal, molluscicidal, insecticidal, or allelopathic (Quarles, 1992). *Chenopodium ambrosioides* (wormseed), *C. quinoa,* and *C. album* have been found effective against plant-parasitic nematodes but the active component in each species remains to be identified. *Chenopodium ambrosioides* was cultivated in the USA to control hookworm and roundworm infestations until effective synthetic compounds became available. Essential oils, saponins, flavonoids, and steroids have been identified in this species and are all active against nematodes. The essential oils have a unique component, peroxide ascaridole, which is anthelmintic. However, a possible limitation on the use of *C. ambrosioides* essential oil for nematode control is that it also contains limonene, a substance toxic to earthworms (Karr, et al., 1990). Methanol extracts of stems and leaves of *C. album* and *C. murale*, species with no essential oils, were effective against *M. javanica* and *T. semipenetrans*. Although beta-sitosterol is known to be responsible for some of their nematicidal action, this could be mostly an effect of a mixture of fatty acids. Palmitic, palmitoleic, stearic, oleic, linoleic, and linolenic acids are the major components of *Chaenopodium* oil (Malik, et al., 1985).

Calotropis procera (family Asclepiadaceae) has attracted the attention of many researchers lately. Soil amendments in the form of chopped leaves resulted in significant reductions in the population build-up of many nematode species (Ahmad, et al., 1996). Leaf extracts usually showed a strong effect on egg hatching, juvenile and adult mortality, and decreased root penetration by endoparasitic nematodes. Bare root dip treatments with leaf extract significantly reduced gall development caused by *M. incognita* in tomato and *Capsicum annuum* (Akhtar, et al., 1992). Seed dressing with latex of *C. procera* resulted in significant control of *M. incognita* and *Rotylenchulus reniformis*, and increased plant growth parameters, chlorophyll content of leaves, water absorption capacity of roots, and root nodulation of pigeon pea (*Cajanus cajan*) and chickpea (*Cicer arietinum*) (Anver and Alam, 1992). Seeds of okra, cabbage, cauliflower, and aubergine dressed with latex from *C. procera* or *C. gigantea* also resulted in lower nematode numbers and improvement of plant growth (Sebastian and Gupta, 1996; Siddiqui and Alam, 1988b, c; Wani, et al., 1994).

19.8 New Approaches on the Use of Antagonistic Plants

Bare root dipping in plant extracts is a relatively new way of inducing resistance against nematodes (Owino and Waudo, 1992; Siddiqui and Alam, 1988a, 1989a; Tiyagi, et al., 1990). A significant control of *M. incognita*, for instance, in tomato and *Capsicum annuum* came about after root-dipping seedlings of these crops in extracts of *Azadirachta indica*, *Ricinnus comunis*, *Eruca sativa*, *Brassica juncea*, or *Calotropis procera*. Neem-based industrial products such as Nimin were also tested with good results (Akhtar, et al., 1992; Akhtar and Mahmood, 1993, 1994a, b).

Seed dressing with plant latices of *Calotropis procera*, *C. gigantea*, *Euphorbia pulcherrima*, *E. milii*, *E. nerifolia*, *E. tirucalli*, *E. caducifolia*, or *Carica papaya* has shown to be a promising technique with good results found against nematode species such as *M. incognita*, *R. reniformis*, *Pratylenchus thornei*, and *T. brassicae* (Anver and Alam, 1992; Siddiqui and Alam, 1988b, c; Wani, et al., 1994). Neem-based products (e.g., Achook, Suneem, Repelin, and Welgrow) also showed some effect (Akhtar and Mahmood, 1995, 1997; Dash and Padhi, 1990).

Some neem-based pesticides, developed mainly for insect control (Jawan, Neenguard, Neemark, Margoside, Nimbecidine, and others), have been tried under laboratory and field conditions for nematode control (Gnanapragasam, et al., 1993; Paruthi, et al., 1996; Reshmi and Vijayalakshmi, 1998). Even though the control efficacy has usually not been very high, they may be valuable in a scheme of integrated pest management.

Another important field explored in the last decade is the association of antagonistic plants with nematophagous fungi to control nematodes. Organic matter from *Tagetes minuta*, *Ricinus communis*, and *Datura stramonium* stimulated egg parasitism of *M. incognita* and *M. javanica* by *Paecilomyces lilacinus*. Nagesh and Reddy (1995) studied the management of *M. incognita*, infecting crossandra, by integrating the use of *P. lilacinus* and *Verticillium lecanii* with leaf extracts of *R. communis* and *A. indica* as bare root dip and soil drench. Combinations of the antagonistic plants and the nematophagous fungi resulted in significantly higher plant growth parameters, flower yield, and lowered root gall index. *Tagetes patula* leaf extracts in pot test increased *M. incognita* egg parasitism by *Fusarium solani* and *F. oxysporum* (Owino, 1992; Owino and Sikora, 1992).

The interaction of antagonistic plants with mycorrhiza has also been investigated. The growth of tomato seedlings was synergistically enhanced after the vesicular arbuscular mycorrhizal (VAM) fungus *Glomus fasciculatum* was introduced into *M. incognita*-infested nursery beds amended with *Calotropis procera* leaves. The interactive effect of these components also resulted in a significant reduction of root-knot galls and a significant decrease in the number of eggs per egg mass. Colonization by mycorrhiza was significantly higher in the roots of tomato seedlings where *G. fasciculatum* was introduced into nursery beds amended with *Calotropis* leaves, a sure indication of their complementary interaction (Rao, et al., 1996a). In an assay intended to protect black gram (*Vigna mungo*) against *M. incognita*, Sankaranarayanan and Sundarababu (1996) tested *Glomus fasciculatum*, leaf extracts of *Calotropis procera, Tagetes erecta, Catharanthus roseus,* and *Bougainvillea spectabilis,* as well as carbofuran and phorate. The significant increases in all growth parameters established the favorable effect of all leaf extracts. These led to a higher spore population and a more intense mycorrhizal colonization than those observed with the nematicides, but gall indices and the nematode populations were lower with these chemicals than with the leaf extracts.

Intercropping antagonistic plants with commercial crops is controversial. Although benefits truely exist, like soil improvement and lessening of pests and diseases, it is a disappointment to find, quite often, no yield increase and income improvement to offset the increased production costs. A promising exception with *Muhlenbergia schreberi* has been reported (Meyer, et al., 1992). This plant emerged from their experiments as the most promising candidate for ground cover in commercial peach orchards. According to Meyer et al. (1992), this short-stature perennial grass tolerated drought, grew well in partial shade, did not harbor *Tetranychus urticae*, pentatomids, or the mirid *Lygus lineolaris*, inhibited populations of ring nematodes (*Criconemella xenoplax*), and survived the winter weather with little injury. *Muhlenbergia schreberi* successfully crowded out most weed species and it did not appear to be highly competitive with peach trees for water and nutrients, even when growing directly under the tree canopy.

Systemic nematicidal compounds would be highly desirable for obvious reasons. Encouraging reports have appeared on the subject. Sitaramaiah and Pathak (1981) reported that tomato yield increased and root-knot incidence (*M. incognita* and *M. javanica*) decreased after tomato plants had been sprayed under field conditions with catechol at 0.001 or 0.01 mol/l. Another promising compound is (2R, 5R)-dihydroxymethyl-(3R, 4R)-dihydroxypyrrolidine [DMDP], a polyhydroxy alkaloid isolated from tropical legume seeds and leaves of genera *Lonchocarpus* and *Derris*. It is a fructose analogue with insect antifeedant and glucosidase enzyme inhibition activities (Birch, et al., 1993b). It showed systemic activity against *M. javanica* in tomato and also when applied as a drench or seed coating. Some natural products, such as

2-furfuraldehyde, a fumigant commercially produced from sugarcane bagasse and known to have insecticidal, fungicidal, and nematicidal properties (Daneel and de Jager, 1996), deserve more in-depth studies.

Natural nematicides may also be found even under the sea. Twelve seaweed species were screened against *M. javanica*. Exposure to crude extracts of *Jolina laminarioides, Cystoseira trinodis,* and *Zoanthid* species caused over 50% mortality in *M. javanica*. Zoanthamine, a white crystalline compound isolated from *Zoanthid* species, was effective against the juveniles (Atta ur, et al., 1997).

19.9 Concluding Remarks

The farmers' reluctance to adopt antagonistic plants to control nematodes is the most important constraint for their use as a sound agronomic conservationist practice. Since caution seems to be inherent to the farmers' nature, it is paramount to dispel any doubts about the efficiency of the methods, assuring him of good profits by means of the research data and good extension work. Another difficulty is that most antagonistic plant species are not cash crops. Thus, a more convincing research is strongly needed and the generated data have to be made available to growers by extension scientists who should know how to explain them and change the growers' point of view. It is also desirable that research finds additional applications for the selected antagonist as compensations for the expenses to grow it. As an illustration, consider what happened with the velvet bean. When it appeared in the USA it was considered as "one of the most important crops of recent introduction" (Tracy and Coe, 1918). Soon the area under cultivation with this plant species reached about 2,000,000 ha, in 1917 (Coe, 1918) due to its soil improving qualities and for cattle and hog feeding. Velvet bean use started to decline somewhat in the early 1920s but the crop continued to be important in the South until the mid-1940s. By 1965, velvet bean had almost disappeared as a crop from the USA. This can probably be explained by a sharp decrease in mineral fertilizer prices and to the increasing popularity of soybean as a commercial crop (Buckles, 1995). This author states "the history of velvet bean's uses in various settings suggests that although not all old technologies can become new again, but changing conditions may provide fresh opportunities for building on older practices." Interesting velvet bean peculiarities that remain to be explored are: (1) it has strong allelopathic properties and control important weeds; (2) more than 50 chemicals have been isolated from the plant, such as alanine, arachidic acid, arginine, aspartic acid, beta-sitosterol, bufotenine, cystine, L-dopa, gallic acid, glutamic acid, glycine, histidine, lecithin, leucine, linoleic acid, mucunadine, mucunain, mucunine, myristic acid, nicotine, oleic acid, palmitic acid, proline, saponins, serine, serotonin, tyrosine, and valine; (3) it is considered an excellent cover crop; (4) it is a high-protein fodder for most animals; (5) in Mexico, the beans are toasted and ground to make a tasty high-protein coffee or used to "stretch" real coffee, although in excess it can lead to stomach pains; (6) it has attributes as analgesic, anti-inflammatory, antipyretic, carminative, diuretic, hypocholesterolemic, hypotensive, hypoglycemic, vermifuge, and others; and (7) the beans may also be used for human consumption, but it is necessary to eliminate the non-nutritive factors they contain such as anti-trypsin factor and L-dopa. Additionally, *Mucuna pruriens* seeds have been used in Unani Medicine, in India, and other countries for treating male sexual dysfunction in humans. The effect of the powdered seeds of *M. pruriens* on the mating behavior, libido and potency of sexually normal male rats was investigated. The drug produced a striking and sustained increase of sexual activity (Amim, et al., 1996). Adoption of velvet bean as antagonist may also suffer from constraints peculiar to a given region. In Benin, Africa, for instance, where they grow it mainly for human and animal consumption, farmers are unable to seed the crop at appropriate dates to get good biomass production and maximal benefits and also have risk of loss because of bush fires in the dry season and of the presence of snakes within its canopy. Nevertheless, after 5 years of extension work, *Mucuna* has been accepted as an efficient technology for sustainable agriculture there (Galiba, et al., 1998). In Central America, where intercropping velvet bean and corn is common, rats may become a nuisance; they use the velvet bean stems to climb up and eat the corn.

Researchers have been looking for new approaches on the management of plant-parasitic nematodes, such as the discovery of effective nematode antagonists, the development of techniques for better utilization of antagonistic plants and discovery of other properties of these plant parts so they can become marketable, which would stimulate their cultivation by the farmers. In the search for alternatives to chemical control of plant-parasitic nematodes, the potential nematicide value of a number of plant parts, by-products, and residues have been studied. However, much research has dealt with the nematostatic and/or nematicidal effects of crude plant extracts on certain nematode species, mainly *M. incognita*. In only a few cases the active substance has been identified and even in these cases the mode of action is unknown. It follows that these are attractive and promising research areas lying wide open ahead; and the industry people is watching developments closely. It remains necessary: (1) to continue screening additional plant species in a quest for new antagonists and nematicide properties; (2) to increase in number and in depth evaluation studies of potential and known antagonists on to what extent their use is safe in regard to natural enemies of nematodes and to non-target organisms; (3) to develop from plants new, stable, and low cost nematicidal or nematostatic formulations.

References

Abid, M., and M. A. Maqbool. 1990. Effects of inter-cropping of *Tagetes erecta* on root-knot disease and growth of tomato. International Nematology Network Newsletter 7:41 –42.

Abid, M., and M. A. Maqbool. 1991. Effect of bare-root dip treatment in oil-cakes and neem leaf extract on the root-knot development and growth of tomato and eggplant. Pakistan Journal of Nematology 9:13 – 16.

Abid, M., M. J. Zaki, and M. A. Maqbool. 1995. Neem derivatives for the control of root-knot nematode (*Meloidogyne javanica*) on okra. Pakistan Journal of Phytopathology 7:212 –214.

Ahmad, R., A. Ali, S. T. Sahi, N. Javed, and A. S. Shakir. 1993. Control of root-knot nematode (*Meloidogyne javanica*) by organic soil amendments. Pakistan Journal of Nematology 11:25 –29.

Ahmad, R., M. Tariq, N. Javed, and M. Inam ul Haq. 1991. Effect of extract of various plants on egg hatching and larval mortality of root-knot nematode, *Meloidogyne incognita*. Pakistan Journal of Phytopathology 3:38 –42.

Ahmad, R., M. Z. Shahab, M. Inam ul Haq, N. Javed, M. A. Dogar, and M. Y. Khan. 1996. Effect of soil amendment with *Calotropis procera* for the control of *Meloidogyne javanica* infection on egg plant. Pakistan Journal of Nematology 14:55 –59.

Ahmed, S., and M. Grainge. 1986. Potential of the neem tree (*Azadirachta indica*) for pest control and rural development. Economic Botany 40:201 –209.

Ajayi, V. A., C. N. Akem, and S. O. Adesiyan. 1993. Comparison of nematicidal potential of *Vernonia amygdalina* leaf extract and carbofuran on the growth and yield of root-knot nematode infested soyabean. Afro Asian Journal of Nematology 3:119 – 127.

Ajith, K., and M. S. Sheela. 1996. Utilization of green leaves of neem and *Eupatorium* for the management of soil organisms in bhindi and cowpea. Indian Journal of Nematology 26:139 – 143 (publ. 1997).

Akhtar, M. 1998. Biological control of plant-parasitic nematodes by neem products in agricultural soils. Applied Soil Ecology 7:219 –223.

Akhtar, M., and I. Mahmood. 1993. Control of plant-parasitic nematodes with "Nimin" and some plant oils by bare-root dip treatment. Nematologia Mediterranea 21:89 –92.

Akhtar, M., and I. Mahmood. 1994a. Control of root-knot nematode by bare-root dip in undecomposed and decomposed extracts of neem cake and leaf. Nematologia Mediterranea 22:55 –57.

Akhtar, M., and I. Mahmood. 1994b. Prophylactic and therapeutic use of oilcakes and leaves of neem and castor extracts for the control of root-knot nematode on chilli. Nematologia Mediterranea 22: 127 – 129.

Akhtar, M., and I. Mahmood. 1995. Control of root-knot nematode *Meloidogyne incognita* in tomato plants by seed-coating with Achook and neem oil. International Pest Control 37: 86 – 87.

Akhtar, M., and I. Mahmood. 1997. Control of root-knot nematode *Meloidogyne incognita* in tomato plants by seed coating with suneem and neem oil. Journal of Pesticide Science 22: 37 – 38.

Akhtar, M., A. H. Wani, and M. M. Alam. 1992. Control of root-knot nematode with bare-root dip in leaf extracts of Persian lilac and *Calotropis*. Current Nematology 3: 41 – 44.

Alam, M. M. 1986. Possible use of weeds as soil amendment for the management of root-knot and stunt nematodes attacking eggplant. Agricultural Wastes 16: 97 – 102.

Alam, M. M., A. M. Khan, and S. K. Saxena. 1979. Mechanism of control of plant parasitic nematodes as a result of the application of organic amendments to the soil. V. Role of phenolic compounds. Indian Journal of Nematology 9: 136 – 142 (publ. 1980).

Alam, M. M., M. A. Siddiqui, and A. Ahmad. 1990. Antagonistic plants. Pp. 41-49. In M. S. Jairajpuri, M. M. Alam, and I. Ahmad, eds. Nematode Biocontrol: Aspects and Prospects. Delhi, India: CBS Publishers and Distributors.

Alen, Y., S. Nakajima, T. Nitoda, N. Baba, H. Kanzaki, and K. Kawazu. 2000. Two antinematodal phenolics from *Knema hookeriana*, a Sumatran rainforest plant. Zeitschrift-fur-Naturforschung. Section C, Biosciences. 55: 300 – 303.

Allen, E. H., and J. Feldmesser. 1970. Nematicidal effect of alpha-tomatine on *Panagrellus redivivus*. Phytopathology 60: 1013.

Allen, E. H., and J. Feldmesser. 1971. Nematicidal activity of alfa-chaconine: effect of hydrogenion concentration. Journal of Nematology 3: 58 – 61.

Amim, K. M. Y., M. N. Khan, S. Sillur Rehman, and N. A. Khan. 1996. Sexual function improving effect of *Mucuna pruriens* in sexually normal male rats. Fitoterapia 67: 53 – 58.

Anver, S., and M. M. Alam. 1992. Effect of latex seed dressing on interacting root-knot and reniform nematodes. Afro Asian Journal of Nematology 2: 17 – 20.

Asha, M. K., D. Prashanth, B. Murali, R. Padmaja, and A. Amit. 2001. Anthelmintic activity of essential oil of *Ocimum sanctum* and eugenol. Fitoterapia 72: 669 – 670.

Asmus, R. M. F., and S. Ferraz. 1988. Antagonismo de algumas espécies vegetais, principalmente leguminosas, *Meloidogyne javanica*. Fitopatologia Brasileira 13: 20 – 24.

Atta ur, R., A. M. Khan, M. Shabbir, M. Abid, M. I. Chaudhary, A. Nasreen, M. A. Maqbool, M. Shameel, and R. Sualeh. 1997. Nematicidal activity of marine organisms. Pakistan Journal of Nematology 15: 95 – 100.

Awan, M. N., N. Javed, R. Ahmad, and M. Inam ul Haq. 1992. Effect of leaf extract of four plant species on larval mortality of citrus nematode (*Tylenchulus semipenetrans* Cobb) and citrus plant growth. Pakistan Journal of Phytopathology 4: 41 – 45.

Babu, S. P. S., D. Roy, N. C. Sukul, S. B. Mahato, P. Tauro, and S. S. Narwal. 1992. Nematicidal principle from *Acacia auriculiformis*. Proceedings, February 12 – 14,

1992, held at CCS Haryana Agricultural University, Hisar-125 004, India. Pp. 181 – 183.

Babu, S. P. S., and N. C. Sukul. 1990. Essential oils as nematicidal principles. Environment and Ecology 8: 1118 – 1120.

Barbosa, L. C. A., F. F. Barcelos, A. J. Demuner, and M. A. dos Santos. 1999. Chemical constituents from *Mucuna aterrima* with activity against *Meloidogyne incognita* and *Heterodera glycines*. Nematropica 29: 81 – 88.

Barrons, K. C. 1939. Studies of the nature of root-knot resistance. Journal of Agricultural Research 58: 262 – 271.

Basu, S. D., and S. K. Roy. 1975. *Rotylenchulus* sp., a new ectoparasitic nematode in tea soil. Two and a Bud 22: 17.

Belcher, J. V., and R. S. Hussey. 1977. Influence of *Tagetes patula* and *Arachis hypogaea* on *Meloidogyne incognita*. Plant Disease Reporter 61: 525 – 528.

Berner, D. K., A. S. Killani, E. Aigbokhan, and D. C. Couper. 1992. *Macrophomina phaseolina* on the tropical cover crop *Mucuna pruriens* var. utilis. Plant Disease 76: 1283.

Bijloo, J. D. 1966. The "pisum test": a simple method for the screening of substances on their therapeutic nematicidal activity. Nematologica 11: 643 – 644.

Birch, A. N. E., W. M. Robertson, and L. E. Fellows. 1993a. Plant products to control plant parasitic nematodes. Pesticide Science 39: 141 – 145.

Birch, A. N. E., W. M. Robertson, I. E. Geoghegan, W. J. McGavin, T. J. W. Alphey, L. E. Fellows, A. A. Watson, M. S. J. Simmonds, and E. A. Porter. 1992. Activity of the natural plant product (2R, 3R, 4R, 5R)-2,5-bis (hydroxymethyl) pyrrolidine-3,4-diol (DMDP) as an anti-nematode agent. Proceedings Brighton Crop Protection Conference Pest and Diseases, November 23-26, 1992, Brighton, England. Pp. 67 – 72.

Birch, A. N. E., W. M. Robertson, I. E. Geoghegan, W. J. McGavin, T. J. W. Alphey, M. S. Phillips, L. E. Fellows, A. A. Watson, M. S. J. Simmonds, and E. A. Porter. 1993b. DMDP – a plant-derived sugar analogue with systemic activity against plant parasitic nematodes. Nematologica 39: 521 – 535.

Brito, J. A. de, and S. Ferraz. 1987a. Antagonismo de *Brachiaria decumbens* e *Panicum maximum* cv. guiné *Meloidogyne javanica*. Nematologia Brasileira 11: 270 – 285.

Brito, J. A. de, and S. Ferraz. 1987b. Seleção de gramíneas antagonistas a *Meloidogyne javanica*. Nematologia Brasileira 11: 260 – 269.

Buckles, D. 1995. Velvetbean: a new plant with a history. Economic botany 49: 13 – 25.

Castro, C. O., and C. L. Munoz. 1982. Natural thiophene derivatives in the roots of *Tagetes jalisciencis*. Revista Latinoamericana de Quimica 13: 36 – 37.

Chandravadana, M. V., E. S. J. Nidiry, R. M. Khan, and M. S. Rao. 1994. Nematicidal activity of serpentine against *Meloidogyne incognita*. Fundamental and Applied Nematology 17: 185 – 186.

Chatterjee, A., N. C. Sukul, S. Laskar, and S. Ghoshmajumdar. 1982. Nematicidal principles from two species of Lamiaceae. Journal of Nematology 14: 118 – 120.

Chavda, J. C., B. A. Patel, and D. J. Patel. 1988. Screening of pigeonpea lines to *Rotylenchulus reniformis*. International Nematology Network Newsletter 5: 28 – 29.

Chitwood, D. J. 1992. Nematicidal compounds from plants. Pp. 185 – 204. In H. N. Nigg and D. Seigler, eds. Phytochemical Resources for Medicine and Agriculture. New York: Plenum Press.

Chitwood, D. J. 1993. Naturally occurring nematicides. Pp. 300-315. In D. J. Chitwood, S. O. Duke, J. J. Menn, and J. R. Plimmer, eds. Pest control with enhanced environmental safety. ACS Symposium Series No. 524. Washington, D. C. : American Chemical Society.

Chitwood, D. J. 2002. Phytochemical based strategies for nematode control. Annual Review of Phytopathology 40: 221 – 249.

Coe, H. S. 1918. Origin of the Georgia and Alabama varieties of velvet bean. Journal of the American Society of Agronomy 12: 175 – 179.

Cohn, E. 1971. Biological control of the citrus nematode. Scientific Activities 127: 127.

Conijn, C. G. M. , L. P. G. Molendijk, M. Schepman, A. T. Koster, A. M. E. Schenk, B. Kroonen Backbier, F. J. Gommers, and H. Brinkman. 1996. *Tagetes* spp. and root lesion nematodes. Gewasbescherming 27: 106 – 110.

Cook, C. G. , and G. A. White. 1996. *Crotalaria juncea*: a potencial multi-purpose fiber crop. Pp. 389 – 394. In C. G. Cook and G. A. White, eds. Progress in New Crops. Arlington, VA: ASHS Press.

Costa, D. C. , and S. Ferraz. 1990. Avaliação do efeito antagônico de algumas plantas, principalmente de inverno, *à Meloidogyne javanica*. Nematologia Brasileira 14: 61 – 70.

Daneel, M. , and K. de Jager. 1996. A naturally occurring fumigant for nematode control. Inligtingsbulletin Instituut vir Tropiese en Subtropiese Gewasse, No. 288, Pp. 9.

Dash, B. K. , and N. N. Padhi. 1990. Response of root-knot nematode infecting tomato to seed treatment by some neem products. Orissa Journal of Agricultural Research 2: 148 – 150.

Davis, E. L. , and J. R. Rich. 1987. Nicotine content of tobacco roots and toxicity to *Meloidogyne incognita*. Journal of Nematology 19: 23 – 29.

Debprasad, R. , D. Prasad, R. P. Singh, and D. Ray. 2000. Chemical examination and antinemic activity of marigold (*Tagetes erecta* L.) flower. Annals of Plant Protection Sciences 8: 212 – 217.

Devakumar, C. , B. K. Goswami, and S. K. Mukerjee. 1985. Nematicidal principles from neem (*Azadirachta indica* A. Juss). Part 1. screening of neem kernel fractions against *Meloidogyne incognita*. Indian Journal of Nematology 15: 121 – 124.

Dhangar, D. S. , D. C. Gupta, and R. K. Jain. 1995. Effect of marigold (*Tagetes erecta*) intercropped with brinjal in different soil types on disease intensity of root-knot nematode (*Meloidogyne javanica*). Indian Journal of Nematology publ. 1996, 25: 181 – 186.

Dias, C. R. , A. V. Schwan, D. P. Ezequiel, M. C. Sarmento, and S. Ferraz. 2000. Efeito de extratos aquosos de plantas medicinais na sobrevivência de juvenis de *Meloidogyne incognita*. Nematologia Brasileira 24: 203 – 210

Duke, J. A. 1981. Handbook of legumes of world economic importance. New York: Plenum Press.

Edwards, L. , T. Vrain, and R. S. Utkhede. 1994. Effect of antagonistic plants on apple replant disease. Acta Horticulturae 363: 135 – 140.

Egunjobi, O. A. , and S. O. Afolami. 1975. Effects of water-soluble extracts of neem (*Azadirachta indica*) on *Pratylenchus brachyurus* and on maize. Journal of Nematology

7: 321.

Eisenback, J. D. 1987. Reproduction of northern root-knot nematode (*Meloidogyne hapla*) on marigolds. Plant Disease 71: 281.

Ellenby, C. 1951. Mustard oils and control of the potato-root eelworm, *Heterodera rostochiensis* Wollenweber: further field experiments. Annals of Applied Biology 1: 287 – 288.

Enwerem, N. M., J. I. Okogun, C. O. Wambebe, D. A. Okorie, and P. A. Akah. 2001. Anthelmintic activity of the stem bark extracts of *Berlinia grandiflora* and one of its active principles, betulinic acid. Phytomedicine 8: 112 – 114.

Evans, P. H., W. S. Bowers, and E. J. Funk. 1984. Identification of fungicidal and nematicidal components in the leaves of *Piper betle* (Piperaceae). Journal of Agricultural and Food Chemistry 32: 1254 – 1256.

Fassuliotis, G., and G. P. Skucas. 1969. The effect of pyrrolizidine alkaloid ester and plants containing pyrrolizidine on *Meloidogyne incognita acrita*. Journal of Nematology 1: 287 – 288.

Ferraz, C. A. M., E. Cia, and N. P. Sabino. 1977. Effect of Bengal beans and groundnuts in rotation with cotton. Bragantia 36: 1 – 9.

Firoza, K., and M. A. Maqbool. 1996. Effect of plant extracts in the control of *Helicotylenchus dihystera* (Cobb, 1893) Sher, 1961. Pakistan Journal of Nematology 14: 61 – 66.

Galiba, M., P. Vissoh, G. Dagbenonbakin, and F. Fagbohoun. 1998. The reactions and fears of farmers utilizing velvet bean (*Mucuna pruriens*). Pp. 55 – 64. In D. Buckles, A. Etèka, O. Osiname, M. Galiba, and N. Galiano, eds. Cover Crops in West Africa: Contributing to Sustainable Agriculture. Otawa, Canada: International Maize and Wheat Improvement Center.

Gavilano, L., C. J. Li, F. J. Schimitz, and K. Schubert. 1998. Natural products from *Ambrosia psilostachya*, *Solidago missouriensis* and *Lespedeza stuevei* against *Meloidogyne incognita* nematodes. Journal of Nematology 30: 497.

Ghosal, S., S. Singh, and S. K. Bhattacharya. 1971. Alkaloids of *Mucuna pruriens*: chemistry and pharmacology. Planta Medical 19: 279 – 284.

Gnanapragasam, N. C. 1981. The influence of cultivating *Eragrostis curvula* in nematode-infested soils, on the subsequent build-up of populations in replanted tea. Tea Quarterly 50: 160 – 162.

Gnanapragasam, N. C., M. Mohotti, B. Sureshkumar, and G. P. Udamulla. 1993. Effect of "Jawan" a neem based natural pesticide in controlling nematode pests of tea. Sri Lanka Journal of Tea Science 62: 47 – 52.

Gokte, N., M. L. Maheshwari, and V. K. Mathur. 1991. Nematicidal activity of new essential oils against root-knot and cyst nematode species. Indian Journal of Nematology 21: 123 – 127.

Gokte, N., and G. Swarup. 1988. Effect of neem (*Azadirachta indica*) seed kernel powder treatment on penetration and development of *Anguina tritici* on wheat. Indian Journal of Nematology 18: 149 (publ. 1989).

Gommers, F. J. 1971. A nematicidal principle from the roots of a *Helenium* hybrid. Phytochemistry 10: 1945 – 1946.

Gommers, F. J. 1973. Nematicidal principles in Compositae. Mededelingen

Landbouwhogeschool Wageningen 73(17): ii + 71 pp.
Gommers, F. J. 1981. Biochemical interactions between nematodes and plants and their relevance to control. Helminthological Abstracts, Series B 50: 9 – 24.
Gommers, F. J., and J. Bakker. 1988. Physiological diseases induced by plant responses or products. Pp. 3 – 22. In F. J. Gommers and J. Bakker, eds. Diseases of Nematodes, Vol. 1 Boca Raton, FL: CRC Press.
Gommers, F. J., and D. J. M. Voor in't Holt. 1976. Chemotaxonomy of Compositae related to their host suitability for *Pratylenchus penetrans*. Netherlands Journal of Plant Pathology 82: 1 – 8.
Gonzalez, J. A., A. Estevez Braun, R. Estevez Reyes, L. Bazzocchi, L. Moujir, I. A. Jimenez, A. G. Ravelo, and A. G. Gonzalez. 1995. Biological activity of secondary metabolites from *Bupleurum salicifolium* (Umbelliferae). Experientia 51: 35 – 39.
Gonzalez, J. A., A. Estevez Braun, R. Estevez Reyes, and A. G. Ravelo. 1994. Inhibition of potato cyst nematode hatch by lignans from *Bupleurum salicifolium* (Umbelliferae). Journal of Chemical Ecology 20: 517 – 524.
Goswami, B. K., and K. Vijayalakshmi. 1986. Nematicidal properties of some indigenous plant materials against root-knot nematodes, *Meloidogyne incognita* on tomato. Indian Journal of Nematology 16: 65 – 68.
Grewal, P. S. 1988. Stimulation of fungal control of mycophagous nematodes and weed moulds in mushroom compost. Brighton Crop Protection Conference. Pests and Diseases 3: 1199 – 1204.
Grewal, P. S. 1989. Effects of leaf-matter incorporation on *Aphelenchoides composticola* (Nematoda), mycoflora composition, mushroom compost quality and yield of *Agaricus bisporus*. Annals of Applied Biology 115: 299 – 312.
Hanawa, F., T. Yamada, and T. Nakashima. 2001. Phytoalexins from *Pinus strobus* bark infected with pinewood nematode, *Bursaphelenchus xylophilus*. Phytochemistry 57: 223 – 228.
Haroon, S., and G. C. Smart, Jr. 1983a. Development of *Meloidogyne incognita* inhibited by *Digitaria decumbens* cv. Pangola. Journal of Nematology 15: 102 – 105.
Haroon, S., and G. C. Smart, Jr. 1983b. Effects of Pangola digitgrass on *Meloidogyne arenaria*, *M. javanica* and *M. hapla*. Journal of Nematology 15: 649 – 650.
Haroon, S., and G. C. Smart, Jr. 1983c. Root extracts of Pangola digitgrass affect egg hatch and larval survival of *Meloidogyne incognita*. Journal of Nematology 15: 646 – 649.
Haroon, S. A., and S. H. Abadir. 1989. The effect of four summer legume cover crops on the population level of *Meloidogyne incognita*, *Pratylenchus penetrans* and *Trichodorus christiei*. Assiut Journal of Agricultural Sciences 20: 25 – 35.
Harry, O. K. R. E., H. Mojtahedi, S. F. Vaughn, P. F. Dowd, G. S. Santo, R. A. Holser, and T. P. Abbott. 1999. Milkweed seedmeal: a control for *Meloidogyne chitwoodi* on potatoes. Industrial Crops and Products 9: 145 – 150.
Harvey, S. G., H. N. Hannahan, and C. E. Sams. 2002. Indian mustard and allyl isothiocyanate inhibit *Sclerotium rolfsii*. Journal of the American Society for Horticultural Science 127: 27 – 31.
Haseeb, A., and F. Butool. 1990. Evaluation of nematicidal properties in some members of the family Lamiaceae. International Nematology Network Newsletter 7: 24 – 26.

Hatakeda, K., N. Saito, S. Ito, Y. Ikusima, and T. Asano. 1985. A new nematicidal compound in french marigold. Japanese Journal of Nematology 15: 11 – 13.

Huang, C. S., and E. F. S. Mota e Silva. 1980. Interruption of the life cycle of *Meloidogyne incognita* by *Crotalaria* spp. Fitopatologia Brasileira 5: 402 – 403.

Hussaini, S. S., R. Rao, and H. K. Pandu. 1996. Toxicity of water soluble leaf extracts against larvae and egg masses of three *Meloidogyne* species. Indian Journal of Nematology 26: 23 – 31 (publ. 1997).

Imran, M., and S. K. Saxena. 1993. Effect of plant materials on the population of tomato root-knot nematode (*Meloidogyne incognita*). Indian Journal of Mycology and Plant Pathology 23: 180 – 182.

Insunza, V. B., and A. A. Valenzuela. 1995. Control of *Ditylenchus dipsaci* on garlic (*Allium sativum*) with extracts of medicinal plants from Chile. Nematropica 25: 35 – 41.

Karr, L. L., C. D. Drewes, and J. R. Coats. 1990. Toxic effects of d-limonene in the earthworm *Eisenia fetida* {Sauvigny}. Pesticide Biochemistry and Physiology 36: 175 – 186.

Kathirvel, M., M. Balasubramanian, M. Gopalan, and C. V. Sivakumar. 1992. Effect of seed treatment with botanicals and chemicals for the control of root-knot nematode, *Meloidogyne incognita* infesting okra, *Abelmoschus esculentus* L. Indian Journal of Plant Protection 20: 191 – 194.

Kawazu, K., Y. Nishii, and S. Nakajima. 1980. Two nematicidal substances from roots of *Cirsium japonicum*. Agricultural and Biological Chemistry 44: 903 – 906.

Khan, A. 1992. Effect of neem leaves and carbofuran on *Pratylenchus thornei* associated with three wheat varieties. Pakistan Journal of Scientific and Industrial Research 35: 459 – 460.

Khan, A. M., S. K. Saxena, and Z. A. Siddiqi. 1971. Efficacy of *Tagetes erecta* in reducing root infesting nematodes of tomato and okra. Indian Phytopathology 24: 166 – 169.

Khan, H. A., F. Qamar, M. Saeed, and S. A. Khan. 1990. The nematicidal properties of compounds of plant origin with emphasis on polyphenols – a review. Proceedings of Parasitology 9: 87 – 91.

Khan, M. W., M. M. Alam, A. M. Khan, and S. K. Saxena. 1974. Effect of water soluble fractions of oil-cakes and bitter principles of neem on some fungi and nematodes. Acta Botanica Indica 2: 120 – 128.

Khanna, A. S., and N. K. Sharma. 1994. Efficiency of neem leaf powder as an additive in management of mushroom nematodes. Indian Journal of Hill Farming 7: 114 – 115.

Kimura, Y., M. Mori, A. Suzuki, and A. Kobayashi. 1981. Isolation and identification of two nematicidal substances from roots of *Erigeron philadelphicus* L. and nematicidal activities of their related compounds. Agricultural and Biological Chemistry 45: 2915 – 2917.

Kloepper, J. W., R. Rodríguez-Kábana, J. A. McInroy, and D. J. Collins. 1991. Analysis of populations and physiological characterization of microorganisms in rhizospheres of plants with antagonistic properties to phytopathogenic nematodes. Plant and Soil 136: 95 – 102.

Kloepper, J. W., R. Rodríguez-Kábana, J. A. McInroy, and R. W. Young. 1992.

Rhizosphere bacteria antagonistic to soybean cyst (*Heterodera glycines*) and root-knot (*Meloidogyne incognita*) nematodes: identification by fatty acid analysis and frequency of biological control activity. Plant and Soil 139: 75 – 84.

Kogiso, S., K. Wada, and K. Munakata. 1976a. Nematicidal polyacetylenes, 3z, 11 E-and 3E, 11E-trideca-1, 211-triene-5, 7, 9-triyne from *Carthamus tinctorius* L. Tetrahedron Letters 2: 109 – 110.

Kogiso, S., K. Wada, and K. Munakata. 1976b. Isolation of nematicidal polyacetylenes from *Carthamus tinctorius*. Agricultural and Biological Chemistry 40: 2085 – 2089.

Kogiso, S., K. Wada, and K. Munakata. 1976c. Odoracin, a nematicidal constituent from *Daphne odora*. Agricultural and Biological Chemistry 40: 2119 – 2120.

Krishnamurthy, G. V. G., P. S. N. Murthy, M. S. Chari, and G. Ramaprasad. 1993. Further studies with plant extracts on root-knot nematode (*Meloidogyne javanica*) larvae. Botanical pesticides in integrated pest management: Proceedings of National Symposium held on January 21 – 22, 1993, Rajahmundry, India. Pp. 438 – 448.

Kuiper, K. 1963. Enige bijzondere aaltjesaantastingen in 1962. Netherlands Journal of Plant Pathology 69: 153 – 154.

Kurundkar, B. P., and P. V. Jadhav. 1993. Efficacy of some oil seed cakes and plant extracts in managing root-knot of okra. Indian Phytopathology 46: 254 – 256.

Lawrence, B. M. 1985. A review of the world production of essential oils (for 1984). Perfumer and Flavorist 10(5): 1 – 16.

Leela, N. K., R. M. Khan, P. P. Reddy, and E. S. J. Nidiry. 1992. Nematicidal activity of essential oil of *Pelargonium graveolens* against the root-knot nematode *Meloidogyne incognita*. Nematologia Mediterranea 20: 57 – 58.

Leela, N. K., and K. V. Ramana. 2000. Nematicidal activity of the essential oil of allspice (*Pimenta dioica* L. Merr.). Journal of Plant Biology 27: 75 – 76.

Lorenzetti, F., S. MacIsaac, J. T. Arnason, D. V. C. Awang, and D. Buckles. 1998. The phytochemistry, toxicology, and food potential of velvetbean (*Mucuna* Adans. spp., Fabaceae). Pp. 67 – 84. In D. Buckles, A. Etèka, O. Osiname, M. Galiba, and N. Galiano, eds. Cover Crops in West Africa Contributing to Sustainable Agriculture. Otawa, Canada: International Maize and Wheat Improvement Center.

Macedo, M. E., R. Consoli, T. S. M. Grandi, A. M. G. dos. Anjos, A. B. de. Oliveira, N. M. Mendes, R. O. Queiroz, and C. L. Zani. 1997. Screening of Asteraceae (Compositae) plant extracts for larvicidal activity against *Aedes fluviatilis* (Diptera: Culicidae). Memorias do Instituto Oswaldo Cruz 92: 565 – 570.

Mahajan, R., P. Singh, and P. S. Kalsi. 1986. Nematicial activity of some sesquiterpenoids against root-knot nematode (*Meloidogyne incognita*). Nematologica 32: 119 – 123.

Malik, M. S., B. S. Verma, N. K. Sangwan, D. S. Bhatti, and K. S. Dhindsa. 1985. Phytochemical and nematicidal studies on *Chaenopodium* species. Plant Physiology and Biochemistry 12: 133 – 138.

Mani, A., and K. C. Chitra. 1989. Toxicity of certain plant extracts to *Meloidogyne incognita*. Nematologia Mediterranea 17: 43 – 44.

Mareggiani, G., A. Pelicano, A. Fraschina, G. Bilotti, N. Gorosito, and G. Zipeto. 1997. "In vitro" activity of natural plant products on *Meloidogyne incognita* larvae (Nematoda: Meloidogynidae). Revista de la Facultad de Agronomia, Universidad de

Buenos Aires 16: 141 – 145.

Martowo, B., and D. Rohana. 1987. The effect of intercropping of pepper (*Capsicum annum* L.) with some vegetable crops on pepper yield and disease incidence caused by Meloidogyne spp. Buletin Penelitian Hortikultura 15(4): 55 – 59.

Matsuda, K., M. Kimura, K. Komai, and M. Hamada. 1989. Nematicidal activities of (-)-N-methylcystisine and (-)-anagyrine from *Sophora flavescens* against pine wood nematodes. Agricultural and Biological Chemistry 53: 2287 – 2288.

Matsuda, K., K. Yamada, M. Kimura, and M. Hamada. 1991. Nematicidal activity of matrine and its derivatives against pine wood nematodes. Journal of Agricultural and Food Chemistry 39: 189 – 191.

Medina, A. L., J. N. BeMiller, J. Janick, and J. E. Simon. 1993. Marigold flower meal as a source of an emulsifying gum. New crops exploration: research and commercialization, Indianapolis, Indiana, October 6 – 9, 1991, Pp. 389 – 393 (publ. 1993).

Mejia, E. G. de, G. Loarca Pina and M. Ramos Gomez. 1997. Antimutagenicity of xanthophylls present in Aztec Marigold (*Tagetes erecta*) against 1-nitropyrene. Mutation Research, Genetic Toxicology and Environmental Mutagenesis 389: 219 – 226.

Meyer, J. R., E. I. Zehr, R. L. Meagher, Jr., and S. K. Salvo. 1992. Survival and growth of peach trees and pest populations in orchard plots managed with experimental ground covers. Agriculture, Ecosystems and Environment 41: 353 – 363.

Miller, P. M. 1978. Toxicity of homogenized leaves of woody and herbaceous plants to root lesion nematodes in water and in soil. Journal of the American Society for Horticultural Science 103: 78 – 81.

Milman, I. A. 1990. Alanto- and isoalantolactones. Chemistry of Natural Compounds 26(3): 251 – 262.

Mitarai, M., J. C. V. Sicat, Y. Uchida, Y. Okada, and M. Nagata. 1997. Effect of AITC on the control of nematodes and its optimum application method. Bulletin of the Faculty of Agriculture, Miyazaki University. 44: 35 – 44.

Mohamed, A. S. A., T. Mori, S. Q. Islam, M. Sato, and T. Yamasaki. 2000. Lethal activity of gallo- and condensed tannins against the free-living soil-inhabiting nematode, *Caenorhabditis elegans*. Journal of Pesticide Science 25: 410 – 415.

Mojumder, V., and S. D. Mishra. 1990. In vitro studies on toxicity of embelin and its haloderivatives to *Meloidogyne incognita* juveniles. Current Nematology 1: 47 – 48.

Mojumder, V., and S. D. Mishra. 1991. Nematicidal efficacy of some plant products and management of *Meloidogyne incognita* pulse crops by soaking seeds in their aqueous extracts. Current Nematology 2: 27 – 32.

Momin, R. A., R. S. Ramsewak, and M. G. Nair. 2000. Bioactive compounds and 1,3-di[(cis)-9-octadecenoyl]-2-[(cis, cis)-9, 12-octadecadienoyl] glycerol from Apium graveolens L. seeds. Journal of Agricultural and Food Chemistry. 48: 3785 – 3788.

Mori, T., A. S. A. Mohamed, M. Sato, and T. Yamasaki. 2000. Ellagitannin toxicity in the free-living soil-inhabiting nematode, *Caenorhabditis elegans*. Journal of Pesticide Science 25: 405 – 409.

Motsinger, R. E., E. H. Moody, and C. M. Gay. 1977. Reaction of certain french marigold (*Tagetes patula*) cultivars to three *Meloidogyne* spp. Journal of Nematology 9: 278 (abstr.).

Munakata, K. 1983. Nematicidal natural products. Pp. 299-310. In D. L. Whitehead and W. S. Bowers, eds. Natural Products for Innovative Pest Management. Pergamon Press, Oxford, UK.

Naganathan, T. G., R. Arumugam, M. Kulasekaran, and S. Vadivelu. 1988. Effect of antagonistic crops as intercrops on the control of banana nematodes. South Indian Horticulture 36: 268 – 269.

Nagesh, M., and P. P. Reddy. 1995. Comparative efficacy of *Paecilomyces lilacinus* (Thom.) Samson and *Verticillium lecanii* (A. Zimmerman) Viegas in combination with botanicals against *Meloidogyne incognita* (Kofoid and White, 1919) Chitwood, 1949 infecting *Crossandra undulaefolia* L. Journal of Biological Control 9: 109 – 112.

Nagesh, M., P. P. Reddy, and M. S. Rao. 1997. Integrated management of *Meloidogyne incognita* on tuberose using *Paecilomyces lilacinus* in combination with plant extracts. Nematologia Mediterranea 25: 3 – 7.

Nandal, S. N., and D. S. Bhatti. 1986a. The effect of certain edaphic factors on the nematicidal activity of plant extracts. Nematologia Mediterranea 14: 295 – 298.

Nandal, S. N., and D. S. Bhatti. 1986b. Influence of four plant extracts on the hatching of *Meloidogyne javanica* and invasion of host roots. Nematologia Mediterranea 14: 291 – 294.

Nandal, S. N., and D. S. Bhatti. 1990. Efficacy, persistence and field application potential of some weeds/shrubs for controlling *Meloidogyne javanica* on brinjal. Nematologia Mediterranea 18: 113 – 115.

National Research Council. 1992. Neem: a tree for solving global problems. Washington, D. C.: National Academy Press. 141.

Nidiry, E. S. J., R. M. Khan, and P. P. Reddy. 1993. In vitro nematicidal activity of *Gloriosa superba* seed extract against *Meloidogyne incognita*. Nematologia Mediterranea 21: 127 – 128.

Njoku, C. J., D. C. Hopp, F. Alali, I. U. Asuzu, and J. L. McLaughlin. 1997a. Dihydroguaiaretic acid: a bioactive component of the stem bark of *Pycnanthus angolensis*. Planta Medica 63: 580 – 581.

Njoku, C. J., L. Zeng, I. U. Asuzu, N. H. Oberlies, J. L. McLaughlin, and L. Zeng. 1997b. Oleanolic acid, a bioactive component of the leaves of *Ocimum gratissimum* (Lamiaceae). International Journal of Pharmacognosy 35: 134 – 137.

Nogueira, M. A., J. S. de Oliveira, and S. Ferraz. 1996a. Nematicidal hydrocarbons from *Mucuna aterrima*. Phytochemistry 42(4): 997 – 998.

Nogueira, M. A., J. S. de Oliveira, S. Ferraz, and M. A. dos Santos. 1996b. Nematicidal constituents in *Mucuna aterrima* and its activity on *Meloidogyne incognita* race 3. Nematologia Mediterranea 24: 249 – 252.

Nuttall, Z. 1920. Los jardines del antiguo Mexico. Memoires de la Sociedad Cientifica "Antonio Alazate." 37: 193 – 213.

Onda, M., K. Abe, K. Yonezawa, N. Esumi, and T. Suzuki. 1970. Studies on the constituents of *Bocconia cordata*. II Bocconine. Chemical and Pharmaceutical Bulletin 18: 435 – 439.

Onda, M., K. Takiguchi, M. Hirakura, H. Fukushima, M. Akagawa, and F. Naoi. 1965. Studies on the constituents of *Bocconia cordata*. Part I. On nematicidal alkaloids. Nippon Nageikaguku Kaishi 39: 168 – 170.

Oostenbrink, M. 1960. *Tagetes patula* L. als voorvrucht van enkele land-en tuinbowwgewassen op zand-en dalgrond. Meded. LandbHoogesch. OpzoekStns Gent 25: 1065 – 1075.

Oostenbrink, M., K. Kuiper, and J. J. s' Jacob. 1957. *Tagetes* als feindpflanzen von *Pratylenchus*-arten. Nematologica 2: 424 – 433.

Owino, P. O. 1992. Effect of marigold leaf extract and captafol on fungal parasitism of root knot nematode eggs - Kenyan isolates. Nematologia Mediterranea 20: 211 – 213.

Owino, P. O., and R. A. Sikora. 1992. Influence of leaf extracts of marigold and the fungicide Benlate on fungal parasitism of *Meloidogyne javanica* eggs in Kenyan soil. Pakistan Journal of Nematology 10: 57 – 67.

Owino, P. O., and S. W. Waudo. 1992. Medicinal plants of Kenya: effects on *Meloidogyne incognita* and the growth of okra. Afro Asian Journal of Nematology 2: 64 – 66.

Padma, V., K. Suman, S. Satyawati, P. Vasudevan, S. Kashyap, and S. Sharma. 1997. *Tagetes*: a multipurpose plant. Bioresource Technology 62: 29 – 35.

Paruthi, I. J., R. S. Kanwar, and D. C. Gupta. 1996. In vitro evaluation of achook – a neem based pesticide against root-knot nematode, *Meloidogyne incognita*. Indian Journal of Nematology 26: 270 – 272 (publ. 1997).

Perich, M. J., C. Wells, W. Bertsch, and K. E. Tredway. 1994. Toxicity of extracts from three *Tagetes* against adults and larvae of yellowfever mosquito and *Anopheles stephensi* (Diptera: Culicidae). Journal of Medical Entomology 31: 833 – 837.

Potter, M. J., K. Davies, and A. J. Rathjen. 1998. Suppressive impact of glucosinolates in *Brassica* vegetative tissues on root lesion nematode *Pratylenchus neglectus*. Journal of Chemical Ecology 24: 67 – 80.

Prasad, D., and M. M. Haque. 1982. Reaction on varieties of marigold against root-knot nematode, *Meloidogyne incognita*. Indian Journal of Nematology 12: 418 – 419.

Prasad, D., D. K. Nagia, S. Kumar, and M. L. Saini. 1994. Integrated approach for management of plant parasitic nematodes in brinjal, *Solanum melongena* L. Plant Protection Bulletin Faridabad 46: 31 – 33.

Pritts, M. P. 1992. Weed control in strawberries: some new approaches. Pennsylvania Fruit News 72: 97 – 102.

Prot, J. C., and J. M. Kornprobst. 1985. Effects of quassinoids extracted from *Hannoa undulata* seed on the penetration and reproduction of *Meloidogyne javanica* on tomato. Revue de Nematologie 8: 383 – 389.

Qamar, F., Z. Kapadia, S. A. Khan, and Y. Badar. 1995. *Datura metel* L. a plant with nematicidal potential. Pakistan Journal of Scientific and Industrial Research 38: 319 – 321.

Qamar, F., M. Saeed, A. Khan, and N. Seema. 1989. Control of anguinosis of wheat using plant extracts. Pakistan Journal of Scientific and Industrial Research 32: 260 – 261.

Quarles, W. 1992. Botanical pesticides from *Chenopodium*. IPM Practitioner 14 (2): 1 – 11.

Rajaram, N., and K. Janardhanan. 1991. The biochemical composition and nutritional potential of the pulse, *Mucuna gigantea* (Willd) DC. Plant Foods for Human Nutrition 44: 45 – 51.

Rangaswamy, S. D., P. P. Reddy, and S. Joshi. 1993. Histopathological and histochemical investigations on antagonistic trap crops (marigold and mustard) and susceptible tomato infested with *Meloidogyne incognita*. Current Nematology 4: 203 - 206.

Rao, M. S., P. P. Reddy, and S. M. Das. 1996a. Effect of integration of *Calotropis procera* leaf and *Glomus fasciculatum* on the management of *Meloidogyne incognita* infesting tomato. Nematologia Mediterranea 24: 59 - 61.

Rao, M. S., P. P. Reddy, A. Mittal, M. V. Chandravadana, and M. Nagesh. 1996b. Effect of some secondary plant metabolites as seed treatment agents against *Meloidogyne incognita* on tomato. Nematologia Mediterranea 24: 49 - 51.

Rao, M. S., P. P. Reddy, and M. Nagesh. 1997. Integration of *Paecilomyces lilacinus* with neem leaf suspension for the management of root-knot nematodes on egg plant. Nematologia Mediterranea 25: 249 - 252.

Reddi, C. K. 1983. Nitrogen fixation and nematode resistance of 13 tropical legumes. Dissertation Abstracts International, B 44(2): 380b.

Resende, I. C., S. Ferraz, and A. R. Condé. 1987. Efeito de seis variedades de mucuna (*Stizolobium* spp.) sobre *Meloidogyne incognita* raça 3 e *M. javanica*. Fitopatologia Brasileira 12: 310 - 313.

Reshmi, B., and M. Vijayalakshmi. 1998. Effect of aqueous extracts of neem based pesticides on the mortality of second-stage juveniles of *Meloidogyne incognita*. Annals of Plant Protection Sciences 6: 53 - 55.

Reynolds, L. B., J. W. Potter, and B. R. Ball-Coelho. 2000. Crop rotation with *Tagetes* sp. is an alternative to chemical fumigation for control of root-lesion nematodes. Agronomy Journal 92: 957 - 966.

Rhoades, H. L. 1980. Relative susceptibility of *Tagetes patula and Aeschynomene americana* to plant nematodes in Florida. Nematropica 10: 116 - 120.

Rich, J. R., G. S. Rahi, C. H. Opperman, and E. L. Davis. 1989. Influence of the castor bean (*Ricinus communis*) lectin (ricin) on motility of *Meloidogyne incognita*. Nematropica 19: 99 - 103.

Rodríguez-Kábana, R. and E. G. Ingram. 1978. Susceptibility of pigeon pea to plant parasitic nematodes in Alabama. Nematropica 8(1): 32 - 35.

Rodríguez-Kábana, R., P. S. King, D. G. Robertson, and C. F. Weaver. 1988a. Potential of crops uncommon to Alabama for management of root-knot and soybean cyst nematodes. Annals of Applied Nematology 2: 116 - 120.

Rodríguez-Kábana, R., P. S. King, D. G. Robertson, C. F. Weaver, and E. L. Carden. 1988b. New crops with potential for management of soybean nematodes. Nematropica 18(1): 45 - 52.

Rodríguez-Kábana, R., N. Kokalis Burelle, D. G. Robertson, P. S. King, and L. W. Wells. 1994. Rotations with coastal bermudagrass, cotton, and bahiagrass for management of *Meloidogyne arenaria* and Southern blight in peanut. Supplement to Journal of Nematology 26: 665 - 668.

Rodríguez-Kábana, R., J. Pinochet, D. G. Robertson, and L. Wells. 1992. Crop rotation studies with velvetbean (*Mucuna deeringiana*) for the management of *Meloidogyne* spp. Supplement to Journal of Nematology 24: 662 - 668.

Rodríguez-Kábana, R., D. G. Robertson, C. F. Weaver, and L. Wells. 1991a.

Rotations of bahiagrass and castorbean with peanut for the management of *Meloidogyne arenaria*. Supplement to Journal of Nematology 23: 658 – 661.

Rodríguez-Kábana, R., D. G. Robertson, L. Wells, P. S. King, and C. F. Weaver. 1989a. Crops uncommon to Alabama for the management of *Meloidogyne arenaria* in peanut. Supplement to Journal of Nematology 21: 712 – 716.

Rodríguez-Kábana, R., C. F. Weaver, D. G. Robertson, and H. Ivey. 1988c. Bahiagrass for the management of *Meloidogyne arenaria* in peanut. Annals of Applied Nematology 2: 110 – 114.

Rodríguez-Kábana, R., D. B. Weaver, R. Garcia, D. G. Robertson, and E. L. Carden. 1989b. Bahiagrass for the management of root-knot and cyst nematodes in soyabean. Nematropica 19: 185 – 193.

Rodríguez-Kábana, R., D. B. Weaver, D. G. Robertson, E. L. Carden, and M. L. Pegues. 1991b. Additional studies on the use of bahiagrass for the management of root-knot and cyst nematodes in soybean. Nematropica 21: 203 – 210.

Rohde, R. A., and W. R. Jenkins. 1958. The chemical basis of resistence of *Asparagus* to the nematode *Trichodorus christiei*. Phytopathology 48: 463 (abstr.).

Sabira, B., W. Aneela, B. S. Siddiqui, and Q. Fatima. 2000. Nematicidal constituents of the aerial parts of *Lantana camara*. Journal of Natural Products. 63: 765 – 767.

Sadhana, S., and D. S. Walia. 1996. Fungitoxicity test of certain essential oils against storage fungi. International Journal of Tropical Plant Diseases 14: 227 – 228.

Salawu, E. O. 1992. Effect of neem leaf extract and ethoprop singly and in combination on *Meloidogyne incognita* and growth of sugarcane. Pakistan Journal of Nematology 10: 51 – 56.

Saleh, M. A., F. H. Abdel Rahman, N. A. Ibrahim, and N. M. Taha. 1987. Isolation and struture determination of new nematicidal triglyceride from *Argemone mexicana*. Journal of Chemical Ecology 13: 1361 – 1370.

Sangwan, N. K., B. S. Verma, K. K. Verma, and K. S. Dhindsa. 1990. Nematicidal activity of some essential plant oils. Pesticide Science 28: 331 – 335.

Sangwan, N. K., K. K. Verma, B. S. Verma, M. S. Mali, and K. S. Dhindsa. 1985. Nematicidal activity of essential oils of *Cymbopogon grasses*. Nematologica 31: 93 – 99.

Sanjay, S., D. Prasad, R. P. Singh, and S. Srivastava. 2000. Antinemic studies of the chemical constituents of the leaves of *Murraya koenigii* (L.) Spreng. Annals of Plant Protection Sciences. 8: 183 – 186.

Sankaranarayanan, C., and R. Sundarababu. 1996. Effect of leaf extracts on growth of blackgram inoculated with vesicular arbuscular mycorrhiza (*Glomus fasciculatum*) and root- knot nematode (*Meloidogyne incognita*). Indian Journal of Nematology 26: 144 – 147 (publ. 1997).

Sano, Z., and K. Nakasono. 1986. Histological responses of three leguminous enemy plants to the penetration and development of *Meloidogyne incognita*. Japanese Journal of Nematology 16: 48 – 55.

Sano, Z. I., K. Nakasono, and M. Araki. 1983. Penetration and development of *Meloidogyne incognita* in some enemy and host plants. Proceedings of the Association for Plant Protection of Kyushu 29: 132 – 136.

Sarah, J. L., R. Fogain, and C. Valette. 1997. Nematode resistance in bananas: varietal screening and resistance mechanisms. Fruits Paris 52: 267 – 271.

Sasanelli, N. 1992. Nematicidal activity of aqueous extracts from leaves of *Ruta graveolens* on *Xiphinema index*. Nematologia Mediterranea 20: 53 – 55.

Sasanelli, N. 1995. Prospects for the use of some plants with nematicide action. Informatore Agrario 51: 55 – 56.

Sasanelli, N., and T. D'Addabbo. 1993. Effect of *Cineraria maritima*, *Ruta graveolens* and *Tagetes erecta* leaf and root extracts on Italian populations of *Meloidogyne* species. Nematologia Mediterranea 21: 21 – 25.

Sasanelli, N., and T. D'Addabbo. 1995. Effect of aqueous solutions of rutin on the beet cyst nematode *Heterodera schachtii*. Nematologia Mediterranea 23(1): 31 – 34.

Saxena, D. B., B. K. Goswami, and S. S. Tomar. 1987. Nematicidal properties of some plant extracts against second stage juveniles of *Meloidogyne incognita* (Kofoid and White) Chitwood. Journal of Research Assam Agricultural University 8: 59 – 62 (publ. 1990).

Scheffer, F., R. Kickuth, and J. H. Visser. 1962. Die Wurzelausscheidungen von *Eragrotis curvula*, Nees und ihr Einfluss auf Wurzelknoten-Nematoden. Zeitschrift fur Pflanzenahrung und Bodenkunde 98: 114 – 120.

Schmutterer, H. 1997. Side-effects of neem (*Azadirachta indica*) products on insect pathogens and natural enemies of spider, mites and insects. Journal of Applied Entomology 121: 121 – 128.

Scott, J. M. 1910. The velvet bean. Gainesville, FL, University of Florida Agricultural Experiment Station.

Sebastian, S., and P. Gupta. 1996. Evaluation of three plant latices as seed treatment against *Pratylenchus thornei* on chickpea. International Chickpea and Pigeonpea Newsletter No. 3: 39 – 40.

Seinhorst, J. W., and C. H. Klinkenberg. 1963. De invloed van *Tagetes* op de bevolkingsdichtheid van aaltjes en op de opbrengst van cultuurgewassen. Netherlands Journal of Plant Pathology 69: 153.

Sellami, S., and A. Mouffarrah. 1994. Effect of some aqueous plant extracts on juvenile hatching and larval mortality against *Meloidogyne incognita*. Mededelingen Faculteit Landbouwkundige en Toegepaste Biologische Wetenschappen, Universiteit Gent 59: 813 – 816.

Sharma, R., and P. C. Trivedi. 1992a. Effect of root extract of some plants on larval hatching of *Meloidogyne incognita*. Current Nematology 3: 31 – 34.

Sharma, R., and P. C. Trivedi. 1992b. Reduction of root-knot disease of brinjal using dry root powder amendment. Current Nematology 3: 133 – 138.

Sherif, A., R. G. Hall, and M. El Amamy. 1987. Drugs, insecticides and other agents from *Artemisia*. Medical Hypotheses 23: 187 – 193.

Siddiqui, M. A., and M. M. Alam. 1987a. Control of plant parasitic nematodes by intercropping with *Tagetes minuta*. Nematologia Mediterranea 15: 205 – 211.

Siddiqui, M. A., and M. M. Alam. 1987b. Efficacy of seed dressing with extracts of neem and Persian lilac against *Meloidogyne incognita* and *Rotylenchulus reniformis*. Nematologia Mediterranea 15: 399 – 403.

Siddiqui, M. A., and M. M. Alam. 1988a. Control of root-knot and reniform nematodes by bare-root dip in leaf extracts of margosa and Persian lilac. Zeitschrift fur Pflanzenkrankheiten und Pflanzenschutz 95: 138 – 142.

Siddiqui, M. A., and M. M. Alam. 1988b. Effect of latex seed dressing on *Rotylenchulus reniformis* and plant growth of some vegetables. Nematologia Mediterranea 16: 129 - 130.

Siddiqui, M. A., and M. M. Alam. 1988c. Effect of seed dressing with plant latex on *Tylenchorhynchus brassicae* and plant growth of cabbage and cauliflower. Pakistan Journal of Nematology 6: 65 - 71.

Siddiqui, M. A., and M. M. Alam. 1989a. Control of stunt nematode by bare-root dip in leaf extracts of margosa and Persian lilac. Pakistan Journal of Nematology 7: 33 - 38.

Siddiqui, M. A., and M. M. Alam. 1989b. Effect of root-exudates of neem and Persian lilac on plant parasitic nematodes. Anzeiger fur Schadlingskunde, Pflanzenschutz, Umweltschutz 62: 33 - 35.

Silva, G. S., S. Ferraz, and J. M. Santos. 1990a. Efeito de *Crotalaria* spp. sobre *Meloidogyne javanica*, *M. incognita* raça 3 e *M. exigua*. Fitopatologia Brasileira 15: 94 - 96.

Silva, G. S. da, S. Ferraz, and J. M. dos Santos. 1990b. Histopatologia de raízes de *Crotalaria* parasitadas por *Meloidogyne javanica*. Fitopatologia Brasileira 15: 46 - 48.

Silva, G. S. da, S. Ferraz, and J. M. dos Santos. 1989a. Atração, penetração e desenvolvimento de larvas de *Meloidogyne javanica* em raizes de *Crotalaria* spp. Nematologia Brasileira 13: 151 - 163.

Silva, G. S. da, S. Ferraz, and J. M. dos Santos. 1989b. Resistência de espécies de *Crotalaria* a *Pratylenchus brachyurus* e *P. zeae*. Nematologia Brasileira 13: 81 - 86.

Silva, G. S. da, S. Ferraz, and J. M. dos Santos. 1989c. Resistência de espécies de *Crotalaria* a *Rotylenchulus reniformis*. Nematologia Brasileira 13: 87 - 92.

Singh, R. P., S. S. Tomar, C. Devakumar, B. K. Goswami, and D. B. Saxena. 1991. Nematicidal efficacy of some essential oils against *Meloidogyne incognita*. Indian Perfumer 35: 35 - 37.

Singh, R. P., S. S. Tomar, C. Devakumar, and D. B. Saxena. 1990. Plant products - a versatile source of agro-chemicals. Pesticide Research Journal 2: 61 - 68.

Sitaramaiah, K., and K. N. Pathak. 1979. Effect of phenolics and an aromatic acid on *Meloidogyne javanica* infecting tomato. Nematologica 25: 281 - 287.

Sitaramaiah, K., and K. N. Pathak. 1981. Effect of growth regulators, phenolics and an aromatic acid on root-knot severity (*Meloidogyne incognita* and *M. javanica*) on tomato. Zeitschrift fur Pflanzenkrankheiten und Pflanzenschutz 88: 651 - 654.

Slootweg, A. F. G. 1956. Root rot of bulbs caused by *Pratylenchus* and *Hoplolaimus* spp. Nematologica 1: 192 - 201.

Soule, J. A., and J. Janick. 1996. Novel annual and perennial *Tagetes*. Progress in new crops: Proceedings of the Third National Symposium, Indiana, USA, 22 - 25 October, 1996. Pp. 546 - 551.

Starr, J. L. 1979. Peroxidase isozymes from *Meloidogyne* spp. and their origin. Journal of Nematology 11: 1 - 5.

Starr, J. L., and M. C. Black. 1995. Reproduction of *Meloidogyne arenaria*, *M. incognita,* and M. javanica on sesame. Supplement to Journal of Nematology 27: 624 - 627.

Steiner, G. 1941. Nematodes parasitic on and associated with roots of marigolds (*Tagetes* hybrids). Proceedings of the Biological Society of Washington 54: 31 - 34.

Stephan, Z. A., A. A. Al Askari, and B. G. Antoon. 1989. Effect of *Haplophyllum tuberculatum* plant extract on root-knot nematode. International Nematology Network Newsletter 6: 31 –32.

Stoyanov, D. 1971. Control of plant-parasitic nematodes by rotation and by fallowing. Revista de Agricultura, Cuba 4: 75 –80.

Stoyanov, D. 1973. Control of the parasitic nematodes of bananas by means of crop rotation and their survival in soil without a host plant. Serie Agricultura, Academia de Ciencias de Cuba 20: 1 –8.

Sturhan, D. 1963. Der pflanzenparasitische nematode *Longidorus maximus*, seine biologie und oekologie, mit untersuchungen an *L. elongatus* und *Xiphinema diversicaudatum*. Zeitschrift fur Angewandte Zoologie 50: 129 –193.

Suatmadji, R. W. 1969. Studies on the effect of *Tagetes* species on plant parasitic nematodes. Dissertation, Wageningen Agricultural University. The Netherlands. 133 pp.

Sukul, N. C. 1970. Nematocidal properties of two species of polygonaceous plants. Proceedings of the Zoological Society of Calcutta 23: 139 –146.

Sukul, N. C., S. P. Sinhababu, S. C. Datta, B. Nandi, and A. Sukul. 2001. Nematotoxic effect of *Acacia auriculiformis* and *Artemisia nilagirica* against root-knot nematodes. Allelopathy Journal 8: 65 –72.

Sundarababu, R., C. Sankaranarayana, and S. Vadivelu. 1990. Nematode management with plant products. Indian Journal of Nematology 20: 177 –178.

Supratoyo, M. 1993. Studies on the effect of *Tagetes erecta* and *Tagetes patula* for controlling plant parasitic nematodes on banana. Ilmu Pertanian 5: 681 –691.

Sweelam, M. E. 1989. The potential use of some ornamental plants for nematode control in Egypt. Bulletin of Faculty of Agriculture, University of Cairo 40: 391 –393.

Sweet, R. 1817. *Tagetes florida*. Brit. Fl. Gard., Ser. I. 5. t. 35.

Tada, M., Y. Hiroe, S. Kiyohara, and S. Suzuki. 1988. Nematicidal and antimicrobial constituents from *Allium grayi* Regel and *Allium fistulosum* L. var. *caespitosum*. Agricultural and Biological Chemistry 52(9): 2383 –2385.

Takasugi, M., Y. Yachida, M. Anetai, T. Masamune, and K. Kegasawa. 1975. Identification of asparagusic acid as a nematicide occurring naturally in the roots of asparagus. Chemical letters: 43 –44.

Tanda, A. S., A. S. Atwal, and Y. P. S. Bajaj. 1988. Antagonism of sesame to the root-knot nematode (*Meloidogyne incognita*) on okra in tissue culture. Nematologica 34: 78 –87 (publ. 1989).

Tanda, A. S., A. S. Atwal, and Y. P. S. Bajaj. 1989. In vitro inhibition of root-knot nematode *Meloidogyne incognita* by sesame root exudate and its amino acids. Nematologica 35: 115 –124.

Taylor, C. E., and a. F. Murant. 1966. Nematicidal activity of aqueous extracts from raspberry canes and roots. Nematologica 12: 488 –494.

Tenente, R. C. V., L. G. E. Lordello, and J. F. S. Dias. 1982. Estudo com a excreção radicular de mucuna preta na eclosão de larvas, porcentagem de penetração e crescimento de *Meloidogyne incognita* race 4. Trabalhos apresentados à V Reunião de Nematologia 5:271 –284.

Timchenko, L. S., and T. K. Maiko. 1989. Nematicidal properties of plants-antagonists of nematodes of decorative plants. Byulleten' Vsesoyuznogo Instituta Gel'mintologii im. K. I. Skryabina No. 50: 81 – 84.

Tiyagi, S. A., A. Ahmad, and M. M. Alam. 1990. Control of root-knot, reniform and stunt nematodes by root dip in leaf extract of lemongrass. International Pest Control 32: 70 – 71.

Tiyagi, S. A., M. A. Siddiqui, and M. M. Alam. 1986. Toxicity of an insect-repellent plant to plant-parasitic nematodes. International Nematology Network Newsletter 3: 16 – 17.

Topp, E., S. Millar, H. Bork, and M. Welsh. 1998. Effects of marigold (*Tagetes* sp.) roots on soil microorganisms. Biology and Fertility of Soils 27: 149 – 154.

Tracy, S. M., and H. S. Coe. 1918. Velvet beans. Washington, D. C.: United States Department of Agriculture. Farmers bulletin n°962.

Tsai, B. Y., J. West, S. D. Van Gundi, and E. Rodrigues. 1991. Screening plants for nematicidal agents. Pp. 1 – 26. In I. Kudu and M Jacobson, eds. Phytochemical Pesticides. Oxford: CRC Press.

Tsao, R., Q. Yu, I. Friesen, J. Potter, and M. Chiba. 2000. Factors affecting the dissolution and degradation of oriental mustard-derived sinigrin and allyl isothiocyanate in aqueous media. Journal of Agricultural and Food Chemistry 48: 1898 – 1902.

Uhlenbroek, J. H., and J. D. Bijloo. 1958. Investigations on nematicides. I. Isolation and structure of a second nematicidal principle occurring in *Tagetes* roots. Recueil des Travaux Chimiques des Pays-Bas 77: 1004 – 1008.

Uhlenbroek, J. H., and J. D. Bijloo. 1959. Investigations on nematicides. II. Structure of a second nematicidal principle isolated from *Tagetes* roots. Recueil des Travaux Chimiques des Pays-Bas 78: 382 – 390.

Valle, L. A. C. do, W. P. Dias, and S. Ferraz. 1996a. Reação de algumas espécies vegetais, principalmente leguminosas, ao nematóide de cisto da soja, *Heterodera glycines* Ichinohe. Nematologia Brasileira 20: 30 – 40.

Valle, L. A. C. do, S. Ferraz, W. P. Dias, and D. A. Teixeira. 1996b. Controle do nematóide de cisto da soja, *Heterodera glycines* Ichinohe, com gramíneas forrageiras. Nematologia Brasileira 20: 1 – 11.

Varma, M. K., H. C. Sharma, and V. N. Pathak. 1978. Efficacy of *Tagetes patula* and *Sesamum orientale* against root knot of eggplant. Plant Disease Reporter 62: 274 – 275.

Vats, R., and S. N. Nandal. 1993a. Effect of different concentrations of leaf extracts of neem and *Eucalyptus* used as bare-root-dip treatment of tomato seedlings against *Meloidogyne javanica*. Current Nematology 4: 15 – 18.

Vats, R., and S. N. Nandal. 1993b. Preliminary studies on efficiency of aqueous plant parts extracts as bare-root-dip treatment of tomato seedlings against *Meloidogyne javanica*. Current Nematology 4: 181 – 185.

Vats, R., and S. N. Nandal. 1995. Evaluation of differently prepared plant extracts as bare-root dip treatment of tomato seedlings against *Meloidogyne javanica*. Crop Research Hisar 9: 123 – 128.

Vernon Carter, E. J., J. T. Ponce Palafox, and R. Pedroza Islas. 1996. Pigmentation of Pacific white shrimp (*Penaeus vannamei*) using Aztec marigold (*Tagetes erecta*) extracts as the carotenoid source. Archivos Latinoamericanos de Nutricion 46: 243 – 246.

Viala, S. S. de, B. B. Brodie, E. Rodriguez, and D. M. Gibson. 1998. The potential of Thiarubrine C as a nematicidal agent against plant-parasitic nematodes. Journal of Nematology 30: 192 – 200.

Villar, E. M. J., and E. Zavaleta-Mejía. 1990. Effect of *Crotolaria longirostrata* Hook y Arnott on root galling nematodes (*Meloidogyne* spp.). Revista Mexicana de Fitopatologia 8: 166 – 172.

Walia, K. K., and D. C. Gupta. 1995. Neem - an effective biocide against *Meloidogyne javanica* attacking vegetable crops. Plant Disease Research 10: 59 – 61.

Wani, A. H. 1992. Control of root-knot nematode on okra with seed soaking in neem leaf extract. Current Nematology 3: 39 – 40.

Wani, A. H., A. P. Ansari, and M. M. Alam. 1994. Nematicidal potential of latex extracts of *Calotropis procera* and *Euphorbia caducifolia* on root-knot nematodes infecting okra. Annals of Plant Protection Sciences 2: 72 – 74.

Welle, H. B. A. 1964. Physostigmine and structural analogues as direct and systemic nematicides. Dissertation, University of Utrecht, The Netherlands, 187 pp.

White, G. A., and J. R. Haun. 1965. Growing *Crotalaria juncea*, a multi-purpose fiber legume, for paper pulp. Economic Botany 19: 175 – 183.

Whittington, D. P., and E. I. Zehr. 1992. Populations of *Criconemella xenoplax* on peach interplanted with certain herbaceous plants. Supplement to Journal of Nematology 24: 688 – 692.

Widhi, S., and P. C. Trivedi. 1995. Nematicidal and nematostatic response of aqueous extracts of certain plants of semi-arid niche. Current Nematology 6: 43 – 53.

Widmer, T. L., and G. S. Abawi. 2000. Mechanism of suppression of *Meloidogyne hapla* and its damage by a green manure of Sudan grass. Plant Disease 84: 562 – 568.

Zavaleta-Mejía, E., and R. O. Gomez. 1995. Effect of *Tagetes erecta* L.-tomato (*Lycopersicon esculentum* Mill.) intercropping on some tomato pests. Fitopatologia 30: 35 – 46.

Zhao, B., and B. G. Zhang. 1996. Nematicidal activity of aloperine against pine wood nematodes. Scientia Silvae Sinicae 32: 243 – 247.

Zinovieva, S. V., Z. V. Udalova, I. S. Vasiljeva, and V. A. Paseschnichenko. 1997. Action of sterol glycosides on *Meloidogyne incognita* infecting tomato and cucumber roots. Russian Journal of Nematology 5: 77 – 80.

Zygadlo, J. A., C. A. Guzman, and N. R. Grosso. 1994. Antifungal properties of the leaf oils of *Tagetes minuta* L. and *T. filifolia* Lag. Journal of Essential Oil Research 6: 617 – 621.

Acknowledgments The authors especially thank Dr. Murilo Geraldo de Carvalho and Dr. Onkar Dev Dhingra for reviewing the manuscript, and to all the nematologists who have generated the information presented in this chapter.

20 Biological Control of Nematodes by Fungal Antagonists

Senyu Chen and Donald W. Dickson

20.1 Introduction

All organisms in an ecosystem are influenced by abiotic and biotic factors. Nematodes are not exceptions. In an undisturbed ecosystem, many nematode populations might be at equilibrium. When humans introduced agriculture into an ecosystem, the equilibrium might be broken and community structure might be changed dramatically so that some nematodes became severe pests of cultivated crops. Nevertheless, these agricultural pests are continuously subjected to attack by a number of natural enemies. The organisms that have adverse effects on nematode populations are collectively called nematode antagonists.

The action of antagonists in maintaining nematode population density at a lower average than would occur in their absence can be called biological control of nematodes. Nematode antagonists have been observed in a wide range of organisms including fungi, bacteria, viruses, richettseae, plants, protozoans, turbellarians, tardigrades, enchytraeids, mites, insects, and nematodes. Among these, fungal antagonists have been most extensively studied and appeared to be the most important in regulating nematode populations in soil. This chapter emphasizes nematode suppression by fungal antagonists with recent advances of the research, and on perspectives on biological control of nematodes using potential fungal agents. More information concerning various aspects is provided in a number of previous reviews, book chapters, and books (Barron, 1977; Dackman, et al., 1992; Davis, 1998; Duddington, 1957; Gray, 1987, 1988; Jatala, 1986; Kerry, 1984, 1986, 1987, 1988, 1990, 1993; Kerry and Jaffee, 1997; Li et al., 2000; Mankau, 1980; Morgan-Jones and Rodríguez-Kábana, 1985, 1987, 1988; Rodríguez-Kábana and Morgan-Jones, 1988; Sikora, 1992; Stirling, 1988, 1991; Tribe, 1977a, 1980).

Fungal antagonists of nematodes have been studied since the first observation of the nematophagous habit of the fungus, *Harposporium anguillulae*, by Lohde in 1874. Linford's (Linford, 1937; Linford, et al., 1938) efforts to use predacious fungi to control plant-parasitic nematodes stimulated interest in the nematode-predacious fungi. Early research done in France, the USA, England, and the former USSR was substantial contributions to our understanding of the taxonomy and ecology of the nematode-predacious fungi, but there has been only limited success in using them as biological control agents against nematodes (Stirling, 1991). The lack of success of biological control of nematodes using fungi was contrasted to the development of highly effective nematicides of fumigants in the 1940 – 1950s and organophosphates and carbamates in the 1950 – 1970s. Consequently, the interests in biological control declined during these periods. The resurgence of interest in biological control occurred in the mid-1970s. This resulted from both the continuing environmental problems associated with the use of nematicides (Kerry, 1993; Stirling, 1991) and demonstrations of suppression of nematodes by fungal parasites. Some efforts have been made to market biological control agents (Cayrol, et al., 1978; Liu, et al., 1996; Timm, 1987; Warrior, et al., 1999) for nematode management, but the products generally have not been accepted or they were used only at a small scale. Recently, more and more evidence showed that biological control of plant-parasitic nematodes with fungal antagonists is promising. Biological control of plant-parasitic nematodes has been at a crucial stage where successful examples of using nematode antagonists in management are needed to warrant continuous supports from public and industries.

20.2 Types of Fungal Antagonists of Nematodes

Fungal antagonists of nematodes are those fungi, including nematophagous and non-nematophagous fungi that have some adverse effect on nematodes. Numerous fungi have been isolated from nematodes. Cooke and Godfrey (1964) published a key that included 97 predacious and endoparasitic fungal species attacking vermiform nematodes. Crump (1991) listed 129 species of fungi isolated from beet, cereal, and potato cyst nematodes. Rodríguez-Kábana and Morgan-Jones (1988) summarized fungal species isolated from root-knot and cyst nematodes. Many more fungal species have been reported recently from nematodes throughout the world. Li et al. (2000) did an extensive taxonomic review of nematophagous fungi.

Nematophagous fungi use nematodes for nutrition. In early literature, nematophagous fungi are generally referred to as predacious (trapping) and endoparasitic fungi. Barron (1977) provided a detailed description of the predacious and endoparasitic fungi of vermiform nematodes. After more knowledge of fungi associated with eggs, females, and cysts of sedentary endoparasitic nematodes, such as Heteroderidae was accumulated, a third group of nematophagous fungi was proposed: opportunistic fungi (Morgan-Jones and Rodríguez-Kábana, 1987). According to the mode of action, the fungal antagonists of nematodes can be grouped into: (1) predacious fungi; (2) endoparasites of vermiform nematodes; (3) parasites of sedentary females and eggs; (4) fungi producing antibiotic substances; and (5) vesicular-arbuscular mycorrhizal (VAM) fungi. There is no clear-cut between these categories. Some individual fungi can belong to different categories.

20.2.1 Predacious Fungi

Predacious organisms capture, kill, and then consume their prey. Like some omnivorous plants, predacious fungi have evolved special devices for capturing animals such as vermiform nematodes. These devices are adhesive hyphae, adhesive branches, adhesive nets, adhesive knobs, non-constricting rings, constricting rings (Fig. 20.1) and stephanocysts (Barron, 1977; Liou and Tzean, 1992). However, killing of nematodes by some predacious fungi may be a slow process and fungi may undergo parasitism for a long period. Based on this point of view, these fungi are considered parasites of nematodes as well.

20.2.1.1 Fungi with Adhesive Hyphae

Adhesive hyphae are usually produced by lower fungi most of which are in the genera *Stylopage* and *Cystopage*, Zygomycetes. Because they have no septa, they cannot produce complex devices for capturing nematodes (Barron, 1977). However, some septate fungi in Deuteromycetes, such as *Arthrobotrys botryospora*, *Dactylaria psychrophila*, and *Arthrobotrys superba*, also can capture nematodes with adhesive hyphae under certain conditions. These fungi produce adhesive materials that are deposited on some points of hyphal surfaces. When a nematode touches these points, it is captured. The hyphae produce appressoria that penetrate through the nematode body wall, and the mycelium grows throughout the nematode body. After consuming all of the nematode contents, the fungus draws all plasma back outside of the nematode body for the development of sporangia and spores.

20.2.1.2 Fungi with Adhesive Branches

Adhesive branches are produced by Deuteromycetes and lower fungi such as *Stylopage*. The behavior of these fungi is similar to that of adhesive hyphae. Adhesive branches are usually made up of one to three cells and normally

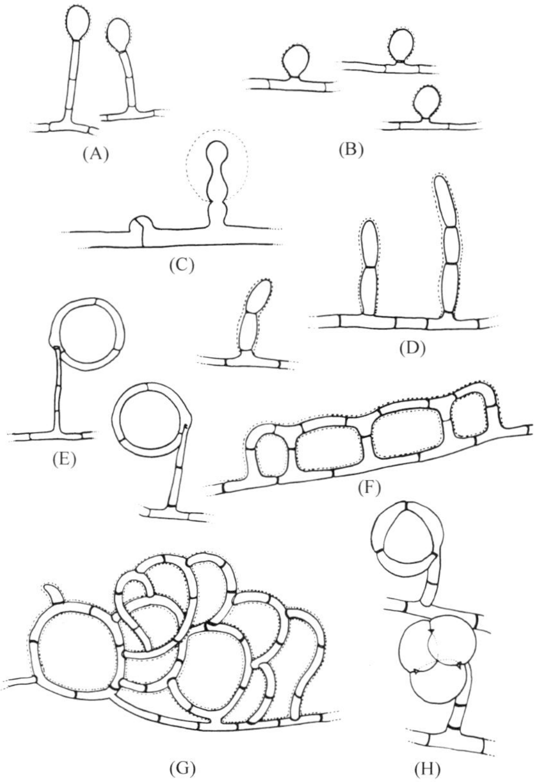

Figure 20.1 Organs of capture in predacious nematophagous fungi. (A) Stalked adhesive knobs. (B) Sessile adhesive knobs. (C) Hour-glass adhesive knob of *Nematoctonus*. (D) Adhesive branches. (E) Non-constricting rings. (F) Two-dimensional scalariform adhesive net. (G) Three-dimensional adhesive net. (H) Constricting rings (Barron, 1977).

anastomose to form simple adhesive hoops or two-dimensional networks. A thin film of adhesive materials is secreted over the entire surface of the branch. The trap is elevated above the substrate. Nematodes that become attached to branches are held fast, and they are quickly penetrated and consumed. The adhesive branches are probably primitive traps from which more complex organs of capture have evolved (Barron, 1977).

20.2.1.3 Fungi with Adhesive Nets

Adhesive nets are common trapping devices of predacious fungi and usually found in the Deuteromycetes. These fungi have been studied intensively to determine their mechanism of infection of nematodes and their efficacy in reducing nematode densities. Nordbring-Hertz (1972) demonstrated the presence of an adhesive substance coating the surface of the capture organs. The adhesive nets are effective; once a nematode is captured, the prey is held fast. If the nematode struggles, it becomes attached to another net and may become attached to several nets or branches (Barron, 1977; Gray, 1988). Large nematodes as well as small nematodes can be captured by this device. Once a nematode is captured, a penetration hypha enters the nematode within 1 h and then swells to form an infection bulb. From this structure assimilative hyphae develop.

20.2.1.4 Fungi with Adhesive Knobs

Adhesive knobs are morphologically distinct cells, and they are considered to be highly specialized trapping devices. These knobs are covered by a thin layer of adhesive materials and are either sessile or borne at the apex of a short non-adhesive stalk. When a nematode touches the knobs, a flattened mass of adhesive is produced at the point of attachment, forming a thick pad. This increases the area of attachment many folds, ensuring that the nematode is firmly held (Gray, 1988). Adhesive knobs contain numerous electron-dense spherical bodies close to the part of the cell wall where is likely to come into contact with a nematode (Wimble and Young, 1984). Adhesive knobs are normally quite closely spaced along hyphae, and when a nematode is caught it is quite normal for it to become attached to several other knobs as it struggles to free itself, making escape impossible (Gray, 1988). In the struggle to escape, the nematode usually breaks the knob off and carries the knob on its body surface. The knob then penetrates the host and the nematode is parasitized. The adhesive knobs are found in Deuteromycetes and Basidiomycetes. Deuteromycetes, such as *Dactylaria candida*, produce a globose infection bulb after penetration. From this bulb, the assimilative hyphae arise to colonize and digest the host contents. Basidiomycetes, such as *Nematoctonus* sp., do not form an infection bulb after penetration. This group of fungi is unique in forming the clamp connections on the secondary hyphae.

20.2.1.5 Fungi with non-constricting rings

Non-constricting rings are produced by erect, lateral branches arising from the prostrate septate hyphae of Deuteromycetes, such as *Dactylaria*. Fungi that produce non-constricting rings, as exemplified by *D. candida* and *Dactylaria lysipaga*, often produce adhesive knobs as well (Drechsler, 1937). Knobs and rings are commonly found alternating on the same hyphal strand (Gray,

1988). Also, an adhesive layer, similar to those associated with adhesive nets and adhesive knob traps, has been observed on the inside of the ring (Dowsett and Reid, 1977). The ring is composed of three cells. When a nematode enters the ring and if its diameter is only a little larger than the inner diameter of the ring, the ring will hold the nematode. The connection between the ring and stalk is very delicate and the nematode usually will break off the ring in its struggle to escape. One nematode may carry several rings around its body. Finally the fungus penetrates the nematode body wall and consumes the nematode.

20.2.1.6 Fungi with Constricting Rings

The most sophisticated device for predacious fungi is the constricting ring formed by species of *Dactylella, Dactylaria*, *Arthrobotrys*, and a few other Deuteromycetes. The constricting rings, which also consist of three cells, are similar to non-constricting rings in morphology. However, unlike the non-constricting ring, the constricting ring is borne on a strong stalk that is not easily broken. The size of the ring varies between and within species, but generally falls within a range of 20 μm to 40 μm (Gray, 1988). When the constricting ring is stimulated, the three cells expand rapidly inward. If a nematode moves into the ring, the inner walls of the ring cells will be triggered to close, and the nematode will be pinched and have no chance to escape (Fig. 20.2). The mechanisms involved in closure of the rings, however, are debatable. Both mechanical and hot water, but not chemical stimulation have

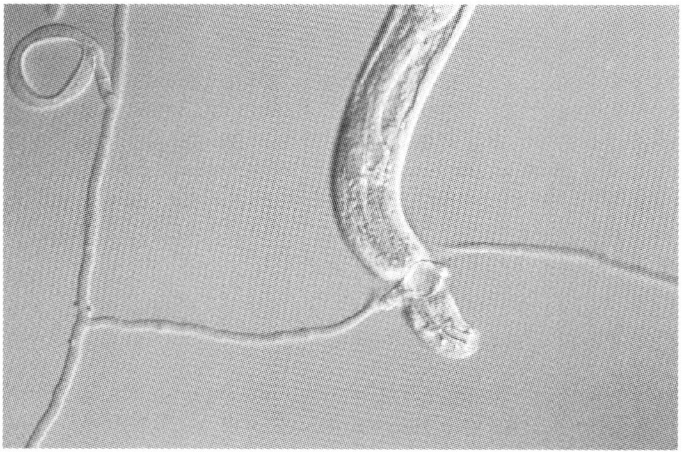

Figure 20.2 Nematode trapped in a constricting ring of *Arthrobotrys dactyloides*. Courtesy of B. A. Jaffee.

triggered the swelling of the constricting rings (Couch, 1937; Müller, 1958). Couch (1937) believed that the closure was affected by swelling of a colloidal substance. The swelling may be due to a rearrangement of molecules of water and colloidal material already in the cell, or it might be that additional water is imbibed from the stalk and thread cells. Müller (1958) argued that colloidal material could not account for ring closure but rather stimulation of the inner walls of the three cells changed the permeability of the cell membranes and the cells absorbed water from the surrounding medium, causing the ring to close.

20.2.1.7 Stephanocysts

Stephanocysts are unique reporductive structures, apparently a kind of conidia or spores, present in Basidiomycetes, especially in the genus *Hyphoderma* (Liou and Tzean, 1992). A typical stephanocyst consists of a cup-like basal cell and a terminal globose cell. At the juncture of the two cells, a row of spines surrounds the circumference. Liou and Tzean (1992) observed that the stephanocysts also function as nematode-trapping devices. The two-celled stephanocysts are adhesive, and can attach to passing nematodes while attached to or detached from the hyphae or short stalks. Once a nematode is captured, the fungus forms infective peg to penetrate, and then consumes the prey nematode.

20.2.2 Endoparasites of Vermiform Nematodes

Endoparasitic fungi are different from the predacious fungi in that they have no special trapping devices. Most endoparasitic fungi of vermiform nematodes are obligate parasites or have limited saprophytic ability. They have no extensive hyphal development outside the body of nematodes. Some fungi attacking vermiform nematodes, however, are facultative parasites and can undergo saprophytic activity without nematodes.

20.2.2.1 Obligate Endoparasites of Vermiform Nematodes

The obligate endoparasitic fungi that attack vermiform nematodes can be found in lower and higher fungi. They can be placed into four groups based on their mechanism of infection: (1) encysting species; (2) species forming adhesive conidia; (3) species with conidia that may be ingested (Barron, 1977; Gray, 1988); and (4) species with gun cells (Barron, 1987; Robb and Barron, 1982).

Encysting species Encysting species that produce zoospores belong to Chytridiomycetes and Oomycetes. The zoospores can swarm for a short time, and if they reach their host nematode they encyst on nematode cuticles. *Catenaria* spp., which belong to Chytridiomycetes, are the most common endoparasitic fungi encountered on vermiform nematodes. Their zoospores have one posterior whiplash flagellum, and sporangia are formed in chains within the nematode body. *Myzocytium* spp. (Oomycetes), which also can parasitize nematodes, are different from *Catenaria* spp. in that they produce oospores and zoospores with two flagella, one tinsel and the other whiplash.

Species forming adhesive conidia The endoparasitic fungi forming adhesive conidia are found in Zygomycetes, Basidiomycetes, and Deuteromycetes. Species of *Meristacrum* (Zygomycetes) form conidia that are forcibly discharged at their maturity. If the primary conidia fail to contact a nematode, they may germinate and produce a secondary conidium. Both primary and secondary conidia have a layer of sticky material on their surfaces. The endoparasitic Basidiomycetes such as some species of *Nematoctonus* also infect nematodes using adhesive conidia, but the conidia formed by species of *Nematoctonus* are similar to the adhesive knobs formed on hyphae by the predacious species. Most endoparasitic fungi forming adhesive conidia were found in Deuteromycetes such as many well-known species in the genera *Verticillium*, *Cephalosporium*, *Acrostolagmus*, and *Hirsutella* (Fig. 20. 3). The infective conidia produced by these fungi either have adhesive mucilage in parts of the surface of the conidia or have mucous sheath enveloping conidium. Although the saprophytic abilities of these fungi in natural soil are not well studied, most of these fungi can be extensively cultured on artificial media and produce infective conidia. Therefore they may be considered facultative endoparasites. *Hirsutella rhossiliensis*, however, behaves as an obligate endoparasite in nature soil because it will go extinct without supplying a minimum number of host nematodes (Jaffee and Zehr, 1985).

Species with conidia that may be ingested Some species of endoparasites have developed morphologically adapted conidia, which, when ingested by a nematode, become lodged in either its buccal cavity or esophagus. The species belong almost exclusively to the genus *Harposporium* (Deuteromycetes). The conidia are crescent-shaped, helicoid, or irregular-shaped, and they are well adapted to infection. These fungi cannot infect plant-parasitic nematodes because the lumen of the stylet is too small to ingest the conidia.

Species with gun cells Species of *Haptoglossa* (Oomycetes) have evolved a special infective cell called "gun" cell to attack nematode or rotifer hosts (Barron, 1987; Beakes and Glockling, 1998; Robb and Barron, 1982). The biflagelate zoospores are discharged from zoosporangium, swim a short distance, and encyst on substratum within a few minutes (Fig. 20. 4(A), (B)). After 2 –3 h of quiescence, the encysted spore undergoes synchronous germination, initially producing a narrow germ tube (Fig. 20. 4(C)). The germ tube soon begins to swell and within an hour or so forms the ovoid to

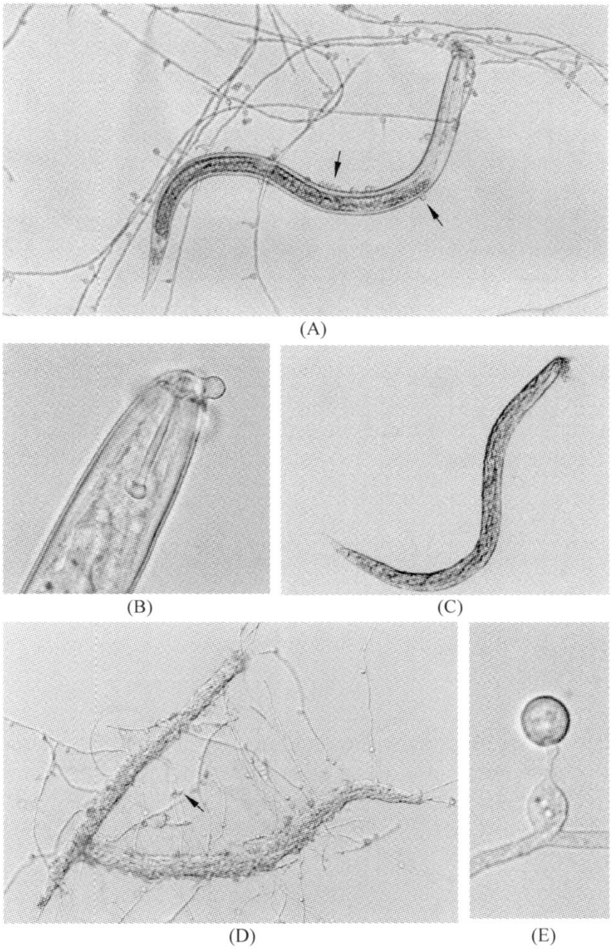

Figure 20.3 Parasitism of second-stage juveniles (J2) of *Heterodera glycines* by the endoparasitic fungus *Hirsutella minnesotensis*. (A) J2 with conidia attached to the cuticle on cornmeal agar, arrows. (B) J2 with attached conidium that is penetrating the nematode cuticle. (C) J2 filled with mycelium. (D) The fungus growing from the nematode body and sporulate; a conidiogenous cell with an attached conidium, arrow. (E) A conidiogenous cell with a conidium. Modified from Chen et al. (2000b).

slightly tapering gun cell, which is delimited from the early empty cyst by the thin cross wall (Fig. 20.4(D)). The infective cell takes about another 2–3 h to become mature, and by then the original cyst often appears collapsed and the gun cell has acquired a prominent apical beak, often held at an acute angle to the main cell body (Fig. 20.4(E)) (Beakes and Glockling, 1998). The gun cell often anchors on substrate in the manner of a cannon. Fig. 20.5 illustrates a cannon-like gun cell with descriptions of structures (Barron, 1987). Once a nematode contacts the gun cell, it triggers the cell to fire a "missile" (projectile) through the host cuticle and then serve as a hypodermic conduit through which the parasite cytoplasm is injected into the host (Fig. 20.4(F), (G)) (Beakes and Glockling, 1998).

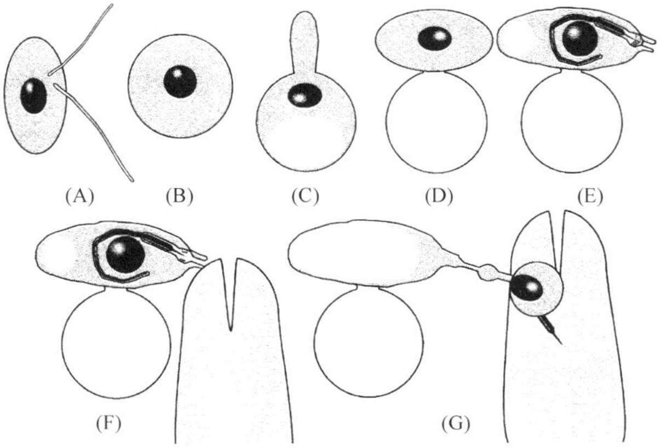

Figure 20.4 Schematic diagrammatic summary of gun cell differentiation in a zoosporic *Haptoglossa*. (A) Zoospore. (B) Cyst. (C) Cyst germination. (D) Delimited infection cell. (E) Mature infection cell. (F) Triggered infection cell. (G) Fired infection cell and infected nematode (Beakes and Glockling, 1998).

20.2.2.2 Facultative Parasites of Vermiform Nematodes

Whereas the saprophytic ability of many fungal endoparasites of vermiform nematodes in nature is unclear, a few fungi such as the species *Catenaria anguillulae* can grow saprophytically on dead or injured nematodes (Sayre and Keeley, 1969). As discussed above, many fungi of Deuteromycetes that produce adhesive conidia can grow in various substrates and produce infective conidia. It is probably appropriate to group these fungi with facultative parasites. A recently described fungus *Esteya vermicola* that was isolated from the pinewood nematode *Bursaphelenchus xylophilus* produced infective, adhesive conidia when cultured on poor nutrition media or grown from parasitic

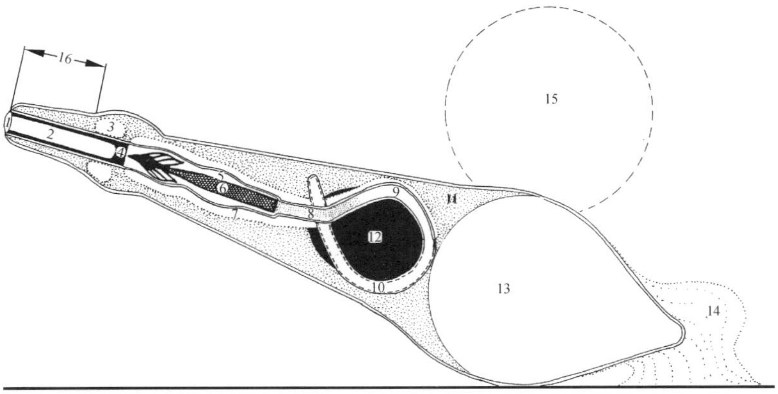

Figure 20.5 Diagrammatic representation of the gun cell of *Haptoglossa mirabilis*. 1. Muzzle. 2. Bore with extra wall layer. 3. Apical vacuolar system. 4. Plug. 5. Projectile chamber. 6. Projectile. 7. Lateral vacuolar system. 8–10. Lower flexuous tube. 11. Protoplasm. 12. Basal vacuole. 14. Adhesive pad. 15. Empty zoospore cyst. 16. Apical cone (Barron, 1987).

nematode, and produced non-infective, non-adhesive conidia when cultured on enriched media (Liou, et al., 1999). Like predacious fungi, *E. vermicola* might have adapted to both saprophytic and parasitic growth. Some predacious fungi, such as those with adhesive knobs, act very similarly to these endoparasitic fungi that form adhesive conidia. Consequently, the predacious fungi with adhesive knobs can be considered facultative endoparasites of vermiform nematodes. Many fungi, such as *Verticillium chlamydosporium*, that colonize cysts or egg masses also can attack second-stage juveniles (J2) within the cysts or egg masses. Although the mode of infection of juveniles in cysts or egg masses is poorly understood, these fungi are probably facultative parasites of vermiform juveniles.

20.2.3 Parasites of Sedentary Females and Eggs

Sedentary nematode females, cysts, eggs, and egg masses are also subjected to the attack by fungi (Fig. 20.6). Unlike motile vermiform nematodes that may actively move toward and contact predacious or endoparasitic fungi, sedentary stages of nematodes may have no chance to contact parasitic fungi unless the fungi have some mechanism to reach the nematodes. Fungi attaching sedentary nematodes diversify biologically and ecologically. While a few of them are obligate parasites, most fungi in this group can live in soil as saprophytes. Mechanisms of attacking nematodes differ between obligate parasites and facultative parasites.

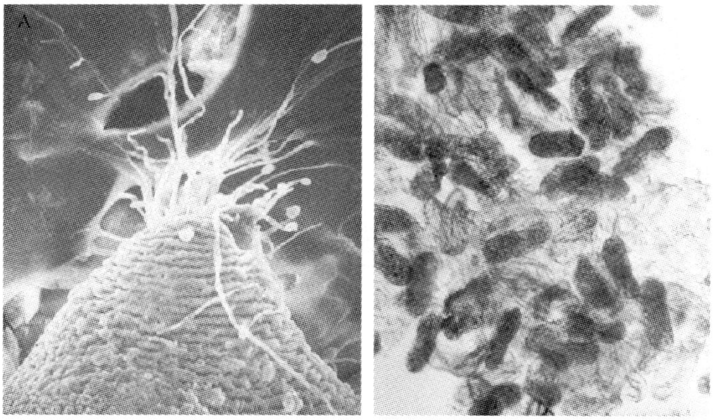

(A) (B)

Figure 20.6 Nematode cysts (A) and eggs (B) colonized by fungi.

20.2.3.1 Obligate Parasites of Sedentary Females and Eggs

Obligate parasites of sedentary females and eggs cannot develop mycelium in soil. Therefore, these fungi must have a motile or disseminative stage in order to reach the target nematode. Three species, *Catenaria auxiliaris*, *Nematophthora gynophila*, and an undescribed lagenidiaceous fungus, have been reported on cyst nematodes (Kerry and Crump, 1980). All of these fungi produce thick-walled resting spores that remain in soil from one season to the next. Zoospores are probably the infective stage (Stirling, 1991; Tribe, 1977b). The zoospores move toward nematode females, and when they contact the target nematode, they penetrate the body wall and assimilate the nematode body contents. *Catenaria auxiliaris* and *N. gynophila* also attack nematode eggs within the female body (Tribe, 1977b).

20.2.3.2 Facultative Parasites of Eggs and Sedentary Females

Numerous fungi have been isolated from eggs, sedentary females, egg masses, or cysts of Heteroderidae. Some of them may not be parasitic species, while others are facultative parasites. The facultative parasitic fungi produce an expanding mycelium growth that enables them to reach the nematode targets of sessile stages. These fungi belong to taxonomic divergent groups of Chytridiomycetes, Basidiomycetes, and Deuteromycetes, but they are encountered more frequently in Deuteromycetes.

The existence of mycofloras associated with eggs, egg masses, females, or cysts of nematodes belonging to the Heteroderidae has been known for more than a century. As early as 1877, Kühn reported a fungus, *Tarichium auxiliarum*, that parasitized females of *Heterodera schachtii* (Morgan-Jones and Rodríguez-Kábana, 1988). During the 1920s and 1930s, a large number of cysts in several surveys were examined in Europe for fungal parasites of the cyst nematodes, *H. schachtii* and *Heterodera avenae* (Morgan-Jones and Rodríguez-Kábana, 1988). Recently, studies on the species and frequency of fungi associated with cyst and root-knot nematodes have increased. Rodríguez-Kábana and Morgan-Jones (1988) reviewed the fungi isolated up to 1988 from Heteroderidae collected from Australia, Europe, North America, and South America. Since then, many more fungal species have been reported from these nematodes collected in various geographic locations. The fungal species most commonly isolated are limited to a few genera, indicating there is a certain degree of specialization towards the nematodes. These fungi include *Acremonium*, *Alternaria*, *Catenaria*, *Cylindrocarpon*, *Exophiala*, *Fusarium*, *Gliocladium*, *Nematophthora*, *Paecilomyces*, *Penicillium*, *Phoma*, and *Verticillium*. Besides these fungal genera, *Neocosmospora vasinfecta* and *Stagonospora heteroderae* were frequently isolated from *Heterodera glycines* in warm climatic conditions (Chen, et al., 1994).

20.2.4 Fungi Producing Antibiotic Substances

Many fungi isolated from cysts and egg masses are probably saprophytic. Their effects on nematodes are not clear. Presumably, some of these fungi produce substances toxic to nematodes or their existence in egg masses or cysts inhibits or stimulates hatching of juveniles from eggs. Toxic effects of fungal culture filtrates on vermiform nematodes and eggs have been reported in a number of studies from several fungi such as species of *Paecilomyces*, *Verticillium*, *Fusarium*, *Aspergillus*, *Trichoderma*, *Myrothecium* and *Penicillium*.

A few studies have been done to characterize toxic compounds produced by fungi. *Paecilomyces lilacinus* releases chitinase and protease, which can cause undifferentiated *Meloidogyne hapla* eggs to become deformed and vacuolate (Fitters, et al., 1992). Meyer et al. (2000) reported that non-enzymic factors produced by *Trichoderma virens* (syn. *Gliocladium virens*) inhibited *Meloidogyne incognita* egg hatch and J2 mobility. Toxins produced by *Fusarium* spp. were tested on *M. incognita* and some were highly toxic to the nematode (Ciancio, et al., 1988). An antibiotic from *Cylindrocarpon olidum* was isolated, purified and characterized (Coosemans, 1991). It showed good nematicidal activity and had a low toxicity towards vertebrates. Purified extracts of *Penicillium* sp., *Penicillium oxalicum*, *Penicillium anatolicum* and *Aspergillus niger* showed high nematicidal activity at 100 ppm and 200 ppm (Molina and Davide, 1986).

Fungi that produce antibiotic substances may be common in soil. Many more soil fungi that are antagonistic to nematodes through the release of toxins, antibiotics or enzymes remain to be discovered. Nematode density has been correlated negatively with activities of chitinase, collagenase, and proteinase of some soil microorganisms (Mian, et al., 1982; Rodríguez-Kábana, et al., 1989), including fungi such as *Cunninghamella elegans* (Galper, et al., 1991). Kloepper et al. (1991) observed that plants with properties antagonistic to plant-parasitic nematodes have rhizosphere microflora distinct from those of host plants, and they also found that greater numbers of microorganisms in the rhizosphere of antagonistic plants were chitinolytic. Whether the mycofloras are involved in the antagonism of the plants to nematodes through the release of substances is worthy of study.

20.2.5 Vesicular-arbuscular Mycorrhizal Fungi

The vesicular-arbuscular mycorrhizal (VAM) fungi are endomycorrhizal fungi that invade deeply into roots. All VAM fungi belong to the order Glomales (Zygomycetes). The symbiotic association is obligate for these fungi, and they have not been successfully cultured apart from their hosts. During the past two decades many studies on effects of VAM fungi on nematodes have been reported. The roles of VAM fungi in regulating nematode populations and their modes of action have not been elucidated fully. The response of nematodes to VAM fungi varies and may depend on the specific association, soil nutrient level, and the timing of the observation (Ingham, 1988). Both antagonistic and beneficial effects of VAM on nematode populations have been reported. The VAM fungi may compete for nutrition and root space, modify root exudates, alter plant physiology, colonize nematode feeding sites, reduce the number of giant cells, or release nematoxin or antibiotics (Ingham, 1988; Surech, et al., 1985; Bagyaraj, et al., 1979). On the other hand, VAM fungi may improve plant growth and offset the yield loss normally caused by nematode parasitism, and increase food sources for nematodes thereby increase nematode populations. Francl and Dropkin (1985) reported that *Glomus fasciculatum* could parasitize *H. glycines* eggs but was not sufficient to effectively reduce nematode population density. As a group, the VAM fungi cannot be placed in any group discussed above and thus we consider them a unique type of fungal antagonists. More detailed reviews on interaction between VAM fungi and nematode populations have been provided by Ingham (1988).

20.3 Modes of Infection of Nematodes by Fungi

The knowledge of mode of infection of nematodes by fungi is important for using them as biological control agents. Most research on mode of infection of nematodes was conducted on predacious and endoparasitic fungi that attack vermiform nematodes. The relationships of *Arthrobotrys oligospora* or *Drechmeria coniospora* with nematodes have been used as models for studying the recognition mechanism of fungal infection. Various aspects of modes of infection have been intensively reviewed (Dackman, et al., 1992; Dijksterhuis, et al., 1994; Nordbring-Hertz, 1988; Nordbring-Hertz, et al., 1981; Tunlid, et al., 1992).

20.3.1 Attraction

Attraction between nematodes and nematophagous fungi was observed in many studies. The attraction of fungi to nematodes increased with increasing dependence of the fungi on nematodes for nutrients (Jansson and Nordbring-Hertz, 1979, 1980). Several factors, such as carbon dioxide and various organic and inorganic substances, have been thought to be involved in nematode attraction (Gray, 1988). A volatile or a small rapidly diffusing compound was continuously produced in non-spontaneous trap formers (fungi insensitive in trap formation in response to organic matter, including net-forming species). Spontaneous trap formers (fungi sensitive in trap formation in response to organic matter, including branch-, knob- and ring-forming species) and endoparasites produced larger, or less volatile, slowly diffusing compounds, which attract nematodes (Gray, 1988; Jansson and Nordbring-Hertz, 1979). Sialic acid located on the head region of nematodes was found to be involved in chemotaxis of zoospores to nematodes (Jansson and Nordbring-Hertz, 1984). The motile zoospores of endoparasitic fungi are thought to be attracted to nematodes as a result of the chemical gradients formed by exudates released from nematode body orifices. It is not clear if there is a chemotropic response of facultative fungal parasites of eggs or females towards their sedentary hosts. Zoospores of *C. auxiliaris* (Tribe, 1977b) and *N. gynophila* (Kerry, 1974), however, are thought to move toward nematode females.

20.3.2 Attachment

When nematodes come in contact with nematophagous fungi, some morphological changes and chemical interactions may occur. The binding sites vary among combinations between nematodes and fungi (Durschner-Pelz and Atkinson, 1988). Three patterns of binding sites of nematodes with fungi have been observed: (1) specifically to the head and tail; (2) all over the body; and (3) very sparse or no binding (Jansson, et al., 1985). Zoospores usually attach near natural body openings. Molecular moieties are responsible for recognition between nematodes and nematophagous fungi (Tunlid, et al., 1992). Although it had been suggested previously that predacious fungi were not nematode-specific (Rosenzweig, et al., 1985), the results obtained by Gaspard and Mankau (1987) indicated that host-specificity does exist in the interaction between predacious fungi and nematodes. Binding of lectin-carbohydrates is involved in recognition between predacious fungi and nematodes (Wharton and Murray, 1990). The qualitative and quantitative variation in carbohydrates on the surface of different nematodes may provide more opportunity for ecological specialization of endoparasitic fungi that use lectins in host recognition. Attachment and penetration of nematodes by fungi are two separate events. Blocking the binding sites by saccharides prevented penetration of *Trichostrongylus colubriformis* by *A. oligospora,* but had no effect on capture of the nematode; thus capture and penetration may be two distinct processes, with capture being less specific than penetration. Freeze-dried conidia of *D. coniospora* attached to *Caenorhabditis elegans* and *Panagrellus redivivus* as readily as did untreated conidia, but infection was significantly delayed (Zuckerman, et al., 1988).

20.3.3 Penetration

Many light and electron microscopic studies have been published showing the penetration of nematode cuticle, mainly vermiform nematodes, by different nematophagous fungi (Jansson and Nordbring-Hertz, 1988). Whether penetration of the cuticle is enzymatic or mechanical, or both, is not fully understood. Organelles such as endoplasmic reticulum in infective hyphae proliferated at a high rate during the infection process, which suggests the involvement of enzymatic activities (Veenhuis, et al., 1989). Some studies, however, have shown that these organelles are not involved in either the secretion of adhesives or the lysis of the cuticle (Jansson and Nordbring-Hertz, 1988). Protein changes of *Caenorhabditis elegans* and *D. coniospora* were observed during infection (Coles, et al., 1989). *Hirsutella rhossiliensis* spores adhere to vermiform nematode cuticle before they penetrate the cuticle. A bulbous infection hypha formed from the spore and secondary hyphae developed from the hypha (Jaffee and Zehr, 1982). Some observations suggested that mechanical force is involved in the penetration of the nematode cuticle (Veenhuis, et al., 1985; Wharton and Murray, 1990). The species of

Haptoglossa penetrates cuticle solely with mechanical force from the gun cell (see Section 20.2.2.1).

Females and cysts may be colonized by one of the following ways: (1) directly through the cuticle; (2) by penetration through the natural body openings of the vulval cone or oral opening; and (3) after colonization of the feeding cell within the root, which may itself cause the death of the female (Kerry, 1988). Only a few fungi have been shown to penetrate the cuticular wall of nematode cysts. Probably, the first study of penetration of cyst wall by fungi was reported by Hanssler and Hermanns (1981) using transmission electron microscopy. In their study, *Verticillium lecanii* penetrated cyst wall of *H. schachtii* only 48 hours after inoculation. It was concluded that lytic enzymes secreted by the fungus played a major role in its penetration of the cyst wall. Inside the cyst cavity, the fungus penetrated eggshells and colonized the juveniles (Hanssler, 1990). Kim et al. (1992) observed that a sterile fungus, designated as Arkansas Fungus 18 (ARF18), penetrated cyst wall and eggs of *H. glycines*. Chen and Dickson (1996) examined 12 fungal species with the aid of light, and scanning and transmission election microscopy for their ability to penetrate the cyst wall of *H. glycines*. Nine of them penetrated the cyst wall from inside the cyst. At least three species, *Exophiala pisciphila*, *Pyrenochaeta terrestris*, and *Fusarium oxysporum*, also penetrated the cyst wall from the outside. Generally, penetration could be completed by a single hypha (Fig. 20.7). The penetration hyphal pegs within the cyst wall were generally less that 1 μm (usually about 0.5 μm) in diameter, which is much less than the diameter of a regular hypha. Numerous organelles were observed in penetration hyphae of *P. terrestris*. When penetration of the cyst wall was initiated from outside to inside, hypha was observed attached to the cyst wall and appeared to dissolve a hole through the wall. The contents of the fungal cell with condensed organelles extended into the cyst wall, from which several branches of hyphae may have developed. At this stage, the wall of the penetrating hypha was not obvious. The penetrating hyphae of *V. chlamydosporium* in the cyst wall, however, appeared to have cell walls. The ability to penetrate the cyst wall from the outside may not be important for a fungus to colonize a cyst because fungi can readily enter cysts through natural body openings (Kerry, 1988). In fact, fungi were usually found inside cyst cavities within 1 day after inoculation, before any direct penetration of the cyst wall was observed. The cuticular penetration may be important for the infection process to occur on young females. If a fungus can invade young females, the fungus may have more chance to destroy eggs within the cysts because eggs in the early developmental stage are more vulnerable to fungal infection (Chen and Chen, 2003; Irving and Kerry, 1986).

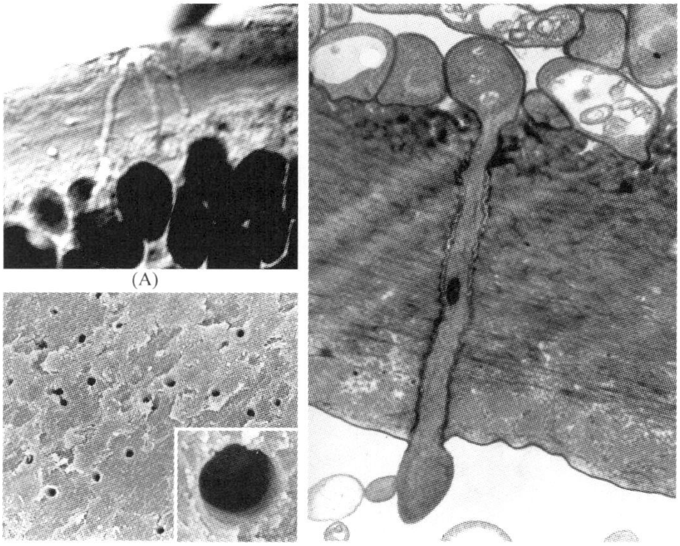

Figure 20. 7 Penetration of cyst wall of *Heterodera glycines* by fungi. (A) *Pyrenochaeta terrestris* initiated penetration at a point from which penetration hyphae developed. (B) *Verticillium chlamydosporium* penetrated cyst wall from inside. (C) and (D) Holes in cyst wall resulting from penetration by *Fusarium oxysporum* (Chen and Dickson, 1996).

Penetration of nematode egg by fungi is completed by either an appressorium or a simple hypha. Appressoria involved in penetration of nematode eggs have been observed in *V. suchlasporium*, *V. chlamydosporium* (Lopez-Llorca and Claugher, 1990; Stirling and Mankau, 1979), *Dactylella oviparasitica* (Stirling and Mankau, 1979), and *P. lilacinus* (Dunn, et al., 1982; Holland, et al., 1999). *Paecilomyces lilacinus* was also able to colonize nematodes through simple hyphal penetration (Dunn, et al., 1982). Zoospores of *C. anguillulae* encysted on nematode eggs by chance and germinated in such a way that the germ tube penetrated the lipid layer.

Enzymatic and mechanical activities may be involved in the penetration of eggs, although the processes are poorly understood. Fungi without any chitinolytic activity probably penetrate only mechanically damaged eggs (Kunert and Lysek, 1987), while chitinolytic activity is probably important in dissolving the eggshell. Recent studies done by Segers et al. (1996, 1999) demonstrated that a subtilisin-like protease designated as VCP1 was involved in infection of nematode eggs. The subtilisin was originally isolated from *V. chlamydosporium,* later also isolated from *Metarhizium anisopliae*, an insect pathogen, and *V. lecanii*, pathogens of nematodes and insects, but not from plant-pathogenic species of *Verticillium*, suggesting its role in the infection of invertebrates. Kunert and Lysek (1987) showed no evidence for the involvement of lipolytic enzymes in the penetration of the nematode egg shell by parasitic fungi.

20.3.4 Pathogenicity of Fungi to Nematodes

Once a fungus penetrates a nematode, the fungus usually proliferates within the nematode, eventually kills the nematode, and consumes the nematode body contents. The pathogenicity, however, varies among nematophagous fungi. Among the predacious fungi, those that form unmodified adhesive hyphae or branches are considered primitive types, whereas fungi forming constricting rings are more evolved and highly aggressive (Gray, 1988). Ingestion of nematodes by predacious fungi is often completed within 48 h to 72 h (Jansson and Nordbring-Hertz, 1988). Most endoparasites of vermiform nematodes are obligate parasites, and they are dependent on nematodes for nutrition. Cayrol and Frankowski (1986) demonstrated that a single *H. rhossiliensis* conidium attached to *Ditylenchus dipsaci* was sufficient for infection and causing death of the nematode. Some fungi isolated from cysts and eggs of Heteroderidae are saprophytic or weak pathogens. Many, however, are capable of parasitizing sedentary nematode females and eggs. Chen et al. (1996b) studied 21 isolates of 18 fungal species for their pathogenicity to eggs of the *H. glycines*. The pathogenicity to the nematode eggs varied among the fungal species.

Variation of host specificity and pathogenicity to nematodes has been observed among isolates within a species of nematophagous fungi in many studies. Tedford et al. (1994) tested 25 isolates of *H. rhossiliensis* and divided the isolates into four groups according to their host preference. Liu and Chen (2001) tested 93 isolates of *H. rhossiliensis* for their pathogenicity to *H. glycines* J2 in the laboratory and in the greenhouse. Although most isolates parasitized high percentages of J2 on water agar and in soil in the laboratory, some of them were weak parasites of the nematode. Isolates from bacteria-feeding nematodes did not parasitize *H. glycines*. Variations in pathogenicity were also observed among strains of egg-parasitic fungi such as *V. chlamydosporium* (Kerry, 1987), *P. lilacinus* (e.g., Stirling and West, 1991), *Cylindrocarpon destructans* (Rodríguez-Kábana and Morgan-Jones, 1988), *F. oxysporum*, *Fusarium solani* (Godoy, et al., 1982), *Gliocladium roseum* (Rodríguez-Kábana, et al., 1984b), *C. anguillulae* (Voss, et al., 1992), and *Acremonium strictum* (Nigh, et al., 1980). Variation in pathogenicity among isolates of a fungal species against nematodes appears to be the rule rather than the exception.

Only a few studies have been reported on biochemical processes of assimilation of nematode body contents by fungi. Hydrolytic enzymes are not likely to have a function in the digestion of nematode body contents. Oxidation of lipids, however, occurs later or during development of new vegetative mycelium (Jansson and Nordbring-Hertz, 1988).

20.4 Suppressive Soils Associated with Fungal Antagonists

Nematodes and their fungal antagonists in soil have co-evolved for a long period of time in natural history. The equilibrium between nematode density and some nematophagous fungi may exist in undisturbed soil or in soil with continuing monoculture. A soil in which a nematode population on a susceptible cultivar maintains at a low level compared to an average infection level in a region can be called nematode suppressive soil. A nematode suppressive soil can be caused by biotic or abiotic factors. A low nematode population is often attributed to some factors such as resistance of host plants or soil environments. Only a few surveys have demonstrated the prevalence of natural biological control agents. The mechanisms resulting in nematode suppressive soils are poorly understood and extent and frequency of natural occurrence are yet to be determined. It is difficult to quantify fungal parasitism of nematodes. Consequently, many reports on suppressive soils associated with fungi were observational rather than quantitative, and the conclusions made from some studies were superficial. A good example of nematode suppressive soils associated with fungal parasites is the decline of cereal cyst nematode in Europe. Reports from other nematodes and regions suggest suppressive soils are common in various agricultural systems throughout the world.

20.4.1 Decline of *Heterodera avenae* in Europe

The decline of the cereal cyst nematode *H. avenae* has been well documented in Europe. This phenomenon was first demonstrated by Gair et al. (1969) in Britain. In a field infested by *H. avenae*, cereals were grown continuously for 13 years. The nematode population density peaked two years after the start of the experiment, then declined rapidly, and maintained low levels for the rest of years of the experiment. Similar population dynamics of *H. avenae* was observed in a microplot study (Kerry and Crump, 1998). This phenomenon was also observed from Germany (Ohnesorge, et al., 1974) and Denmark

(Jakobsen, 1974). Application of formalin at 3000 liters/ha removed the suppressiveness, suggesting some biological factors were responsible for the decline (Kerry and Crump, 1998).

The decline of the cereal cyst nematode population has been attributed to fungal parasites of eggs and females. In Britain, major fungal species included *N. gynophila*, *V. chlamydosporium*, *C. destructans*, and *C. auxiliaris*, but *N. gynophila* and *V. chlamydosporium* were the most important, killing large numbers of females and eggs in many soils (Kerry, 1975; Kerry, et al., 1982a). In eastern Scotland, *Verticillium* sp., *N. gynophila*, *Paecilomyces carneus*, and *C. destructans* were the major species for controlling the population densities of *H. avenae* below economic threshold levels (Boag and Lopez-Llorca, 1989).

20.4.2 Suppression of *Heterodera glycines* by Fungi

The soybean cyst nematode occurs in most soybean producing regions. Decline of the soybean cyst nematode has been reported from the USA and China. Hartwig (1981) observed decline of *H. glycines* population density in breeding microplots and field plots after 5 years of monoculture. He believed pathogens might suppress the nematode populations, but types of pathogens involved were not determined. Carris et al. (1989) surveyed population changes of the soybean cyst nematode over 3 years in two Illinois fields. In one field the nematode maintained a low level without significant damage to the soybean even a susceptible soybean cultivar was grown continuously for 4 years before the study. In the other field, the nematode increased rapidly during the soybean-growing season. In the field with suppressed nematode population, the mycota was more diverse, and frequency of fungal colonization of cysts and frequency of pathogenic fungi were higher than that in the field with high nematode population densities. The data suggested that the fungi might be involved in nematode suppression.

Soil suppressive to the soybean cyst nematode was also observed in a Florida field (Chen, et al., 1994, 1996a). The soybean cyst nematode was introduced into the field in 1985 and the nematode population developed poorly (about one cyst/100 cm^3 soil in 1992) over 7 years of monoculture of soybean, suggesting the soil was suppressive to the soybean cyst nematode. Microwave-heating treatment removed the suppression from the soil in a greenhouse assay. The nematode density was fourfold higher and the number of eggs produced per female was 73% higher in microwave-heating-treated soil compared with untreated soil. The nematode population density was negatively correlated with both percentage of cysts colonized by fungi and fungal parasitism of eggs. A black, yeast-like fungus was frequently isolated from the soil and might be the major fungal parasite but not the only fungus responsible for the suppression of the nematode (Chen, et al., 1996a).

In a continuously cropped soybean field in Arkansas, the *H. glycines* population density dropped late in the soybean growing season rather than increasing as expected. Nematode females, eggs, and juveniles in the soil were infected by the sterile fungus ARF18. Subsequent surveys revealed that the fungus was frequently isolated from fields in Arkansas (Kim and Riggs, 1991). In a field with ARF18, *H. glycines* population density declined from 140 eggs/g soil in November to 4 eggs/g soil the next July, a reduction of 97%, while percentage reduction of eggs in another area under comparable conditions was 42% to 66% (Kim and Riggs, 1994). Application of 37% formalin solution to the soil at the fungicidal but not nematicidal rate of 0.4 mL/100 g soil resulted in 23- to 60-fold increase in the number of nematode eggs compared with untreated soil in a greenhouse test. When a field soil that was naturally infested with ARF18 was autoclaved and inoculated with the *H. glycines* eggs and J2, the nematode egg density increased more than 3.6-fold compared with the nematode density in soil without heating treatment. These results suggest ARF18 can suppress the soybean cyst nematode population density in some fields in Arkansas.

Decline of the soybean cyst nematode populations has also been observed in several locations in China (Liu and Wu, 1993). The nematode population densities increased in the first few years of monoculture and then declined thereafter to a level that had no significant damage to soybean. Three soil samples have been assayed in the greenhouse to determine suppressiveness to the nematode. The nematode female numbers on soybean roots increased 2.4- to 3.2-fold in soil treated with a fungicide, compared to untreated soil. Autoclaved soil produced 36% to 133% higher nematode females than did the autoclaved soil mixed with untreated field soil at a ratio of 9:1, indicating the suppressiveness can be transferred to the autoclaved soil. The results suggest that fungal antagonists of the nematode might be involved in the suppression of the nematode populations. While several fungi were common in cysts from two soils, a single predominant fungus, *P. lilacinus*, was observed in one soil. Other factors, however, may also be partially responsible for the suppression but have not yet been determined.

20.4.3 Suppression of *Heterodera schachtii* by Fungi

Suppression of *H. schachtii* by fungal parasites of eggs and juveniles has been observed in several locations in Europe and the USA. In a field in The Netherlands, sugar beet was grown continuously from 1965 to 1982 (Heijbroek, 1983). At first the population of *H. schachtii* was high but declined during the first few years of the trial. Although there was a later increase in number of nematodes, they did not reach the initial density. The most important fungal parasites in the field were *V. chlamydosporium* and *C. destructans*. The fungal parasitism of eggs reduced nematode population density. Suppression of *H. schachtii* population by *H. rhossiliensis* was reported in oil radish fields in Germany (Müller, 1982). The fungus parasitized as high as 90% of *H. schachtii* J2. Soil treated with fungicide reduced the fungal parasitism and increased the nematode population density. Although sugar beet cyst nematodes were suppressed by fungal pathogens, mainly *C. destructans*, *V. chlamydosporium*, *N. gynophila*, and *C. auxiliaris* (Crump and Kerry, 1983), the soil containing these fungal pathogens did not cause *H. schachtii* population densities to fall below the damage threshold (Crump and Kerry, 1987). It appears that *H. schachtii* is not commonly under natural control by fungi as *H. avenae* is in Europe. Westphal and Becker (1999) demonstrated a soil suppressive to *H. schachtii* in a field in California. Treatment with fumigants or aerated steam reduced suppressiveness of the soil against *H. schachtii*. More fungal-infested females and cysts were found in the suppressive soil than a conducive soil (Westphal and Becker, 2001). The most common fungi isolated from the infested cysts were *Fusarium oxysporum*, *Fusarium* sp. nov., and *Dactylella oviparasitica*.

20.4.4 Suppression of *Meloidogyne* spp. by *Dactylella oviparasitica*

Ferris et al. (1976) observed that root-knot nematode populations were low in some old peach orchards in California, despite the climate and host were suitable for the nematode developments. A subsequent survey led to the discovery of *D. oviparasitica*, a fungus parasitic to the eggs of root-knot nematode (Stirling and Mankau, 1978). The fungus parasitized 20% to 60% of the eggs in the orchard throughout the year and reduced nematode numbers and root galling in a sterile soil in greenhouse (Stirling, et al., 1979). A greenhouse study and the field observations suggested that *D. oviparasitica* was responsible for the suppression of root-knot nematode in the peach orchards.

20.4.5 Suppression of *Mesocriconema xenoplax* by *Hirsutella rhossiliensis*

Mesocriconema xenoplax is one of several factors contributing to peach tree short life disease complex in southeastern USA. In some orchards, the nematode population declined unexpectedly even though the weather patterns and farm practices were suitable for the nematode. The observation of this phenomenon led to suggestion that some biological factors may regulate the nematode population. A survey conducted in 22 peach orchards in South Carolina and one orchard in Georgia resulted in a discovery of frequent parasitism of the nematode by *H. rhossiliensis* (Jaffee and Zehr, 1982). The nematodes parasitized by the fungus were often with brown heads and distorted hyphae-filled bodies. A greenhouse study showed that the fungus suppressed the nematode multiplication (Eayre, et al., 1987). The fungal parasitism of the nematode may be partially responsible for the suppression of the nematode population in some orchards in southern USA (Zehr, 1985). However, even in the presence of the fungus, nematode population densities were generally far above damage levels. Further studies on population biology of the nematode in California peach orchards suggested that the fungus might be a week regulator of *M. xenoplax* (Jaffee, et al., 1989).

20.5 Potential Fungal Agents for Control of Nematodes

Numerous fungi including various types of fungal antagonists of nematodes have been tested for their efficacy in control of plant-parasitic nematodes. However, only a few fungi have been commercialized (Table 20.1), and the products are not accepted or only used at a small scale. Before the 1970s, most fungi tested for control of nematodes were predacious fungi and a few endoparasites of vermiform nematodes. It seems the predacious fungi are unlikely to be effective in control of nematodes as introduced agents into an agricultural cropping fields unless more efficient organisms are identified and

Table 20.1 Commercial biological control agents developed from fungal antagonists.

Product name	Fungal species	Type of fungi	Country	Reference
Royal 300	*Arthrobotrys superba*	Predacious	France	Cayrol, 1981; Cayrol, et al., 1978
Royal 350	*Arthrobotrys irregular*	Predacious	France	Cayrol, 1981; Cayrol, et al., 1978
Biocon	*Paecilomyces lilacinus*	Egg-parasitic, producing antibiotics	The Philippines	Timm, 1987
Soybean Root Bio-Protectant	*Paecilomyces lilacinus*	Egg-parasitic, producing antibiotic	China	Liu, et al., 1996
DiTera	*Myrothecium*	Producing antibiotics	USA	Warrior, et al., 1999
Nemout	Fungi		Saudi Arab	Al-Hazmi, et al., 1993

better delivering systems developed. Recently, studies on nematode egg parasites and the fungi that produce toxins to nematodes have been increased. Some of them have been shown to be promising in control of various nematodes. The followings are a few fungi as examples that recently have been more or less extensively tested and have shown some potential in control of plant-parasitic nematodes.

20.5.1 *Paecilomyces lilacinus*

Paecilomyces lilacinus is a typical soil-borne fungus that has been reported from numerous parts of the world, but it seems to be most frequent in warmer regions (Domsch, et al., 1980). The fungus has been found from various types of habitats. Since Jatala et al. (1979) discovered the infection of eggs of *M. incognita* and *Globodera pallida* and females of *M. incognita*, the fungus has been isolated from eggs, egg masses, females, and cysts of many plant-parasitic nematodes throughout the world. The fungus first colonizes the gelatinous matrix of *Meloidogyne*, *Tylenchulus*, and *Nacobbus*, and cysts of *Heterodera* and *Globodera*; eventually a mycelial network develops and engulfs the nematode eggs. Penetration of nematode eggs is completed with an appressorium or simple hyphae (Holland, et al., 1999; Jatala, 1986). Both mechanical and enzymatic activities may be involved in the penetration. Morgan-Jones et al. (1984) reported the fungal hyphae penetrated the eggshell of *Meloidogyne arenaria* through small pores dissolved in the vitelline layer. The fungus penetrates the eggs of *Meloidogyne* faster than the eggs of *Globodera* and *Nacobbus* because the eggshell of *Meloidogyne* is simpler than the eggshell of *Globodera* and *Nacobbus* (Rogers, 1966). Following penetration, the fungus grows and proliferates in the eggs in early embryonic development. After depleting all nutrients in the egg, the mycelium may penetrate and rupture the cuticle of the infected egg from within and then emerge to infect other eggs in the vicinity. The fungus may also colonize the juveniles within the eggshell, and the 3rd and 4th stages of juveniles on water agar (Holland, et al., 1999). Culture filtrates of *P. lilacinus* were toxic to nematodes (Cayrol, et al., 1989; Chen, et al., 2000a; Khan and Goswami, 2000). Cuticles of nematodes were ruptured, and the nematodes were killed within a few hours after exposure to the culture filtrates. A peptidal antibiotic P-168 has been isolated from culture of *P. lilacinus* and characterized (Isogai, et al., 1980). This substance has anti-microbial activity against fungi, yeast, and gram-positive bacteria, and therefore may enable the fungus to compete with soil microorganisms. *Paecilomyces lilacinus* appears to be a good root colonizer (Cabanillas, et al., 1988) and rhizosphere competitor. Its depth of distribution in sandy soils, however, appears to be limited to the upper 15cm (Hewlett, et al., 1988). The fungus can grow well over a wide range of temperature and pH and on various plant and animal substrates (Alam, 1990; Jatala, 1986). The fungus is also a parasite of insects.

Paecilomyces lilacinus has been tested widely for its potential as a biological control agent and shown to suppress nematode population densities and increase plant yields. Not all tests with *P. lilacinus*, however, resulted in successful control of nematodes (Hewlett, et al., 1988). The variation of experimental results may be attributed to the variation of virulence among isolates (Stirling and West, 1991) and experimental conditions. A strain of *P. lilacinus* was shown to suppress *M. incognita* and *M. arenaria* populations but not *Meloidogyne javanica* (Wu, et al., 1990). Combination of *P. lilacinus* with *Pasteuria penetrans* increased efficacy in control of *M. incognita* (Sosamma and Koshy, 1997; Dube and Smart, 1987).

A formulation called "Biocon", containing *P. lilacinus*, was sold for control of root-knot and cyst nematodes in the Philippines (Timm, 1987). A biocontrol agent named "Soybean Root Bio-Protectant" has been developed and used to control the soybean cyst nematode over 12,600 hectares in China (Liu, et al., 1996). The agent, which contains *P. lilacinus*, organic materials, nematicidal plant broth, and mineral fertilizers, reduced *H. glycines* density and increased soybean yields (Wang, et al., 1997). Although *P. lilacinus* has been isolated from human eyes and sinuses (Agrawal, et al., 1979; Rockhill and Klein, 1980), there is no evidence that the isolates from nematodes are pathogenic to human.

20.5.2 *Verticillium chlamydosporium*

Verticillium chlamydosporium (syn. *Diheterospora chlamydosporia*) has been known as a soil fungus since 1913 and is worldwide in distribution (Domsch, et al., 1980). Since Willcox and Tribe (1974) discovered its parasitism of nematode eggs, the fungus has been found on various nematodes but mainly species of *Heterodera* and *Meloidogyne*. Gams (1988) reclassified the fungus into two species and two varieties of each species: *Verticillium chlamydosporium* Goddard var. *chlamydosporium*, *Verticillium chlamydosporium* var. *catenulatum*, *Verticillium suchlasporium* var. *suchlasporium*, and *V. suchlasporium* var *catenatum*. Cluster analysis of the enzyme activities indicated that some subspecific groupings of isolates may exist, but it did not support their division into two species (Carder, et al., 1993). The name *V. chlamydosporium* discussed herein represents the two species and four varieties. The ecology of

V. chlamydosporium and biological control of cyst and root-knot nematodes were recently reviewed by Kerry and Jaffee (1997).

Verticillium chlamydosporium enters nematode cysts through natural openings or penetrates cyst wall directly (Kerry, 1988). The fungus forms branched mycelial network and penetrates eggs by simple branches of hyphae or by formation of appressoria (Lopez-Llorca and Claugher, 1990). Enzymatic activities are involved in penetration. Electron microscopy study showed that the fungus could disintegrate eggshell vitelline layer and partially dissolve chitin and lipid layers. A 32 kDa protease has been isolated from the infection of *H. avenae* eggs by *V. suchlasporium* and considered to be involved in the pathogenicity of the fungus to nematode eggs (Lopez-Llorca and Robertson, 1992). An enzyme designated as VCP1 from *V. chlamydosporium* can hydrolyze proteins from the outer layer of the eggshell of *M. incognita* and expose its chitin layer (Segers, et al., 1996). *Verticillium chlamydosporium* may produce toxins that inhibit hatching or kill eggs of nematodes (Caroppo, et al., 1990; Meyer, et al., 1990). Juveniles within a cyst or egg mass may be colonized by *V. chlamydosporium*. Whether or not the juveniles were killed by toxin before the colonization is not clear. *Verticillium chlamydosporium* also can parasitize females of cyst and root-knot nematodes (Kerry, et al., 1982b; Morgan-Jones, et al., 1981) and other invertebrate animals such as snails.

The ability of colonization of plant roots by *V. chlamydosporium* varies among fungal isolates and plant species (Bourne and Kerry, 1999; Bourne, et al., 1994). Some studies showed that *V. chlamydosporium* are able to colonize plant roots (Kerry, 1984; Stiles and Glawe, 1989), while others indicated that *V. chlamydosporium* cannot invade the root cortex and is confined to the rhizoplane (De Leij and Kerry, 1991). The fungus appears to be non-pathogenic to plants (Kerry, 1984; Stiles and Glawe, 1989) and has no pathogenic tendency to higher animals and humans.

Verticillium chlamydosporium is one of major fungal parasites responsible for suppression of *H. avenae* in Europe (Kerry, 1975). The potential of the fungus in biological control of nematodes has been evaluated in many studies in greenhouse, microplots, and fields. The efficacy in control of nematodes is affected by several factors. Host plants have great influence on the growth of the fungus in the rhizosphere and the control efficacy (Borrebaeck, et al., 1984). The fungus is more efficient in control of nematodes at lower nematode densities than at higher densities and in a poor host of nematode than in a susceptible host (Kerry and Jaffee, 1997). Root-knot nematodes in large galls may escape from the fungus attack, and control efficacy may be limited. Different isolates varied in their pathogenicity to nematode eggs (Irving and

Kerry, 1986). A combination of the fungus with *Pasteuria penetrans* increased efficacy of reducing nematode population of *M. incognita* on tomato (De Leij, et al., 1992a). *Verticillium chlamydosporium* is a promising biocontrol agent, and efforts have been made to develop commercially acceptable formulation (Stirling, et al., 1998). No commercial product from this fungus, however, has yet been on the market.

20.5.3 *Verticillium lecanii*

Verticillium lecanii has been frequently isolated from soils in various geographic locations (Domsch, et al., 1980). Besides its wide range of substrates of dead plants and animals, the fungus is a catholic hyperparasite and parasitizes arthropods, rust fungi, powdery mildews, and many other fungi. It has been used commercially for control of greenhouse aphids. The fungus could penetrate cyst wall and colonize eggs of *H. schachtii* within 60 h (Hanssler, 1990; Hanssler and Hermanns, 1981). Lytic enzymes secreted by the fungus played a major role during its penetration of the cyst wall and eggshell. Gintis et al. (1983) observed chitinase activity of the fungus on chitin agar and the ability to invade *H. glycines* eggs. Meyer et al. (1990), however, reported that one strain of *V. lecanii* reduced viability of *H. glycines* eggs without colonization of the eggs, indicating that the fungus produced toxins that killed the nematode eggs. The fungus has been evaluated as a biological control agent against the soybean cyst nematode in the laboratory, greenhouse, and fields for several years. Benomyl-tolerant mutants were induced and a mutant was more efficient in reducing nematode populations of *H. glycines* and *M. incognita* in the greenhouse (Meyer, 1994; Meyer and Huettel, 1991; Meyer and Meyer, 1995, 1996). Application of alginate prills containing *V. lecanii* mutant strains at 5 g prills per pot (530 g soil) significantly suppressed nematode population of *H. glycines* in non-sterilized soil, but no reduction of the nematode population was observed with 0.5 g prills per pot (Meyer and Meyer, 1996). Microplot tests showed a significant control of *H. glycines* population with *V. lecanii* at 340 kg alginate prills/hectare; no control of the nematode, however, was obtained in field plots (Meyer, et al., 1997). More research is needed to determine whether or not the fungus has potential as a biological control agent against the soybean cyst nematode and other nematodes. The wide range of target pests and plant pathogens ensures the commercial value of the fungus as a biological control agent. If the efficacy of control for nematodes is warranted, the fungus may be promising for control of nematodes at a large scale of field use.

20.5.4 *Hirsutella rhossiliensis*

Hirsutella rhossiliensis was first described in 1980 (Minter and Brady, 1980) based on a specimen collected from Wales in 1953. Sturhan and Schneider (1980) reported this fungus from the hop cyst nematode, *Heterodera humuli*, and named it as *Hirsutella heteroderae* (syn. *H. rhossiliensis*). The fungus has a wide range of hosts including plant-parasitic nematodes, free-living nematodes, entomopathogenic nematodes, and mites, although different isolates may have different host preferences.

Hirsutella rhossiliensis is a hyphomycete with simple erect phialides, which are swollen at base and tapering towards the apex. When a host nematode contacts the conidia on the phialides, the conidia may attach to the nematode cuticle, and infect the host nematode within a few days. Following penetration, the fungus forms an infection bulb in the nematode cavity, from which assimilative hyphae are developed. After converting nematode body contents to mycelial mass, the fungus may emerge from the nematode cadaver, produce spores, and infect other nematodes. An average of 112 conidia may be formed from mycelium developed from a single juvenile of *H. schachtii* at 20℃ (Jaffee, et al., 1990). KCl increased infection of nematodes by the fungus (Jaffee and Zehr, 1983). Conidia detached from the phialides may loss infectivity. Some conidia died shortly after sporulation and others may be viable and virulent for at least 200 days (Jaffee, et al., 1990). Variability of morphology, pathogenicity, and genetics was observed among isolates (Liu and Chen, 2001; Tedford, et al., 1994).

Parasitism of nematodes by *H. rhossiliensis* is dependent on nematode density; the percentage of nematodes parasitized by the fungus is correlated positively with host nematode density (Jaffee, et al., 1992). The number of conidia attached to the cuticle of nematodes by *H. rhossiliensis* is correlated with the amount of conidia in the soil. Since the fungus is a poor soil competitor, local populations of the fungus may go extinct unless supplied with some minimum number of nematodes (the host threshold density) (Jaffee and Zehr, 1985). Natural epidemics of this fungus among populations of nematodes develop slowly and only after long periods of high host densities. Transmission of spores was greater in loamy sand than in coarse sand (Jaffee, et al., 1990). In contrast to the theory that addition of organic matter may enhance activities of nematophagous fungi, addition of organic matter to soil decreased parasitism of *M. xenoplax* by *H. rhossiliensis* (Jaffee, et al., 1994).

The potential of the fungus as a biological control agent has been controversial. Müller (1982) reported that the fungus might suppress cyst nematodes in some sugar beet fields in Germany. The fungus was considered partially responsible for suppression of *M. xenoplax* population in some orchards in southern USA (Zehr, 1985). High numbers and percentages of *M. xenoplax* parasitized by *H. rhossiliensis* were also found in some California peach orchards (Jaffee, et al., 1989). In greenhouse studies, *H. rhossiliensis* suppressed *G. pallida* on potato (Velvis and Kamp, 1996), *H. schachtii* on cabbage (Jaffee and Muldoon, 1989), *Pratylenchus penetrans* on potato (Timper and Brodie, 1994), and *H. glycines* on soybean (Liu and Chen, 2001). Results obtained by Tedford et al. (1993) indicated that long-term interactions between populations of *H. rhossiliensis* and cyst or root-knot nematodes did not result in biological control. In a field microplot test, *H. rhossiliensis* failed in suppression of *H. schachtii* (Jaffee, et al., 1996). The fungus, however, significantly reduced *H. glycines* in field plots (S. Chen, et al., unpublished). *Hirsutella rhossiliensis* has been formulated in alginate pellets and used in control of nematodes in laboratory or greenhouse studies (Jaffee, et al., 1996; Lackey, et al., 1993). More research, however, is needed to determine whether or not the fungus has potential as a commercial biological control agent.

20.5.5 *Fusarium* spp.

Fusarium is a large genus including many species with a wide range of trophic adaptations. A number of *Fusarium* species have been isolated from females, cysts, egg masses, and eggs of nematodes. *Fusarium oxysporum* and *F. solani* are the most commonly encountered species. Strains of these two species are either phytopathogenic or non-phytopathogenic, but in general they are highly competitive in soil. Only a few species of *Fusarium* have been tested in laboratory and greenhouse for their potential as biological control agents on nematodes. Nigh et al. (1980) demonstrated that a high percentage of *H. schachtii* eggs were parasitized by *F. oxysporum* in California sugar beet fields. Laboratory and greenhouse bioassays indicated that isolates from the fields are highly pathogenic to the nematode eggs. Chen et al. (1996b) reported that one isolate of *F. oxysporum* and one isolate of *F. solani* could colonize more than 30% and 20% of the eggs, respectively, in yellow females to light brown cysts of *H. glycines* on water agar. The same isolate of *F. oxysporum* colonized more than 70% of the eggs in newly formed females on roots in sterile soil in greenhouse pots. Species of *Fusarium* produce a large range of toxins, which are antagonistic to streptomyces, bacteria, fungi, and nematodes (Ciancio, et al., 1988). Hallmann and Sikora (1994) reported that isolates of non-phytopathogenic endophytes, *F. oxysporum*, reduced root galls of tomato induced by root-knot nematodes by 52% to 75%. Culture filtrates of the non-phytopathogenic, endophytic isolates of *F. oxysporum* killed juveniles of *M. incognita* within 8 hours (Hallmann and Sikora, 1994). Nematicidal effect of culture filtrates was also observed from *F. solani* on *M. incognita* (Mani and Sethi, 1984). Toxins that are thermostable and independent of pH are responsible for the nematicidal effects. It seems that strains of non-phytopathogenic *Fusarium* species, which are either highly pathogenic to nematode eggs or produce metabolites toxic to nematodes, may exist in natural soil. Such strains plus their high competence in soil and rhizosphere may be effective biological control agents.

20.5.6 Arkansas Fungus 18

Arkansas Fungus 18 (ARF18) is an unidentified sterile fungus, which has been frequently isolated from soybean fields in Arkansas and other southern states in the USA. Electron microscopy study indicated that the fungus is a species of ascomycetes since it has simple pores on septa and Woronin bodies near the pores (Kim, et al., 1992). ARF18 can parasitize *H. glycines* females, eggs, juveniles within eggs and cysts, and young stages of the nematode on plant roots. The fungus enters cysts either through natural openings or by direct penetration through cyst wall. Sclerotium-like structures were frequently observed on the surface of cyst wall and numerous pores were found below the structure. It appears that enzymatic activity is involved in the penetration of the cyst wall. In the cysts, the fungus proliferates and penetrates eggs in early embryogenesis or first-stage juveniles (Kim, et al., 1992). In a test on water agar, the fungus parasitized 89% of *H. glycines* eggs in cysts (Kim and Riggs, 1991). In a greenhouse study, the fungus reduced the nematode density by 96% when 1 g freshly made 2.6% of mycelium pellets/100 g soil was used in an autoclaved soil with addition of 1500 eggs (Kim and Riggs, 1995). Variation may exist among isolates (Kim, et al., 1998). Based on appearance of their sclerotium-like structure, the fungus can be separated into two groups: ARF18-C (compact) and ARF18-L (loose). The ARF18-C group parasitized more eggs on water agar than the ARF18-L group. In soil, however, isolates of ARF18-L parasitized higher percentage of eggs, juveniles, and young females on soybean roots than isolates of ARF18-C (Timper and Riggs, 1998; Timper, et al., 1999). It appears that ARF18-L isolates are better rhizosphere colonizers than ARF18-C isolates. More information is needed to determine whether ARF18 can be developed as a commercial biological control agent or not.

20.6 Ecological Aspects of Nematode Suppression by Fungi

20.6.1 Distribution and Habitat of Nematophagous Fungi

Nematophagous fungi are worldwide in distribution and have been reported from many countries (Gray, 1987). Virtually, nematophagous fungi occur wherever nematodes can live. The distributions, however, differ among species of nematophagous fungi. Many predacious fungi and facultative parasitic fungi are widely distributed in the world. Some species, however, are distributed in a limited geographic area.

In Britain and Ireland, the most frequently recorded endoparasites of nematodes were *Acrostalagmus obovatus* and *H. anguillulae*, and the most common predacious fungi were *Dactylella bembicoides*, *Dactylella ellipsospora* and *Dactylella cionopaga* (Duddington, 1951; Gray, 1983). The species of endoparasitic fungus, *Harposporium helicoides*, was frequently isolated in Ontario, Canada (Barron, 1978), but rarely in Ireland (Kerry and Andersson, 1983). While the distributions of *Dactylella mammillata* in Britain, *A. oligospora* in Ireland and *Arthrobotrys musiformis* in both Ireland and Britain appear highly restricted, all species are locally abundant (Gray, 1987). Liu et al. (1993) reported that the predacious fungi that were the most widely distributed in China include *Monacrosporium thaumasium*, *Monacrosporium eudermatum*, *A. oligospora*, *Arthrobotrys conoides*, *Arthrobotrys robusta*, *Arthrobotrys oviformis*, and *Arthrobotrys brochopaga*. *Monacrosporium thaumasium* was frequently encountered in the temperate areas in China. *Arthrobotrys oligospora* was most abundant in temperate soils in some surveys but not in other studies (Gray, 1983; Liu, et al., 1993).

Studies of the species and frequency of fungal parasites of females and eggs of sedentary endoparasitic nematodes have been increased recently. Although numerous species have been isolated from cysts and females, only a few species have been determined as parasites of nematodes. The most common species encountered in females and eggs are similar, but the taxonomically diverse mycofloras exist in different geographic locations. Although *V. chlamydosporium* is worldwide in distribution (Domsch, et al., 1980) and has been reported as a major fungal parasite of *H. avenae* in Europe, the fungus was encountered with a low frequency in the egg masses of the *Meloidogyne* spp., and it was not found in cysts of *H. glycines* in subtropical Florida (Chen, et al., 1996c). The fungus appeared to be more adapted to temperate climates than to subtropical climates. Stirling and Kerry (1983) investigated fungi colonizing females of *H. avenae* in Australia. The most widely distributed parasite was *V. chlamydosporium*, but its abundance in soil was lower than the suppressive level as detected in cereal fields in Europe.

In contrast to *V. chlamydosporium*, *N. vasinfecta* is probably adapted to high temperatures of tropical or subtropical climates (Domsch, et al., 1980). It was encountered at a high frequency in cysts of *H. glycines* in Florida (Chen, et al., 1994), but with a low frequency in Illinois (Carris, et al., 1989). *Fusarium oxysporum* has been encountered in cysts and females of nematodes at a high frequency at both high and low temperature climates.

Nematophagous fungi have been recovered from a variety of habitats. Predacious fungi were isolated often from leaf mold, decaying wood, partly decayed plants, dung, living bryophytes, and soil (Duddington, 1951). Gray and Bailey (1985) surveyed the distribution of nematophagous fungi in a deciduous woodland. They found that nematophagous fungi were isolated throughout the soil profile and down to a maximum of 35 cm deep. Predators forming constricting rings, adhesive branches, and adhesive knobs were restricted to the upper litter and humus layers. The net-forming predators and endoparasites were isolated throughout the soil profile at all depths, although they were more abundant in the deeper mineral rich soil. In another survey, Gray (1983) found that nematophagous fungi were abundant in all habitats examined, although most widely in temporary agricultural pasture, coniferous leaf litter, and coastal vegetation. A number of species showed distinct habitat associations, in particular *A. robusta*, *A. musiformis*, and *D. cionopaga*.

Most egg-parasitic fungi reported in literature were isolated from females, eggs, egg masses, and cysts of Heteroderidae. Egg-parasitic fungi can be found in various habitats. The fungal species isolated from cysts collected from pastures (Hay and Skipp, 1993) were similar to those isolated from cysts collected from cultivated agricultural soils. Cysts and egg masses of the sedentary nematodes may provide unique niches to the egg-parasitic fungi. The eggs of these nematodes are aggregated. When these cysts and egg masses are exposed to soil, they will be vulnerable to attack by fungi. The substrates of the cysts and egg masses may provide important nutrients for the fungi before the fungi parasitize the eggs. Migratory and free-living nematodes deposit eggs separately, making the observation of their egg-parasitic fungi difficult. *Rhopalomyces elegans*, one of the earliest studied parasites of eggs of many nematodes including migratory nematodes, is worldwide in distribution and can be found in soil, rotting plant debris, organic matters, and animal dung.

20.6.2 Nutritional Adaptation of Nematophagous Fungi

Nematophagous fungi are capable of using nematodes as the only or a part of their nutrient source. This ability may have developed over a long period of time in the evolution process. Limited attempts have been made to understand the evolution process of nematophagous fungi. Cooke (1962a, 1962b) hypothesized that competition stimulates the predatory activity of predacious fungi as an adaptation to overcome their low competitive saprophytic ability. This hypothesis was supported by some observations (Cooke, 1963; Quinn, 1987). It appears that the development of predacious efficiency has been accompanied by the tendency to lose characters such as rapid growth rate and good competitive saprophytic ability that are associated with an efficient saprophytic existence in the soil (Cooke, 1963). The positive correlation between ability of nematophagous fungi to destroy nematodes and attraction of the fungi to nematodes also indicates the evolution toward dependence of the fungi on nematodes for nutrients. Belder and Jansen (1994a) observed that formation of trapping devices was less temperature- and nutrient-dependent in simple adhesive hyphae than in complex adhesive network.

Barron (1992) argued that many nematophagous fungi evolved from lignolytic and cellulolytic fungi. The evolutionary process is a response of these fungi to nutrient divergence in N-limiting habitats. This argument was based mainly on a series of observations by Cooke (1962b, 1963). Although the teleomorphs of most predacious Hyphomycetes are not known, some predacious fungi are members of the gilled fungi (Saikawa and Wada, 1986). For example, the teleomorphs of *Nematoctonus* spp. are gilled fungi, *Hohenbuehelia* spp. Many fungi occurring in rotting wood prey on nematodes. The adhesive knob-forming fungus, *Pleurotus ostreatus* (oyster mushroom), has developed a unique method of inactivating and eventually colonizing its nematode victims. Nematodes are paralyzed by a toxin release from minute spatulate secretory cells of the fungus. The gilled fungi *Resupinatus* and *Stignatolemma* also possess secretory-like cells on their hyphae that closely resemble those of *P. ostreatus*. It is probable that the method of capturing nematodes developed by *P. ostreatus* is more common in lignolytic fungi than currently known.

Nutritional adaptation may also have evolved in opportunistic fungi associated with females, cysts, and eggs of sedentary nematodes. Even though a wide range of taxonomically diverse mycofloras have been encountered in females, cysts, egg masses, and eggs, the most common fungi are limited to a few genera, including *Verticillium, Paecilomyces*, *Exophiala*, *Fusarium*, *Gliocladium*, *Cylindrocarpon*, and *Phoma*. Only a few fungal species have been associated with more than one species of nematodes, and the common fungal species isolated from cysts collected from different geographic locations are similar, suggesting that a degree of specialization may be a prerequisite for successful colonization of the ecological niches represented by eggs, females, or cysts of heteroderid nematodes (Rodríguez-Kábana and Morgan-Jones, 1988). Many of these fungi are root colonizers and some are parasites of plants. *Pyrenochaeta terrestris* was commonly encountered in cysts of *H. glycines* and was pathogenic to the nematode eggs and soybean plants (Chen, et al., 1996b). Probably, the fungus was evolved from a plant-parasitic form to an organism capable of using both plants and nematodes as nutrient sources. Some non-phytopathogenic isolates of *F. oxysporum* are egg parasites of cyst and root-knot nematodes.

Chen et al. (1996c) compared the mycoflora of cysts of the soybean cyst nematode, *H. glycines*, with the mycoflora of egg masses of root-knot nematodes, *Meloidogyne* spp., in two adjacent fields on the same farm with similar soil texture and climate conditions. The similarity index for fungi colonizing *Meloidogyne* egg masses and *H. glycines* cysts was low (0.10 – 0.18). In contrast, the average similarity index between different sample dates for fungi in brown cysts of *H. glycines* was 0.54, and the average similarity index between locations and sampling dates for fungi in cysts of *H. glycines* collected from different locations in the southeastern USA at different times was 0.48 (Chen, et al., 1994). This suggested that the mycoflora in egg masses of *Meloidogyne* spp. differed from that in the brown cysts of *H. glycines*. Although some fungi encountered in the cysts of *H. glycines* were also found in the egg masses of the *Meloidogyne* spp., the frequencies of most fungi in the egg masses differed from that in the cysts of *H. glycines*. *Paecilomyces lilacinus* colonized only 1 of 1711 brown cysts of *H. glycines* (Chen, et al., 1994) that was compared with 26% of the egg masses of *Meloidogyne* spp. colonized by this fungus. Probably, the Floridian strain of *P. lilacinus* is more adapted to root-knot nematodes than to cyst nematodes. Mycofloras in cysts of *H. glycines* were compared among 45 fields in Minnesota (Chen and Chen, 2002). The mycofloras in the cysts diversified, and no single highly pathogenic fungus dominated.

20.6.3 Interaction of Nematophagous Fungi with Host Nematodes

Plant-parasitic nematodes spend a part of their life in soil and are subjected to attack by fungal antagonists. Nematode species and their density are a major factor influencing types and abundance of their predators or parasites. Migratory nematodes may be more vulnerable to attack by predacious fungi than the sedentary endoparasitic nematodes. Egg-parasitic or female-parasitic fungi may be more effective in regulating nematode population of sedentary endoparasitic nematodes, which deposit eggs in egg sac or retain the eggs within the female body, than the migratory nematodes, which deposit eggs singly. Egg-parasitic fungi may be more efficient in control of root-knot nematodes forming small galls than those forming large galls.

Linford et al. (1937) demonstrated that addition of organic matter to soil resulted in marked changes in the indigenous nematode populations. Soil amendments increased free-living nematodes and decreased plant-parasitic nematodes (Gray, 1987). It was suggested that the organic supplement, which led to the increase in nematode density, also resulted in increased nematophagous activity of fungi in the soil. The enhanced activity of the nematophagous species present in the soil would then stabilize the rate of increase of the soil nematodes, and eventually reduce their population densities, including those of plant-parasitic nematodes (Gray, 1987). This hypothesis has been used as the basis for a number of experiments on the biological control of plant-parasitic nematodes in soil. Cooke's (1962a, 1962b, 1968) series of experiments suggested that in nature a simple labile equilibrium does not exist. Recent work indicates that equilibria do exist between nematophagous fungi and soil nematodes (Jaffee, 1993; Gray, 1987). The dependency on nematode density and response to soil organic amendments vary among the predacious fungi (Cooke, 1963). Species that form either a branch, knob, or constricting ring are sensitive in forming traps in soil amended with organic matter, whereas net-forming fungi are not (Gray, 1987). The mode of action of organic amendments against nematodes consists of more than the direct effects on nematophagous fungi. Organic amendments can improve soil structure and soil fertility, alter the level of plant resistance, release toxic compounds, and stimulate antagonistic microorganisms (Stirling, 1991). All of these changes by organic amendments may affect nematodes dramatically.

While the population of predacious fungi is related more to the nutritional value of the organic matter in soil than to the number of nematodes present, the population of obligate parasitic fungi is generally dependent on the density of host nematodes. The relationship between *H. rhossiliensis* and host nematodes is a good example. Jaffee et al. (1989) observed a spatial density-dependent parasitism of *M. xenoplax* by *H. rhossiliensis* in peach orchards in California. In a laboratory assay, the disease dynamics of *H. schachtii* exhibited temporal density-dependent parasitism (Jaffee, et al., 1992). The parasitism increased to nearly 100% with repeated addition of many hosts and declined to nearly 0% unless a minimum number of hosts (the host threshold density) was supplied.

The decline of cereal cyst nematode in Europe was also a phenomenon of density-dependent parasitism of nematodes by fungi. In the first two years of monoculture, the cereal cyst nematode increased and reduced crop yield. With continuous monoculture, the nematode declined to less than damaging levels due to the increase of parasitism of the nematode by fungi, mainly *N. gynophila* and *V. chlamydosporium*. *Nematophthora gynophila* is an obligate parasite and its density is dependent on the density of host nematodes (Perry, 1978). Facultative-parasitic fungi associated with eggs, sedentary females, and cysts are mainly influenced by environmental factors and independent or weakly dependent on nematode density (Chen and Chen, 2002). *Verticillium chlamydosporium* is a facultative parasite, and its density may be less dependent on the nematode populations than the obligate parasite *N. gynophila*. In a greenhouse study, however, *V. chlamydosporium* density increased with increasing nematode density (Bourne and Kerry, 1999). Presumably, the ability to parasitize nematode females and eggs gives the fungus a competitive advantage in soil.

20.6.4 Interaction of Fungal Antagonists of Nematodes with Other Soil Organisms

Soil organisms represent almost all kinds of animals, fungi, and bacteria. The interaction among the soil organisms is complicated but may be critical in determining the effectiveness of suppression of nematodes by their antagonists. The biggest problem in biological control of nematodes using nematophagous fungi is probably the difficulty of overcoming the competition from other soil organisms. The knowledge of interaction of the nematophagous fungi with other soil organisms, especially in the rhizosphere, is useful in exploiting potential biological control agents, but limited studies have been done in this area.

Fungistasis is a common phenomenon in soil ecosystem. Fungistasis may be removed by soil sterilization or the addition of a nutrient source. Pioneer colonizers of roots and the rhizosphere were less sensitive to fungistasis than secondary or non-colonizers (Dix, 1967). Conidium germination of predacious fungi was inhibited by field soil, but the inhibition was temporarily overcome by organic amendments, air-drying, or soil sterilization (Mankau, 1962a, b). Cooke (1968) demonstrated that sensitivity to mycostatic factors was not related to the type of trap produced and that germination of conidia did not necessarily result in establishment of the fungus. The fungistasis is a major problem in use of predacious fungi in agricultural soils for the purposes of biological control of plant-parasitic nematodes.

Some fungi are highly pathogenic to nematodes in the laboratory, but poor soil competitors. *Arthrobotrys dactyloides* is a predacious fungus forming constricting rings and also is able to colonize eggs of *H. glycines* (Chen, et al., 1996b). The fungus is ubiquitous in distribution but a poor competitor in soil. In sterile soil, the fungus colonized a large portion of cysts and eggs, and reduced nematode populations. The fungus, however, was not found in cysts and eggs in untreated natural field soil (Chen, 1994). Similarly, *Hirsutella rhossiliensis* can grow and sporulate well in artificial medium, but the fungus seems to be a poor competitor in natural field soil (Jaffee and Zehr, 1985).

Some soil microscopic animals may feed on fungi. Jaffee et al. (1997) observed enchytraeid worms fed on *H. rhossiliensis*, *Monacrosporium gephyropagum*, *Arthrobotrys thaumasia* and *Arthrobotrys haptotyla*, and reduced the fungal population densities. In a later experiment, however, exclusion of enchytraeids, collembolans, and mites did not affect the populations of *H. rhossiliensis* and *M. gephyropagum* (Jaffee, 1999). *Hirsutella rhossiliensis* was quite sensitive to biotic inhibition when formulated as pelletized hyphae but insensitive when formulated as parasitized nematodes (Jaffee, 2000). In contrast, *A. haptotyla* was more sensitive to biotic inhibition when added to soil as fungus-colonized nematodes than as pelletized hyphae.

As hypothesized by Linford et al. (1937), soil bacterial population may affect predators and parasites of nematodes. Bacteria provide food for free-living nematodes, and these nematodes also serve as a food source for many predacious fungi. Bacteria, particularly *Pseudomonas* spp., are the most abundant microorganisms in the rhizosphere. Actinomycetes can comprise up to more than one third of the total microflora associated with plant roots (Stirling, 1991). These organisms frequently produce antibiotics and inhibit fungal growth. Fungi such as *Fusarium*, *Penicillium*, *Trichoderma*, *Aspergillus*, *Mortierella*, *Pythium* and some species of the Mucorales are commonly isolated from the rhizosphere and bulk soil. Most of these fungi may not be pathogenic to plant-parasitic nematodes. Mycorrhizal fungi are important components of the rhizosphere mycoflora. Some mycorrhizal fungi are also egg colonizers (Francl and Dropkin, 1985). Mycorrhizal fungi can either stimulate or inhibit nematophagous fungi in the rhizosphere. Because the egg-parasitic fungi are generally closely associated with plant roots, mycorrhizal fungi may have great effects on this group of nematophagous fungi. Their colonization of the rhizosphere may prevent colonization of roots by nematophagous fungi.

Competition for nutrients between nematophagous fungi and saprophytic fungi may also occur in cysts and egg masses. Many fungi isolated from cysts and egg masses are saprophytic. While Hay and Skipp (1993) reported that multi-species of fungi are commonly encountered in cysts of *Heterodera trifolii* collected from pasture soil in New Zealand, many studies (Carris, et al.,

1989; Chen, et al., 1994; Gintis, et al., 1983; Morgan-Jones and Rodríguez-Kábana, 1985) suggest that cysts colonized by one fungus are not readily colonized by other fungi. Colonization of cysts by non-antagonistic, saprophytic fungi may inhibit egg-parasitic fungi and therefore protect nematode eggs from being destroyed by the egg-parasitic fungi (Chen and Chen, 2003).

20.6.5 Effects of Plants and Cropping System

Crop species and cultivars influence microflora and fauna in soil, especially in the rhizosphere. A number of factors affect the root exudates; the principal factors are the species and developmental stage of the plants. There is some evidence that fungal growth is affected by root exudates (Vancura, 1988). Only limited studies, however, have been done to investigate the effect of plant species on the nematophagous fungi (Bourne and Kerry, 1999, 2000). Plant species differed in their ability to support growth of *V. chlamydosporium* in their roots or rhizosphere at different nematode densities; fungal density increased more with increased nematode density on the roots of some species than others (Bourne and Kerry, 1999; Bourne, et al., 1994). In nutrient poor sandy soil, the rhizosphere of tomato roots was able to support a larger population density of the egg-parasitic fungus, *V. chlamydosporium*, than the bulk soil (Bourne and Kerry, 2000). Similarly, the population densities of the predacious fungi, *A. dactyloides*, *A. superba*, and *Monacrosporium ellipsosporum*, were greater in tomato rhizosphere than in root-free soil (Persson and Jansson, 1999). The ability of colonization of roots is essential for effective control of root-knot nematodes by fungi. The locations and size of egg masses of root-knot nematodes vary among the species and their hosts. Egg masses on root surfaces may be more vulnerable to attack by egg-parasitic fungi, while egg masses formed inside roots may be able to escape the fungal attack. *Dactylella oviparasitica* suppressed populations of *M. incognita* on peach trees in a California orchard but not on grape plants in adjacent vineyards. A greenhouse experiment showed that the nematode produced smaller egg masses on peach than on grape; *D. oviparasitica* parasitized most eggs formed on peach and only about 50% of eggs on grape (Stirling and Mankau, 1979; Stirling, et al., 1979). It was suggested that the size of egg masses determines the suppression of root-knot nematode by *D. oviparasitica*.

A limited number of studies have been reported on the effects of crop sequences on nematophagous fungi. Steudel et al. (1990) studied the effect of sugar beet-cereals-green manure rotation on the *H. schachtii* population densities and the frequency of its fungal parasites of eggs. The highest level of parasitism of the nematode by fungi was observed after years with sugar beet.

In the interim, parasitism decreased to about 10%. No difference of fungal parasitism was observed among plots planted to oil radish, white mustard, "Pegletta," or fallow. A high nematode multiplication on susceptible catch crops promoted antagonistic fungi; consequently the nematode population density decreased more quickly following sugar beet. Chen and Reese (1999) demonstrated the effect of crop sequence on parasitism of *H. glycines* by the endoparasite *H. rhossiliensis*. The percentage of *H. glycines* J2 parasitized by the fungus increased in the first year of soybean after 5 years of corn. Fungal parasitism was similar in plots of 2nd through 5th years of soybean after 5 years of corn and in plots of soybean monoculture. When corn was planted, parasitism of J2 decreased. The effect of crop sequence on the fungal parasitism of J2 may be attributed to a density-dependent relationship between the parasite and its host nematode.

Bernard et al. (1997) investigated the incidence of fungi invading females, cysts, and eggs of *H. glycines* in various cropping-tillage plots. No significant difference in percentage of cysts or females colonized by fungi was observed among six cropping-tillage treatments. The diversity of egg-parasitic fungi was similar among the treatments. Although the percentage of eggs parasitized by total fungi did not differ among the cropping-tillage treatments, *P. lilacinus* parasitized more eggs in the disc treatment than in any of the no-till treatments and *V. chlamydosporium* occurred more frequently in the moldboard plow treatment than in any other treatment.

20.6.6 Effects of Soil Abiotic Factors

Soil abiotic variables, such as soil texture, temperature, moisture, nutrients, organic matter, and pH may determine microflora and fauna in the soil (Gray, 1987; Kerry, 1993). These soil factors affect activities of fungal antagonists of nematodes directly or indirectly. It may be impossible to use microbial agents efficiently for control of nematodes without understanding how the agents interact with soil abiotic factors.

20.6.6.1 Soil Temperature

Optimum temperature for growth and development differs among nematophagous fungi. Predacious fungi may require a low temperature for their growth and predacious activities (Belder and Jansen, 1994b). Isolates of *M. ellipsosporum* from Antarctica adapted to a lower temperature than did isolates from UK (Gray, 1982). The endoparasitic fungus, *H. rhossiliensis*, infects J2 of *H. schachtii* fastest at 25℃ to 30℃ (Tedford, et al., 1995). The optimum temperature for control of root-knot nematodes by *V. chlamydosporium* is 25℃ (De Leij, et al., 1992b). When the temperature exceeded 25℃, the percentage of eggs of the nematodes parasitized by the fungus decreased; at these temperatures the nematodes developed fast and hatched before they were colonized by the fungus. Difference in temperature requirement exists among isolates of *V. chlamydosporium* (Kerry, 1981). *Nematophthora* infected more females of *H. schachtii* at 10℃ and 15℃ than at 20℃ (Kerry, 1984). *Dactylella oviparasitica* grew most rapidly at 24℃ – 27℃, but colonized more *Meloidogyne* eggs at 15℃ than at 27℃ (Stirling, 1979). In contrast, *C. anguillulae* infected more nematodes at 24℃ – 28℃ than at 12℃ (Sayre and Keeley, 1969).

20.6.6.2 Soil Moisture

Nematodes are aquatic animals and need a film of water in soil micropores for moving and living. Nematodes are most active when soil moisture is at or close to field capacity (Jones, 1975). At low soil water potentials from –0.1 bar to –0.25 bar, nematodes may be immobilized. Mobility of nematodes determines the chance of contacting their natural enemies, especially predacious fungi and some other fungi that do not actively move in soil. Movement of soil water may disseminate fungal spores. Gray (1985) found that species producing adhesive networks were isolated more frequently from soils with low moisture. In contrast, ring-forming species were isolated significantly more frequently from soil with relatively high moisture and organic matter contents. Fungi can grow and develop in relatively dry soils; their growth is not restricted when the soil water potential is above –30 bar. Nematophagous fungi that produce motile zoospores are more active in soil with higher water potentials. Parasitism of *H. avenae* by *N. gynophila* and *C. auxiliaris* was higher in wet seasons than in dry seasons (Kerry, et al., 1982b; Stirling and Kerry, 1983). Another effect of moisture on fungi is related to soil aeration. Establishment of *V. chlamydosporium* was better, and control of *M. incognita* with the fungus was greater in well aerated soil than in soil that was less aerated (Kerry, et al., 1993). Similarly, infection of *H. schachtii* by *H. rhossiliensis* was higher at low water potentials than at high water potentials (Tedford, et al., 1992).

20.6.6.3 Soil Texture

Soil texture refers to the relative proportion of particles of different sizes. The size of soil particles determines the total surface, with which soil chemical compounds, microorganisms, and plant roots interact. Light soils are generally favorable to large populations of nematodes. The size of soil micropores influences both dispersal of nematophagous fungi and the movement of nematodes. Soil texture affects the proliferation and survival of fungi in soil.

Tedford et al. (1992) reported that transmission of *H. rhossiliensis* to *H. schachtii* was greatest in loamy sand, intermediate in loam, and lowest in sand. Water drainage and aeration are generally better in sandy soils than in heavy clay soils; hence a sandy soil may favor fungal growth if adequate nutrients are provided. Also, nematode movement may be better in soil with better drainage and aeration. Therefore, it may be expected that a light sandy soil favors fungal activity antagonistic to nematodes. However, most sandy soils have low organic matter contents, and consequently saprophytic activity of fungi may be low. Combination of fungal agents and organic amendments may be a good practice for control of nematodes in a light soil with some facultative parasites. The bulk soil is a major barrier for biological control agents to reach target nematodes. Generally, a large dosage is needed to ensure effectiveness of an immediate control of nematodes.

20.6.6.4 Effects of Soil pH, Nutrients, Organic Matter, and Agricultural Chemicals

Soil chemical characteristics are important factors influencing nematophagous fungi. Different fungi may have different requirements of soil pH. While the ring- and knob-forming species are associated with soils of a low pH, species with unmodified hyphae were isolated significantly more frequently from soils with a higher pH. Conidium-forming endoparasites were isolated from samples with relatively low soil pH (Gray, 1987). Optimal pH for *P. lilacinus* sporulation occurred at pH 4 – 6 (Villanueva and Davide, 1984). Activity of pelletized *H. rhossiliensis* was negatively correlated with soil pH (Jaffee and Zasoski, 2001). Maximum activity occurred at pH 4.5, and activity gradually declined to near zero as the pH increased to 6.5 and rapidly declined to near zero as the pH dropped below 4.0. It is concluded that low soil pH suppresses soil organisms that otherwise interfere with growth of *H. rhossiliensis* from alginate pellets.

Gray (1987) studied the effects of soil nutrients N, P, and K on predacious fungi and endoparasitic fungi. Endoparasites with adhesive conidia were independent of soil nutrients, while those species with ingested conidia were isolated from soils with high concentrations of nutrients. Knob-forming predators, which rely on their ability to produce traps spontaneously, were isolated from soils with low concentration of nutrients, while those species with constricting rings were isolated from richer soils, which contain greater densities of nematodes. Net-forming species were largely independent of soil fertility, although generally they were isolated from soils with limited nutrients, especially low K. Culbreath et al. (1984) observed that the level of fungal colonization of *M. arenaria* eggs was higher in soils deficient in N than in soils with adequate N fertilization. Natural concentration of heavy metals in the soil examined by Gray (1987) generally restricted the distribution of nematophagous fungi, with their presence being significantly associated with soils containing low concentrations of Cr, Ni, and Cu. The parasitism of *M. xenoplax* by *H. rhossiliensis* was greatly influenced by the kind and concentration of ionic solutes in the ambient solution; KCl increased infection of the nematode by the fungus (Jaffee and Zehr, 1983).

Gray (1985) studied the effects of organic matter on endoparasitic fungi and predacious fungi of vermiform nematodes. Those endoparasites producing conidia were significantly associated with soils with high organic matter. Ring-forming species were isolated more frequently from soil with relatively high organic matter contents, while species producing adhesive networks were isolated more frequently from soils with low organic matter contents. Since Linford's demonstrations that addition of organic matter to soil reduced nematode population, numerous studies have been done to use organic matters to control plant-parasitic nematodes. Most of these early efforts failed. The effects of the organic amendments on nematophagous fungi are complicated, and there is no simple principle. Predacious fungi may increase in response to the decomposition of organic matter. Timm et al. (2001) demonstrated that frequencies and population densities of most species of predacious fungi were similar in conventionally and organically managed plots. In contrast, parasitism of *M. xenoplax* by *H. rhossiliensis* was decreased by addition of organic matter (Jaffee, et al., 1994). Mankau (1962b) compared the effects of various types of organic manure on a soil nematode community. Dung and green manure appeared to promote the greatest activity of predacious fungi.

Recently, studies of the effects of soil amendments on egg and female parasites have increased. The effect of organic matter on egg-parasitic fungi may depend on the type of organic matter and the soil biological communities and environmental conditions. Pyrowolakis et al. (1999) demonstrated that green manure increased egg-parasitic activities of fungi on *H. schachtii* in soil from a mangold-monoculture field and reduced the activities in soil from a field in sugar beet-wheat rotation. Straw fertilizer favored more fungal parasitism of *H. schachtii* eggs than manure (Sosnowska and Banaszak, 2000). Some studies have demonstrated that amendments of chitin or collagen into soil suppressed plant-parasitic nematode populations. Some studies, however, disproved the idea that organic amendments stimulated the fungal parasitism of eggs or females (Mankau, 1961; Rodríguez-Kábana et al., 1984a).

Although the mechanisms of the suppression of nematodes by soil amendments are not fully understood, chitinolytic or collagenolytic and other enzymatic activities of soil organisms, including fungi, increased after addition of the amendments; the changes of enzymatic activities of soil microorganism are thought to be responsible for the suppression of nematode populations (Galper, et al., 1991; Godoy, et al., 1983; Mian, et al., 1982; Rodríguez-Kábana, et al., 1989; Spiegel, et al., 1986). Mycofloras previously associated with parasitism of eggs of *H. glycines* and *M. arenaria* were isolated from chitin-amended soil, but the exact relationship between nematodes and the microorganisms stimulated by chitin remained obscure (Stirling, 1991).

Effect of agricultural chemicals on nematophagous fungus is poorly understood. Crump and Kerry (1986) tested the effects of 25 chemicals on the growth of *V. chlamydosporium*. Of the nematicides tested, oxamyl was the only one that decreased growth of *C. destructans* and *V. chlamydosporium*. The insecticides and herbicides tested had little effect on the fungal growth. Fungicides used to control plant diseases may also suppress fungal parasites of nematodes but it has not been well studied.

20.7 Prospects

In the 20th century, scientists have made tremendous progress in exploring the possibility of using fungal antagonists to control plant-parasitic nematodes. Numerous papers have been published. However, only a few successful examples in the use of fungal antagonists to control plant-parasitic nematodes at a small scale have been reported, and no widely accepted commercial biological control agents for nematode control have yet been developed. At the beginning of the new century and millenium, it is important to reassess the possibility of using fungal antagonists to control plant-parasitic nematodes, and to identify problems, challenges, opportunities, and directions.

Biological control of nematodes is not an easy task because of the complex of agricultural soil ecosystems. Many more greenhouse and field evaluations are needed to determine efficacy of a biological agent than a chemical nematicide. Consequently, more investment may be needed for development of a biological agent than a chemical nematicide that generally needs millions of dollars. Such a large investment used in development of a chemical nematicide has never been used in development of a biological agent against nematodes. Most research on biological control has been conducted by individuals in public research institutes or universities. These research projects are usually short-term, rarely more than 5 years. In such a short period, it is difficult to plan continuous research from searching for potential fungi to commercializing a biological control product. Involvement of industry is very important in speeding up the development of biological control agents. Although there has been an increase of involvement, industries are still in the "wait and see mode" and skeptical of biological control of nematodes using fungal antagonists. Government or public funding agencies will continue to be the major source of support for research of biological control of nematodes, especially in those aspects such as basic understanding of biology, ecology, and epidemiology of nematophagous fungi. However, without input from industry, a successful exploitation and development of biological agents for nematode control may not be achieved in the near future.

The evolution of molecular biology and biotechnology provides opportunities and also challenges for the research of biological control of nematodes. Enzymatic and DNA analysis are useful in identifying fungal strains with various degrees of pathogenicity to nematodes. Genetic engineering and gene transfer technology have been used in the development of biological agents for control of insects, plant pathogens, and weeds. Biotechnology will be highly useful in construction of a nematophagous fungus with combination of the best attributes from different species or strains so that a high efficient control agent may be obtained. On the other hand, with a rapid increase of application of molecular biology and biotechnology in isolation of pest- or pathogen-resistant genes and in transformation of plants, there is a trend that public institutions and industries will become more interested in the development of resistant cultivars than the development of biological control agents. This challenge may be as great as the challenge from development of cheap nematicides in the 1940s and 1950s.

Nematode control agents using fungal antagonists may not be able to replace nematicides or any other control methods. Rather, biological control should be a part of an integrated management program. Some fungi may not be able to suppress nematode densities to below a damage threshold. However, by reducing nematode inoculum, a fungus may reduce yield loss caused by nematodes. With combination of biological control, a resistant cultivar or crop rotation may be used more efficiently than otherwise without biological control agents. Nematologists continue to debate whether manipulation of natural enemies in soil is likely to be more successful than the addition of selected agents used to inundate pests (Kerry, 1993). Kerry believed that both approaches appear valid and much can be learned from either.

Although there have been many surveys of nematophagous fungi, we may expect that a large portion of fungal antagonists in nature is yet to be discovered. More efficient agents may be found with more extensive investigations. Suppression of nematodes by fungal antagonists occurs in many agricultural soils. Understanding the mechanisms of the suppression may lead to a success in biological control through manipulation of soil environments or through introduction of biological agents. Variations of virulence among strains are common in nematophagous fungi. Different strains may be adapted to different environmental conditions and therefore may be useful in certain regions or fields under certain agricultural systems but not in others. A system should be developed to rapidly identify strains with certain attributes, to maintain virulence or pathogenicity of strains, and to formulate easily with different strains according to the use in different regions and fields.

Nematode targets should be considered when a fungus is used as a biological control agent (Kerry, 1992, 1993). Facultative parasites of eggs and females, such as *V. chlamydosporium*, may be effective for some cyst nematodes (Kerry, 1987). Nematode eggs are more susceptible when they are immature than when they contain J2 (Chen and Chen, 2003; Irving and Kerry, 1986). It is important that fungi infect the nematode soon after a female or egg mass is present on the root surface. A successful parasite of eggs of sedentary endoparasitic nematodes should prove to be a good root colonizer. This is especially important for root-knot nematodes because the females and a large portion of gelatinous matrices containing eggs remain embedded in the root. Only a good root colonizer, such as *P. lilacinus* (Cabanillas, et al., 1988), may attack young females and the embedded egg masses. Fungi that infect mobile stages of nematodes by producing traps or adhesive spores may be more effective against ectoparasitic than endoparasitic nematodes (Eayre, et al., 1987). Fecundity of nematodes and the rate of embryonic development are likely to affect control efficiency (Kerry, 1992).

Inoculum level is an important factor affecting the efficacy of a fungal agent on nematodes. In many greenhouse and field studies, high dosages were used. Much more effort is needed to reduce the effective dosage so that a practical, cheap agent may be developed. A good rhizosphere competitor and root colonizer may be used as a seed-coating agent so that a low effective dosage may be obtained.

An ideal biological control agent should attack different developmental stages of nematodes. This generally cannot be obtained by a single species of fungus because most nematophagous fungi cannot attack all developmental stages. Multiple organisms may be formulated into a single agent or developed in separated formulations and applied separately. But the costs may be a concern to use multiple agents in control of nematodes, especially the nematodes of field crops. Furthermore, interaction among agents may not always be synergistic. A fungal agent may be integrated with some organic or inorganic compounds, which may improve soil environments for fungal antagonists of nematodes, or improve plant root development and increase plant tolerance to the nematode attack. A fungus tolerant to fungicide, for example benomyl-tolerant mutants of *V. lecanii* (Meyer, 1994), may be formulated with a low dosage of fungicide to inhibit fungal competitor and encourage establishment of the biocontrol agent in soil.

Whether a biological agent can be commercialized or not depends on its biological effectiveness and market value. Biological effectiveness is largely determined by natural factors, whereas the market value is greatly influenced by social components and may be changed over time. High costs of biological control agents may be acceptable for use on some ornamental, fruit, and other plants with high economic values. For use on field crops, a product must be relatively cheap and can compete with nematicides and resistant cultivars. It is difficult for products of nematophagous fungi to meet these requirements.

The value of environmental benefit with the use of biological control has not been taken into account by consumers. With increasing concern of food quality, this may be changed in future. In the 20th century, crop production has much emphasized on quantity, especially in the developing countries where the increase of food demands has been intensive. This may be changed in the 21st century. With more consideration of the quality of food products and environment, biological control of plant-parasitic nematodes may become more realistic. Use of fungal antagonists for control of nematodes may be an important component in an integrated pest management program in sustainable agriculture and a significant contribution to the safe and friendly use of natural resources on the planet.

References

Agrawal, P. K., B. Lai, S. Wahab, O. P. Srivastava, and S. C. Misra. 1979. Orbital Paecilomycosis due to *Paecilomyces lilacinus* (Thom) Samson. Sabouraudia 17: 363 – 370.

Al-Hazmi, A. S., A. A. M. Ibrahim, and A. T. Abdul-Raziq. 1993. Evaluation of a nematode-encapsulating fungi complex for control of *Meloidogyne javanica* on potato. Pakistan Journal of Nematology 11: 139 – 149.

Alam, M. M. 1990. *Paecilomyces lilacinus* – A nematode biocontrol agent. Pp. 71 – 82. In M. S. Jairajpuri, M. M. Alam, and I. Ahmad, eds. Nematode Bio-control (aspects and prospects). Delhi: CBS Publishers and Distributors.

Bagyaraj, D. J., A. Manjunath, and D. D. R. Reddy. 1979. Interactions of vesicular-arbuscular mycorrhizae with root knot nematodes in tomato. Plant and Soil 51: 397 – 403.

Barron, G. L. 1977. The Nematode-destroying Fungi. Lancaster, PA: Lancaster Press.

Barron, G. L. 1978. Nematophagous fungi: endoparasites in Ontario and their ability to parasitize *Rhabditis tericola*. Microbial Ecology 4: 157 – 163.

Barron, G. L. 1987. The gun cell of *Haptoglossa mirabilis*. Mycologia 79: 877.

Barron, G. L. 1992. Lignolytic and cellulolytic fungi as predators and parasites. Pp. 311 – 326. In G. C. Carroll and D. T. Wicklow, eds. The Fungal Community: Its Organization and Role in the Ecosystem. New York: Marcel Dekker.

Beakes, G. W., and S. L. Glockling. 1998. Injection tube differentiation in gun cells of a *Haptoglossa* species which infects nematodes. Fungal Genetics and Biology 24: 45 – 68.

Belder, E., and E. Jansen. 1994a. The influence of temperature, nutrition, light and the growth time of the mycelium on capture and infection of *Meloidogyne hapla* and *Arthrobotrys oligospora*. Fundamental and Applied Nematology 17: 57 – 66.

Belder, E. D., and E. Jansen. 1994b. Saprophytic and predacious abilities in *Arthrobotrys oligospora* in relation to dead and living root-knot nematodes. Fundamental and Applied Nematology 17: 432.

Bernard, E. C., L. H. Self, and D. D. Tyler. 1997. Fungal parasitism of soybean cyst nematode, *Heterodera glycines* (Nematoda: Heteroderidae), in differing cropping-tillage regimes. Applied Soil Ecology 5: 57 – 70.

Boag, B., and L. V. Lopez-Llorca. 1989. Nematodes and nematophagous fungi associated with cereal fields and permanent pasture in eastern Scotland. Crop Research 29: 1 – 10.

Borrebaeck, C. A. K., B. Mattiasson, and B. Nordbring-Hertz. 1984. Isolation and partial characterization of a carbohydrate-binding protein from a nematode-trapping fungus. Journal of Bacteriology 159: 53 – 56.

Bourne, J. M., and B. R. Kerry. 1999. Effect of the host plant on the efficacy of *Verticillium chlamydosporium* as a biological control agent of root-knot nematodes at different nematode densities and fungal application rates. Soil Biology and Biochemistry 31: 75 – 84.

Bourne, J. M., and B. R. Kerry. 2000. Observations on the survival and competitive ability of the nematophagous fungus *Verticillium chlamydosporium* in soil. International Journal of Nematology 10: 9 – 18.

Bourne, J. M., B. R. Kerry, and F. A. A. M. De Leij. 1994. Methods for the study of *Verticillium chlamydosporium* in the rhizosphere. Journal of Nematology 26: 587 – 591.

Cabanillas, E., K. R. Barker, and M. E. Daykin. 1988. Histology of the interactions of *Paecilomyces lilacinus* with *Meloidogyne incognita* on tomato. Journal of Nematology 20: 362 – 265.

Carder, J. H., R. Segers, T. M. Butt, D. J. Barbara, N. Mende, and J. Coosemans. 1993. Taxonomy of the nematophagous fungi *Verticillium chlamydosporium* and *V. suchlasporium* based on secreted enzyme activities and RFLP analysis. Journal of Invertebrate Pathology 62: 178 – 184.

Caroppo, S., B. Perito, and O. Pelagatti. 1990. In vitro evaluation of nematicide activity by several fungi against *Meloidogyne incognita* eggs. Redia 73: 451 – 462.

Carris, L. M., D. A. Glawe, C. A. Smyth, and D. I. Edwards. 1989. Fungi associated with populations of *Heterodera glycines* in two Illinois soybean fields. Mycologia 81: 66 – 75.

Cayrol, J. C. 1981. Possibilities of biological control of *Meloidogyne* nematode crop pest with the use of the "R 350" nematophagous fungus (*Arthrobotrys irregularis*). La Defense des Vegetaux 35: 279 – 282.

Cayrol, J. C., C. Djian, and L. Pijarowski. 1989. Study of the nematocidal properties of the culture filtrate of the nematophagous fungus *Paecilomyces lilacinus*. Revue de Nematologie 12: 331 – 336.

Cayrol, J. C., J. P. Frankowski, A. Laniece, G. D. Hardemare, and J. P. Talon. 1978. Control of nematodes in mushroom cultures. Statement of a method of biological control using a predatory Hyphomycetes: *Arthrobotrys robusta* strain Antipolis (Royal 300). Pepinieristes Horticulteurs Maraichers, Revue Horticole 184: 23 – 30.

Cayrol, J. C., and J. P. Frankowski. 1986. Influence of the number of parasitizing conidia of *Hirsutella rhossiliensis* on the mortality of *Ditylenchus dipsaci*. Revue de Nematologie 9: 411 – 412.

Chen, F. J., and S. Y. Chen. 2002. Mycofloras in cysts, females, and eggs of the soybean cyst nematode in Minnesota. Applied Soil Ecology 19: 35 – 50.

Chen, S. Y. 1994. Fungal antagonists of the soybean cyst nematode, *Heterodera glycines*. Ph. D. dissertation, University of Florida, Gainesville, Florida.

Chen, S. Y., and F. J. Chen. 2003. Fungal parasitism of *Heterodera glycines* eggs as influenced by egg age and pre-colonization of cysts by other fungi. Journal of Nematology 35: 271 – 277.

Chen, S. Y., and D. W. Dickson. 1996. Fungal penetration of the cyst wall of *Heterodera glycines*. Phytopathology 86: 391 – 427.

Chen, S. Y., D. W. Dickson, J. W. Kimbrough, R. McSorley, and D. J. Mitchell. 1994. Fungi associated with females and cysts of *Heterodera glycines* in a Florida soybean field. Journal of Nematology 26: 296 – 303.

Chen, S. Y., D. W. Dickson, and D. J. Mitchell. 1996a. Population development of *Heterodera glycines* in response to mycoflora in soil from Florida. Biological Control 6: 226 – 231.

Chen, S. Y., D. W. Dickson, and D. J. Mitchell. 1996b. Pathogenicity of fungi to *Heterodera glycines*. Journal of Nematology 28: 145 – 158.

Chen, S. Y., D. W. Dickson, and D. J. Mitchell. 2000a. Viability of *Heterodera glycines* exposed to fungal filtrates. Journal of Nematology 32: 190 – 197.

Chen, S. Y., D. W. Dickson, and E. B. Whitty. 1996c. Fungi associated with egg masses of *Meloidogyne incognita* and *M. javanica* in a Florida tobacco field. Nematropica 26: 153 – 157.

Chen, S. Y., X. Z. Liu, and F. J. Chen. 2000b. *Hirsutella minnesotensis* sp. nov. – a new pathogen of the soybean cyst nematode. Mycologia 92: 819 – 824.

Chen, S. Y., and D. C. Reese. 1999. Parasitism of the nematode *Heterodera glycines* by the fungus *Hirsutella rhossiliensis* as influenced by crop sequence. Journal of Nematology 31: 437 – 444.

Ciancio, A., A. Logrieco, F. Lamberti, and A. Bottalico. 1988. Nematicidal effects of some *Fusarium* toxins. Nematologia Mediterranea 16: 137 – 138.

Coles, G. C., M. B. Dicklow, and B. M. Zuckerman. 1989. Protein changes associated with the infection of the nematode *Caenorhabditis elegans* by the nematophagous fungus *Drechmeria coniospora*. International Journal for Parasitology 19: 733 – 736.

Cooke, R. C. 1962a. Behavior of nematode-trapping fungi during decomposition of organic matter in the soil. Transactions British Mycological Society 45: 314 – 320.

Cooke, R. C. 1962b. The ecology of nematode-trapping fungi in the soil. Annals of Applied Biology 50: 507 – 513.

Cooke, R. C. 1963. Ecological characteristics of nematode-trapping Hyphomycetes. Annals of Applied Biology 52: 431 – 437.

Cooke, R. C. 1968. Relationships between nematode-destroying fungi and soil-borne phytonematodes. Phytopathology 58: 909 – 913.

Cooke, R. C., and B. E. Godfrey. 1964. A key to the nematode-destroying fungi. Transactions British Mycological Society 47: 61 – 74.

Coosemans, J. 1991. Antagonistic activity of a *Cylindrocarpon olidum* (Wollenw.) isolate against nematodes. Mededelingen van de Faculteit Landbouwwetenschappen Rijksuniversiteit Gent 56: 223 – 228.

Couch, J. N. 1937. The formation and operation of the traps in the nematode-catching fungus, *Dactylella bembicodes* Drechsler. Journal of the Elisha Mitchell Scientific Society 53: 301 – 309.

Crump, D. H. 1991. Fungal species isolated from beet (BCN), cereal (CCN) and potato (PCN) cyst nematodes. Bulletin Section Regionale Ouest Palearctique (SROP) 14: 58 – 64.

Crump, D. H., and B. R. Kerry. 1983. Possibilities for biological control of beet cyst-nematode with parasitic fungi. Aspects of Applied Biology 24: 59 – 64.

Crump, D. H., and B. R. Kerry. 1986. Effects of some agrochemicals on the growth of two nematophagous fungi. Nematologica 32: 363 – 366.

Crump, D. H., and B. R. Kerry. 1987. Studies on the population dynamics and fungal parasitism of *Heterodera schachtii* in soil from a sugar-beet monoculture. Crop Protection 6: 49 – 55.

Culbreath, A. K., R. Rodríguez-Kábana, and G. Morgan-Jones. 1984. An agar disc method for isolation of fungi colonizing nematode eggs. Nematropica 14: 145 – 154.

Dackman, C., H.-B. Jansson, and B. Nordbring-Hertz. 1992. Nematophagous fungi and their activities in soil. Pp. 95 – 130. In G. Stotzky and J. M. Bollag, eds. Soil biochemistry, Vol. 7. New York: Dekker.

Davis, K. G., 1998. Natural parasites and biological control. Pp. 369 – 387. In S. B. Sharma, ed. The Cyst Nematodes. Dordrecht: Kluwer Academic Publishers.

De Leij, F. A. A. M., K. G. Davis, and B. R. Kerry. 1992a. The use of *Verticillium chlamydosporium* Goddard and *Pasteuria penetrans* (Thorne) Sayre & Starr alone and in combination to control *Meloidogyne incognita* on tomato plants. Fundamental and Applied Nematology 15: 235 – 242.

De Leij, F. A. A. M., J. A. Dennehy, and B. R. Kerry. 1992b. The effect of temperature and nematode species on interactions between the nematophagous fungus *Verticillium chlamydosporium* and root-knot nematodes (*Meloidogyne* spp.). Nematologica 38: 65 – 79.

De Leij, F. A. A. M., and B. R. Kerry. 1991. The nematophagous fungus *Verticillium chlamydosporium* as a potential biological control agent for *Meloidogyne arenaria*. Revue de Nematologie 14: 157 – 164.

Dijksterhuis, J., M. Veenhuis, and W. Harder. 1994. Nematophagous fungi: Structure-function relationships and physiological aspects. Advances in Microbial and Physiology 36: 111 – 143.

Dix, N. J. 1967. Mycostasis and root exudation: factors influencing the colonization of bean roots by fungi. Transactions of the British Mycological Society 50: 23 – 31.

Domsch, K. H., W. Gams, and T. H. Anderson. 1980. Compendium of Soil Fungi, vols. 1 and 2. London: Academic Press.

Dowsett, J. A., and J. Reid. 1977. Transmission and scanning electron microscope observations on the trapping of nematodes by *Dactylaria candida*. Canadian Journal of Botany 55: 2963 – 2970.

Drechsler, C. 1937. Some Hyphomycetes that prey on free-living tericolous nematodes. Mycologia 29: 447 – 552.

Dube, B., and G. C. Smart, Jr. 1987. Biological control of *Meloidogyne incognita* by *Paecilomyces lilacinus* and *Pasteuria penetrans*. Journal of Nematology 19: 222 – 227.

Duddington, C. L. 1951. The ecology of predacious fungi. I. Preliminary survey. Transactions British Mycological Society 34: 322 – 331.

Duddington, C. L. 1957. The friendly fungi: a new approach to the eelworm problem. London: Faber and Faber.

Dunn, M. T., R. M. Sayre, A. Carrell, and W. P. Wergin. 1982. Colonization of nematode eggs by *Paecilomyces lilacinus* (Thom) Samson as observed with scanning election microscope. Scanning Electron Microscopy 3: 1351 – 1357.

Durschner-Pelz, V. V., and H. J. Atkinson. 1988. Recognition of *Ditylenchus* and other nematodes by spores of the endoparasitic fungus *Verticillium balanoides*. Journal of Invertebrate Pathology 51: 97 – 106.

Eayre, C. G., B. A. Jaffee, and E. I. Zehr. 1987. Suppression of *Criconemella xenoplax* by the nematophagous fungus *Hirsutella rhossiliensis*. Plant Disease 71: 832 – 834.

Ferris, H., M. V. McKenry, and H. E. McKinney. 1976. Spatial distribution of nematodes in peach orchards. Plant Disease Reporter 60: 18 – 22.

Fitters, P. F. L., E. Belder, and E. Den-Belder. 1992. Effect of fungal products on nematodes. Gewasbescherming 23: 16 (Abstr.).

Francl, L. J., and V. H. Dropkin. 1985. *Glomus fasciculatum*, a weak pathogen of *Heterodera glycines*. Journal of Nematology 17: 470 – 475.

Gair, R., P. L. Mathias, and P. N. Harvey. 1969. Studies of cereal nematode populations and cereal yields under continuous or intensive culture. Annals of Applied Biology 63: 503 – 512.

Galper, S., E. Cohn, Y. Spiegel, and I. Chet. 1991. A collagenolytic fungus, *Cunninghamella elegans*, for biological control of plant-parasitic nematodes. Journal of Nematology 23: 269 – 274.

Gams, W. 1988. A contribution to the knowledge of nematophagous species of *Verticillium*. Netherlands Journal of Plant Pathology 94: 123 – 148.

Gaspard, J. T., and R. Mankau. 1987. Density-dependence and host-specificity of the nematode-trapping fungus *Monacrosporium ellipsosporum*. Revue de Nematologie 10: 241 – 246.

Gintis, B. O., G. Morgan-Jones, and R. Rodríguez-Kábana. 1983. Fungi associated with several developmental stages of *Heterodera glycines* from an Alabama soybean field soil. Nematropica 13: 181 – 200.

Godoy, G., R. Rodríguez-Kábana, and G. Morgan-Jones. 1982. Parasitism of eggs of *Heterodera glycines* and *Meloidogyne arenaria* by fungi isolated from cysts of *H. glycines*. Nematropica 12: 111 – 119.

Godoy, G., R. Rodríguez-Kábana, R. A. Shelby, and G. Morgan-Jones. 1983. Chitin amendments for control of *Meloidogyne arenaria* in infested soil. II. Effects on microbial population. Nematropica 13: 63 – 74.

Gray, N. F. 1982. Psychro tolerant nematophagous fungi from the maritime Antarctic. Plant and Soil 64: 431 – 436.

Gray, N. F. 1983. Ecology of nematophagous fungi: Distribution and habitat. Annals of Applied Biology 102: 501 – 509.

Gray, N. F. 1985. Ecology of nematophagous fungi: effect of soil moisture, organic matter, pH and nematode density on distribution. Soil Biology and Biochemistry 17: 499 – 507.

Gray, N. F. 1987. Nematophagous fungi with particular reference to their ecology. Biological Revue 62: 305 – 338.

Gray, N. F. 1988. Fungi attacking vermiform nematodes. Pp. 3 – 38. In G. O. Poinar, Jr. and H. -B. Jansson, eds. Diseases of Nematodes vol. II. Boca Raton, Florida: CRC Press.

Gray, N. F., and F. Bailey. 1985. Ecology of nematophagous fungi: vertical distribution in a deciduous woodland. Plant and Soil 86: 217 – 223.

Hallmann, J., and R. A. Sikora. 1994. In vitro and in vivo control of *Meloidogyne incognita* with culture filtrates from nonpathogenic *Fusarium oxysporum* on tomato. Journal of Nematology 26: 102 (Abstr.).

Hanssler, G. 1990. Parasitism of *Verticillium lecanii* on cysts of *Heterodera schachtii*. Zeitschrift fur Pflanzenkrankheiten und Pflanzenschutz 97: 194 – 201.

Hanssler, G., and M. Hermanns. 1981. *Verticillium lecanii* as a parasite on cysts of *Heterodera schachtii*. Zeitschrift fur Pflanzenkrankheiten und Pflanzenschutz 88: 678 – 681.

Hartwig, E. E. 1981. Breeding productive soybean cultivars resistant to the soybean cyst nematode for the Southern United States. Plant Disease 65: 303 – 307.

Hay, F. S., and R. A. Skipp. 1993. Fungi and actinomycetes associated with cysts of *Heterodera trifolii* Goffart (Nematoda: Tylenchida) in pasture soils in New Zealand. Nematologica 39: 376 – 384.

Heijbroek, W. 1983. Some effects of fungal parasites on the population development of the beet cyst nematode (*Heterodera schachtii* Schm.). Mededelingen van de Faculteit Landbouwwetenschappen Rijksuniversiteit Gent 48: 433 – 439.

Hewlett, T. E., D. W. Dickson, D. J. Mitchell, and M. E. Kannwischer-Mitchell. 1988. Evaluation of *Paecilomyces lilacinus* as a biocontrol agent of *Meloidogyne javanica* on tobacco. Journal of Nematology 20: 578 – 584.

Holland, R. J., K. L. Williams, K. Alamgir, and A. Khan. 1999. Infection of *Meloidogyne javanica* by *Paecilomyces lilacinus*. Nematology 1: 131 – 139.

Ingham, R. E. 1988. Interactions between nematodes and vesicular arbuscular mycorrhizae. Agriculture, Ecosystems and Environment 24: 169 – 182.

Irving, F., and B. R. Kerry. 1986. Variation between strains of the nematophagous fungus, *Verticillium chlamydosporium* Goddard. II. Factors affecting parasitism of cyst nematode eggs. Nematologica 32: 474 – 485.

Isogai, A., A. Suzuki, S. Higashikawa, and S. Tamura. 1980. Structure of peptidal antibiotic P-168 produced by *Paecilomyces lilacinus*. Agricultural and Biological Chemistry 44: 3033 – 3036.

Jaffee, B. A. 1993. Density-dependent parasitism in biological control of soil-borne insects, nematodes, fungi and bacteria. Biocontrol Science and Technology 3: 235 – 246.

Jaffee, B. A. 1999. Enchytraeids and nematophagous fungi in tomato fields and vineyards. Phytopathology 89: 398 – 406.

Jaffee, B. A. 2000. Augmentation of soil with the nematophagous fungi *Hirsutella rhossiliensis* and *Arthrobotrys haptotyla*. Phytopathology 90: 498 – 504.

Jaffee, B. A., H. Ferris, J. J. Stapleton, M. V. K. Norton, and A. E. Muldoon. 1994. Parasitism of nematodes by the fungus *Hirsutella rhossiliensis* as affected by certain organic amendments. Journal of Nematology 26: 152 – 161.

Jaffee, B. A., J. T. Gaspard, and H. Ferris. 1989. Density-dependent parasitism of the soil-borne nematode *Criconemella xenoplax* by *Hirsutella rhossiliensis*. Microbial Ecology 17: 193 – 200.

Jaffee, B. A., and A. E. Muldoon. 1989. Suppression of cyst nematode by natural infestation of a nematophagous fungus. Journal of Nematology 21: 505 – 510.

Jaffee, B. A., A. E. Muldoon, and W. A. M. Didden. 1997. Enchytraeids and nematophagous fungi in soil microcosms. Biology and Fertility of Soils 25: 382 – 388.

Jaffee, B. A., A. E. Muldoon, R. Phillips, and M. Mangel. 1990. Rates of spore transmission, mortality, and production for the nematophagous fungus *Hirsutella rhossiliensis*. Phytopathology 80: 1083 – 1088.

Jaffee, B. A., A. E. Muldoon, and B. B. Westerdahl. 1996. Failure of a mycelial formulation of the nematophagous fungus *Hirsutella rhossiliensis* to suppress the nematode *Heterodera schachtii*. Biological Control 6: 340 – 346.

Jaffee, B. A., R. Phillips, A. Muldoon, and M. Mangel. 1992. Density-dependent host-pathogen dynamics in soil microcosms. Ecology 73: 495 – 506.

Jaffee, B. A., P. F. Santos, and A. E. Muldoon. 1997. Suppression of nematophagous fungi by enchytraeid worms: A field exclosure experiment. Oecologia 112: 412 – 423.

Jaffee, B. A., and R. J. Zasoski. 2001. Soil pH and the activity of a pelletized nematophagous fungus. Phytopathology 91: 324 – 330.

Jaffee, B. A., and E. I. Zehr. 1982. Parasitism of the nematode *Criconemella xenoplax* by the fungus *Hirsutella rhossiliensis*. Phytopathology 72: 1378 – 1381.

Jaffee, B. A., and E. I. Zehr. 1983. Effects of certain solutes, osmotic potential, and soil solutions on parasitism of *Criconemella xenoplax* by *Hirsutella rhossiliensis*. Phytopathology 73: 544 – 546.

Jaffee, B. A., and E. I. Zehr. 1985. Parasitic and saprophytic abilities of the nematode-attacking fungus *Hirsutella rhossiliensis*. Journal of Nematology 17: 341 – 345.

Jakobsen, J. 1974. The importance of monocultures of various host plants for the population density of *Heterodera avenae*. Tidsskrift for Planteavl 78: 697 – 700.

Jansson, H. -B., A. Jeyaprakash, and B. M. Zuckerman. 1985. Differential adhesion and infection of nematodes by the endoparasitic fungus *Meria coniospora* (Deuteromycetes). Applied and Environmental Microbiology 49: 552 – 555.

Jansson, H. -B., and B. Nordbring-Hertz. 1979. Attraction of nematodes to living mycelium of nematophagous fungi. Journal of General Microbiology 112: 89 – 93.

Jansson, H. -B., and B. Nordbring-Hertz. 1980. Interactions between nematophagous fungi and plant-parasitic nematodes: attraction, induction of trap formation and capture.

Nematologica 26: 383 – 389.

Jansson, H.-B., and B. Nordbring-Hertz. 1984. Involvement of sialic acid in nematode chemotaxis and infection by an endoparasitic nematophagous fungus. Journal of General Microbiology 130: 39 – 43.

Jansson, H.-B., and B. Nordbring-Hertz. 1988. Infection events in the fungus-nematode system. Pp. 59 – 72. In G. O. Poinar, Jr. and H.-B. Jansson, eds. Diseases of Nematodes, vol. 2. Boca Raton, Florida: CRC Press.

Jatala, P. 1986. Biological control of plant-parasitic nematodes. Annual Review of Phytopathology 24: 453 – 489.

Jatala, P., R. Kaltenbach, and M. Bocangel. 1979. Biological control of *Meloidogyne incognita acrita* and *Globodera pallida* on potato. Journal of Nematology 11: 303. (Abstr.).

Jones, F. G. W. 1975. The soil as an environment for plant parasitic nematodes. Annals of Applied Biology 79: 113 – 139.

Kerry, B. R. 1974. A fungus associated with young females of the cereal cyst-nematode, *Heterodera avenae*. Nematologica 20: 259 – 260.

Kerry, B. R. 1975. Fungi and the decrease of cereal cyst-nematode populations in cereal monoculture. Bulletin of European Mediterranean Plant Protection Organization 5: 353 – 361.

Kerry, B. R. 1981. Fungal parasites: a weapon against cyst nematodes. Plant Disease 65: 390 – 393.

Kerry, B. R. 1984. Nematophagous fungi and the regulation of nematode populations in soil. Helminthological Abstracts Series B Plant Nematology 53: 1 – 14.

Kerry, B. R. 1986. An assessment of the role of parasites and predators in the regulation of cyst nematode populations. Pp. 433 – 450. In F. Lamberti and C. E. Taylor, eds. Cyst Nematodes. New York: Plenum Press.

Kerry, B. R. 1987. Biological control. Pp. 233 – 263 In R. H. Brown and B. R. Kerry, eds. Principles and Practice of Nematode Control in Crops. Sydney: Academic Press.

Kerry, B. R. 1988. Fungal parasites of cyst nematodes. Agriculture, Ecosystems and Environment 24: 293 – 305.

Kerry, B. R. 1990. An assessment of progress toward microbial control of plant-parasitic nematodes. Journal of Nematology 22: 621 – 631.

Kerry, B. R. 1992. Key factors in the development of fungal agents for the control of cyst and root-knot nematodes. Pp. 139 – 144. In E. C. Tjamos, G. C. Papavizas, and R. J. Cook, eds. Biological Control of Plant Diseases. New York: Plenum Press.

Kerry, B. R. 1993. The use of microbial agents for the biological control of plant parasitic nematodes. Pp. 81 – 104. In D. G. Jones, ed. Exploitation of Microorganisms. London: Chapman and Hall.

Kerry, B. R., and S. Andersson. 1983. *Nematophthora gynophila* and *Verticillium chlamydosporium*, fungal parasites of cyst-nematodes, occur frequently in Swedish soils infested by cereal cyst-nematodes. Vaxtskyddsnotiser 47: 79 – 80, 88.

Kerry, B. R., and D. H. Crump. 1980. Two fungi parasitic on females of cyst-nematodes (*Heterodera* spp.). Transactions British Mycological Society 74: 119 – 125.

Kerry, B. R., and D. H. Crump. 1998. The dynamics of the decline of the cereal cyst nematode, *Heterodera avenae*, in four soils under intensive cereal production.

Fundamental and Applied Nematology 21:617 – 625.

Kerry, B. R., D. H. Crump, and L. A. Mullen. 1982a. Natural control of the cereal cyst nematode, *Heterodera avenae* Woll., by soil fungi at three sites. Crop Protection 1: 99 – 109.

Kerry, B. R., D. H. Crump, and L. A. Mullen. 1982b. Studies of the cereal cyst-nematode, *Heterodera avenae* under continuous cereals, 1975-1978. II. Fungal parasitism of nematode females and eggs. Annals of Applied Biology 100:489 – 499.

Kerry, B. R., and B. A. Jaffee. 1997. Fungi as biological control agents for plant parasitic nematodes. Pp. 203 – 218. In D. T. Wicklow and B. E. Soderstrom, eds. The Mycota: a Comprehensive Treatise on Fungi as Experimental Systems for Basic Applied Research: Volume 4 Environmental and Microbial Relationships. Berlin Herdelberg: Springer-Verlag.

Kerry, B. R., I. A. Kirkwood, F. A. A. M. De Leij, J. Barba, M. B. Leijdens, and P. C. Brookes. 1993. Growth and survival of *Verticillium chlamydosporium* Goddard, a parasite of nematodes, in soil. Biocontrol Science and Technology 3:335 – 365.

Khan, M. R., and B. K. Goswami. 2000. Effect of culture filtrates of *Paecilomyces lilacinus* isolates on hatching of *Meloidogyne incognita* eggs. Annals of Plant Protection Sciences 8:62 – 65.

Kim, D. G., and R. D. Riggs. 1991. Characteristics and efficacy of a sterile hyphomycete (ARF18), a new biocontrol agent for *Heterodera glycines* and other nematodes. Journal of Nematology 23:275 – 282.

Kim, D. G., and R. D. Riggs. 1994. Techniques for isolation and evaluation of fungal parasites of *Heterodera glycines*. Journal of Nematology 26:592 – 595.

Kim, D. G., and R. D. Riggs. 1995. Efficacy of the nematophagous fungus ARF18 in alginate-clay pellet formulations against *Heterodera glycines*. Journal of Nematology 602 – 608.

Kim, D. G., R. D. Riggs, and J. C. Correll. 1998. Isolation, characterization, and distribution of a biocontrol fungus from cysts of *Heterodera glycines*. Phytopathology 88:465 – 471.

Kim, D. G., R. D. Riggs, and K. S. Kim. 1992. Ultrastructure of *Heterodera glycines* parasitized by Arkansas fungus 18. Phytopathology 82:429 – 433.

Kloepper, J. W., R. Rodríguez-Kábana, J. A. McInroy, and D. J. Collins. 1991. Analysis of populations and physiological characterisation of microorganisms in rhizospheres of plant with antagonistic properties to phytopathogenic nematodes. Plant and Soil 136:95 – 102.

Kunert, J., and H. Lysek. 1987. Lipolytic activity of ovicidal soil fungi. Biologia Czechoslovakia 42:285 – 293.

Lackey, B. A., A. E. Muldoon, and B. A. Jaffee. 1993. Alginate pellet formulation of *Hirsutella rhossiliensis* for biological control of plant-parasitic nematodes. Biological Control 3:155 – 160.

Li, T. F., K. Q. Zhang, and X. Z. Liu. 2000. Taxonomy of nematophagous fungi. Beijing: Chinese Scientific and Technological Publication.

Linford, M. B. 1937. Stimulated activity of natural enemies of nematodes. Science 85: 123 – 124.

Linford, M. B., F. Yap, and J. M. Oliveira. 1937. Reduction of soil populations of the root-knot nematode during decomposition of organic matter. Soil Science 45: 127 – 141.

Liou, J. Y., J. Y. Shih, and S. S. Tzean. 1999. *Exteya*, a new nematophagous genus from Taiwan, attacking the pinewood nematode (*Bursaphelenchus xylophilus*). Mycological Research 103: 241 – 248.

Liou, J. Y., and S. S. Tzean. 1992. Stephanocysts as nematode-trapping and infecting propagules. Mycologia 84: 786 – 790.

Liu, X. Z., and S. Y. Chen. 2001. Screening isolates of *Hirsutella* species for biological control of *Heterodera glycines*. Biocontrol Science and Technology 11: 151 – 160.

Liu, X. Z., W. F. Qiu, and Z. Q. Miao. 1993. Distribution of nematode-trapping fungi in China. Acta Mycologica Sinica 12: 253 – 256.

Liu, X. Z., M. H. Sun, R. J. Guo, X. D. Zhang, Y. Q. Xie, and W. F. Qiu. 1996. Biological control of soybean cyst nematode in China. Pp. 11 – 14. In W. Tang, R. J. Cook, and A. Rovira, eds. Proceeding of the International Workshop on Biological Control of Plant Diseases, Beijing, May 22 – 27, 1996. Beijing: China Agricultural University Press.

Liu, X. Z., and X. Y. Wu. 1993. Decline of soybean cyst nematode: a preliminary result. Proceeding of Second International Workshop on Plant Nematology, University of Karachi, Pakistan. Pp. 61 – 69.

Lopez-Llorca, L. V., and D. Claugher. 1990. Appressoria of the nematophagous fungus *Verticillium suchlasporium*. Micron and Microscopica Acta 21: 125 – 130.

Lopez-Llorca, L. V., and W. M. Robertson. 1992. Immunocytochemical localization of a 32 kDa protease from the nematophagous fungus *Verticillium suchlasporium* in infected nematode eggs. Experimental Mycology 16: 261 – 267.

Mani, A., and C. L. Sethi. 1984. Some characteristic of culture filtrate of *Fusarium solani* toxic to *Meloidogyne incognita*. Nematropica 14: 121 – 129.

Mankau, R. 1961. An attempt to control root-knot nematodes with *Dactylaria thaumasia* Drechsler and *Arthrobotrys arthrobotryoides* Lindau. Plant Disease Reporter 45: 164 – 166.

Mankau, R. 1962a. Soil fungistasis and nematophagous fungi. Phytopathology 52: 611 – 615.

Mankau, R. 1962b. The effect of some organic additives upon a soil nematode population and associated natural enemies. Nematologica 7: 65 – 73.

Mankau, R. 1980. Biocontrol: Fungi as nematode control agents. Journal of Nematology 12: 244 – 252.

Meyer, S. L. F. 1994. Effects of a wild type strain and a mutant strain of the fungus *Verticillium lecanii* on *Meloidogyne incognita* populations in greenhouse studies. Fundamental and Applied Nematology 17: 563 – 567.

Meyer, S. L. F., and R. N. Huettel. 1991. Comparisons of fungi and fungus-bioregulator combinations for control of *Heterodera glycines*, the soybean cyst nematode. Journal of Nematology 23: 540 (Abstr.).

Meyer, S. L. F., R. N. Huettel, and R. M. Sayre. 1990. Isolation of fungi from *Heterodera glycines* and in vitro bioassays for their antagonism to eggs. Journal of Nematology 22: 532 – 537.

Meyer, S. L. F., G. Johnson, M. Dimock, J. W. Gahey, and R. N. Huettel. 1997. Field efficacy of *Verticillium lecanii*, sex pheromone, and pheromone analogs as potential management agents for soybean cyst nematode. Journal of Nematology 29: 282 – 288.

Meyer, S. L. F., S. I. Massoud, D. J. Chitwood, and D. P. Roberts. 2000. Evaluation of *Trichoderma virens* and *Burkholderia cepacia* for antagonistic activity against root-knot nematode, *Meloidogyne incognita*. Nematology 2: 871 – 879.

Meyer, S. L. F., and R. J. Meyer. 1995. Effects of a mutant strain and a wild type strain of *Verticillium lecanii* on *Heterodera glycines* populations in the greenhouse. Journal of Nematology 27: 409 – 417.

Meyer, S. L. F., and R. J. Meyer. 1996. Greenhouse studies comparing strains of the fungus *Verticillium lecanii* for activity against the nematode *Heterodera glycines*. Fundamental and Applied Nematology 19: 305 – 308.

Mian, I. H., G. Godoy, R. A. Shelby, R. Rodríguez-Kábana, and G. Morgan-Jones. 1982. Chitin amendments for control of *Meloidogyne arenaria* in infested soil. Nematologica 12: 71 – 84.

Minter, D. W., and B. L. Brady. 1980. Mononematous species of *Hirsutella*. Transactions of the British Mycological Society 74: 271 – 282.

Molina, G. C., and R. G. Davide. 1986. Evaluation of microbial extracts for nematicidal activity against plant parasitic nematodes *Meloidogyne incognita* and *Radopholus similis*. Philippine Agriculturist 69: 173 – 186.

Morgan-Jones, G., G. Godoy, and R. Rodríguez-Kábana. 1981. *Verticillium chlamydosporium*, fungal parasite of *Meloidogyne arenaria* females. Nematologica 11: 115 – 120.

Morgan-Jones, G., and R. Rodríguez-Kábana. 1985. Phytonematode pathology: Fungal modes of action. A perspective. Nematropica 15: 107 – 114.

Morgan-Jones, G., and R. Rodríguez-Kábana. 1987. Fungal biocontrol for the management of nematodes. Pp. 94 – 99. In J. A. Veech and D. W. Dickson, eds. Vistas on Nematology. Hyattsville, MD: Society of Nematologists.

Morgan-Jones, G., and R. Rodríguez-Kábana. 1988. Fungi colonizing cysts and eggs. Pp. 39 – 58. In G. O. Poinar, Jr. and H. -B. Jansson, eds. Diseases of Nematodes, Vol. 2. Boca Raton, FL: CRC Press.

Morgan-Jones, G., J. F. White, and R. Rodríguez-Kábana. 1984. Phytonematode pathology: ultrastructural studies. II. Parasitism of *Meloidogyne arenaria* eggs and larvae by *Paecilomyces lilacinus*. Nematropica 14: 57 – 71.

Müller, H. G. 1958. The constricting ring mechanism of two predacious hyphomycetes. Transactions British Mycological Society 41: 341 – 364.

Müller, J. 1982. The influence of fungal parasites on the population dynamics of *Heterodera schachtii*. Nematologica 28: 161.

Nigh, E. A., I. J. Thomason, and S. D. Van Gundy. 1980. Identification and distribution of fungal parasites of *Heterodera schachtii* eggs in California. Phytopathology 70: 884 – 889.

Nordbring-Hertz, B. 1972. Scanning electron microscopy of the nematode-trapping organs in *Arthrobotrys oligospora*. Physiology of Plant 26: 279 – 284.

Nordbring-Hertz, B. 1988. Nematophagous fungi: strategies for nematode exploitation and for survival. Microbiological Sciences 5: 108 – 116.

Nordbring-Hertz, B., E. Friman, P. R. Johansson, and B. Mattiasson. 1981.

Host-microorganism interaction: Developmentally regulated lectin-mediated capture of nematodes by nematode-trapping fungi. Biology, Biochemistry, Clinical Biochemistry 1:43 –90.

Ohnesorge, B., J. Friedel, and U. Oesterlin. 1974. Investigations on the distribution pattern of *Heterodera avenae* and its changes in a field under continuous cereal cultivation. Zeitschrift fur Pflanzenkrankheiten und Pflanzenschutz 81:356 –363.

Perry, J. N. 1978. A population model for the effect of parasitic fungi on numbers of the cereal cyst-nematode, *Heterodera avenae*. Journal of Applied Ecology 15: 781 –787.

Persson, C., and H. B. Jansson. 1999. Rhizosphere colonization and control of *Meloidogyne* spp. by nematode-trapping fungi. Journal of Nematology 31: 164 –171.

Pyrowolakis, A., R. P. Schuster, and R. A. Sikora. 1999. Effect of cropping pattern and green manure on the antagonistic potential and the diversity of egg pathogenic fungi in fields with *Heterodera schachtii* infection. Nematology 1: 165 –171.

Quinn, M. A. 1987. The influence of saprophytic competition on nematode predation by nematode-trapping fungi. Journal of Invertebrate Pathology 49: 170 –174.

Rockhill, R. C., and M. D. Klein. 1980. *Paecilomyces lilacinus* as the cause of chronic maxillary simusitis. Journal of Clinical Microbiology 11: 737 –739.

Robb, E. J., and G. L. Barron. 1982. Nature's ballistic missile. Science 218: 1221 – 1222.

Rodríguez-Kábana, R., D. Boube, and R. W. Young. 1989. Chitinous materials from blue crab for control of root-knot nematode. 1. Effect of urea and enzymatic studies. Nematropica 19: 53 –74.

Rodríguez-Kábana, R., and G. Morgan-Jones. 1988. Potential for nematode control by mycofloras endemic in the tropics. Journal of Nematology 20: 191 –203.

Rodríguez-Kábana, R., G. Morgan-Jones, and B. O. Gintis. 1984a. Effects of chitin amendments to soil on *Heterodera glycines*, microbial populations, and colonization of cysts by fungi. Nematropica 14: 10 –25.

Rodríguez-Kábana, R., G. Morgan-Jones, G. Godoy, and B. O. Gintis. 1984b. Effectiveness of species of *Gliocladium, Paecilomyces, and Verticillium* for control of *Meloidogyne arenaria* in field soil. Nematropica 14: 155 –170.

Rogers, W. P. 1966. The reversible inhibition of exsheathment in some parasitic nematodes. Comparative Biochemistry and Physiology 17: 1103 –1110.

Rosenzweig, W. D., D. Premachandran, and D. Pramer. 1985. Role of trap lectins in the specificity of nematode capture of fungi. Canadian Journal of Microbiology 31: 693 – 695.

Saikawa, M., and N. Wada. 1986. Adhesive knobs in *Pleurotus ostreatus* (the oyster mushroom), as trapping organs for nematodes. Transactions of the Mycological Society of Japan 27: 113 –118.

Sayre, R. M., and L. S. Keeley. 1969. Factors influencing *Catenaria anguilulae* infections in a free-living and a plant parasitic nematode. Nematologica 15: 492 –502.

Segers, R., T. M. Butt, J. H. Carder, J. N. Keen, B. R. Kerry, and J. F. Peberdy. 1999. The subtilisins of fungal pathogens of insects, nematodes and plants: Distribution and variation. Mycological Research 103: 395 –402.

Segers, R., T. M. Butt, B. R. Kerry, A. Beckett, and J. F. Pedberdy. 1996. The role of the proteinase VCP1 produced by the nematophagous *Verticillium chlamydosporium* in the infection process of nematode eggs. Mycological Research 100: 421 – 428.

Sikora, R. A. 1992. Management of the antagonistic potential in agricultural ecosystems for the biological control of plant parasitic nematodes. Annual Review of Phytopathology 30: 45 – 70.

Sosamma, V. K., and P. K. Koshy. 1997. Biological control of *Meloidogyne incognita* on black pepper by *Pasteuria penetrans* and *Paecilomyces lilacinus*. Journal of Plantation Crops 25: 72 – 76.

Sosnowska, D., and H. Banaszak. 2000. The effect of organic fertilizers on fungi parasitization of beet cyst nematode (*Heterodera schachtii* Schmidt) eggs in sugar beet cultivated in a three years rotation. Journal of Plant Protection Research 40: 73 – 79.

Spiegel, Y., E. Cohn, and I. Chet. 1986. Use of chitin for controlling plant parasitic nematodes. 1. Direct effects on nematode reproduction and plant performance. Plant and Soil 95: 87 – 95.

Steudel, W., J. Müller, and J. Schlang. 1990. Investigations on fungal egg parasites of the sugar beet nematode (*Heterodera schachtii* Schmidt) in two sugar beet-cereal-rotations. Journal of Phytopathology 129: 316 – 326.

Stiles, C. M., and D. A. Glawe. 1989. Colonization of soybean roots by fungi isolated from cysts of *Heterodera glycines*. Mycologia 81: 797 – 799.

Stirling, G. R. 1979. Effect of temperature on parasitism of *Meloidogyne incognita* eggs by *Dactylella oviparasitica*. Nematologica 25: 104 – 110.

Stirling, G. R. 1988. Biological control of plant-parasitic nematodes. Pp. 93 – 139. In G. O. Poinar, Jr. and H. -B. Jansson, eds. Diseases of Nematodes, Vol. II. Boca Roton, FL: CRC Press.

Stirling, G. R. 1991. Biological control of plant parasitic nematodes: Progress, problems and prospects. Wallingford, UK: CAB International.

Stirling, G. R., and B. R. Kerry. 1983. Antagonists of the cereal cyst nematode *Heterodera avenae* Woll. in Australian soils. Australian Journal of Experimental Agriculture and Animal Husbandry 23: 318 – 324.

Stirling, G. R., K. A. Licastro, L. M. West, and L. J. Smith. 1998. Development of commercially acceptable formulations of the nematophagous fungus *Verticillium chlamydosporium*. Biological Control 11: 217 – 223.

Stirling, G. R., and R. Mankau. 1978. Parasitism of *Meloidogyne* eggs by a new fungal parasite. Journal of Nematology 10: 236 – 240.

Stirling, G. R., and R. Mankau. 1979. Mode of parasitism of *Meloidogyne* and other nematode eggs by *Dactylella oviparasitica*. Journal of Nematology 11: 282 – 288.

Stirling, G. R., M. V. McKenry, and R. Mankau. 1979. Biological control of root-knot nematodes (*Meloidogyne* spp.) on peaches. Phytopathology 69: 806 – 809.

Stirling, G. R., and L. M. West. 1991. Fungal parasites of root-knot nematode eggs from tropical and sub-tropical regions of Australia. Australasian Plant Pathology 20: 149 – 154.

Sturhan, D., and R. Schneider. 1980. *Hirsutella heteroderae*, a new nematode-parasitic fungus. Phytopathologische Zeitschrift 99: 105 – 115.

Surech, C. K., D. J. Bagyaraj, and D. D. R. Reddy. 1985. Effect of vesicular-arbuscular mycorrhizae on survival, penetration and development of root-knot nematode in tomato. Plant and Soil 87: 305 – 308.

Tedford, E. C., B. A. Jaffee, and A. E. Muldoon. 1992. Effect of soil moisture and texture on transmission of the nematophagous fungus *Hirsutella rhossiliensis* to cyst and root-knot nematodes. Phytopathology 82: 1002 – 1007.

Tedford, E. C., B. A. Jaffee, and A. E. Muldoon. 1994. Variability among isolates of the nematophagous fungus *Hirsutella rhossiliensis*. Mycological Research 98: 1127 – 1136.

Tedford, E. C., B. A. Jaffee, and A. E. Muldoon. 1995. Effect of temperature on infection of the cyst nematode *Heterodera schachtii* by the nematophagous fungus *Hirstuella rhossiliensis*. Journal of Invertebrate Pathology 66: 6 – 10.

Tedford, E. C., B. A. Jaffee, A. E. Muldoon, C. E. Anderson, and B. B. Westerdahl. 1993. Parasitism of *Heterodera schachtii* and *Meloidogyne javanica* by *Hirsutella rhossiliensis* in microplots over two growing seasons. Journal of Nematology 25: 427 – 433.

Timm, M. 1987. 'Biocon' controls nematodes biologically. Bio-Technology 5: 772 – 774.

Timm, L., D. Pearson, and B. Jaffee. 2001. Nematode-trapping fungi in conventionally and organically managed corn-tomato rotations. Mycologia 93: 25 – 29.

Timper, P., and B. B. Brodie. 1994. Effect of *Hirsutella rhossiliensis* on infection of potato by *Pratylenchus penetrans*. Journal of Nematology 26: 304 – 307.

Timper, P., and R. D. Riggs. 1998. Variation in efficacy of isolates of the fungus ARF against the soybean cyst nematode *Heterodera glycines*. Journal of Nematology 30: 461 – 467.

Timper, P., R. D. Riggs, and D. L. Crippen. 1999. Parasitism of sedentary stages of *Heterodera glycines* by isolates of a sterile nematophagous fungus. Phytopathology 89: 1193 – 1199.

Tribe, H. T. 1977a. Pathology of cyst nematodes. Biological Reviews 52: 477 – 507.

Tribe, H. T. 1977b. A parasite of white cysts of *Heterodera*: *Catenaria auxiliaris*. Transactions British Mycological Society 69: 367 – 376.

Tribe, H. T. 1980. Prospects for the biological control of plant-parasitic nematodes. Parasitology 81: 619 – 639.

Tunlid, A., H. -B. Jansson, and B. Nordbring-Hertz. 1992. Fungal attachment to nematodes. Mycological Research 96: 401 – 412.

Vancura, V. 1988. Microorganisms, their mutual relations and functions in the rhizosphere. Pp. 191 – 280. In V. Vancura and F. Kunc, eds. Soil Microbial Associations. Amserdam: Elsevier.

Veenhuis, M., W. Harder, and B. Nordbring-Hertz. 1989. Occurrence and metabolic significance of microbodies in trophic hyphae of the nematophagous fungus *Arthrobotrys oligospora*. Antonie van Leeuwenhoek 56: 241 – 249.

Veenhuis, M., B. Nordbring-Hertz, and W. Harder. 1985. An electron microscopical analysis of capture and initial stages of penetration of nematodes by *Arthrobotrys oligospora*. Antonie van Leeuwenhoek 51: 385 – 398.

Velvis, H., and P. Kamp. 1996. Suppression of potato cyst nematode root penetration by the endoparasitic nematophagous fungi *Hirsutella rhossiliensis*. European Journal of Plant Pathology 102: 115 – 122.

Villanueva, L. M., and R. G. Davide. 1984. Influence of pH, temperature, light and agar medium on the growth and sporulation of a nematophagous fungus, *Paecilomyces lilacinus* (Thom.) Samson. Philippine Agriculturist 67: 223 – 231.

Voss, B., P. Utermohl, and U. Wyss. 1992. Variation between strains of the nematophagous endoparasitic fungus *Catenaria anguillulae* Sorokin. II. Attempts to achieve parasitism of *Heterodera schachtii* Schmidt in pot trials. Zeitschrift fur Pflanzenkrankheiten und Pflanzenschutz 99: 311 – 318.

Warrior, P., L. A. Rehberger, M. Beach, P. A. Grau, G. W. Kirfman, and J. M. Conley. 1999. Commercial development and introduction of DiTera, a new nematicide. Pesticide Science 55: 376 – 379.

Wang, C. J., C. Y. Song, X. D. Zhang, Y. Q. Xie, X. Z. Liu, and M. Z. Wang. 1997. Sustainable control efficacy of *Paecilomyces lilacinus* against *Heterodera glycines*. Journal of Biological Control (China) 13: 26 – 28.

Westphal, A., and J. O. Becker. 1999. Biological suppression and natural population decline of *Heterodera schachtii* in a California field. Phytopathology 89: 434 – 440.

Westphal, A., and J. O. Becker. 2001. Components of soil suppressiveness against *Heterodera schachtii*. Soil Biology and Biochemistry 33: 9 – 16.

Wharton, D. A., and D. S. Murray. 1990. Carbohydrate/lectin interactions between the nematophagous fungus, *Arthrobotrys oligospora*, and the infective juveniles of *Trichostrongylus colubriformis* (Nematoda). Parasitology 101: 101 – 106.

Willcox, J., and H. T. Tribe. 1974. Fungal parasitism of cysts of *Heterodera*. I. Preliminary investigations. Transactions British Mycological Society 62: 585 – 594.

Wimble, D. B., and T. W. K. Young. 1984. Ultrastructure of the infection of nematodes by *Dactylella lysipaga*. Nova Hedwigia 40: 9 – 29.

Wu, J. Q., M. C. Chen, and S. D. Xue. 1990. Preliminary study on the control of root cyst nematodes in kenaf by *Paecilomyces lilacinus*. China's Fiber Crops 2: 19 – 21.

Zehr, E. I. 1985. Evaluation of parasites and predators of plant parasitic nematodes. Journal of Agricultural Entomology 2: 130 – 134.

Zuckerman, B. M., M. B. Dicklow, G. C. Coles, and H.-B. Jansson. 1988. Cryopreservation studies on the nematophagous fungus *Drechmeria coniospora*. Revue de Nematologie 11: 327 – 331.

21 Biological Control of Nematodes with Bacterial Antagonists

Z. X. Chen and D. W. Dickson

21.1 Introduction

Nematodes and bacteria are both ubiquitous organisms in soil. Bacteria are numerically the most abundant organisms in arable soil, typically ranging from 10^6 to 10^8 colony-forming units per gram dry weight of soil measured by soil dilutions plating on nutrient media (Clark, 1967). Direct examination of soil by microscopy usually detects population density at least 100 times greater than this. The total bacterial biomass is less than that of fungi, but exceeds that of algae, protozoa, and nematodes combined. Bacteria and nematodes have coexisted in almost every conceivable environmental circumstance over the millennia. A full spectrum of ecological interactions has evolved between these organisms, ranging from neutralism, mutualism, protocooperation, commensalism, competition, and amensalism, to parasitism (Sayre, 1988).

Biological control of nematodes was proposed as soon as plant-parasitic nematodes were recognized as devastating plant pathogens. Cobb reported the first bacterial disease of nematodes in 1906, though he erroneously identified it as a monad disease. The bacterium was recognized as *Pasteuria* sp. some 75 years later. Bacterial diseases of animal-parasitic nematodes were first observed in *Parascaris equorum* during the early 20th century (Weinberg and Keilin, 1912; cited in Bird and Bird, 1991). These bacteria parasitize externally and cause lesions on the nematode cuticle.

Biological control of nematodes was largely neglected during most time of the 20th century. The discovery of chemical nematicides provided a cost-effective and often spectacular means for nematode control. Consequently, chemical nematicides dominated nematode control in the past century even though they were environmentally and ecologically dangerous. Increased environmental awareness and concerns of hazardous nematicides warranted suspension and prohibition of several nematicides in the past. Several fumigants and other nematicides are now under critical review by the Environmental Protection Agency (EPA) of the USA and are planned to be phased out in the near future. Currently, farmers have a shortened list of nematicides available for nematode control. Nematodes cause annual losses to agriculture on the order of several billion dollars in the USA alone. Therefore, control of nematodes is important to maintain economic and social stability. With the economical globalization, the competition of agriculture production has intensified. These scenarios have provided the great economical, environmental, and social impetus for development of biological control of nematodes.

The notation of nematode diseases in the early age provided valuable guidance to searchers of potential candidates for biological control of plant-parasitic nematodes. Among all organisms studied, fungi, bacteria, and viruses appear to be promising agents for nematode biocontrol. More than a hundred species of fungi have been characterized for their nematode-destroying abilities, a few of them have been extensively evaluated for biological control potentials. Chen and Dickson review the fungal antagonists in detail in a separate chapter. A few viral diseases of nematodes have been noted, but the biological control potential has never been evaluated. Readers may find more information on the viral diseases of nematodes in the review by Hess and Poinar (1988).

Stirling (1991) defined biological control of nematodes as "a reduction of nematode populations that is accomplished through the action of living organisms other than nematode-resistant host plants, which occurs naturally or through the manipulation of the environment or the introduction of antagonists." Biological control is achieved through mechanisms such as parasitism, predation, competition, and antibiosis. Predation of nematodes is considered a function of soil fungi, protozoa, nematodes, insects, mites, etc., and thus is not discussed in this chapter. The chapter will focus on bacteria that cause parasitism, competition, and antibiosis which adversely affect the fitness, survival, and reproduction of nematodes.

Bacterial antagonists received little attention in the past. Only a few species have been identified as biocontrol candidates for plant-parasitic nematodes. One species, *Pasteuria penetrans*, however, has received much attention in recent years. It is highlighted here for its promising biological control potential against root-knot nematodes.

Several reviews and publications must be mentioned. They are: Biological Control of Plant Parasitic Nematodes: Progress, Problems and Prospects by Stirling (1991); Diseases of Nematodes, Vol. I and II, edited by Poinar and Jansson (1988); Chapter of nematode pathology in the book of the Structure of

Nematodes, by Bird and Bird (1991); and reviews by Chen and Dickson (1998), Sayre (1980, 1988), and Siddiqui and Mahmood (1999). For literature in pre-computer era, readers should consult reviews by Dollfus (1946), Bergman and van Duuren (1959), and Esser and Sobers (1964). The chapter "surface adhesion to nematodes and its consequences" by Bird in this book also should be consulted for understanding of the biological control of nematodes using bacterial antagonists.

21.2 Epiphytic Nematicidal Bacteria

A number of rhizobacteria or plant growth (health)-promoting rhizobacteria have been reported to have nematicidal effects. The bacteria produce metabolites, excretory enzymes, and antibiotics that are detrimental to nematodes (Hallmann, et al., 1999; Kloepper, et al., 1991, 1999; Sikora, 1997). They are non-parasitic and usually non-specific to plant-parasitic nematodes.

Many soil bacteria are able to break down protein and nitrogenous compounds and release ammonia and nitrates to soil, which kill nematodes through direct exposure and contact. A few bacteria are reported to produce excretory enzymes and antibiotics that degrade nematode cuticles and kill nematodes. Nematodes are affected in egg, juvenile, and adult stages that lodge in soil. Endoparasitic nematodes are usually little affected as they live inside the root tissues and are protected from the direct contact with soil metabolites.

A diversified group of bacteria have been reported to be nematicidal. They include genera *Acinebacter*, *Agrobacterium*, *Bacillus*, *Burkholderia*, *Chromobacterium*, *Enterobacter*, *Pseudomonas*, *Serratia*, *Stenotrophomonas*, and *Streptomyces*. All of them are common soil bacteria and usually have ubiquitous distribution and are abundant in soils.

Acinebacter – two *Acinebacter* species were reported to cause greater than 90% mortality in *Panagrellus redivivus* in Petri dish tests after 24 h inoculation at 22℃ (Tian, et al., 1995). Further tests including *Xiphinema thornei*, *Trichodorus pakistanensis*, *Scutellonema clathricaudatum*, *Criconemoides profuses*, and *Meloidogyne hapla* juveniles showed that the strains resulted in 100% death of the nematodes after 6 h inoculation at 22℃. A greenhouse experiment demonstrated they are effective against *M. hapla*.

Agrobacterium – *Agrobacterium radiobacter* strain G12, a plant health-promoting rhizobacterium isolated from rhizosphere, was able to induce systemic resistance of potato against the potato cyst nematode *Globodera pallida* (Hackenberg and Sikora, 1994; Hackenberg, et al., 1999; Hallmann, et al., 1998; Hasky, et al., 1998). The bacterium inhibited penetration and establishment of juveniles in potato root system (Hallmann, et al., 1998) and also reduced the hatch of *G. pallida in vitro* (Hackenberg and Sikora, 1994). It did not affect the juveniles that formed a syncytium inside roots and the sex ratio or number of eggs/cyst. In a split-root test, both the vital and heat-killed bacterial cells had the ability to induce systemic resistance against *G. pallida*, suggesting that it resulted from some bacterial metabolites (Hallmann, et al., 1998). Potato yields were increased in field plots planted with the bacterium treated tubers (Racke and Sikora, 1992).

Bacillus – this is a large group of bacteria that have shown diversified effects on both free-living and plant-parasitic nematodes. Various strains of *Bacillus thuringiensis* (Bt) were reported to have nematicidal effects against free-living nematodes *Caenorhabditis elegans* (Borgonie, et al., 1996a, b; Leyns, et al., 1995; Zuckerman, et al., 1993), *Panagrellus redivivus* (Tian, et al., 1995), and *Turbatrix aceti* (Meadows, et al., 1990), as well as plant-parasitic nematodes *Heterodera glycines* (Sharma, 1995; Sharma and Gomes, 1996a, 1996b), *M. hapla* (Chen, et al., 2000), *M. incognita* (Chahal and Chahal, 1993; Sharma, 1994; Zuckerman, et al., 1993), *M. javanica* (Carneiro, et al., 1998; Osman, et al., 1988), *Meloidogyne* spp. (Ivanova, et al., 1996; Rai and Rana, 1979), *Radopholus similis* (Mena, et al., 1997), *Rotylenchulus reniformis* (Zuckerman, et al., 1993), and *Tylenchulus semipenetrans* (Osman, et al., 1988).

Although the physiological mode of Bt toxin action in insects is known, little information is available in nematicidal action. Bt toxins take several steps in killing insects. Bt toxins must be ingested by a susceptible insect, where they are solubilized and proteolytically activated in the gut. Activated toxin binds to the midgut, inserts into the membrane, and forms a pore, resulting in intestinal lysis and insect death. Free-living nematodes are able to graze on the bacteria and toxins so it is possible that toxins may be activated and act on the nematode intestines resulting in nematode death. How Bt kills plant-parasitic nematodes is unclear. They are obligate plant parasites, feeding on plant cells by establishing a sophisticated feeding tube. Therefore, it is difficult to interpret the experimental effects of soil application of Bt toxins and bacteria on plant-parasitic nematodes. It has been suggested that extracellular toxins cause the deaths of the nematodes (Carneiro, et al., 1998; Chahal and Chahal, 1993; Meadows, et al., 1990; Rai and Rana, 1979), but supportive scientific data are scarce.

Recently, Marroquin et al. (2000) demonstrated that the action of Bt toxins on *C. elegans* is similar to toxin effects in insects. Endotoxins Cry5B and

Cry6A, when tested alone, caused extensive damage to the intestines, a decrease in fertility, and death. Bt preparations that lack Cry5B crystals do not adversely affect the growth, morphology, or fertility, suggesting that the toxic effects were due to the Cry5B toxin and not the Bt bacterial culture. The toxin damaged the intestine (Borgonie, et al., 1996a, b; Marroquin, et al., 2000), the same target tissue of Bt toxins in insects. Nematodes fed Cry5B-containing Bt preparations showed extensive changes in their gut morphology. These changes include the formation of vacuole-like structures in gut cells, a shrinking of the gut away form the body wall, tight constrictions at various points along the gut, and an overall pitted appearance. Cry5B resistant mutants were identified and five genes were defined (Marroquin, et al., 2000; Griffitts, et al., 2001). All mutants resisted the effects of the Cry5B toxin on the intestine, but remained sensitive to Cry6A toxin. Therefore, Cry6A probably utilized a different toxicity pathway.

Other *Bacillus* species that demonstrated nematicidal effects include: *B. laterosporus* (Carneiro, et al., 1998), *B. circulans* (Carneiro, et al., 1998); *B. subtilis* (Ehteshamul, et al., 1997; Gokte and Swarup, 1988; Siddiqui and Mahmood, 1993, 1995a, b), *B. pumilis* (Gokte and Swarup, 1988), *B. pumilis* (Gokte and Swarup, 1988), *B. cereus* (Gokte and Swarup, 1988; Kempster, et al., 2001), *B. sphaericus* (Hasky, et al., 1998), and *B. licheniformis* (Siddiqui and Mahmood, 1992). The bacterial stock solutions and their dilutions showed varied degrees of ovicidal, larvicidal, and nematicidal activities, or caused immobilization of juveniles *in vitro* tests. *Bacillus sphaericus* was reported to induce systemic resistance of potato against *Globodera pallida* (Hasky, et al., 1998), whereas *B. cereus* induced resistance of white clover against clover cyst nematode *H. trifolii* (Kempster, et al., 2001).

Largely stimulated by the success of Bt controlling various insect pests, *Bacillus* species were widely investigated and evaluated for their biological control potential. Some experiments successfully demonstrated yield increases and nematode suppression after application of the bacterium culture stocks alone or in combination with other organisms and organic amendments in greenhouse and field tests (Ehteshamul, et al., 1997). The nematicidal activities were not specific and a wide range of nematode genera, both free-living and plant-parasitic, was affected. With the absence of nematodes, the bacteria usually were regarded as plant health-promoting rhizobacteria that improved plant growth and increased yield, besides their insecticidal and biocidal properties (Hasky, et al., 1998). The bacteria were known to be able to degrade organic matter and produce volatile compounds such as ammonia and nitrites (Oka, et al., 1993). Therefore, their efficacy usually was enhanced by adding green manure and chitin amendment (Ehteshamul, et al., 1997; Hallmann, et al., 1999). Sikora (1997) concluded that production of allelopathic chemicals, alteration of root exudates, induced resistance, and increased plant tolerance were the major mechanisms of actions for nematode suppression. Nevertheless, exotoxins also were reported to be effective in nematode control (Ivanova, et al., 1996; Meadows, et al., 1990; Rai and Rana, 1979). An extracellular enzyme isolated from *B. cereus* had collagenolytic and proteolytic properties that were able to damage the cuticle of *M. javanica* juveniles (Sela, et al., 1998).

Burkholderia – *Burkholderia cepacia* has recently been tested against *M. incognita* on pepper (Meyer, et al., 2000, 2001). Culture filtrates of *B. cepacia* inhibited egg hatch and J2 mobility *in vitro*. In greenhouse tests, heat-killed *B. cepacia* suppressed eggs and J2 per g of roots on tomato (Meyer, et al., 2000), whereas viable preparations suppressed eggs and J2 per g of roots on bell pepper (Meyer, et al., 2001). An important note that must be made here is that *B. cepacia* also is an emerging opportunistic pathogen that causes fatal infections in patients suffering from cystic fibrosis and chronic granulomatous diseases. Current technology was unable to differentiate the environmental and biological control isolates from the human pathogens (LiPuma, 2001). An excellent review has been published recently (Parke and Gurian-Sherman, 2001), and one may also visit website http://go.to/cepacia for updating information.

Chromobacterium – *Chromobacterium* sp. produces chitinase in soil, which is a known inhibitor of egg hatch of many nematodes. A species of *Chromobacterium* was isolated from a nematode suppressive soil and demonstrated the ability of inhibiting egg hatch of *Globodera rostochiensis in vitro* as well as in soil microcosm tests (Cronin, et al., 1997a, b).

Enterobacter – *Enterobacter* species also occurs as the endophytic and rhizospheric plant growth-promoting bacteria. *Enterobacter cloacae* reportedly stimulated tomato plant growth and inhibited reproduction of *M. incognita* (Duponnois, et al., 1999). In that same report, *E. cloacae* reportedly increased the endospore attachment of *Pasteuria penetrans* to the nematode cuticles *in vitro* tests, and subsequently increased the reproduction of *P. penetrans* in plant roots.

Pseudomonas – Various species of *Pseudomonas* have been frequently reported to affect the viability and pathenogenicity of a wide range of nematodes. *Pseudomonas fluorescens* was the most studied bacterium and reportedly increased plant growth and reduced nematode damage caused by *Globodera rostochiensis* (Cronin, et al., 1997b), *Heterodera* spp. (Gokte and Swarup, 1998; Oostendorp and Sikora, 1989; Siddiqui, et al., 1998),

Hirschmanniella gracilis (Ramakrishnan, et al., 1998), *Meloidogyne* spp. (Aalten, et al., 1998; Eapen, et al., 1996; Hanna, et al., 1999; Hoffmann, et al., 1998; Gokte and Swarup, 1998; Santhi and Sivakumar, 1995; Siddiqui, et al., 2001), *Radopholus similis* (Aalten, et al., 1998), and *Tylenchulus semipenetrans* (Santhi, et al., 1999). *Pseudomonas fluorescens* produced a wide range of secondary metabolites and hydrolytic enzymes, which were believed to be the source of nematicidal activity. For example, a strain of *P. fluorescens,* which produced 2,4-diacetylphloroglucinol (DAPG) and lytic enzymes, inhibited egg hatch of *G. rostochiensis* similarly to that inoculated with purified DAPG and lytic enzymes (Cronin, et al., 1997a, b; Dunne, et al., 1998). A mutant lacking the ability of producing DAPG was not effective, suggesting that DAPG and DAPG-producing *P. fluorescens* were promising as potential biological control inoculants against potato cyst nematodes (Cronin, et al., 1997a, b).

Other investigations involved *P. mendocina* and *P. putida* against root-knot nematodes and *Rodopholus similis* (Aalten, et al., 1998; Duponnois and Mateille, 1999), *P. aureofaciens* – *Criconemalla xenoplax* (Westcott and Kluepfel, 1993), *P. aeruginosa* – *M. javanica* (Ara, et al., 1997; Ehteshamul, et al., 1997; Perveen, et al., 1998, Siddiqui, 2000; Siddiqui and Ehteshamul, 2000; Siddiqui, et al., 2001), and *P. chlororaphis* – *Pratylenchus penetrans* on strawberry (Hackenberg, et al., 2000). *Pseudomonas stutzeri* reportedly improved plant growth and yield of tomato but increased nematode galling and egg mass production of *M. incognita* as well (Khan and Tarannum, 1999). Interestingly, the bacterial symbiont *P. oryzihabitans* from *Steinernema abbasi* and *Xenorhabdus nematophilus* from *S. carpocapsae* reduced egg hatch of *M. javanica* when the bacteria were introduced to egg masses (Samaliev, et al., 2000).

Serratia – *Serratia marcessens* reduced *M. incognita* population and gall formation on table grape in a greenhouse test. Further field tests showed that *S. marcessens* alone or in combination with fenamiphos achieved the best nematode control and improved plant growth (Ali and Kamal, 1998).

Stenotrophomonas – *Stenotrophomonas maltophilia* was isolated from a field suppressive to potato cyst nematodes for the ability of producing chitinase. *Stenotrophomonas maltophilia* reduced the ability of *G. rostochiensis* to hatch in soil microcosms planted with potato seed tubers. It was proposed that the inhibition of egg hatch by *S. maltophilia* and other chitinase-producing bacteria was responsible for the nematode suppressiveness (Cronin, et al., 1997a, b).

Streptomyces – *Streptomyces* spp. are a large group of actinomycetes, which are well-known for their ability to produce various antibiotics (Campbell, 1981; Campbell, et al., 1983; Quarles, 1991). The avermectins are macrocyclic lactones produced by *S. avermitilis*. They have potent, broad-spectrum anthelminthic, insecticidal and miticidal activities (Cayrol, et al., 1993; Quarles, 1991). Consequently, abamectin B1 was commercialized under the name Vertimec for the control of root-knot nematodes (Cayrol, et al., 1993). Some *Streptomyces* species improved plant growth and yield following applications (Ali and Kamal, 1998; Esnard, et al., 1998). The affected nematodes included *Belonolaimus longicaudatus* (Nair, et al., 1995), *Bursaphelenchus xylophilus* (Omura, et al., 1987), *Helicotylenchus multicinctus* (Esnard, et al., 1998), *Heterodera glycines* (Nair, et al., 1995), *M. hapla* (Chen, et al., 1999, 2000), *M. incognita* (Ali and Kamal, 1998; Dicklow, et al., 1993), *Radopholus similis* (Esnard, et al., 1998), *Rotylenchulus reniformis* (Dicklow, et al., 1993), *Pratylenchus penetrans* (Dicklow, et al., 1993), and free-living nematodes (Dicklow, et al., 1993; Esnard, et al., 1998; Nair, et al., 1995).

A new species, *S. costaricanus*, isolated from nematode suppressive soils in Costa Rica have shown promise against plant-parasitic nematodes (Dicklow, et al., 1993). In greenhouse and field experiments, *S. costaricanus* reduced root galling on tomato and pepper caused by *M. incognita*. Additionally, it was effectively shown to improve crop growth and yield and suppressing nematode populations of *Helicotylenchus multicinctus*, *M. hapla*, *Pratylenchus penetrans*, *Radopholus similis*, and *Rotylenchulus reniformis* on various crops (Chen, et al., 2000; Dicklow, et al., 1993; Esnard, et al., 1998). Studies on sterile filtrates from *S. costaricanus* cultures suggested that thermostable macromolecules were the major determinant of the nematicidal activity (Dicklow, et al., 1993).

In contrast to the success of rhizobacteria controlling nematodes, a few studies demonstrated that the efficacy was often inconsistent, insignificant, and unrepeatable (Becker, et al., 1988; Tian and Riggs, 2000). For example, Tian and Riggs (2000) tested 201 bacterial isolates from at least 11 genera in a greenhouse test using pasteurized silt loam soil. Among tested isolates, 139 isolates had no influence, 36 isolates reduced (suppressive), and 27 isolates increased the numbers of cysts, and eggs + juveniles. The retest in the same soil using 20 of the 36 suppressive isolates showed the results were highly variable and inconclusive (Tian and Riggs, 2000).

21.3 Endophytic Bacteria

Hallmann et al. (1997) defined bacterial endophytes as those that can be isolated from surface-disinfected plant tissue or extracted from within the plant and that do not visibly harm the plant. Therefore, the definition is based on current practical technology and the physiological function of the bacteria. It excludes non-extractable endophytic bacteria and endopathogenic bacteria. Endophytic bacteria received great emphasis in recent years for their ability of inducing systemic resistance of plants and suppression of various plant disease pathogens including plant-parasitic nematodes.

Generally speaking, there are not many differences between the endophytic bacteria and epiphytic bacteria, the bacterial group described in the previous section. The endophytic bacteria originated from the epiphytic bacterial communities of the rhizosphere and rhizoplane. However, the function and suppression mechanism of the endophytes must be different from the epiphytes. In the rhizosphere, the major function of the bacteria is decomposing, and the yielded secondary metabolites, lytic enzymes, antibiotics, and changes of chemical and physical properties of the microcosm are the major suppressive sources. In endophytic phase, the decomposing function was minimized. In contrast, endophytic bacteria established a symbiont and mutual relationship with the plant (Hallmann, et al., 1997, 2001). The endophytic bacteria usually maintain an optimum population density in plant tissue, depending on the bacteria and plant species, plant age, and other environmental conditions (Chen, et al., 1995; Frommel, et al., 1991; Hallmann, et al., 1997). In general, the average population densities of endophytic bacteria ranged from 1×10^3 to 1×10^5 colony-forming units (cfu) per gram of plant tissue (Dong, et al., 1994; Frommel, et al., 1991; Quadt-Hallmann and Kloepper, 1996). The numbers were remarkably lower than those of most pathogenic bacteria, which could reach 1×10^7 to 1×10^{10} cfu/g of plant tissue under severe disease pressure (Grimault and Prior, 1994). Also, endophytes used intercellular exudates rather than breaking down the plant root cells and tissues as the major nutrient source (Foster, et al., 1983). Some enzymes, which were produced to break down cell walls when the bacteria existed at the epiphytic phase, were inhibited after the bacteria become endophytes (Benhamou, et al., 1996). Therefore, there is a drastic change of bacterial biological activity and population dynamics that were modified to resolve the mutual relationship between the bacteria and the plant.

The beneficial effects of bacterial endophytes included biological control of pathogens, plant growth promotion, and induced systemic resistance (Barker and Koenning, 1998; Hallmann et al., 1997; Hasky, et al., 1998; Kloepper, et al., 1999; Reitz, et al., 2001). The mode of action of endophytic bacteria against plant-parasitic nematodes remained largely unknown and speculative. Since endophytic bacteria colonized the similar ecological niche to that of endoparasitic nematodes, it was imaginable that the niche competition and direct antagonism might contribute to the biological control of nematodes. The suppression of *Meloidogyne* spp. by *Rhizobium* bacteria was an example of niche competition (Siddiqui, et al., 1995). Interestingly, plant-parasitic nematodes were generally considered as a carrying agent that facilitated the penetration and establishment of endophytic bacteria in plant roots (Hallmann, et al., 1997, 1998). In a study, *Alcaligenes piechaudii* and *Burkholderia pickettii* occurred only in cotton plants infected by *M. incognita* (Hallmann, et al., 1998).

Plant growth promotion and systemic resistance induced by bacterial endophytes are promising for plant-parasitic nematode control (Barker and Koenning, 1998; Reitz, et al., 2000; Kloepper, et al., 1999; Ramamoorthy, et al., 2001). *Rhizobium etli* strain G12 significantly reduced root penetration caused by potato cyst nematode juveniles (Reitz, et al., 2000). The split-root tests showed that both live and heat-killed bacterial cells had the property to induce systemic resistance on potato. Further investigations indicated the heat-stable bacterial surface carbohydrates – lipopolysaccharides were the chemical inducers (Reitz, et al., 2000). However, the signal transduction pathways for the systemic resistance against nematodes are still unknown. Systemic resistance that is induced by other pathogens is usually characterized by enhanced accumulation of pathogenesis-related proteins (i. e., chitinase and beta-1, 3-glucanase), increased peroxidase activity, and cell wall modifications such as lignification (Reitz, et al., 2001). The strain G12 was observed preferentially colonizing galled tissues of potato and *Arabidopsis* roots caused by *M. incognita* infections (Hallmann, et al., 2001). Apparantly, future studies should investigate the biological control potential and interactions between endophytic bacteria and endoparasitic nematodes.

21.4 Ectoparasitic Bacteria

The first external association between bacteria and nematodes was reported in 1851 by Leidy (cited in Sayre, 1988). The relationship between the bacteria *Arthromitus cristatus* and nematodes *Thelastoma attenuatum*, *Aorucus agile*, and *Rhigonema infectum* was considered a commensalism (Sayre, 1988).

A few bacteria cause lesions on nematode cuticle and has been reviewed (Bird, 1980; cited in Bird and Bird, 1991). The nematodes under attack include *Parascaris equorum*, *Heterakis*, *Oxyuris*, *Toxascaris*, *Strongylus*, and *Thelastoma pterygoton*. All of them are animal-parasitic nematodes. These unidentified bacteria are capable of dissolving part of the nematode cuticle and establishing colonies thereon.

Additionally, four types of cuticular lesions from *Strongylus edentatus* parasitizing swines and horses were characterized as (1) filamentous, (2) flat, (3) cratered, and (4) proliferate (Anderson, et al., 1973, 1978). *Enterobacter aerogenes*, *Escherichia coli*, *Micrococcus* sp., *Streptococcus* sp., and *S. faecalis* occurred in the filamentous, flat, and proliferate lesions. Filamentous lesions additionally contained *E. coli*, whereas flat lesions additionally contained *Proteus vulgaris* and *E. coli*. Cratered lesions contained only *S. faecalis* and *Enterobacter aerogenes*. Both *E. coli* and *E. aerogenes* were recovered from worm surfaces having no lesions, but healthy cuticle specimens generally did not contain *S. faecalis* (Anderson, et al., 1978). Some of these bacteria also existed in nematode excretory glands, ovaries, and intestines (Anderson, et al., 1973).

So far, ectoparasitic bacteria on plant-parasitic nematodes have not been discovered yet. The only similar case is the association between bacterial genus *Clavibacter* and nematode *Anguina*. The bacterium, nematode, and a virus interacted with each other in the host plant rye grass (*Lolium rigidium*), which causes a disease of livestock known as annual rye grass toxicity. The bacterial attachment results in the pathological changes in the fine structure of nematode cuticle. Heavy infestation of the bacteria could harm the ability of nematodes penetrating host plants. The biological control potential of such associations has been reviewed in the chapter contributed by Bird in this book. It must be stressed, however, that the bacterium is not a true parasite of the nematode.

21.5 Endoparasitic Bacteria

Due to the chemical nature of the nematode cuticle, few bacteria and fungi could utilize and degrade the chitin-free extracellular exoskeleton of the nematode. The physical body design of the plant-parasitic nematodes prevents bacteria from entering through body openings, such as stylet, excretory/secretory pore, and anus. To become an endoparasite of nematodes, the bacterium must overcome the exoskeleton barrier.

A few groups of bacteria were found inside and parasitizing juveniles and adults of nematodes (Sayre and Starr, 1988). Adams and Eichmuller (1963) reported the bacterium *Pseudomonas denitrificans* parasitizing *Xiphinema americanum*. The bacterium was found throughout the body of juveniles and females of the nematode. In females, the bacteria were concentrated in the intestines and ovaries, which lead them to suggest that the infection might be transmitted transovarially.

An actinomycete, probably a *Streptomyces* species, was observed attacking

Ditylenchus triformis (Dürschner, 1984). The actinomycete produces two forms of conidia, spiraliform and spiral. The spiral conidia attach to the nematode cuticle and eventually kill the nematode in 3 to 7 days. The nematode is filled with the bacterial mycelium consisting of fine hyphae (less than 1.0 μm). The mycelium emerges from the nematode carcass, producing new infective conidia. The bacterium also attacks *D. dipsaci*, *Aphlenchoides composticola*, and *Aphlenchus avenae*, but not *Rhabditis* sp.

Pasteuria spp. are another group of bacteria attacking a wide range of nematodes. The bacteria have germ tubes penetrating the nematode cuticle and entering nematode body cavity to establish parasitism. This unusual group of bacteria has received quite a lot of research attention due to its enormous biological control potential. The rest of the chapter will introduce the important aspects of these bacteria and their biological control potential.

A few recent reviews must be mentioned before introducing *Pasteuria* spp. The historical background of the discovery of *Pasteuria ramosa*, the type species of this genus, was reviewed by Sayre and Starr (1985). The biological control was reviewed by Stirling (1991), Bird and Bird (1991), and most recently by Chen and Dickson (1998). The chapter by Bird in this book has covered some aspects of *Pasteuria penetrans* associated with root-knot nematodes. Therefore, it is obviously unnecessary to present every detail of *Pasteuria* spp. in this chapter. The authors are more interested in analyzing the information floating in public domains and in providing an insightful overview.

21.6 *Pasteuria*

21.6.1 Taxonomy

Pasteuria spp. were differentiated by host preference, developmental characteristics, and size and shape of sporangia and endospores (Fig. 21.1) (Sayre and Starr, 1989). Four species of *Pasteuria* have been described. *Pasteuria ramosa,* which parasitizes water fleas of the genera *Daphnia* and *Moina*, is the type species of the genus. The other three species of *Pasteuria* are parasites of plant-parasitic nematodes: *P. penetrans* on *Meloidogyne* spp., *P. thornei* on *Pratylenchus* spp., and *P. nishizawae* on cyst nematodes of the genera *Heterodera* and *Globodera* (Sayre and Starr, 1989). Two other proposed new species of *Pasteuria* have been isolated from *Heterodera goettingiana* in Münster, Germany (Sturhan et al., 1994), and from *B. longicaudatus* in Florida (Bekal et al., 2001; Giblin-Davis et al., 1990, 1995).

Chen and Dickson (1998) recently summarized the host and temperature

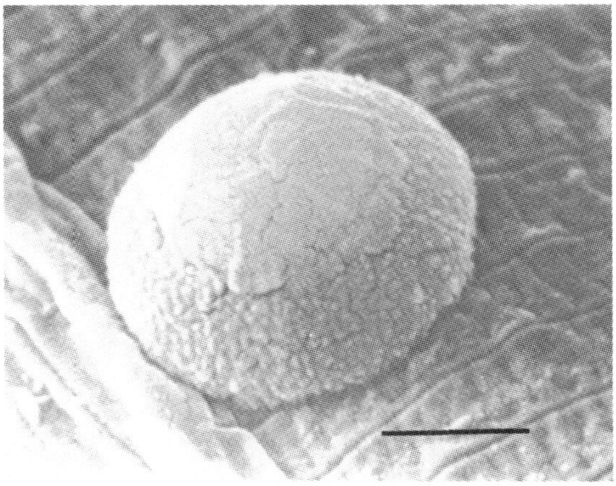

(A)

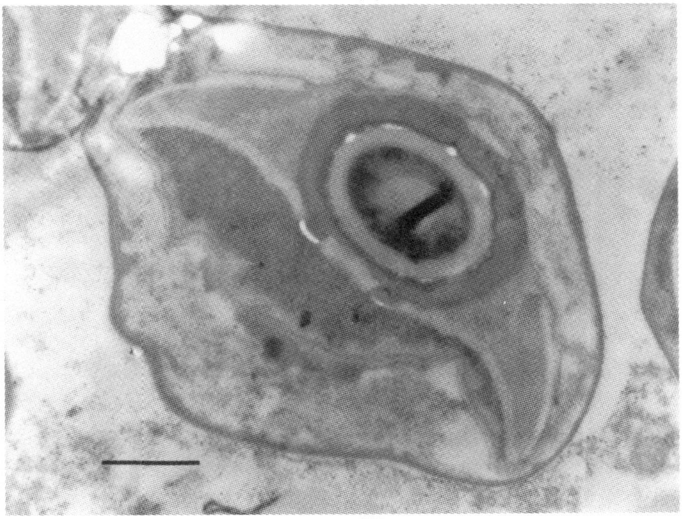

(B)

Figure 21.1 Ultrastructure morphology of *Pasteuria penetrans* endospores. (A) SEM photomicrograph of an endospore attached to the cuticle of second-stage juvenile of *Meloidogyne* sp., bar = 1 μm. (B) TEM photomicrograph of mature endospores, bar = 0.5 μm.

preference, pathogenicity, and occurrence of various isolates reported from all over the world. *Pasteuria* spp. have been reported from more than 116 genera of nematodes distributed worldwide (Fig. 21.2). Endospores of *Pasteuria* spp. have shown remarkable size variations, ranging from approximately 2 to

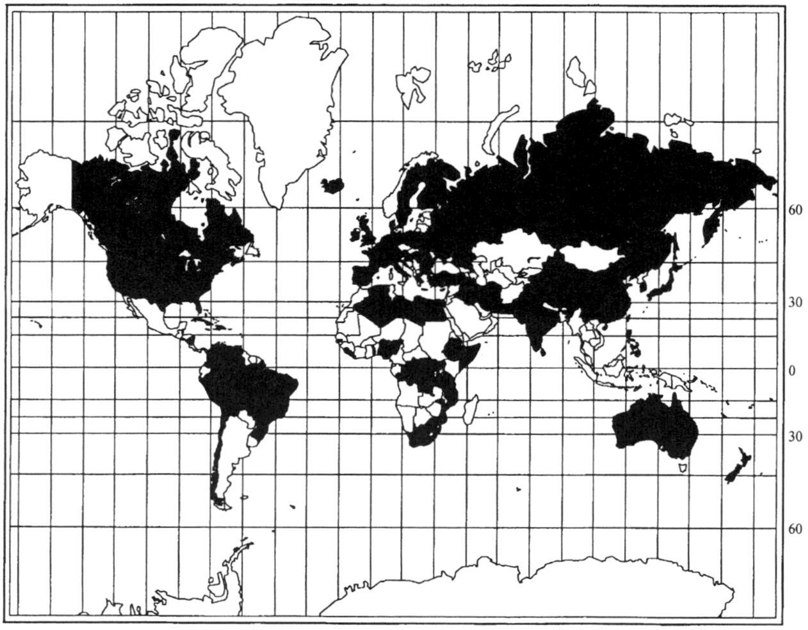

Figure 21. 2 Worldwide distribution of *Pasteuria* spp. Countries with gray color indicated that at least one strain of *Pasteuria* spp. had been reported. See Chen and Dickson (1998) for the list of countries.

8 μm (Table 21. 1). This wide range of host nematodes, endospore dimensions, and host specificity of *Pasteuria* spp. caused considerable confusion in the taxonomy of the genus *Pasteuria*. With only three species of *Pasteuria* having been established that are associated with nematodes, the rest of isolates have not been speciated. It is unlikely that there will be a new species for each nematode genus or species parasitized. Especially since it is known that some isolates are able to parasitize nematodes accross genera (Bhattacharya and Swarup, 1988; Mankau and Prasad, 1977; Oostendorp, et al., 1990; Pan, et al., 1993; Sharma and Davies, 1996; Vargas and Acosta, 1990). Molecular techniques have added valuable tools to bacterial systematics, which should benefit the taxonomy of this group of bacteria as well. Recently, the 16s rRNA of *P. ramosa* from *Daphnia*, *P. penetrans* from *Meloidogyne*, *Pasteuria* strain S-1 from *Belonolaimus*, and *Pasteuria* strain NA from *Heterodera* were successfully sequenced and compared to support the current taxonomical position (Anderson, et al., 1999; Atibalentja, et al., 2000; Bekal, et al., 2001; Ebert, et al., 1996). It is foreseeable that a similar

approach would be applied to other isolates in the future in an effort to classify this very diverse and complex group of bacteria.

Table 21.1 Variation of endospore diameters of *Pasteuria* spp. originated from different nematode hosts (modified from Ciancio et al., 1994).

Nematode species	Endospore diameter ± SD (μm)
Tylenchulus semipenetrans	2.6[a]
Acrobeloides sp.	2.7 ±0.3 (2.0 –3.0)[b]
Coslenchus turkeyensis	2.7 ±0.3 (2.5 –3.0)
Aphelenchoides rutgersi	2.7 ±0.3 (2.0 –3.0)
Boleodorus cylachtus	3.0 ±0.1 (2.8 –3.5)
Tylenchulus semipenetrans	3.0 ±0.2 (2.5 –4.0)
Meloidogyne spp.	3 –4
Discocriconemella mauritiensis	3.5 ±0.2 (3.3 –4.0)
Longidorella sp.	3.5 ±0.4 (3.0 –4.0)
Aphelenchoides dactylocercus	4.0
Merlinius sp.	4.0 ±0.1 (4.0 –4.5)
Pratylenchus neglectus	4.0 ±0.3 (3.5 –4.5)
Hoplotylus silvaticus	4.1 ±0.2 (4.0 –4.5)
Coslenchus costatus	4.2 ±0.5 (3.0 –5.0)
Helicotylenchus californicus	4.3 ±0.2 (4.0 –4.5)
Longidorelle parva	4.4 ±0.4 (4.0 –5.0)
Cylindrolaimus communis	4.4 ±0.3 (4.0 –5.0)
H. pseudorobustus	4.5 ±0.4 (4.0 –5.0)
H. lobus	4.6 (4.5 –5.0)
Paratrophurus anomalus	4.6 ±0.4 (4.0 –5.0)
H. crenacauda	4.7 ±0.3 (4.5 –5.0)
H. digonicus	4.7 ±0.3 (4.0 –5.0)
H. pseudorobustus	4.7 ±0.3 (4.5 –5.0)
Heterodera fici	4.9 ±0.2 (4.5 –5.0)
Trophonema okamotoi	4.9 (4.7 –5.1)

Continued

Nematode species	Endospore diameter ± SD (μm)
Helicotylenchus digonicus	5.0 ±0.2 (5.0 –5.5)
Merlinius brevidens	5.0
H. dihystera	5.1 ±0.3 (5.0 –5.5)
H. pseudorobustus	5.1 ±0.2 (5.0 –5.5)
Xiphinema pachtaicum	5.1 ±0.2 (5.0 –5.5)
Rotylenchus capensis	5.1 ±0.4 (5.0 –6.0)
Opisthodorylaimus sylphoides	5.2 ±0.7 (5.0 –7.0)
Dorylaimoides mitis	5.3 ±0.3 (5.0 –6.0)
Xiphinema sp.	5.4 ±0.4 (5.0 –6.0)
Xiphinema sp.	5.5 ±0.3 (5.0 –6.0)
H. vulgaris	5.5 ±0.4 (5.0 –6.0)
X. pachtaicum	5.5 ±0.4 (5.0 –6.0)
X. vuittenezi	5.5 ±0.5 (5.0 –6.5)
X. brasiliense	5.6 (5.0 –6.5)
X. americanum	5.7 ± 0.4 (5.5 –6.0)
Xiphinema sp.	5.7 ±0. 3 (5.0 –6.0)
X. setariae	5.7 ±0.3 (5.0 –6.0)
X. longicaudatum	5.7 ±0.4 (5.5 –6.0)
X. insigne	5.8 ±0.3 (5.0 –6.0)
X. elongatum	5.8 ±0.3 (5.0 –6.0)
X. setariae	5.8 ±0.3 (5.0 –6.0)
X. elongatum	5.9 ±0.2 (5.5 –6.0)
X. bergeri	5.9 ±0.2 (5.5 –6.5)
X. brevicolle	5.9 ±0.3 (5.5 –6.0)
Scutellonema clathrycaudatum	5.9 ±0.2 (5.5 –6.0)
Discolaimus major	6.0 ±0.1 (6.0 –6.5)
Longidorus laevicapitatus	6.0 ±0.2 (5.5 –6.5)
Aphasmatylenchus nigeriensis	6.0
Mylonchulus boveyi	6.0

Continued

Nematode species	Endospore diameter ± SD (μm)
L. laevicapicatus	6.1 ±0.2 (6.0 –6.5)
X. ebriense	6.1 ±0.3 (5.0 –6.5)
X. rotundatum	6.1 ±0.3 (5.0 –6.5)
L. attenuatus	6.3 ±0.2 (6.0 –6.5)
X. diversicaudatum	6.5 ±0.5 (6.0 –7.0)
L. euonymus	6.6 ±0.4 (6.0 –7.0)
Rotylenchus laurentinus	6.7 ±0.3 (6.0 –7.0)
X. ifacolum	6.8 ±0.3 (6.0 –7.0)
X. radicicola	6.8 ±0.3 (6.0 –7.0)
X. basiri	6.9 ±0.3 (6.0 –7.5)
X. turcicum	7.0 ±0.3 (6.5 –7.5)
Paralongidorus citri	7.4 ±0.3 (7.0 –8.0)
X. ingens	7.6 ±0.3 (7.0 –8.0)
Longidorus sp.	7.8 ±0.2 (7.5 –8.0)
Axonchium valvulatum	8.0 ±0.1 (8.0 –8.5)

[a] Data of standard deviation are not available.

[b] Numbers in parathesis are the ranges of endospore diameters.

21.6.2 Life Cycle

The life cycle of *P. penetrans* involves four stages: endospores attaching to nematodes, germination, proliferation inside the nematode pseudocoel, and releasing of mature endospores. Endospores are the dominant stage of the bacterium. They attach to the cuticle of *Meloidogyne* spp. J2 when the latter moves through soil searching for host safety. Germination of endospores in root-knot nematodes occurs 4 to 10 days after the endospore-encumbered nematode enters a plant root and begins to feed (Sayre and Wergin, 1977; Serracin, et al., 1997). The germ tube emerges through a central opening in the basal attachment layer of the endospore and penetrates the nematode cuticle. The process of penetration seems to be enzymatic (Mankau, 1975a, b; Mankau, et al., 1976). After entering the pseudocoelom of the nematode, the germ tube develops into a cauliflower-like microcolony consisting of a dichotomously branched septate mycelium (Fig. 21. 3). Daughter colonies form when the intercalary cells in the microcolony lyse (Sayre and Starr,

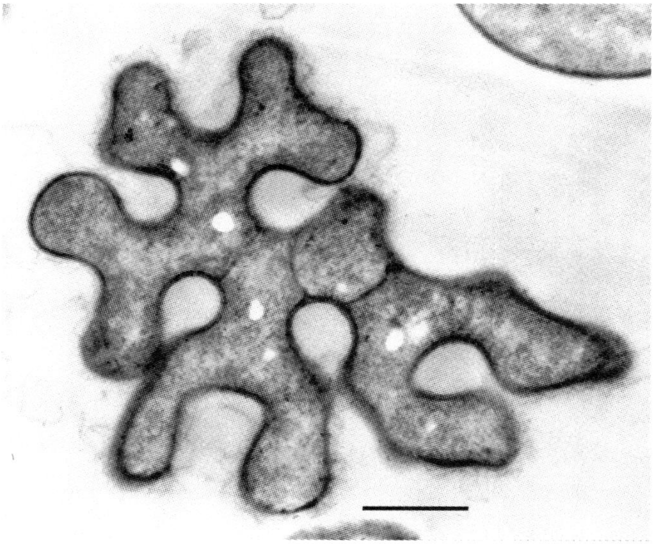

Figure 21.3 Mycelium of *Pasteuria penetrans* consisting of multiple cells. Each cell (top) has several branches, each with dichotomous terminals. Bar = 0.5 μm.

1989). Due to some unknown triggers, the colony forms fragmentations; the terminal cells of the fragmentation enlarge and undergo sporogenesis (Fig. 21.4). Eventually, quartets and doublets of developing sporangia predominate in the nematode body cavity and finally separate into a single sporangium containing an endospore. The mature endospores are released into soil when the plant root with its complement of parasitized root-knot nematode females decomposes.

The life cycle of *P. penetrans* was once considered to be synchronized with the development of root-knot nematodes (Bird, 1986; Sayre and Starr, 1985; Stirling, 1991). This concept must be challenged. *Pasteuria penetrans* has been observed attaching to, developing, and producing mature endospores in J2 (Dickson, et al., 1994; Giblin-Davis, et al., 1990) and males (Page and Bridge, 1985) of *Meloidogyne* spp. At 20℃, *P. penetrans* requires 120 days to complete the life cycle, much longer than that of root-knot nematodes. In both cases, one finishes the life cycle independent of the other. The development of *P. penetrans* is temperature-dependent (Hatz and Dickson, 1992; Serracin, et al., 1997). The minimal growth temperature is 17℃ for *P. penetrans* (Chen and Dickson, 1997a), and 10℃ for the nematodes (Ferris, et al., 1978). The different environmental requirements may prevent

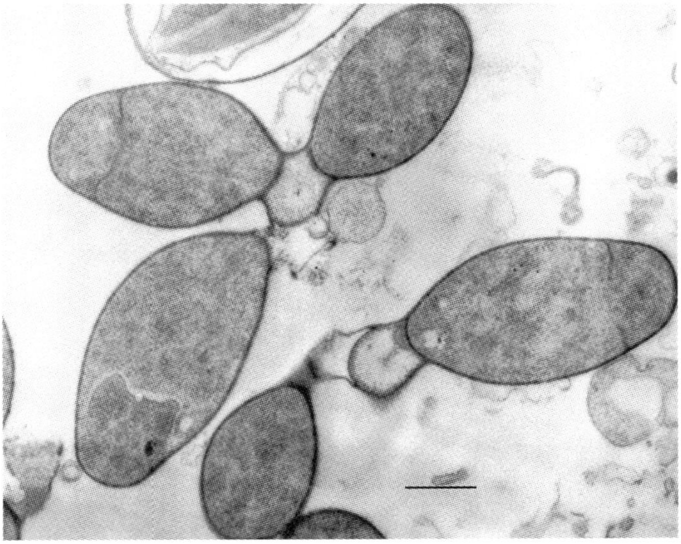

Figure 21.4 Ultrastructure of *Pasteuria penetrans* in early stages of sporogenesis. Bar = 0.5 μm.

synchronized development at certain temperatures. *Pasteuria* spp. apparently do not synchronize with the development of endomigratory, ectoparasitic, or free-living nematodes, because *Pasteuria* spp. have been frequently observed in the pseudocoelom of various developmental stages of host nematodes.

An unusual, multiple germination of endospores has been reported recently (Carneiro, et al., 1999). The report is apparently a hoax. Endospore consists of a single cell, which permits only one time germination at suitable conditions. A single endospore does not have the cellular basis for multiple germinations. The stressed germ tubes look like rod-shaped bacteria that attach to the endospores on nematode cuticles. It is common to have soil bacteria attached to endospores and nematodes when the nematodes are extracted from soil samples (Chen and Dickson, 1998; Bekal, et al., 2001).

21.6.3 Sporogenesis

It appears that different *Pasteuria* spp. share the typical sporogenesis of a Gram-positive, endospore-forming bacterium (Chen, et al., 1997b; Sayre, 1993). Briefly, the sporogenesis process has been divided into seven stages (Fig. 21.5). In stage I, terminal cells of mycelia elongate and the mycelia become fully septate. Stage II is characterized by the formation of a transverse septum that separates the forespore from the endospore mother cell. In stage

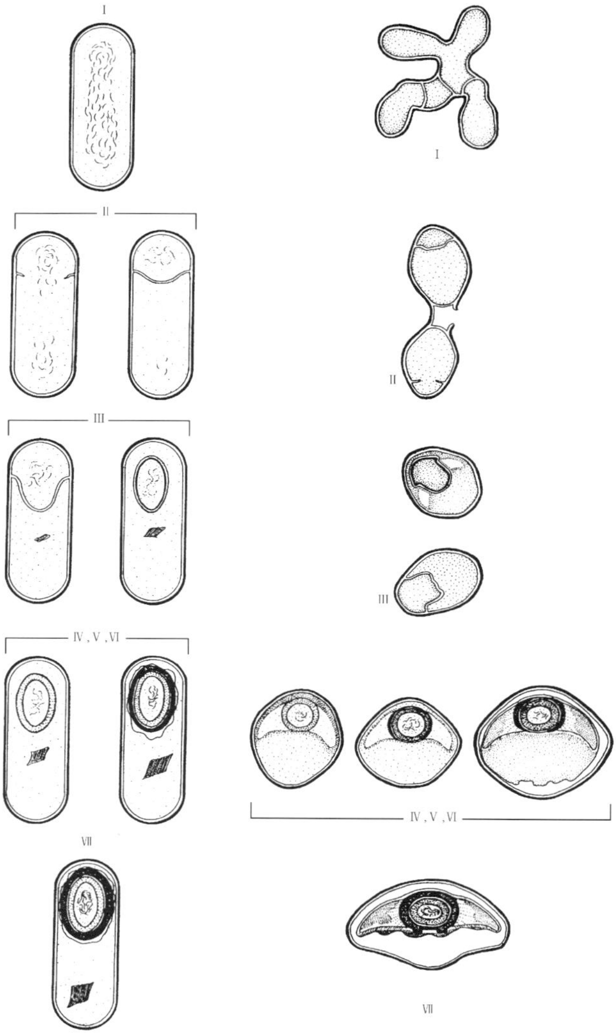

Figure 21. 5 Diagram of sporogenesis of *Pasteuria penetrans* (lower row) compared with sporogenesis of *Bacillus thuringiensis* (upper row). Stage I: Terminal cells of *P. penetrans* elongate and the mycelia become fully septate. Stage II: Formation of a transverse septum. Stage III: Engulfing of the forespore and formation of parasporal fibers. Stages IV to VI: Formation of the cortex, endospore coat, and exosporium. Stage VII: Mature endospore.

III, the forespore is engulfed by the endospore mother cell and condensation of the forespore protoplasm occurs. Parasporal fibers are initiated in stage III. Formation of the cortex, the endospore coat, and the exosporium occurs in stages IV to VI. Mature endospores are formed during stage VII. In accordance with the scheme used to describe the developmental stages of *Bacillus* spp., which assigns the vegetative cells of *Bacillus* spp. as stage 0, the dichotomously branched vegetative mycelium of *P. penetrans* has been assigned as stage 0 (Fig. 21.3).

Interestingly, the vegetative cell of *P. penetrans* has several branches, and each branch has bifurcate terminals (Fig. 21.5). Since each terminal would give rise to an endospore, it is reasonable to postulate that the terminal contains a copy of DNA. This means, however, that each vegetative cell has multiple nucleoids (Chen and Dickson, 1998), equaling to the number of branches times two terminals per branch. The bacterium has multicellular hyphae. When dividing, it was believed that the hyphae separate from each other at the center, resulting in two mutlicellular daughter hyphae (Bishop and Ellar, 1991). Based on this available information, the proliferation of *P. penetrans* mycelium was more like a tissue division rather than a cell division. Whether it was true or not, only further detailed research will resolve this issue. Therefore, the cytology and mycelium proliferation remain one of the most interesting topics to be investigated in future.

21.6.4 Systematics and Phylogeny

Modern bacterial systematics depends on both phenotypic and molecular biological characters. The phenotypic characters are still important in classification and identification of prokaryotes. Currently, endospore-forming bacteria are placed in 13 genera, which are separated based on morphological, physiological, and genetic diversity (Table 21.2). Recently, the 16S rDNA sequences were obtained from endospores of *P. ramosa* (Ebert, et al., 1996), *P. penetrans* (Anderson, et al., 1999), *Pasteuria* sp. from soybean cyst nematodes (Atibalentja et al., 2000), and *Pasteuria* sp. from sting nemaodes (Bekal, et al., 2001). The results suggested that *Bacillus tusciae*, *Alicyclobacillus cycloheptanicus,* and *A. acidocaldarius* are the nearest relatives (Anderson, et al., 1999; Ebert, et al., 1996). The sequence analysis showed that *P. ramosa* does not belong to the Actinomycetales, as was suggested previously (Bird, 1986; Sayre and Wergin, 1977; Sayre et al., 1977). Endospore morphology and sporogenesis of *P. penetrans* was similar to that of a bacterium, except that *P. penetrans* has a complex polymorphic life cycle (Chen, et al., 1997b). Therefore, *Pasteuria* spp. belong to members of the true bacteria (Chen, et al., 1997b; Ebert, et al., 1996).

Table 21.2 Described genera of endospore-forming bacteria and their DNA base composition (from Berkeley and Ali, 1994).

Genus	Mol % GC[a]
Alicyclobacillus	52 – 60
Amphibacillus	36 – 38
Bacillus	32 – 69
Clostridium	22 – 54
Desulfotomaculum	38 – 52
Oscillospira	–[b]
Pasteuria	–
Sporohalobacter	30 – 32
Sporolactobacillus	38 – 40
Sporosarcina	40 – 42
Sulfobacillus	54
Syntrophospora	38
Thermoactinomyces	52 – 55

[a] Mol % GC = mol % guanine (G) plus cytosine (C) content.

[b] – = No information available.

21.6.5 Host Records

Host records of *Pasteuria*-like organisms have been reviewed by Sayre and Starr (1988), Sturhan (1988), and recently updated by Chen and Dickson (1998). *Pasteuria*-like organisms have been reported from 323 species of soilborne nematodes belonging to 116 genera, from 79 countries worldwide. The host nematodes include free-living, predacious, plant-parasitic, and entomopathogenic (*Steinernema glaseri*) nematodes. For detailed information, the reader is referred to the recent review by Chen and Dickson (1998).

21.6.6 Geographical Distribution

Pasteuria spp. have been discovered from both northern and southern hemispheres, from tropical to temperate zones (Fig. 21.2). They have been discovered on five continents and on various islands in the Atlantic, Pacific, and Indian oceans. With their worldwide distribution and reported host specificity, it appears the genus *Pasteuria* may consist of hundreds of species and subspecies, with different host, temperature, and ecological preferences.

21.6.7 Cultivation

Current methods of mass-producing *P. penetrans* rely on the multiplication of the pathogen in its nematode host on greenhouse-grown plants (Stirling and Wachtel, 1980). Such production systems might be improved by culturing the nematode and pathogen in excised or transformed root cultures (Verdejo and Jaffee, 1988; Verdejo and Mankau, 1986), but commercial use of the pathogen will most likely require an *in vitro* method of cultivation. Various media were tested for artificial cultivation of *Pasteuria* spp., but were unsuccessful (Bishop and Ellar, 1991; Reise, et al., 1988; Williams, et al., 1989).

21.6.8 Effect of Temperature, Moisture, and pH

The development of *P. penetrans* in nematode hosts is affected by temperature. Development of *P. penetrans* within females of *M. javanica* and *M. arenaria* was not observed at 10℃ (Hatz and Dickson, 1992). The minimal developmental temperature was recently determined as 17℃ (Chen and Dickson, 1997a), with optimal growth temperature between 28℃ and 35℃ (Hatz and Dickson, 1992; Serracin, et al., 1997). These temperature requirements define *P. penetrans* as a mesophilic bacterium. Different temperature requirements, however, exist for various isolates of the bacterium because of its cosmopolitan distribution. As an example, an Indian isolate of *P. penetrans* that infects both *Heterodera* spp. and *M. incognita* completed its life cycle in *M. incognita* in 49 days at 10℃ to 17℃ (Bhattacharya and Swarup, 1988). Temperature affects endospore attachment (Ahmed, 1990; Hatz and Dickson, 1992; Singh and Dhawan, 1990; Stirling, et al., 1990), germination (Sayre and Wergin, 1977; Serracin et al., 1997), pathogenicity (Stirling, 1981), and endospore production (Chen and Dickson, 1997b, 1998; Hatz and Dickson, 1992).

It is difficult to understand the mesophilic nature of *Pasteuria penetrans*. The host nematodes live in soil or inside plant roots where the temperature usually remains below 25℃. Any temperature above 30℃ become a stress or even detrimental to the plant roots and nematodes. Theoretically, the cospeciation and co-evolution process between the nematodes and bacteria should eliminate the diverged temperature requirements. The great divergence in temperature requirements might suggest that the ancestor of the bacteria have a mesophilic or even thermophilic life and such traits remain largely intact after acquiring the parasitic life form.

Little is known about the effect of soil moisture on endospore attachment and development of *P. penetrans*. It has been reported that the proportion of J2 with attached endospores was greater in moistened soil (Brown and Smart, 1984). The rate of development of *P. penetrans* in infected females was reduced when soils were maintained at or near field capacity (Davies, et al., 1991). Although the reasons for this effect are not known, it is possible that oxygen depletion in wet soil inhibits respiration, resulting in an inhibition of development of both the nematode and the bacterial parasite.

Endospore attachment is reportedly affected by pH. The highest attachment occurred at pH 9 (Ahmed and Gowen, 1991) and decreased at low pH values (Ahmed, 1990). Davies et al. (1988), however, observed that the attachment was higher at pH 7 than pH 4 or pH 9 in tap water, but lower at pH 7 than pH 4 or pH 9 in distilled water. Attachment of sonicated endospores was higher at pH 7 than pH 4 or pH 9 in distilled water and tap water. The sonicated endospores attached in higher numbers per J2 in tap water than in distilled water (Davies, et al., 1988). Recent studies revealed that the endospore surface has a net negative charge, which was greatest at neutral pH and was reduced with a change of pH away from neutral (Afolabi, et al., 1995). Electrostatic forces between the nematode cuticle and the endospore surface oppose attachment because the charges on nematode cuticle also were negative. Reasons for the pH effects remain unclear (Afolabi, et al., 1995).

21.6.9 Endospore Survival in Soil and Natural Enemy

Little is known about the long-term survival of endospores of *P. penetrans* in soil. In a peanut field in Florida, *P. penetrans* endospores maintained suppressive levels for *M. arenaria* over 10 years (Dickson, pers. obs.). Endospores of *P. penetrans* resist various chemicals and environmental conditions (Mani, 1988; Williams, et al., 1989). In the laboratory, endospores of *Pasteuria* sp. were viable for a period of more than 1 year (Mani, 1988). Recent reports indicated that storage of endospores for 5 years and 11 years did not affect the ability of attachment but may decrease the ability of infection (Giannakou, et al., 1997).

Natural enemies of *P. penetrans* endospores in soil have not been reported. Since the endospores are resistant to various environmental conditions, the survival in soil may largely be affected by biotic factors. A bacterium has been observed attached to endospores when the endospores attach to the cuticle of J2 of *Meloidogyne* spp. (Chen and Dickson, 1998). The bacterium is rod-shaped, Gram-negative, and may not be a parasite of the endospores because the ultrastructure and morphology of endospores remained intact when the bacteria were present. We recently observed paramecium engulfing endospores in a counting dish, however whether similar events occur in nature remains to be explored (Hewlett and Dickson, pers. obs.). Apparently, more research should be directed to this important area for understanding the fate of endospores in soil.

21.6.10 Role of *Pasteuria penetrans* in Nematode-suppressive Soil

Pasteuria penetrans has been considered as the primary microorganism responsible for soil suppressiveness to root-knot nematodes in many fields. Stirling and White (1982) observed that *P. penetrans* was widespread in vineyards more than 25 years old, but was difficult to detect in vineyards that were less than 10 years old. In old vineyards that were infested with *P. penetrans* there were fewer root-knot nematodes than in young vineyards without the bacterium. Bird and Brisbane (1988) confirmed the suppressive effect of *P. penetrans* in these vineyard soils. The reproductive capacity of *M. javanica* was much lower in soil infested with *P. penetrans* than in non-infested soil. Autoclaving of soil, which is detrimental to *P. penetrans*, removed the suppressiveness, whereas DBCP and 1, 3-D treatments did not (Stirling, 1984).

Minton and Sayre (1989) observed that population densities of *M. arenaria* declined to levels that only slightly affected plant yields in field plots cropped to hosts of the nematode for more than 20 years. In bioassays where they placed juveniles of *M. arenaria* in "suppressive soil" they found that the juveniles were encumbered with more than 25 endospores/juvenile. Reproduction of *M. arenaria* and root galls were inversely correlated with the number of endospores per juvenile in greenhouse experiments, leading the authors to conclude that the naturally occurring population of *P. penetrans* was keeping the population of *M. arenaria* suppressed.

Nishizawa (1987) observed *Pasteuria* sp. caused population decline of *Heterodera elachista* in monocultured upland rice. In a newly developed experimental field that was reclaimed from a natural forest, a low level of infestation of *H. elachista* was detected at the end of the first season and the population density increased exponentially until the fourth year. High levels of *Pasteuria* sp. were observed during the fifth and sixth years, which coincided with the drastic population reduction of *H. elachista*. It was concluded that *Pasteuria* sp. was the major cause of this decline.

Mankau (1980) studied nematode-suppressive soils that he concluded were induced by *P. penetrans* in West Africa. Root-knot nematodes did not cause severe problems in local areas where the weather conditions and the host crops were ideal for their development. Juveniles recovered from these soils were heavily infested with endospores of *P. penetrans*. When these soils were planted to tomato in a greenhouse, the root-knot nematode populations approached extinction in 4 to 5 generations.

Dickson et al. (1994) observed the sharp decline in the population density of peanut root-knot nematode after several years of monoculture of peanut in a field near Williston, Florida. The field had long been used for peanut nematode research because of the heavy infestation of *M. arenaria* race 1. Over a period of 10 to 15 years, peanut yields gradually increased and root-knot nematode galling on pegs and pods gradually decreased. Juveniles extracted from soil samples were encumbered with endospores of *P. penetrans*. When the soil was subjected to various treatments and planted to tomato, autoclaving of soil removed the suppressiveness and formalin-treated soil was less suppressive than the soil that was air dried or stored at 10℃. Eighty to 100% of juveniles migrating in the soil for 3 days acquired endospores, suggesting that *P. penetrans* is a major factor in suppressing this root-knot nematode population.

The nature of suppressiveness of a nematode-suppressive field in Gainesville, Florida was elucidated recently (Weibelzahl-Fulton, et al., 1996). The field was planted to flue-cured tobacco (*Nicotiana tabacum* L.) for 7 years. Root-knot damages caused by *M. incognita* race 1 and *M. javanica* was initially severe, but decreased over years. Juveniles migrating in soils for 2 days acquired 8 to 13 endospores/juvenile. Soil samples taken from the field were subjected to autoclaving, microwaving, air-drying, and untreated. Autoclaving eliminated both *P. penetrans* endospores and fungi, whereas microwaving did not affect endospores but reduced fungal propagules by 99.9%. When the soils were planted to tobacco, root galls, egg masses, and numbers of eggs were fewer in microwaving, air-drying, and untreated soils than autoclaving soil. The authors concluded that *P. penetrans* was the primary factor suppressing root-knot nematodes in the nematode-suppressive soil.

21.6.11 Biological Control Potential

Pasteuria penetrans is a very promising biological control agent against root-knot nematodes. The role of *P. penetrans* in suppressing plant-parasitic nematodes has been tested on many crops, mostly in greenhouse pots (Chen and Dickson, 1998). *Pasteuria penetrans* suppressed *Meloidogyne* spp. on egg plant, tomato, wheat, tobacco, soybean, bean, pepper, hairy vetch, cucumber, peanut, rye, chickpea, kiwi, grape, brinjal, mung, and okra. Some isolates of *Pasteuria* spp. have been reported to suppress *H. avenae* and *H. zeae* on unspecified crops (Bhattacharya and Swarup, 1988), *Belonolaimus longicaudatus* on bermudagrass turf (Giblin-Davis, 1990), *H. elachista* on rice (Nishizawa, 1987), and *H. cajani* on cowpea (Singh and Dhawan, 1994). A fully successful example of biological control of root-knot nematodes on peanut using *P. penetrans* follows.

Interestingly, cross-genera suppression of nematodes also was observed. Mankau and Prasad (1972) reported that *P. penetrans* reduced tomato root galls induced by *M. javanica* and *M. incognita*. *Pratylenchus scribneri* was reduced 53% in soil and 63% in roots 55 days after the nematodes were inoculated in *P. penetrans* infested soil (Mankau and Prasad, 1972). An Indian isolate of *P. penetrans* parasitized both *Heterodera* spp. and *M. incognita* (Bhattacharya and Swarup, 1988). The bacterium inoculum was mass-produced on *M. incognita* and when the inoculum was mixed into soil, number of cysts of *H. avenae* on wheat roots was reduced.

21.6.12 A Case Study of Biological Control of Peanut Root-knot Nematodes Using *Pasteuria penetrans*

Peanut (*Arachis hypogaea* L.) is one of the most important crops grown in the southern USA and many other countries, such as, Africa, India, and China. There are at least five different kinds of nematodes that infect peanut, including *M. arenaria* race 1, *M. hapla*, *Pratylenchus brachyurus*, *Belonolaimus longicaudatus*, and *Criconemella ornata* (Minton and Baujard, 1990). Evidence that *M. arenaria* is an aggressive pathogen on peanut and causes substantial yield loss at relatively low population densities has been reported. In Texas, initial population densities of 8.8 to 16.6 second-stage juveniles (J2) of *M. arenaria* per 100 cm^3 soil were sufficient to cause 10% suppression of yield (Wheeler and Starr, 1987), whereas the damage threshold in Florida was determined to be as low as 1.0 J2/100 cm^3 soil (McSorley, et al., 1992). The annual loss of peanut yield due to various nematodes was estimated US$1.23 billion in the USA (Koenning, et al., 1999).

The management of *M. arenaria* relies largely on a combination of methods that includes nematicides and crop rotation (Dickson and Hewlett, 1988, 1989; Minton and Baujard, 1990). Economic considerations and potential environmental hazards of nematicides, however, have created a climate of uncertainty regarding their continued use on peanut. It also is documented that nematicides are often unreliable when nematode population densities are high (Dickson and Hewlett, 1988), and some successful rotations must remain in place for more than 4 years (Dickson and Hewlett, 1989). Peanut cultivars resistant to *M. arenaria* have not been developed. Only recently some progress has been made in identifying sources of resistant genotypes (Holbrook and Noe, 1990; Nelson, et al., 1989; also see chapter by Starr and Roberts in this book). Therefore, development of biological control of peanut root-knot nematodes may provide an important alternative or supplemental management tactic for peanut nematode problems.

The biological control potential of *P. penetrans* against *M. arenaria* race 1 was evaluated in central Florida in peanut microplots. Endospores of *P. penetrans* were mass-produced *in vivo in M. arenaria*, which, in turn, was reared on tomato plants in a greenhouse (Chen, et al., 1996a; Stirling and Wachtel, 1980). The endospore concentration was estimated at 79 million endospores/g of root powder by a machine disruption method followed by counting on a hemocytometer (Chen, et al., 1996a). Field microplots were pre-infested with a large number of *M. arenaria* and then treated in 1994 only with 0, 1000, 3000, 10,000, and 100,000 endospores/g of soil. The effects of *P. penetrans* on population dynamics and damage of root-knot nematodes were evaluated in 1994 – 1996 (Chen, and Dickson, unpubl. data).

Pasteuria penetrans remarkably suppressed root and pod galls induced by root-knot nematodes on peanut (Fig. 21. 6). Root and pod galls were significantly reduced at 100, 000 endospores/g of soil in 1994, 10, 000 and 100, 000 endospores/g of soil in 1995, and 1000, 3000, 10, 000, and 100, 000 endospores/g of soil in 1996. There was not a single gall observed on peanut pods at 100, 000 endospores/g of soil in 1996. Furthermore, root galls were not observed in eight of the ten root systems at 100, 000 endospores/g of soil in 1996 (Chen and Dickson, unpubl. data). Apparently, the suppression of root and pod galls was increased year after year.

Figure 21. 6 Biological control of *Meloidogyne arenaria* using *Pasteuria penetrans*. Left: peanut had few galls on pods when microplots was inoculated with 100, 000 endospores/g of soil. Right: peanut from control microplots had few pods and the pods were severely galled by *M. arenaria*.

The increased suppression of root-knot nematodes over years may have resulted from the increased number of endospores of *P. penetrans* in soil. Chen and Dickson (unpubl. data) estimated that endospore densities were increased to 3500, 21,700, 36,000, and 131,100 endospores/g of soil in 1995, and 137,000, 99,000, 136,000, and 133,000 endospores/g of soil in 1996, for treatments of 1000, 3000, 10,000, and 100,000 endospores/g of soil, respectively. Also, the suppression of root and pod galls was quantitatively correlated to the percentage of J2 with endospores attached. These data strongly suggest that the increased suppression of root-knot nematodes was due to the increased numbers of endospores in soil. Effective suppression of root-knot nematodes appears to require 100,000 endospores/g of soil (Chen and Dickson, unpubl. data).

Foliage and pod yields were not significantly different among the treatments in 1994. Increasing of foliage and pod yields was observed in 1995 and 1996. In 1995, pod yields at 10,000 and 100,000 endospores/g of soil were significantly different from the control (Chen, et al., 1996b), whereas in 1996, foliage and pod yields in all the treatments inoculated with *P. penetrans* endospores were 2 to 3 times greater than the control (Chen and Dickson, unpubl. data).

In 1996, there were few numbers of root-knot nematodes present in microplots treated with 100,000 endospores/g of soil. This provided us an opportunity to look for highest densities of endospores in soil. *Pasteuria penetrans* requires host nematodes to complete its life cycle. Without host nematodes, *P. penetrans* cannot amplify itself. We estimated that 133,000 endospores/g of soil is the highest density of *P. penetrans* in soil. At this high level, high numbers of endospores attached to the cuticle that would eventually prevent nematode juveniles from searching host roots and entering and establishing themselves parasitically in roots.

In summary, the 3-year experiment provided much useful information on the biological control potential of *P. penetrans*. First, *P. penetrans* suppressed root and pod galls and increased yields. Approximately 100,000 endospores/g of soil was required for effective suppression of the peanut root-knot nematode on peanut. Secondly, *P. penetrans* had little impact on root-knot nematodes at low endospore densities because only a few nematodes came into contact with endospores and became infected. However, each of these infected nematodes produced 1 to 2 million endospores and the endospores can survive for a long period in soil so that endospore density is amplified and reaches a suppressive level in approximately 3 years when peanut is continuously planted. Thirdly, when the highest density of 133,000 endospores/g of soil is obtained, the nematode population is reduced to non-problematic levels. This provides unambiguous evidence that *P. penetrans* is very promising and has great potential for development as a bionematicide.

21.7 Other Prokaryotic Parasites

Intracellular rickettsia-like organisms have been observed in cyst nematodes, *Globodera rostochiensis*, *Heterodera glycines*, and *H. goettingiana* (Endo, 1979; Shepherd, et al., 1973; Walsh, 1979; Walsh, et al., 1983a, b). They reportedly were transmitted transovarially from generation to generation (Walsh, 1979). Studies showed that the microorganisms presented in the male sperm cells in *H. goettingiana* were not transmittable to the next generation (Walsh, 1979; Walsh, et al., 1983a). The nature of the association, either parasitism or mutualism, was unclear. Little information was available regarding transmission, proliferation, survival, isolation, and culture of these organisms.

21.8 Summary

Given the prodigious numbers of bacteria and nematodes and co-existence over millennia, it would be expected that hundreds of cases of parasitism would be evolved in history. So far, only a few bacterial parasites have been discovered. The huge gap between the theoretical expectation and the reality may suggest that much research effort should be directed to discover bacterial parasites of nematodes in the future. As the world demands on food and fiber supply increase, control of nematodes becomes crucial in agricultural production. With the scheduled phase-out of methyl bromide in 2005 and shrinkage in the number of registered nematicides, chemical control has become a less desirable option. As an alternative to chemical control, biological control of nematodes has received much attention in recent years.

The identification of suppressive soils has stimulated much research recently. Rhizobacteria and bacterial parasites were often considered as the primary factors in suppressing nematode populations. It is becoming increasingly acceptable to scientists and the public that biological control agents can suppress nematodes. The important task now is to gather all viable resources to address current and emerging nematode control challenges and develop biological control options.

P. penetrans acts as a biological nematicide in suppressing root-knot nematodes in the field. The host specificity of *P. penetrans*, the obligate nature of its parasitism, the resistance to various nematicides and adverse environment, the longevity of endospores in soil, and its capacity of amplifying itself and maintaining suppressiveness for years are invaluable traits that make it very promising for development as a commercial biological nematicide. Currently, the lack of an efficient technology for the large-scale production of *P. penetrans* is the greatest single impediment to the use of endospores as biological nematicides. Artificial cultivation of *P. penetrans* must be pursued for producing large quantity of endospores for field application.

References

Aalten, P. M., D. Vitour, D. Blanvillain, S. R. Gowen, and L. Sutra. 1998. Effect of rhizosphere fluorescent *Pseudomonas* strains on plant-parasitic nematodes *Radopholus similis* and *Meloidogyne* spp. Letters in Applied Microbiology 27: 357 – 361.

Adams, R. E., and J. J. Eichmuller. 1963. A bacterial infection of *Xiphinema americanum*. Phytopathology 53: 745 (Abstr.).

Afolabi, P., K. G. Davies, and P. S. Oshea. 1995. The electrostatic nature of the spore of *Pasteuria penetrans*, the bacterial parasite of root-knot nematodes. Journal of Applied Bacteriology 79: 244 – 249.

Ahmed, R. 1990. Studies on the efficacy of *Pasteuria penetrans* for the biological control of *Meloidogyne* species. Ph. D. Dissertation, University of Reading, UK.

Ahmed, R., and S. R. Gowen. 1991. Studies on the infection of *Meloidogyne* spp. with isolates of *Pasteuria penetrans*. Nematologia Mediterranea 19: 229 – 233.

Ali, A. H., and H. M. Kamal. 1998. Biocontrol of the root-knot nematode, *Meloidogyne incognita* on table grape. Bulletin of Faculty of Agriculture, University of Cairo 49: 435 – 452.

Anderson, J. M., J. F. Preston, D. W. Dickson, T. E. Hewlett, N. H. Williams, and J. E. Maruniak. 1999. Phylogenetic analysis of *Pasteuria penetrans* by 16S rRNA gene cloning and sequencing. Journal of Nematology 31: 319 – 325.

Anderson, W. R., F. G. Tromba, D. E. Thompson, and P. A. Madden. 1973. Bacteriologic and histologic examination of *Stephanurus dentatus* parasitizing swine ureters. Journal of Parasitology 59: 765 – 769.

Anderson, W. R., P. A. Madden, and M. L. Colglazier. 1978. Microbial flora of cuticular lesions on *Strongylus edentatus*. Proceedings of the Helminthological Society of Washington 45: 219 – 225.

Ara, J., H. S. Ehteshamul, V. Sultana, A. Ghaffar, and R. Qasim. 1997. Use of *Sargassum* species for the control of *Meloidogyne javanica* in okra. Nematologia Mediterranea 25: 125 – 128.

Atibalentja, N., G. R. Noel, and L. L. Domier. 2000. Phylogenetic position of the North American isolate of *Pasteuria* that parasitizes the soybean cyst nematode, *Heterodera glycines*, as inferred from 16S rDNA sequence analysis. International Journal of Systematic and Evolutionary Microbiology 50: 605 – 613.

Barker, K. R., and S. R. Koenning. 1998. Developing sustainable systems for nematode management. Annual Review of Phytopathology 36: 165 – 205.

Becker, J. O., E. Zavaleta-Mejia, S. F. Colbert, M. N. Schroth, A. R. Weinhold, J. G. Hancock, and S. D. Van-Gundy. 1988. Effects of rhizobacteria on root-knot nematodes and gall formation. Phytopathology 78: 1466 – 1469.

Bekal, S., J. Borneman, M. S. Springer, R. M. Giblin-Davis, and J. O. Becker. 2001. Phenotypic and molecular analysis of a *Pasteuria* strain parasitic to the sting nematode. Journal of Nematology 33: 110 – 115.

Benhamou, N., R. R. Bélanger, and T. Paulitz. 1996. Ultrastructural and cytochemical aspects of the interaction between *Pseudomonas fluorescens* and Ri T-DNA transformed pea roots: Host response to colonization by *Pythium ultimum* Trow. Planta 199: 105 – 117.

Bergman, B. H. H., and A. J. Van Duuren. 1959. Het bietencystenaaltje en zijn bestrijding. VII. De werking van stofwisselingsproducten van sommige micro-organismen op de larven van *Heterodera schachtii*. Mededelingen van het Instituut voor Rationele Suikerproductie. Bergen-op-Zoom. 29(2): 25 – 53.

Berkeley, R. C. W., and N. Ali. 1994. Classification and identification of endospore-forming bacteria. Pp. 1 – 8. In G. W. Gould, A. D. Russell, and D. E. S. Stewart-Tull, eds. Fundamental and applied aspects of bacterial spores. Oxford: Blackwell Scientific Publications.

Bhattacharya, D., and G. Swarup. 1988. *Pasteuria penetrans*, a pathogen of the genus *Heterodera*, its effect on nematode biology and control. Indian Journal of Nematology 18: 61 – 70.

Bird, A. F. 1986. The influence of the actinomycete, *Pasteuria penetrans*, on the host-parasite relationship of the plant-parasitic nematode, *Meloidogyne javanica*. Parasitology 93: 571 – 580.

Bird, A. F., and J. Bird. 1991. The Structure of Nematodes. San Diego, CA: Academic Press.

Bird, A. F., and P. G. Brisbane. 1988. The influence of *Pasteuria penetrans* in field soils on the reproduction of root – knot nematodes. Revue de Nématologie 11: 75 – 81.

Bird, A. F., I. Bonig, and A. Bacic. 1989. Factors affecting the adhesion of microorganisms to the surfaces of plant-parasitic nematodes. Parasitology 98: 155 – 164.

Bishop, A. H., and D. J. Ellar. 1991. Attempts to culture *Pasteuria penetrans* in vitro. Biocontrol Science and Technology 1: 101 – 114.

Borgonie, G., M. Claeys, F. Leyns, G. Arnaut, D. de Waele, and A. Coomans. 1996a. Effect of nematicidal *Bacillus thuringiensis* strains on free-living nematodes. 1. Light microscopic observations, species and biological stage specificity and identification of resistant mutants of *Caenorhabditis elegans*. Fundamental and Applied Nematology 19: 391 – 398.

Borgonie, G., M. Claeys, F. Leyns, G. Arnaut, D. de Waele, and A. Coomans. 1996b. Effect of nematicidal *Bacillus thuringiensis* strains on free-living nematodes. 2. Ultrastructural analysis of the intoxication process in *Caenorhabditis elegans*. Fundamental and Applied Nematology 19: 407 – 414.

Brown, S. M., and G. C. Smart, Jr. 1984. Attachment of *Bacillus penetrans* to *Meloidogyne incognita*. Nematropica 14: 171 – 172.

Campbell, W. C. 1981. An introduction to the avermectins. New Zealand Veterinary Journal 29: 174 – 178.

Campbell, W. C., M. H. Fisher, E. O. Stapley, G. Albers-Schonberg, and T. A. Jacob. 1983. Ivermectin: a potent new antiparasitic agent. Science 221: 823 – 828.

Carneiro, R. G., I. de Souza, and L. C. Belarminoand. 1998. Nematicidal activity of *Bacillus* spp. strains on juveniles of *Meloidogyne javanica*. Nematologia Brasileira 22: 12 – 21.

Carneiro, R. M. D. G., O. Randig, L. G. Freitas, and D. W. Dickson. 1999. Attachment of endospores of *Pasteuria pentrans* to males and juveniles of *Meloidogyne* spp. Nematology 1: 267 – 271.

Cayrol, J. C., C. Djian, and J. P. Frankowski. 1993. Efficacy of abamectin B1 for the control of *Meloidogyne arenaria*. Fundamental and Applied Nematology 16: 239 – 246.

Chahal, P. K., and V. S. Chahal. 1993. Effect of thuricide on the hatching of eggs of root-knot nematode (*Meloidogyne incognita*). Current Nematology 4: 247.

Chen, C., E. M. Bauke, G. Musson, R. Rodríguez-Kábana, and J. W. Kloepper. 1995. Biological control of Fusarium wilt on cotton by use of endophytic bacteria. Biological Control 5: 83 – 91.

Chen, J., G. S. Abawi, and B. M. Zuckerman. 1999. Suppression of *Meloidogyne hapla* and its damage to lettuce grown in a mineral soil amended with chitin and biocontrol organisms. Supplement to Journal of Nematology 31: 719 – 725.

Chen, J., G. S. Abawi, and B. M. Zuckerman. 2000. Efficacy of *Bacillus thuringiensis*, *Paecilomyces marquandii*, and *Streptomyces costaricanus* with and without organic amendments against *Meloidogyne hapla* infecting lettuce. Journal of Nematology 32: 70 – 77.

Chen, Z. X., and D. W. Dickson. 1997a. Minimal developmental temperature of *Pasteuria penetrans*. Supplement to the Journal of Nematology 29: 635 – 639.

Chen, Z. X., and D. W. Dickson. 1997b. Effects of ammonium nitrate and time of harvest on mass production of *Pasteuria penetrans*. Nematropica 27: 53 – 60.

Chen, Z. X., and D. W. Dickson. 1998. A review of *Pasteuria penetrans*: Biology, ecology, and biological control potential. Journal of Nematology 30: 313 – 340.

Chen, Z. X., D. W. Dickson, and T. E. Hewlett. 1996a. Quantification of endospore concentrations of *Pasteuria penetrans* in tomato root material. Journal of Nematology 28: 50 – 55.

Chen, Z. X., D. W. Dickson, D. J. Mitchell, R. McSorley, and T. E. Hewlett. 1997b. Suppression Mechanisms of *Meloidogyne arenaria* race 1 by *Pasteuria penetrans*. Journal of Nematology 29: 1 – 8

Chen, Z. X., D. W. Dickson, R. McSorley, D. J. Mitchell, and T. E. Hewlett. 1996b. Suppression of *Meloidogyne arenaria* race 1 by soil application of endospores of *Pasteuria penetrans*. Journal of Nematology 28: 159 – 168.

Ciancio, A. 1995. Phenotypic adaptations in *Pasteuria* spp. nematode parasites. Journal of Nematology 27: 328 – 338.

Ciancio, A., R. Bonsignore, N. Vovlas, and F. Lamberti. 1994. Host records and spore morphometrics of *Pasteuria penetrans* group parasites of nematodes. Journal of Invertebrate Pathology 63: 260 – 267.

Clark, F. E. 1967. Bacteria in soil. Pp. 15 – 19. In A. Eurges and F. Raw, eds. Soil Biology. London: Academic Press.

Cobb, N. A. 1906. Fungus Maladies of the Sugar Cane, with Notes on Associated Insects and Nematodes. 2nd ed. Honolulu: Hawaiian Sugar Planters Association.

Cronin, D., L. Y. Moenne, C. Dunne, and F. O'Gara. 1997a. Inhibition of egg hatch of the potato cyst nematode *Globodera rostochiensis* by chitinase-producing bacteria. European Journal of Plant Pathology 103: 433 – 440.

Cronin, D., L. Y. Moenne, A. Fenton, C. Dunne, D. N. Dowling, and F. O'Gara. 1997b. Role of 2, 4-diacetylphloroglucinol in the interactions of the biocontrol pseudomonad strain F113 with the potato cyst nematode *Globodera rostochiensis*. Applied and Environmental Microbiology 63: 1357 – 1361.

Davies, K. G., B. R. Kerry, and C. A. Flynn. 1988. Observations on the pathogenicity of *Pasteuria penetrans*, a parasite of root-knot nematodes. Annals of Applied Biology 112: 491 – 501.

Davies, K. G., V. Laird, and B. R. Kerry. 1991. The motility, development and infection of *Meloidogyne incognita* encumbered with spores of the obligate hyperparasite *Pasteuria penetrans*. Revue de Nématologie 14: 611 – 618.

Dicklow, M. B., N. Acosta, and B. M. Zuckerman. 1993. A novel *Streptomyces* species for controlling plant-parasitic nematodes. Journal of Chemical Ecology 19: 159 – 173.

Dickson, D. W., and T. E. Hewlett. 1988. Efficacy of fumigant and nonfumigant nematicides for control of *Meloidogyne arenaria* on peanut. Annals of Applied Nematology 2: 95 – 101.

Dickson, D. W., and T. E. Hewlett. 1989. Effects of bahiagrass and nematicides on *Meloidogyne arenaria* on peanut. Supplement to the Journal of Nematology 21: 671 – 676.

Dickson, D. W., M. Oostendorp, R. M. Giblin-Davis, and D. J. Mitchell. 1994. Control of plant-parasitic nematodes by biological antagonists. Pp. 575 – 601. In D. Rosen, F. D. Bennett, and J. L. Capinera, eds. Pest Management in the Subtropics, Biological Control – a Florida perspective. Hampshire, UK: Intercept.

Dollfus, R. P. 1946. Parasites (Animaux et Végétaux) des Helminthes. Paris: Paul Lechevalier.

Dong, Z., M. J. Canny, M. E. McCully, M. R. Rodoredo, C. F. Cabadilla, E. Ortega, and R. Rodés. 1994. A nitrogen-fixing endophyte of sugarcane stems. Plant Physiology 105: 1139 – 1147.

Duponnois, R., A. M. Ba, and T. Mateille, 1999. Beneficial effects of *Enterobacter cloacae* and *Pseudomonas mendocina* for biocontrol of *Meloidogyne incognita* with the endospore-forming bacterium *Pasteuria penetrans*. Nematology 1: 95 – 101.

Dürschner, U. U. 1984. Observations on a new nematophagous actinomycete. Proceedings of the First International Congress of Nematology, Guelph, Canada, 23 August 1984, Pp 23. Lawrence, KA: Allen Press.

Eapen, S. J., K. V. Ramana, Y. R. Sarma, S. Edison, B. Sasikumar, and K. N. Babu. 1996. Evaluation of *Pseudomonas fluorescens* isolates for control of *Meloidogyne incognita* in black pepper (*Piper nigrum* L.). Proceedings of Biotechnology of spices, medicinal, and aromatic plants, Calicut, India, 24 – 25 April 1996, Pp 129 – 133.

Ebert, D., P. Rainey, T. M. Embley, and D. Scholz. 1996. Development, life cycle, ultrastructure and phylogenetic position of *Pasteuria ramosa* Metchnikoff 1888: Rediscovery of an obligate endoparasite of *Daphnia magna* Straus. Philosophical Transactions of the Royal Society of London Series B – Biological Sciences 351:

1689 – 1701.
Ehteshamul, H. S., V. Sultana, J. Ara, R. Qasim, and A. Ghaffar. 1997. Use of crustacean chitin and plant growth promoting bacteria for the control of *Meloidogyne javanica* root knot nematode in chickpea. Pakistan Journal of Nematology 15: 89 – 93.
Endo, B. Y. 1979. The ultrastructure and distribution of an intracellular bacterium-like microorganism in tissue of larvae of the soybean cyst nematode, *Heterodera glycines*. Journal of Ultrastructure Research 67: 1 – 14.
Esnard, J., M. N. Marban, and B. M. Zuckerman. 1998. Effects of three microbial broth cultures and an organic amendment on growth and populations of free-living and plant-parasitic nematodes on banana. European Journal of Plant Pathology 104: 457 – 463.
Esser, R. P., and E. K. Sobers. 1964. Natural enemies of nematodes. Proceedings of Soil and Crop Science Society of Florida. 24: 326 – 353.
Ferris, H., H. S. Du Vernay, and R. H. Small. 1978. Development of a soil-temperature data base on *Meloidogyne arenaria* for a simulation model. Journal of Nematology 10: 39 – 42.
Foster, R. C., A. D. Rovira, and T. W. Cock. 1983. Ultrastructure of the root – soil interface. St. Paul, MN: American Phytopathological Society.
Frommel, M. I., J. Nowak, and G. Lazarovits. 1991. Growth enhancement and developmental modifications of *in vitro* grown potato (*Solanum tuberosum* ssp. *tuberosum*) as affected by a nonfluorescent *Pseudomonas* sp. Plant Physiology 96: 928 – 936.
Giannakou, I. O., B. Pembroke, S. R. Gowen, and K. G. Davies. 1997. Effects of long term storage and above normal temperatures on spore adhesion of *Pasteuria penetrans* and infection of the root-knot nematode *Meloidogyne javanica*. Nematologica 43: 185 – 192.
Giblin-Davis, R. M. 1990. Potential for biological control of phytoparasitic nematodes in Bermudagrass turf with isolates of the *Pasteuria penetrans* group. Proceedings of the Florida State Horticultural Society 103: 349 – 351.
Giblin-Davis, R. M., D. Williams, T. E. Hewlett, and D. W. Dickson. 1995. Development and host attachment studies using *Pasteuria* from *Belonolaimus longicaudatus* from Florida. Journal of Nematology 27: 500 (Abstr.).
Giblin-Davis, R. M., L. L. McDaniel, and F. G. Bilz. 1990. Isolates of the *Pasteuria penetrans* group from phytoparasitic nematodes in Bermudagrass turf. Journal of Nematology 22: 750 – 762.
Gokte, N., and G. Swarup. 1988. On the potential of some bacterial biocides against root-knot and cyst nematodes. Indian Journal of Nematology 18: 152 – 153 (publ. 1989).
Griffitts, J. S., J. L. Whitacre, D. E. Stevens, and R. V. Aroian. 2001. Bt toxin resistance from loss of a putative carbohydrate-modifying enzyme. Science 293: 860 – 864.
Grimault, V., and P. Prior. 1994. Invasiveness of *Pseudomonas solanacearum* in tomato, eggplant, and pepper: a comparative study. European Journal of Plant Pathology 100: 259 – 267.
Hackenberg, C., A. Muehlchen, T. Forge, and T. Vrain. 2000. *Pseudomonas chlororaphis* strain Sm3, bacterial antagonist of *Pratylenchus penetrans*. Journal of Nematology 32: 183 – 189.

Hackenberg, C., and R. A. Sikora. 1994. Influence of temperature and soil moisture on the biological control of the potato-cyst nematode *Globodera pallida* using the plant-health-promoting rhizobacterium *Agrobacterium radiobacter*. Journal of Phytopathology 142: 338 – 344.

Hackenberg, C., T. C. Vrain, and R. A. Sikora. 1999. Rhizosphere colonization pattern of *Agrobacterium radiobacter* strain G12A, an antagonistic rhizobacterium to the potato cyst nematode *Globodera pallida*. Microbiological Research 154: 57 – 61.

Hallmann, J., A. Quadt-Hallmann, W. F. Mahaffee, and J. W. Kloepper. 1997. Bacterial endophytes in agricultural crops. Canadian Journal of Microbiology 43: 895 – 914.

Hallmann, J., A. Quadt-Hallmann, R. Rodríguez-Kábana, and J. W. Kloepper. 1998. Interactions between *Meloidogyne incognita* and endophytic bacteria in cotton and cucumber. Soil Biology and Biochemistry 30: 925 – 937.

Hallmann, J., A. Quadt-Hallmann, W. G. Miller, R. A. Sikora, and S. E. Lindow. 2001. Endophytic colonization of plants by the biocontrol agent *Rhizobium etli* G12 in relation to *Meloidogyne incognita* infection. Phytopathology 91: 415 – 422.

Hallmann, J., R. Rodríguez-Kábana, and J. W. Kloepper. 1999. Chitin-mediated changes in bacterial communities of the soil, rhizosphere and within roots of cotton in relation to nematode control. Soil Biology and Biochemistry 31: 551 – 560.

Hanna, A. I., F. W. Riad, and A. E. Tawfik. 1999. Efficacy of antagonistic rhizobacteria on the control of root-knot nematode, *Meloidogyne incognita* in tomato plants. Egyptian Journal of Agricultural Research 77: 1467 – 1476.

Hasky, G. K., H. S. Hoffmann, and R. A. Sikora. 1998. Resistance against the potato cyst nematode *Globodera pallida* systemically induced by the rhizobacteria *Agrobacterium radiobacter* (G12) and *Bacillus sphaericus* (B43). Fundamental and Applied Nematology 21: 511 – 517.

Hatz, B., and D. W. Dickson. 1992. Effect of temperature on attachment, development, and interactions of *Pasteuria penetrans* on *Meloidogyne arenaria*. Journal of Nematology 24: 512 – 521.

Hess, R., and G. O. Poinar, Jr. 1988. Viral diseases. Pp. 51 – 67. In G. O. Poinar, Jr., and H. B. Jansson, eds. Diseases of Nematodes, Vol. 1. Boca Raton, FL: CRC Press.

Hoffmann, H. S., M. K. Gulati, and R. A. Sikora. 1998. Yield response and biological control of *Meloidogyne incognita* on lettuce and tomato with rhizobacteria. Zeitschrift fur Pflanzenkrankheiten und Pflanzenschutz 105: 349 – 358.

Holbrook, C. C., and J. P. Noe. 1990. Resistance to *Meloidogyne arenaria* in *Arachis* spp. and implications on development of resistant peanut cultivars. Peanut Science 17: 35 – 38.

Ivanova, T. S., A. K. Vyalykh, M. S. Cokolov, and O. A. Monastyrskii. 1996. Efficiency of biopreparations in control of gall nematode in protected soil. Agrokhimiya (No. 3) 101 – 106.

Kempster, V. N., K. A. Davies, and E. S. Scott. 2001. Chemical and biological induction of resistance to the clover cyst nematode (*Heterodera trifolii*) in white clover (*Trifolium repens*). Nematology 3: 35 – 43.

Khan, M. R., and Z. Tarannum. 1999. Effects of field application of various micro-organisms on *Meloidogyne incognita* on tomato. Nematologia Mediterranea 27: 233 – 238.

Kloepper, J. W., R. Rodríguez-Kábana, G. W. Zehnder, J. F. Murphy, E. Sikora, and C. Fernandez. 1999. Plant root-bacterial interactions in biological control of soilborne diseases and potential extension to systemic and foliar diseases. Australasian Plant Pathology 28: 21 – 26.

Kloepper, J. W., R. Rodríguez-Kábana, J. A. McInroy, and D. J. Collins. 1991. Analysis of populations and physiological characterization of microorganisms in rhizospheres of plants with antagonistic properties to phytopathogenic nematodes. Plant and Soil 136: 95 – 102.

Koenning, S. R., C. Overstreet, J. W. Noling, P. A. Donald, J. O. Becker, and B. A. Fortnum. 1999. Survey of crop losses in response to phytoparasitic nematodes in the United States for 1994. Supplement to the Journal of Nematology 31: 587 – 618.

Leidy, J. 1851. A flora and fauna within living animals. Smithsonian Contributions to Knowledge, Smithsonian Institution, Washington, D. C. 5(2): 67.

Leyns, F., G. Borgonie, G. Arnaut, and D. de Waele. 1995. Nematicidal activity of *Bacillus thuringiensis* isolates. Fundamental and Applied Nematology 18: 211 – 218.

LiPuma, J. J. 2001. *Burkholderia cepacia* complex as human pathogens. Phytopathology (Supplement) 91: S173 (Abstr.).

Mani, A. 1988. Studies on the bacterial parasite *Pasteuria penetrans*: I. Spore viability after storage. II. Culture on citrus nematode *Tylenchulus semipenetrans*. International Nematology Network Newsletter 5: 24 – 25.

Mankau, R. 1975a. Prokaryote affinities of *Duboscqia penetrans*, Thorne. Journal of Protozoology 21: 31 – 34.

Mankau, R. 1975b. *Bacillus penetrans* n. comb. causing a virulent disease of plant-parasitic nematodes. Journal of Invertebrate Pathology 26: 333 – 339.

Mankau, R. 1980. Biological control of *Meloidogyne* populations by *Bacillus penetrans* in West Africa. Journal of Nematology 12: 230 (Abstr.).

Mankau, R., and N. Prasad. 1972. Possibilities and problems in the use of a sporozoan endoparasite for biological control of plant parasitic nematodes. Nematropica 2: 7 – 8.

Mankau, R., and N. Prasad. 1977. Infectivity of *Bacillus penetrans* in plant-parasitic nematodes. Journal of Nematology 9: 40 – 45.

Mankau, R., J. L. Imbriani, and A. H. Bell. 1976. SEM observations on nematode cuticle penetration by *Bacillus penetrans*. Journal of Nematology 8: 179 – 181.

Marroquin, L. D., D. Elyassnia, J. S. Griffitts, J. S. Feitelson, and R. V. Aroian. 2000. *Bacillus thuringiensis* (Bt) toxin susceptibility and isolation of resistance mutants in the nematode *Caenorhabditis elegans*. Genetics 155: 1693 – 1699.

McSorley, R., D. W. Dickson, E. M. Candanedo-Lay, T. E. Hewlett, and J. J. Frederick. 1992. Damage functions for *Meloidogyne arenaria* on peanut. Journal of Nematology 24: 193 – 198.

Meadows, J., S. S. Gill, and L. W. Bone. 1990. *Bacillus thuringiensis* strains affect population growth of the free-living nematode *Turbatrix aceti*. Invertebrate Reproduction and Development 17: 73 – 76.

Mena, J., E. Pimentel, R. Vazquez, R. Garcia, M. Fernandez, R. Moran, L. Perez, M. Garcia, Z. Zaldua, D. Somontes, A. Lopez, M. Gomez, J. D. Mencho. 1997.

Results of the use of *Bacillus thuringiensis* var. *kurstaki* in the control of *Radopholus similis in banana and plantain plantations. Centro Agricola* 24: 41 -49.

Meyer, S. L. F., S. I. Massoud, D. J. Chitwood, and D. P. Roberts. 2000. Evaluation of *Trichoderma virens* and *Burkholderia cepacia* for antagonistic activity against root-knot nemaotde, *Meloidogyne incognita*. Nematology 2: 871 -879.

Meyer, S. L. F., D. P. Roberts, D. J. Chitwood, L. K. Carta, R. D. Lumsden, and W. L. Mao. 2001. Application of *Burkholderia cepacia* and *Trichoderma virens*, alone and in combinations, against *Meloidogyne incognita* on bell pepper. Nematropica 31: 75 -86.

Minton, N. A., and P. Baujard. 1990. Nematode parasites of peanut. Pp. 285 -320. In M. Luc, R. A. Sikora, and J. Bridge, eds. Plant Parasitic Nematodes in Subtropical and Tropical Agriculture. Wallingford, UK: CAB International.

Minton, N. A., and R. M. Sayre. 1989. Suppressive influence of *Pasteuria penetrans* in Georgia soil on reproduction of *Meloidogyne arenaria*. Journal of Nematology 21: 574 -575 (Abstr.).

Nair, M. G., A. Chandra, D. L. Thorogod, and R. M. Giblin-Davis. 1995. Nematicidal and mosquitocidal aromatic nitro compounds produced by *Streptomyces* spp. Extended summaries: 8th International Congress of Pesticide Chemistry (IUPAC) 43: 361 -365.

Nelson, S. C., C. E. Simpson, and J. L. Starr. 1989. Resistance to *Meloidogyne arenaria* in *Arachis* spp. germplasm. Journal of Nematology (Supplement) 21: 654 -660.

Nishizawa, T. 1987. A decline phenomenon in a population of upland cyst nematode, *Heterodera elachista*, caused by bacterial parasite, *Pasteuria penetrans*. Journal of Nematology 19: 546 (Abstr.).

Oka, Y., I. Chet, and Y. Spiegel. 1993. Control of the root-knot nematode *Meloidogyne javanica* by *Bacillus cereus*. Biocontrol Science and Technology 3: 115 -126.

Omura, S., K. Otoguro, N. Imamura, H. Kuga, Y. Takahashi, R. Masuma, Y. Tanaka, and H. Tanaka. 1987. Jietacins A and B, new nematocidal antibiotics from a *Streptomyces* sp. Taxonomy, isolation, and physico-chemical and biological properties. Journal of Antibiotics 40: 623 -629.

Oostendorp, M., and R. A. Sikora. 1989. Seed treatment with antagonistic rhizobacteria for the suppression of *Heterodera schachtii* early root infection of sugar beet. Revue de Nematologie 12: 77 -83.

Oostendorp, M., D. W. Dickson, and D. J. Mitchell. 1990. Host range and ecology of isolates of *Pasteuria* spp. from the southeastern United States. Journal of Nematology 22: 525 -531.

Osman, G. Y., F. M. Salem, and A. Ghattas. 1988. Bio-efficacy of two bacterial insecticide strains of *Bacillus thuringiensis* as a biological control agent in comparison with a nematicide, Nemacur, on certain parasitic Nematoda. Anzeiger fur Schadlingskunde, Pflanzenschutz, Umweltschutz 61: 35 -37.

Page, S. L. J., and J. Bridge. 1985. Observations on *Pasteuria penetrans* as a parasite of *Meloidogyne acronea*. Nematologica 31: 238 -240.

Pan, C., J. Lin, Z. Ni, and S. Wang. 1993. Study on the pathogenic bacteria parasitizing root-knot nematodes discovered in China and their application to biological control. Acta Microbiologica Sinica 33: 313 -316.

Parke, J. L., and D. Gurian-Sherman. 2001. Diversity of the *Burkholderia cepacia* complex and implications for risk assessment of biological control strains. Annual Review of Phytopathology 39: 225 – 258.

Perveen, S., H. S. Ehteshamul, and A. Ghaffar. 1998. Efficacy of *Pseudomonas aeruginosa* and *Paecilomyces lilacinus* in the control of root rot-root knot disease complex on some vegetables. Nematologia Mediterranea 26: 209 – 212.

Poinar, G. O., Jr., and H. -B. Jansson, eds. 1988. Diseases of nematodes, vol. I and II. Boca Raton, FL: CRC Press.

Quadt-Hallmann, A., and J. W. Kloepper. 1996. Immunological detection and localization of the cotton endophyte *Enterobacter asburiae* JM22 in different plant species. Canadian Journal of Microbiology 42: 1144 – 1154.

Quarles, W. 1991. The avermectins: Successful biopesticides. IPM Practitioner 13: 5 – 15.

Racke, J., and R. A. Sikora. 1992. Isolation, formulation and antagonistic activity of rhizobacteria toward the potato cyst nematode *Globodera pallida*. Soil Biology and Biochemistry 24: 521 – 526.

Rai, G. P., and R. S. Rana. 1979. Effectiveness of beta-exotoxin of *Bacillus thuringiensis* var. *thuringiensis* on the ability of *Meloidogyne* sp. from brinjal (*Solanum melongenea* L.) to survive. Journal of Biosciences 1: 271 – 278.

Ramakrishnan, S., C. V. Sivakumar, and K. Poornima. 1998. Management of rice root nematode, *Hirschmanniella gracilis* (de Man) Luc & Goodey with *Pseudomonas fluorescens* Migula. Journal of Biological Control 12: 135 – 141.

Ramamoorthy, V., R. Viswanathan, T. Raguchander, V. Prakasam, and R. Samiyappan. 2001. Induction of systemic resistance by plant growth promoting rhizobacteria in crop plants against pests and diseases. Crop Protection 20: 1 – 11.

Reise, R. W., K. L. Hackett, R. M. Sayre, and R. N. Huettel. 1988. Factors affecting cultivation of three isolates of *Pasteuria* spp. Journal of Nematology 20: 657 (Abstr.).

Reitz, M., H. S. Hoffmann, J. Hallmann, and R. A. Sikora. 2001. Induction of systemic resistance in potato by rhizobacterium *Rhizobium etli* strain G12 is not associated with accumulation of pathogenesis-related proteins and enhanced lignin biosynthesis. Zeitschrift fur Pflanzenkrankheiten und Pflanzenschutz 108: 11 – 20.

Reitz, M., K. Rudolph, I. Schroeder, H. S. Hoffmann, J. Hallmann, and R. A. Sikora. 2000. Lipopolysaccharides of *Rhizobium etli* strain G12 act in potato roots as an inducing agent of systemic resistance to infection by the cyst nematode *Globodera pallida*. Applied and Environmental Microbiology 66: 3515 – 3518.

Samaliev, H. Y., F. I. Andreoglou, S. A. Elawad, N. G. M. Hague, and S. R. Gowen. 2000. The nematicidal effects of the bacteria *Pseudomonas oryzihabitans* and *Xenorhabdus nematophilus* on the root-knot nematode *Meloidogyne javanica*. Nematology 2: 507 – 514.

Santhi, A., and C. V. Sivakumar. 1995. Biocontrol potential of *Pseudomonas fluorescens* (Migula) against root-knot nematode, *Meloidogyne incognita* (Kofoid and White, 1919) Chitwood, 1949 on tomato. Journal of Biological Control 9: 113 – 115.

Santhi, A., R. Sundarababu, C. V. Sivakumar, and R. Sundarababu. 1999. Field evaluation of rhizobacterium, *Pseudomonas fluorescens* for the management of the citrus nematode, *Tylenchulus semipenetrans*. Proceedings of national symposium on rational approaches in nematode management for sustainable agriculture, Anand, India, 23 – 25

November 1998.
Sayre, R. M. 1980. Biocontrol: *Bacillus penetrans* and related parasites of nematodes. Journal of Nematology 12: 260 – 270.
Sayre, R. M. 1988. Bacterial diseases of nematodes and their role in controlling nematode populations. Agriculture, Ecosystems and Environment 24: 263 – 279.
Sayre, R. M. 1993. *Pasteuria*, Metchnikoff, 1888. Pp. 101 – 111. In A. L. Sonenshein, J. A. Hoch, and R. Losick, eds. *Bacillus Subtilis* and Other Gram-positive Bacteria: Biochemistry, Physiology, and Molecular Genetics. Washington, D. C.: American Society for Microbiology.
Sayre, R. M., and M. P. Starr. 1985. *Pasteuria penetrans* (ex Thorne, 1940) nom. rev., comb. n., sp. n., a mycelial and endospore-forming bacterium parasitic in plant-parasitic nematodes. Proceedings of the Helminthological Society of Washington 52: 149 – 165.
Sayre, R. M., and M. P. Starr. 1988. Bacterial diseases and antagonisms of nematodes. Pp. 69 – 101. In G. O. Poinar, Jr., and H. -B. Jansson, eds. Diseases of Nematodes. Boca Raton, Florida: CRC Press.
Sayre, R. M., and M. P. Starr. 1989. Genus *Pasteuria* Metchnikoff, 1888. Pp. 2601 – 2615. In S. T. Williams, M. E. Sharpe, and J. G. Holt, eds. Bergey's Manual of Systematic Bacteriology. Baltimore, MD: Williams and Wilkins.
Sayre, R. M., and W. P. Wergin. 1977. Bacterial parasite of a plant nematode: morphology and ultrastructure. Journal of Bacteriology 129: 1091 – 1101.
Sayre, R. M., W. P. Wergin, and R. E. Davies. 1977. Occurence in *Moina rectirostris* (Cladocera: Daphnidae) of a parasite morphologically similar to *Pasteuria ramosa* (Metchnikoff, 1888). Canadian Journal of Microbiology 23: 1573 – 1579.
Sela, S., H. Schickler, I. Chet, and Y. Spiegel. 1998. Purification and characterization of a *Bacillus cereus* collagenolytic/proteolytic enzyme and its effect on *Meloidogyne javanica* cuticular proteins. European Journal of Plant Pathology 104: 59 – 67.
Serracin, M., A. C. Schuerger, D. W. Dickson, and D. P. Weingartner. 1997. Temperature-dependent development of *Pasteuria penetrans* in *Meloidogyne arenaria*. Journal of Nematology 29: 228 – 238.
Sharma, R. D. 1994. *Bacillus thuringiensis*: a biocontrol agent of *Meloidogyne incognita* on barley. Nematologia Brasileira 18: 79 – 84.
Sharma, R. D. 1995. Efficiency of *Bacillus* spp. toxins to control *Heterodera glycines* on soybean. Nematologia Brasileira 19: 72 – 80.
Sharma, S. B., and K. G. Davies. 1996. Characterization of *Pasteuria* isolated from *Heterodera cajani* using morphology, pathology and serology of endospores. Systematic and Applied Microbiology 19: 106 – 112.
Sharma, R. D., and A. C. Gomes. 1996a. Effect of seed treatment with delta-endotoxins of *Bacillus* spp. on the multiplication of *Heterodera glycines* on soybean and corn. Nematologia Brasileira 20: 21 – 29.
Sharma, R. D., and A. C. Gomes. 1996b. Effects of *Bacillus* spp. toxins on oviposition and juvenile hatching of *Heterodera glycines*. Nematologia Brasileira 20: 53 – 62.
Shepherd, A. M., S. A. Clark, and A. Kempton. 1973. An intracellular microorganism associated with the tissue of *Heterodera* spp. Nematologica 19: 31 – 34.
Siddiqui, I. A. 2000. Effects of cell suspension and cell-free culture filtrate of

Pseudomonas aeruginosa in the control of root rot-root knot disease complex of tomato (*Lycopersicon esculentum* Mill.). Acta Agrobotanica 53: 47 – 55.

Siddiqui, I. A., and H. S. Ehteshamul. 2000. Use of *Pseudomonas aeruginosa* for the control of root rot-root knot disease complex in tomato. Nematologia Mediterranea 28: 189 – 192.

Siddiqui, I. A., H. S. Ehetshamul, and S. S. Shaukat. 2001. Use of rhizobacteria in the control of root rot-root knot disease complex of mungbean. Journal of Phytopathology 149: 337 – 346.

Siddiqui, Z. A., and I. Mahmood. 1992. Biological control of root-rot disease complex of chickpea caused by *Meloidogyne incognita* race 3 and *Macrophomina phaseolina*. Nematologia Mediterranea 20: 199 – 202.

Siddiqui, Z. A., and I. Mahmood. 1993. Biological control of *Meloidogyne incognita* race 3 and *Macrophomina phaseolina* by *Paecilomyces lilacinus* and *Bacillus subtilis* alone and in combination on chickpea. Fundamental and Applied Nematology 16: 215 – 218.

Siddiqui, Z. A., and I. Mahmood. 1995a. Role of plant symbionts in nematode management: A review. Bioresource Technology 54: 217 – 226.

Siddiqui, Z. A., and I. Mahmood, 1995b. Management of *Meloidogyne incognita* race 3 and *Macrophomina phaseolina* by fungus culture filtrates and *Bacillus subtilis* in chickpea. Fundamental and Applied Nematology. 18: 71 – 76.

Siddiqui, Z. A., and I. Mahmood. 1999. Role of bacteria in the management of plant parasitic nematodes: A review. Bioresource Technology 69: 167 – 179

Siddiqui, Z. A., I. Mahmood, and M. A. Ansari. 1995. Effect of different inoculum levels of *Meloidogyne incognita* on the growth of pea in the presence and absence of *Rhizobium*. Nematologia Mediterranea 23: 249 – 251.

Siddiqui, Z. A., I. Mahmood, and S. Hayat. 1998. Biocontrol of *Heterodera cajani* and *Fusarium udum* on pigeonpea using *Glomus mosseae*, *Paecilomyces lilacinus* and *Pseudomonas fluorescens*. Thai Journal of Agricultural Science 31: 310 – 321.

Sikora, R. A. 1997. Biological system management in the rhizosphere an inside-out/outside-in perspective. Mededelingen Faculteit Landbouwkundige en Toegepaste Biologische Wetenschappen, Universiteit Gent. 62: 105 – 112.

Singh, B., and S. C. Dhawan. 1990. A new bacterial strain of *Pasteuria penetrans*, its host range and effect of temperature on spore attachment to second-stage juveniles of pigeon-pea cyst nematode, *Heterodera cajani*. Indian Journal of Nematology 20: 161 – 166.

Singh, B., and S. C. Dhawan. 1994. Effect of *Pasteuria penetrans* on the penetration and multiplication of *Heterodera cajani* in *Vigna unguiculata* roots. Nematologia Mediterranea 22: 159 – 161.

Stirling, G. R. 1981. Effect of temperature on infection of *Meloidogyne javanica* by *Bacillus penetrans*. Nematologica 27: 458 – 462.

Stirling, G. R. 1984. Biological control of *Meloidogyne javanica* with *Bacillus penetrans*. Phytopathology 74: 55 – 60.

Stirling, G. R. 1991. Biological control of plant parasitic nematodes: Progress, problems and prospects. Wallingford, UK: CAB International.

Stirling, G. R., and A. M. White. 1982. Distribution of a parasite of root-knot nematodes in South Australian vineyards. Plant Disease 66: 52 – 53.

Stirling, G. R., and M. F. Wachtel. 1980. Mass production of *Bacillus penetrans* for the biological control of root-knot nematodes. Nematologica 26: 308 – 312.

Stirling, G. R., R. D. Sharma, and J. Perry. 1990. Attachment of *Pasteuria penetrans* spores to the root knot nematode *Meloidogyne javanica* in soil and its effects on infectivity. Nematologica 36: 246 – 252.

Sturhan, D. 1988. New host and geographical records of nematode-parasitic bacteria of the *Pasteuria penetrans* group. Nematologica 34: 350 – 356.

Sturhan, D., R. Winkelheide, R. M. Sayre, and W. P. Wergin. 1994. Light and electron microscopical studies of the life cycle and developmental stages of a *Pasteuria* isolate parasitizing the pea cyst nematode, *Heterodera goettingiana*. Fundamental and Applied Nematology 17: 29 – 42.

Tian, H. L., and R. D. Riggs. 2000. Effects of rhizobacteria on soybean cyst nematode, *Heterodera glycines*. Journal of Nematology 32: 377 – 388.

Tian, H. L., S. H. Wang, and X. Y. Wu. 1995. Screening of rhizobacteria antagonistic to plant parasitic nematodes. Acta Agriculturae Universitatis Pekinensis 21: 209 – 213.

Vargas, R., and N. Acosta. 1990. *Pasteuria penetrans*: a biological agent for the suppression of nematodes in Puerto Rico. Journal of Agriculture of the University of Puerto Rico 74: 319 – 321.

Verdejo, S., and B. A. Jaffee. 1988. Reproduction of *Pasteuria penetrans* in a tissue-culture system containing *Meloidogyne javanica* and *Agrobacterium rhizogenes*-transformed roots. Phytopathology 78: 1284 – 1286.

Verdejo, S., and R. Mankau. 1986. Culture of *Pasteuria penetrans* in *Meloidogyne incognita* on oligoxenic excised tomato root culture. Journal of Nematology 18: 635 (Abstr.).

Walsh, J. A. 1979. An intracellular microorganism in tissues of the potato cyst nematode *Globodera rostochiensis* and pea cyst nematode *Heterodera goettingiana*. Journal of Nematology 11: 317 (Abstr.).

Walsh, J. A., D. L. Lee, and A. M. Shepherd. 1983a. The distribution and effect of intracellular rickettsia-like microorganisms infecting adult males of the potato cyst nematode *Globodera rostochiensis*. Nematologica 29: 227 – 239.

Walsh, J. A., A. M. Shepherd, and D. L. Lee. 1983b. The distribution and effect of intracellular rickettsia-like microorganisms infecting second-stage juveniles of the potato cyst nematode *Globodera rostochiensis*. Journal of Zoology 199: 395 – 419.

Weibelzahl-Fulton, E., D. W. Dickson, and E. B. Whitty. 1996. Suppression of *Meloidogyne incognita* and *M. javanica* by *Pasteuria penetrans* in field soil. Journal of Nematology 28: 43 – 49.

Westcott, S. W., III, and D. A. Kluepfel. 1993. Inhibition of *Criconemella xenoplax* egg hatch by *Pseudomonas aureofaciens*. Phytopathology 83: 1245 – 1249.

Wheeler, T. A., and J. L. Starr. 1987. Incidence and economic importance of plant-parasitic nematodes on peanut in Texas. Peanut Science 14: 94 – 96.

Williams, A. B., G. R. Stirling, A. C. Hayward, and J. Perry. 1989. Properties and attempted culture of *Pasteuria penetrans*, a bacterial parasite of root-knot nematode (*Meloidogyne javanica*). Journal of Applied Bacteriology 67: 145 – 156.

Zuckerman, B. M., M. B. Dicklow, and N. Acosta. 1993. A strain of *Bacillus thuringiensis* for the control of plant-parasitic nematodes. Biocontrol Science and Technology 3: 41 – 46.

22 Biological Control of Insects and Other Invertebrates with Nematodes

Harry K. Kaya and Albrecht M. Koppenhöfer

22.1 Introduction

Of more than 30 nematode families associated in some way with insects, only three – Phaenopsitylenchidae, Steinernematidae, and Heterorhabditidae – are currently being used in biological control of insect pests. A common feature among nematode species in these three families is their association with mutualistic microorganisms that enhance their pathogenicity, mass production capability, or survival in the field. Two other families – Mermithidae and Iotonchiidae – show some degree of biological control potential, and a few species have been used for pest suppression. Other families such as Allantonematidae, Diplogasteridae, Parasitylenchidae, Sphaerulariidae, and Tetradonematidae have species that may have some prospects in pest suppression but have not been seriously pursued because of difficulties in mass production, poor survival, or other features that limit their usefulness in biological control programs. Some nematode species in the families, Carbonematidae, Syrphonematidae, Oxyuridae, Thelastomatidae, Aphelenchoididae, Entaphelenchidae, and Fergusobiidae, are associated with insects, but they show low pathogenicity and have no biological control potential for insects. One species in the family Rhabditidae has been an effective biological control agent against slugs and is available commercially in Europe, and steinernematid and heterorhabditid nematodes and their mutualistic bacteria have been used against plant-parasitic nematodes.

Nematode species may be exploited against invertebrate pests in classical, conservation, or augmentative biological control programs. In classical biological control, the exotic natural enemy (i. e., nematode) is imported to suppress either an exotic or native invertebrate pest. This approach generally follows a sequence of procedures including foreign exploration, quarantine of imported material, mass production of the natural enemy, field colonization, and evaluation (Ehler, 1990). The goal of classical biological control is to obtain long-term suppression of the pest. Host specificity of the nematode to its invertebrate host is of prime concern because of the possible harmful effects of an introduced natural enemy to non-target organisms. Examples of nematodes employed in classical biological control against exotic pests include *Deladenus* (= *Beddingia*) *siricidicola* against the wood wasp *Sirex noctilio* in Tasmania and Victoria, Australia (Bedding, 1993) and *Steinernema scapterisci* against mole crickets *Scapteriscus* spp. in Florida, USA (Parkman and Smart, 1996). There are no examples of classical biological control with nematodes against a native insect pest. However, *Steinernema kushidai* appears to be host specific to scarab beetles (Mamiya, 1989) and could possibly be used for classical biological control of exotic or native scarab pests. The use against native scarab pests would be a new association for this nematode species.

In conservation and augmentative biological control, human intervention is needed to enhance the effectiveness of those natural enemies already in place against native or exotic invertebrate species (Ehler, 1998). In conservation biological control, manipulation of the environment, in the case of nematodes, may simply involve irrigation to enhance nematode survival and infection of the pest. Another approach would be the selective use of pesticides that are compatible with the nematode. In augmentative biological control, direct manipulation of the natural enemy is involved. It may consist of mass production and field releases of the nematode. These field releases represent two points of a continuum from inoculative to inundative. In the inoculative approach, a relatively small number of individuals are released with the intent that the progeny of the released individuals will provide season-long pest suppression. In the inundative approach, a massive number of individuals are released with the intent of obtaining immediate pest suppression. Thus, *Steinernema glaseri* was initially used as an inoculative biological control agent against the Japanese beetle in New Jersey, USA (Gaugler, et al., 1992b), whereas *S. carpocapsae* has been used as a biological insecticide against a number of soil pests (Georgis and Manweiler, 1994; Kaya and Gaugler, 1993).

The focus of this chapter will be on the five nematode families (i. e., Mermithidae, Iotonchiidae, Phaenopsitylenchidae, Steinernematidae, and Heterorhabditidae) that are being used or have the potential to be used in biological control of insects. We will consider each family separately except for Steinernematidae and Heterorhabditidae which possess many similarities and will be covered together. Because of the vast amount of literature on steinernematids and heterorhabditids, a greater emphasis will be placed on these two families. We will discuss the biology, ecology, and field application of nematodes for invertebrate pest control and evaluate their future prospect for pest suppression with emphasis on the more recent literature. In addition, we will discuss the rhabditid nematode for biological control of slugs.

22.2 Mermithidae

These slender nematodes, ranging in length from a few to 405 mm (usually between 50 and 150 mm), parasitize many invertebrate species, but mostly insects (Kaiser, 1991). Mermithids are invariably lethal to their hosts, and a few of them have received consideration as biological control agents. For more detailed information, Kaiser (1991) discusses the bionomics of a number of terrestrial and semi-terrestrial mermithid species, Petersen (1985) covers mermithids that occur in the aquatic environment, and Popiel and Hominick (1992) provide more recent research on mermithids with emphasis on population biology.

A variety of mermithid life cycles occur, but most species typically follow a similar life cycle. A newly hatched, second stage juvenile (J2 or preparasite) emerges, seeks its hosts, penetrates through the host's cuticle using its stylet, and enters the hemocoel where it absorbs nutrients directly through its cuticle and stores them in the trophosome (Kaiser, 1991). During the parasitic phase, the oral cavity becomes highly cuticularized so that the stylet is no longer recognizable, and the stichosome, an extensive system of pharyngeal glands of unknown function, grows into an organ of considerable size. The parasitic juvenile remains free in the hemocoel, but in a few species, the juvenile penetrates the host's brain or nerve ganglion. This behavior is a mechanism to avoid the host's defenses. Upon completion of the parasitic phase, the postparasitic stage emerges from the host causing its death, enters the soil (terrestrial mermithids) or sediment (aquatic mermithids), matures, mates, and oviposits to complete the life cycle.

The number of molts by mermithids remains unclear. There is one molt in the egg from the J1 to J2 stage; whether the mermithid molts during the parasitic stage has not been clearly established except for *Romanomermis culicivorax* (Kaiser, 1991). Two molts seem to occur in the postparasitic stage in some mermithids.

A number of mermithids have been considered for biological control of insects, but only one of them reached commercial production. We will cover four mermithids that have been studied as agents for inundative, augmentation or conservation biological control programs. The most studied mermithid is *R. culicivorax*, a biological control agent of mosquito larvae, that was briefly available as a commercial product in the mid-1970s and early 1980s (Petersen,

1985). Recently, another mermithid of larval mosquitoes, *R. yunanensis*, has been evaluated in the field in China (Peng, et al., 1998). A third mermithid, *Filipjevimermis* (syn. *Oesophagomermis*) *leipsandra*, received some attention as a potential augmentative biological control agent of larvae of the banded cucumber beetle *Diabrotica balteata*. A fourth mermithid, *Agamermis unka*, has been evaluated as an augmentative biological control agent of the brown planthopper, *Nilaparvata lugens*, and the whitebacked planthopper, *Sogatella furcifera*, that are pests in Asian rice fields (Choo, et al., 1998).

22.2.1 *Romanomermis* and Mosquitoes

Romanomermis culicivorax kills its host within 8 days and has a generation time of 1 to 2 months (Fig. 22.1). During the 1970s, it received substantial attention as a biological insecticide and a long-term biological control agent against a number of mosquito species. Most attempts to use this nematode were as a biological insecticide where the J2 stage was applied to obtain immediate control. Petersen (1984) summarized much of these field results showing that larval populations of *Anopheles* and *Culex* mosquitoes could be reduced by 50% to more than 90% at rates between 1200 and 3600 J2s/m^2 of water surface area. Not all field trials have been successful with the preparasites due to polluted waters, high salinity, cold temperatures, or predation of the nematodes by microcrustaceans.

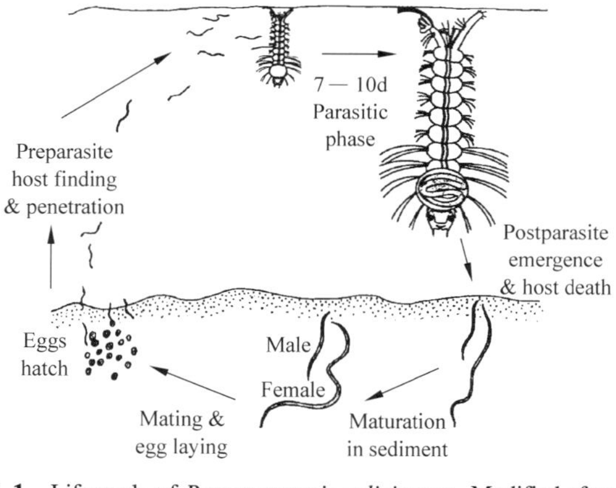

Figure 22.1 Life cycle of *Romanomermis culicivorax*. Modified after Platzer in Tanada and Kaya (1993).

Through recycling of *R. culicivorax,* long-term control has been achieved using the preparasitic and postparasitic stages (Petersen, 1985). For example, *R. culicivorax* recycled in mosquito populations that occurred in permanent and semi-permanent water sites for several years after an initial application of J2s (Petersen and Willis, 1975). Inoculative releases of the postparasitic stage of *R. culicivorax* in rice fields showed partial control of mosquito larvae (Kerwin and Washino, 1985). Hominick and Tingley (1984), however, noted that this mermithid's life cycle precludes it from being a very effective biological control agent because the reproductive potential of mosquitoes greatly exceeds that of the mermithid. Their mosquito-mermithid model predicted that a 90% host reduction required an unrealistic mermithid: mosquito release ratio of about 13:1. The environmental sex determination of the nematode prevented this mermithid from maintaining high rates of parasitism. Moreover, the environmental constraints cited previously reduced their usefulness to pristine waterways (Petersen, 1985). Thus, even though *R. culicivorax* was commercialized during the 1970s, a number of factors operated against it. For example, Song et al. (2001) showed that the J2 could be stored in liquid nitrogen, but the revived nematodes were unable to parasitize mosquitoes. Coupled with poor nematode storage and shipping and poor economics of nematode mass production, commercialization was a high-risk venture. In the early 1980s, the success of *Bacillus thuringiensis* subspecies *israelensis* as a potent bacterial mosquito larvicide triggered the demise of this mermithid as a commercial product.

The isolation of *Romanomermis yunanensis* has initiated resurgence in mosquito biological control in China. This mermithid has been evaluated as a biological control of several larval mosquito species. It parasitizes at least 31 species in six mosquito genera under laboratory conditions (see Peng, et al., 1998). Field tests over a 10 year period (1986 – 1995) in five Chinese provinces against several mosquito species at rates of 2000 – 4000 J2s/m^2 have shown population reductions ranging from 36% to 97% in ponds, rice paddies, and streams (Peng, et al., 1998). When it was applied to small suburban water sites (pots, flowerbeds, and vats) at rates of 400 J2s/pot to 4000 J2s/m^2/flower beds, rates of larval mosquito parasitism ranged from 71% to 100%. Like *R. culicivorax*, *R. yunanensis* does poorly in polluted waters but does occur naturally in rice paddies and streams (Peng, et al., 1998). Perhaps, it can be used in an augmentative release program in less developed countries where mass production may be economically feasible.

22.2.2 *Filipjevimermis* and the Banded Cucumber Beetle

Filipjevimermis leipsandra has a parthenogenetic life cycle. Each female produces several thousand eggs in the soil (Cuthbert, 1968). After hatching from the egg, the J2 actively searches for its host, penetrates through the larval cuticle, enters the hemocoel and develops in the nervous system of the insect.

The mermithid grows for 12 days to 22 days, ruptures the nervous system, and exits from the host, killing it in the process. The generation time is 1 to 2 months.

This nematode has been mass-produced *in vivo*, and 5×10^6 eggs/week can be obtained for augmentative biological control (Creighton and Fassuliotis, 1981, 1982). In small plot tests, J2s from these eggs caused 78% larval parasitism of *D. balteata* (Creighton and Fassuliotis, 1983). Fassuliotis and Creighton (1982) developed the first *in vitro* production of a mermithid, but the development of *F. leipsandra* as a biological control agent suffered a severe reversal when the *in vitro* production of this nematode could not be sustained (Gaugler, 1987).

22.2.3 *Agamermis* and Planthoppers

Agamermis unka parasitizes 50% to 80% of the brown planthopper, *Nilaparvata lugens*, in Korean rice fields (Choo, et al., 1989; Choo and Kaya, 1993). Female *A. unka* can lay as many as 8000 eggs, but most lay between 500 and 3000 eggs (Choo, et al., 1995). In the field, egg deposition occurs from June through early autumn. The eggs hatch in about 3 weeks, and the J2 swims to the water surface, climbs the rice plant, and penetrates directly through the cuticle into the hemocoel of the planthopper nymphs. Although the planthopper reaches the adult stage, the mermithid sterilizes the planthopper. Two or three weeks after parasitism, the postparasitic stage emerges from the adult planthopper, killing it in the process. The postparasite enters the soil, molts to the adult stage, overwinters, and mates the following spring.

Agamermis unka may be practical in conservation and augmentative biological control programs. These approaches are more suited for temperate rice growing regions because *A. unka* does not seem to be an important mortality factor of planthoppers in tropical areas (Cook and Perfect, 1989). The postparasitic and adult mermithids are readily found in temperate Korean rice fields whether they have received pesticide treatments or not (Choo, et al., 1989). In a laboratory study, however, some pesticides were highly toxic to the mermithid (Choo, et al., 1998). If conservation biological control is an approach that will be used for *A. unka*, an insecticide (e. g., fenthion, etofenprox, chlorpyrifos, and imidacloprid) with minimal impact upon the mermithid and other non-target organisms but that is still effective against planthoppers should be used. In addition, proper cultural techniques can increase the effectiveness of *A. unka* (Yan, et al., 1986). Tilling and irrigating the rice fields can increase parasitism of planthoppers by this mermithid. Adult mermithids cannot be grown *in vitro* but can be collected from the field during the non-growing rice period (i. e., fall, winter and early spring). The eggs have been stored at 5℃ for 30 days with 80% hatching 21 days later when they were placed at 25℃ (Choo, et al., 1995). The adults, eggs, or preparasites may be used to augment naturally occurring populations of *A. unka* or introduced into uninfested areas. Such an approach will probably provide acceptable crop protection only when planthopper populations are low.

22.2.4 Evaluation of Mermithids as Biological Control Agents

As a group, mermithids possess many attributes of an ideal biological control agent (Petersen, 1985). They tend to be host specific, are lethal to their hosts, can be easily manipulated in the laboratory, can be mass-produced in their hosts, can be disseminated through standard spray equipment, have the potential for long-term control, and are environmentally safe. However, many of these positive attributes hold true for only a few mermithid species (Petersen, 1985). Many liabilities exist including the variable efficacy, the long life cycles of many mermithids, their obligate parasitism that requires rearing of their hosts, cost of production, poor long-term storage (i. e. for several months), and poor natural dispersal. Hominick and Tingley (1984) stated that the reproductive potential of the host exceeds that of the mermithid and the environmental sex determination of mermithids may prevent maintaining high rates of parasitism. The lack of knowledge about the biology of many mermithid species is an additional constraint in employing them in biological control programs.

22.3 Iotonchiidae

Most nematode species in this family have not been exploited for biological control because they are obligate parasites of insects, have very complex life cycles, are difficult to maintain in culture, and often do not induce significant host damage (Poinar, 1979). However, the iotonchiids, *Paraiotonchium* (syn. *Heterotylenchus*) *autumnale* and *P. muscadomesticae*, have possibilities for biological control. *P. autumnale* has been released to augment naturally occurring nematode populations against the face fly (Poinar, 1979). More recently, *P. muscadomesticae* was isolated from the house fly, *Musca domestica,* in Brazil (Coler and Nguyen, 1994), and it has the potential to be used as a "classical" biological control agent of the ubiquitous house fly (Geden, 1997).

The life cycle of *P. autumnale* and *P. muscadomesticae* involves an alternation of gamogenetic and parthenogenetic generations within their fly host. Except for the brief, free-living, gamogenetic stage in the flies' larval habitat (for face fly, cattle dung and for the house fly, a variety of decomposing organic matter), the life cycle of these nematodes occurs entirely within their fly host. The mated free-living gamogenetic female penetrates into the hemocoel of the larval fly but does not produce eggs until the adult fly emerges. After adult fly emergence, the gamogenetic female lays several eggs that become parthenogenetic females in a few days. These parthenogenetic females produce many eggs that develop into immature gamogenetic nematodes that invade the fly's ovaries, sterilizing the female. The infected female fly "nemaposits" nematodes into the flies' larval habitat during mock oviposition. The male flies are "dead end" hosts because there is no mechanism for the nematodes to infest the flies' larval habitat (Nappi, 1973). To survive the winter months, *P. autumnale* synchronizes its life cycle within the diapausing, female adult fly (Stoffolano, 1967). Because the house fly does not have a diapausing stage, *P. muscadomesticae* would have to survive in adult flies or in developing larvae in colder climatic regions.

The face fly, a major pest of range cattle, was accidentally introduced into North America from Europe in the early 1950s, presumably with its nematode parasite, *P. autumnale* (Stoffolano, 1968). In California, 6% to 40% of flies from the faces of cattle were infected with *P. autumnale* (Kaya and Moon, 1978), but percentage infection varied with the collection site (Kaya, et al., 1979). Flies collected from dung had a higher percentage of parasitism than those collected from the faces of cattle and were physiologically older. Infected flies from dung had ovaries packed with nematodes, whereas those from faces of cattle were young flies containing primarily the gamogenetic or parthenogenetic nematodes. Kaya et al. (1979) concluded that older infected flies became terminal dung seekers and were less likely to pester cattle. Krafsur et al. (1983) showed that in Iowa, the prevalence of infection by *P. autumnale* ranged from 7% to 19% when collected from cattle dung, but averaged only 2.9% when collected from cattle or along fences.

Paraiotonchium autumnale also causes mortality of immature flies. Chirico (1990) noted that larval face flies exposed to *P. autumnale* had an adult emergence rate of 56% compared to the control that had an adult emergence rate of 79%.

Augmentative releases were made with nematode-infected face flies to assist nematode dispersal (Nickle and Welch, 1984). In addition, Chirico (1996) released laboratory-infected face flies into field populations and observed an increased prevalence of infected adult flies in released sites in comparison to control sites. A high infection rate of 83% was obtained under laboratory conditions by allowing the nematode and fly larvae to aggregate at the bottom of the rearing container (Chirico, 1990). Chirico (1996) concluded that it is possible to release infected face flies to induce and establish a higher infection in a recipient host field population, but whether such releases are sufficient to provide a regulatory effect remains to be demonstrated. Because *P. autumnale* appears to operate independently of host density (Kaya and Moon, 1978), it may not regulate face fly densities. The nematode's obligate parasitism and complex life cycle make *in vitro* culture impossible to attain. Despite these drawbacks, the nematode can be maintained in laboratory face fly colonies and introduced into field populations (Chirico, 1990, 1996), especially into new areas of face fly infestations. The nematode will reduce face fly populations through mortality of the immature stages and sterilization of the adults, and in view of the behavioral changes of infected individuals, infected adult flies will be less pestiferous.

The jury is still out on *P. muscadomesticae* as a "classical" biological control agent for the house fly. It is host specific and can be produced in large quantities, infected house fly adults live half as long as uninfected flies, and the nematode can cause mortality of immature house flies (Geden, 1997). However, infection of the house fly by *P. muscadomesticae* in Brazil only ranged from 0.3 to 1.7% (Coler and Nguyen, 1994), suggesting that it may not be effective under natural conditions. Perhaps, through conservation and augmentation, *P. muscadomesticae* may be effective in areas around livestock and poultry farms where house flies breed continuously.

22.4 Phaenopsitylenchidae

This family is characterized by an alternation of a parasitic heterosexual and one or more free-living heterosexual or parthenogenetic generations (Remillet and Laumond, 1991). *Deladenus siricidicola* used in classical biological control against the siricid wood wasp, *Sirex noctilio*, in Australia is in this family.

This wood wasp, at best a secondary pest of stressed trees in Europe and other regions of the world, is a key pest of Monterey pine trees in New Zealand and Australia (Talbot, 1977). The female wood wasp lays eggs along with a toxic mucus and arthrospores of the fungus, *Amylostereum areolatum*, which is mutualistically associated with *S. noctilio*. The combination of toxic mucus, fungus, and infestation of the wood wasps kills the pine tree. The fungus grows throughout the tree providing food for the wood wasp larvae.

Deladenus siricidicola's life cycle is typical for this family (Fig. 22.2). The mated, infective stage female initiates the parasitic cycle by using its spear-like stylet to penetrate the larval cuticle and enter the hemocoel

(Bedding, 1993). The female may increase up to 1000-fold in volume, but reproduction is inhibited until the host pupates which may take up to 3 years. Prior to the end of the wood wasp's pupal stage, the female nematode produces progeny ovoviviparously and the juveniles invade the host's reproductive organs, essentially sterilizing the female but not the male wood wasp. The juvenile nematodes (50 to 200) enter the wood wasp egg which is nemaposited by the female along with the mucus and mutualistic fungus into new trees. The male, a "dead end host" due to the lack of a mechanism for the nematode's release into the tree, is not sterilized because spermatozoa are stored in the vesiculae seminales prior to nematode invasion of the testes.

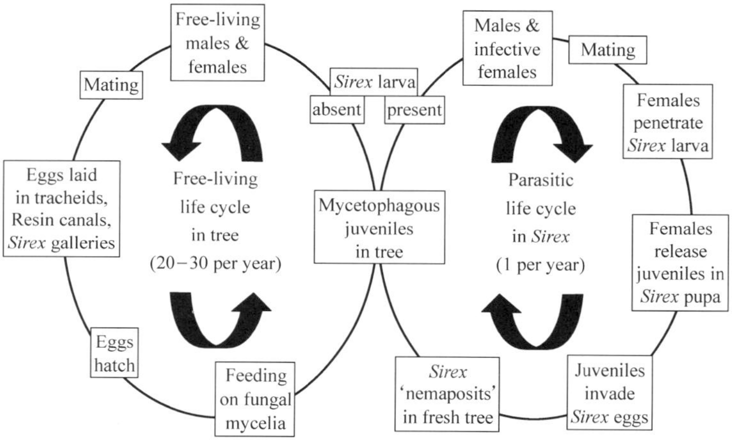

Figure 22.2 Life cycle of *Deladenus siricidicola*.

The nematodes exit from the siricid eggs and enter the free-living, mycetophagous life cycle, feed on the mycelia of *A. areolatum* with their syringe-like stylets, mature, and mate (Bedding, 1984). The female nematodes produce progeny oviparously. The free-living life cycle ends when the mycetophagous nematodes develop adjacent to a siricid larva, initiating the formation of adult males and infective-stage females to renew the parasitic phase. The parasitic and free-living cycles are bound by the specificity of the mycetophagous nematode to the symbiotic fungus as food for both the insect and nematode (Bedding, 1993).

Deladenus siricidicola was found in New Zealand where it caused the collapse of *Sirex* populations (Zondag, 1969). Subsequently, the wood wasp was accidentally introduced into Tasmania and Victoria, Australia causing serious damage in Monterey pine plantations. This created a unique opportunity to conduct classical biological control with this and other closely related nematode species. *D. siricidicola* was released in Tasmania and Victoria in 1970 and within 4 years the number of killed trees dropped spectacularly. In Tasmania, a high rate of parasitism (>90%) was observed in wood wasp populations, and because the wood wasps are effective fliers, they served as a dispersal agent for the nematode to other forests up to 13 km away.

In 1987, outbreaks of *Sirex* occurred in pine plantations in South Australia and Victoria (Haugen, 1990). In response to this threat, an augmentation program with *D. siricidicola* was initiated and 147,000 trees were inoculated with the nematode (Haugen and Underdown, 1990). The augmentation program controlled the *Sirex* population and reduced tree mortality.

The successful biological control of *S. noctilio* can be attributed to a number of factors. (1) The nematodes are easy to produce. The mycetophagous life cycle of the nematode can be maintained in the laboratory without loss of infectivity by the parasitic phase for a number of generations. Bedding and Akhurst (1974) developed an effective mass-rearing program, producing 3 – 10 million nematodes/500 ml flask. Although *D. siricidicola* lacks a resistant stage and loses viability after only 8 weeks of storage, the nematodes were used soon after production by injecting them into *Sirex*-infested trees (Bedding, 1993). (2) The nematode has high infectivity and high transmission efficiency. (3) The nematode is efficiently dispersed by the nemapositing behavior of the infected female host. And (4) the lack of a resistant stage is overcome by having a biphasic life cycle that permits the nematode to increase in numbers and sustain itself until a host becomes available (Gaugler, 1987).

22.5 Steinernematidae and Heterorhabditidae

Nematodes in the family Steinernematidae are represented by the genera *Steinernema* and *Neosteinernema* and in the family Heterorhabditidae are represented by the genus *Heterorhabditis*. They are called entomopathogenic (insect-pathogenic) nematodes because most species have the ability to quickly kill their hosts within 2 days. This rapid mortality is attributed to their association with mutualistic bacteria in the genus *Xenorhabdus* for Steinernematidae and *Photorhabdus* for Heterorhabditidae. Because both families possess many biological similarities, they will be considered together. However, these two nematode families are not very closely related (Blaxter, et al., 1998), even though they belong to the same nematode order (Rhabditida). They have distinctly different reproductive strategies, have different morphological features, and are associated with different bacterial genera (Kaya and Gaugler, 1993; Burnell and Stock, 2000).

Several species in the genera *Steinernema* and *Heterorhabditis* are presently used as microbial insecticides and are produced commercially by various companies around the world. Because of their importance as commercial microbial control agents and widespread distribution, there is a great literature base on these two families. Moreover, the emerging importance of the mutualistic bacteria as a source of metabolites for various uses has generated a significant literature on these microorganisms (Bowen, et al., 1998; Hu, et al., 1999; ffrench-Constant and Bowen, 1999). For additional reading on these two nematode families and their mutualistic bacteria, we refer the reader to books edited by Gaugler and Kaya (1990), Bedding et al. (1993) and Gaugler (2002) and reviews by Kaya and Gaugler (1993), Forst and Nealson (1996), Forst et al. (1997), Lewis et al. (1998), Barbercheck and Millar (2000), Burnell and Stock (2000), and Liu et al. (2000). For techniques on the use of entomopathogenic nematodes, we refer the reader to chapters by Kaya and Stock (1997) and Koppenhöfer (2000), and several chapters in a book edited by Lacey and Kaya (2000). There is an extensive bibliography compiled by Smith et al. (1992) and an updated web-based version is also available at: http://128.146.54.216/nematodes/

22.5.1 Bionomics of the Nematode/Bacterium Complex

Entomopathogenic nematodes are very common soil organisms and have been recovered from soils from various parts of the world (Hominick, 2002; Hominick, et al., 1996). Numerous surveys continuously recover new isolates and many of these constitute new species. Presently, 35 species of entomopathogenic nematodes in two families and three genera (*Steinernema* with 27 species; *Neosteinernema* with one species, and *Heterorhabditis* with seven species) are recognized (Table 22.1). Several taxonomic changes at the generic and species levels have been made, especially in the last 10 years, resulting in confusion in the literature (Adams and Nguyen, 2002). A publication from a meeting of most of the specialists in the area of entomopathogenic nematode taxonomy and systematics gives a synopsis of the current status, protocols, and definitions for the biosystematics of entomopathogenic nematodes (Hominick, et al., 1997). The reader should keep in mind that the systematics and taxonomy of steinernematids and heterorhabditids are in a continual state of flux because new species are being found and described and some of the older described species are being synonymized.

Table 22.1 Described species of entomopathogenic nematode species and their respective symbiotic bacterial species.

Nematode genus	Nematode species	Symbiont species
Steinernema	*abbasi*	undescribed
	affine[1]	*Xenorhabdus bovienii*
	arenarium (= *anomali*)[2]	*Xenorhabdus* sp.
	bicornutum	undescribed
	carpocapsae[3]	*X. nematophilus*
	caudatum	undescribed
	ceratophorum	undescribed
	cubanum	*X. poinarii*
	feltiae (= *bibionis*)	*X. bovienii*
	glaseri	*X. poinarii*
	intermedium	*X. bovienii*
	karii	undescribed
	kraussei	*X. bovienii*
	kushidai	*X. japonicus*
	loci	undescribed
	longicaudum	undescribed
	monticolum	undescribed
	neocurtillae	undescribed
	oregonense	undescribed
	puertoricense	undescribed
	rarum	*Xenorhabdus* sp.
	riobrave	*Xenorhabdus* sp.
	ritteri	*Xenorhabdus* sp.
	sangi	undescribed
	scapterisci	*Xenorhabdus* sp.
	siamkayai	*Xenorhabdus* sp.

Continued

Nematode genus	Nematode species	Symbiont species
Steinernema	*thanhi*	undescribed
	undescribed	*X. beddingii*
Neosteinernema	*longicurvicauda*	undescribed
Heterorhabditis	*argentinensis*	*Photorhabdus luminescens*[5]
	bacteriophora (= *heliothidis*)	*P. luminescens*[5] and *P. temperata*[6]
	brevicaudis	*P. luminescens*[5]
	indica (= *hawaiiensis*[4])	*P. luminescens*
	marelatus (= *hepialus*)	*P. luminescens*[5]
	megidis	*P. temperata*
	zealandica	*P. temperata*

[1] Endings of some specific epithets in *Steinernema* have been corrected to reflect that-*nema* is neuter gender. [2] In brackets previously used names. [3] The species "*carpocapsae*" has been referred to as "*feltiae*" in the literature primarily between 1983 and 1989. The name "*feltiae*" is valid and takes precedent over "*bibionis*." [4] Appears to be conspecific with *H. indica* (Adams, et al., 1998; Hashmi and Gaugler, 1998). [5] Fischer-Le Saux et al. (1999) showed that *Photorhabdus* is a heterogenous group. *P. luminescens* is associated with *H. indica* with some but not all isolates of *H. bacteriophora*. The status of bacterial symbionts with *H. argentinensis*, *H. brevicaudis*, and *H. marelatus* is not clear and we have maintained the bacterial species as *P. luminescens*. This may change at a future date. In addition, one species, *P. asymbiotica*, is not associated with nematodes and has been isolated from human clinical specimens. [6] Some subgroups of *H. bacteriophora* are associated with *P. temperata* (Fischer-Le Saux, et al., 1999).

22.5.1.1 Associated Bacteria

Xenorhabdus and *Photorhabdus* are motile, Gram-negative, facultatively anaerobic rods in the family Enterobacteriaceae. Five species are recognized in the genus *Xenorhabdus* (Table 22.1) (Bonifassi, et al., 1999; Burnell and Stock, 2000). Previously, all *Photorhabdus* were considered as one species (*P. luminescens*) even though they represented a heterogeneous group that included symbionts of *Heterorhabditis* as well as some nonsymbiotic isolates from human wounds (Akhurst, et al., 1996). Recently, Fischer-Le Saux et al. (1999) revised the taxonomy of *Photorhabdus* and proposed the formation of two new species, *P. temperata* and *P. asymbiotica* (Table 22.1). *P. asymbiotica*, isolated from human clinical cases, is not associated with nematodes. In addition, Fischer-Le Saux et al. (1999) proposed that *P. luminescens* be divided into three subspecies.

Major differences between the two genera are that most *Photorhabdus* spp. are luminescent, whereas *Xenorhabdus* spp. have no luminescence and that *Photorhabdus* spp. are catalase positive whereas *Xenorhabdus* spp. are catalase negative. Both bacterial genera also produce phenotypic variant cell types called primary form (phase I) and secondary form (phase II) (Forst and Clarke, 2002). The primary form is the cell type naturally associated with the nematodes, and the secondary form can arise spontaneously when the bacterial cultures are in the stationary nongrowth stage. The secondary form can revert to the primary form but has only been documented with *Xenorhabdus* spp. (Boemare, 2002; Boemare, et al., 1996). A number of differences occur between the two forms. Primary and secondary forms have distinctly different colony morphologies. The primary form produces antibiotics, adsorbs certain dyes, and develops large intracellular inclusions composed of crystal proteins, whereas the secondary form does not produce antibiotics, does not adsorb dyes, and produces intracellular inclusions inefficiently. Differences in pathogenicity between the forms have been documented in some hosts (Volgyi, et al., 1998) but not in others (Jackson, et al., 1995, Nishimura, et al., 1994). The primary form is superior to the secondary form in its ability to support nematode propagation *in vitro*, although recent evidence suggests that this is not always the case (Ehlers, et al., 1990; Volgyi, et al., 1998).

The nematode/bacterium association is highly specific (Akhurst and Boemare, 1990; Boemare, 2002). This specificity is a monoxenic association between the bacterium and nematode and has been attributed to the production of antimicrobial compounds produced by the symbiotic bacterium that prevent the development of other bacteria. Under laboratory conditions, however, other bacterial species have been isolated from the gut of the infective juvenile stage from various steinernematid (Aguillera and Smart, 1993; Aguillera, et al., 1993; Lysenko and Weiser, 1974) and heterorhabditid (Jackson, et al., 1995) species. Recently, the occurrence of natural dixenic associations between the symbiont *P. luminescens* and bacteria related to *Ochrobactrum* spp. from tropical *Heterorhabditis indica* has been reported (Babic, et al., 2000). The *Ochrobactrum* spp. are not pathogenic to insects but are considered human opportunist pathogens. If these nematodes are going to be mass-produced for biological control programs, greater vigilance will be required to ensure that only *Xenorhabdus* or *Photorhabdus* bacteria are present.

22.5.1.2 Biology of Nematode/Bacterium Complex

Steinernematids and heterorhabditids are obligate pathogens in nature. Only the non-feeding third stage infective juvenile (IJ) or dauer juvenile is capable of surviving outside the insect host (Fig. 22.3). The IJ carries cells of its bacterial symbiont in its intestinal tract. After locating a suitable host, the IJ invades it through natural openings (mouth, spiracles, anus) or thin areas of the host's cuticle (common only in heterorhabditids) (Wang and Gaugler, 1998a) and penetrate into the host's hemocoel. The IJ releases its symbiont, and the bacterium and nematode cooperate to overcome the host's immune response. The mutualistic bacterium propagates and produces substances that rapidly kill the host and protect the cadaver from colonization by other microorganisms. The nematode initiates its development, feeding on the bacterial cells and host tissues that have been metabolized by the bacterium and has 1 – 3 generations, depending on host size. As the food resources in the host cadaver are depleted, a new generation of IJs is produced that emerges from the host cadaver in search of a new host. A major difference between steinernematids and heterorhabditids is that species in the former group are amphimictic[1], whilst species in the latter group are hermaphrodites in the first generation but amphimictic in following generations. Therefore, steinernematids require at least two IJs, a male and a female, to invade the host to produce progeny, and heterorhabditids need only one IJ to penetrate into the host as the resulting hermaphroditic adult is self-fertile.

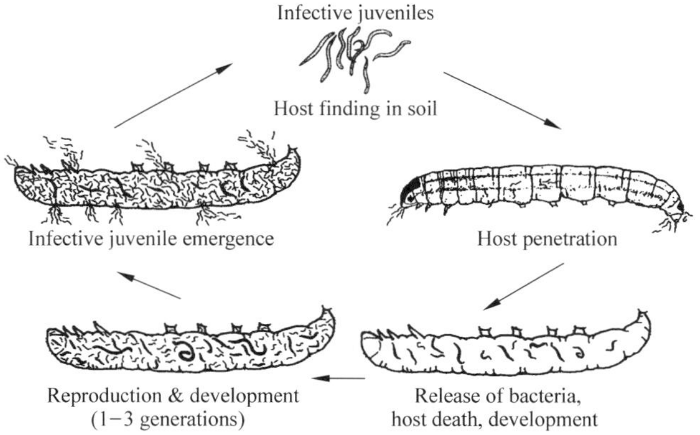

Figure 22.3 Life cycle of steinernematids and heterorhabditids.

Each nematode species is specifically associated with one bacterial symbiotic species (but see recent publication by Babic et al. (2000)); however, a symbiotic species may be associated with more than one nematode species

[1] Hermaphrodites have been found in one undescribed *Steinernema* species from Indonesia (Griffin, et al., 2001). The hermaphrodite is self-fertile, but males are present at a low frequency (1 – 6% of the population).

(Table 22.1). This specificity operates at 2 levels (Akhurst and Boemare, 1990). First, the best nematode reproduction occurs on their natural symbiont even though, in some cases, the nematode can develop on other bacterial species. Second, the natural bacterial symbionts are retained better than other bacterial species. The nematode is dependent upon the mutualistic bacterium for (1) quickly killing its insect host; (2) creating a suitable environment for its development by producing antibiotics that suppress competing microorganisms; and (3) transforming the host tissues into a food source. The bacterium requires the nematode for (1) protection from the external environment; (2) penetration into the host's hemocoel; and (3) inhibition of the host's antibacterial proteins.

22.5.1.3 Infectivity

Laboratory observations under optimal conditions indicate that only a portion of inoculated IJs (usually < 40%) can be recovered from susceptible hosts exposed to them, leading to the "phased infectivity hypothesis" (e.g., Griffin, 1996; Kaya and Gaugler, 1993). Hominick and Reid (1990) who first proposed this hypothesis stated that "... an effective survival strategy might be for infectivity to be phased over time. Thus, upon emergence from a host, some individuals may be immediately infectious, while others become dormant for a time."

This hypothesis suggests that a population of IJs emerging from a host cadaver can be divided into an infectious portion and a non-infectious portion. If the "phased infectivity hypothesis" is correct, the non-infectious portion can consist of a permanently non-infectious and a temporarily non-infectious portion. There is considerable evidence that temporary non-infectivity can be induced in already emerged IJs by environmental factors such as low moisture (Kung, et al., 1991) and low temperature (e.g., Brown and Gaugler, 1996; Griffin, 1996). This is a physiological response to adverse environmental conditions which is different from not infecting a host under optimal conditions. The experimental evidence for an innately non-infectious proportion is limited to one strain of *S. feltiae* (Bohan and Hominick, 1996, 1997). Recent research by Campbell et al. (1999) suggests that "phased infectivity" may only occur under specific conditions and/or in a small proportion of the IJ population of some species.

22.5.1.4 Host Range

Entomopathogenic nematode species, in general, attack a wide spectrum of insects under laboratory conditions where host contact is assured, environmental conditions are optimal, and no ecological or behavioral barriers to infection exist

(Gaugler, et al., 1997; Kaya and Gaugler, 1993). In the field, the range of insects infected by entomopathogenic nematodes is significantly narrower than in the laboratory (Akhurst, 1990; Bathon, 1996; Georgis, et al., 1991; Peters, 1996), adding to their safety as biological control agents. Because isolation of new nematode strains/species is usually done using larvae of the greater wax moth, *Galleria mellonella*, the host range of known species tends to be biased towards generalists or species adapted to lepidopterons. Some species that have been isolated from host cadavers in the field have a restricted host range with *S. kushidai* (Mamiya, 1989) and a putative new *Steinernema* sp. (Stock and Koppenhöfer, in preparation) being specialists against scarab larvae. *S. scapterisci* is adapted to mole crickets and infects other insects poorly (Grewal, et al., 1993, Parkman and Smart, 1996), but recently, Bonifassi et al. (1999) showed that the combination of strain UY61 of *Xenorhabdus* and *S. scapterisci* infects *G. mellonella* larvae very well.

22.5.2 Nematode/Bacterium Interactions with Hosts

The efficacy of entomopathogenic nematodes varies with many biological factors, including nematode species and strain and insect species and their developmental stage (Eidt and Thurston, 1995, Simões and Rosa, 1996). Many soil-dwelling insects have behaviors that reduce host finding, attachment, or penetration of IJs. These behaviors include: (1) a high defecation rate that reduces infection via the anus (scarab grubs); (2) low CO_2 output or CO_2 released in bursts that minimize chemical cues (lepidopteran pupae and scarab grubs); (3) the formation of impenetrable cocoons or soil cells before pupation that serve as physical barriers (many Lepidoptera and Scarabaeidae); (4) walling-off infected individuals that avoids or reduces contamination to other insects in a nest (termites); and (5) aggressive grooming or evasion behavior that reduces IJ contact (scarab grubs) (Gaugler, et al., 1994). IJs can penetrate into insects using several routes, depending on which are accessible (Eidt and Thurston, 1995). In some insects, however, the usual routes of entry may be inaccessible. The mouth may be obstructed by oral filters (wireworms) or too narrow (insects with sucking/piercing mouthparts or small insects with chewing mouthparts). The anus may be constricted by muscles or other structures (wireworms), or the spiracles may be covered with septa (wireworms) or sieve plates (scarab grubs) or simply too narrow for nematode entry (some Diptera and Lepidoptera).

To enter the hemocoel, the IJs have to penetrate through the cuticle (including the trachea) or gut. To achieve this, the nematode may employ physical force such as body thrusting to rupture through the thin trachea or, as with *Heterorhabditis*, use a tooth situated terminally at the anterior end to penetrate directly into the hemocoel. The IJs may also use proteolytic secretions to digest the midgut tissues to gain access into the hemocoel (AbuHatab, et al., 1993; Peters and Ehlers, 1994).

Once inside the body cavity, the nematodes and bacteria need to overcome the host's immune response (Forst, et al., 1997; Simões and Rosa, 1996) that involves interacting humoral and cellular factors. To neutralize the bacterial cells, the insects may use antibacterial proteins and/or phagocytosis followed by nodule formation. To inactivate the nematodes, the insects may encapsulate them followed by melanization. However, the nematodes (e.g., IJs of *S. carpocapsae*) can evade recognition as non-self, but the mechanism of this avoidance is not well established and may involve the structure/chemistry of the nematodes' cuticle. Although *S. glaseri* is initially recognized as non-self and is encapsulated in larvae of the Japanese beetle, *Popillia japonica*, it escapes from the capsules and successfully infects its host (Wang, et al., 1995). In another example, *Heterorhabditis* IJs can avoid encapsulation in tipulid larvae by exsheathing from the J2 cuticle during penetration (Peters, et al., 1997).

By excreting a lipopolysaccharide and inhibiting the activation of the insect enzyme, prophenoloxidase, *Xenorhabdus* bacteria can evade the insects' humoral response. As previously mentioned, the symbiotic bacteria produce antibiotic substances that protect the host resource from colonization by other microorganisms. The nematodes can produce immuno-inhibiting factors that destroy the antibacterial factors produced by the insect; this inhibition allows the mutualistic bacteria to produce insecticidal toxins that rapidly kill the host (Bowen, et al., 1998). Wang and Gaugler (1998b) showed that surface coat proteins of *S. glaseri* suppress the host immune response in Japanese beetle larvae and destroy hemocytes. Nematodes may also produce paralyzing exotoxins and cytotoxic and proteolytic extracellular enzymes. The degree to which any of the above reactions by insect and/or nematode/bacterium occurs is dependent on the insect host and nematode/bacterium complex. This interaction contributes to the variable efficacy of entomopathogenic nematodes against different insects.

22.5.3 Ecology

22.5.3.1 Behavior

Understanding entomopathogenic nematodes' infection process as it relates to host selection, search, attachment, and recognition is critical to successful insect pest control. Another important consideration is to match the pest's biology and ecology with the nematodes' biology and ecology which will lead to successful control of the pest.

One of the major factors restricting nematode host range is the foraging behavior of the IJs. Entomopathogenic nematodes employ different foraging strategies to locate and infect hosts (Gaugler, et al., 1997; Lewis, 2002). Some nematode species forage by using a sit-and-wait strategy (ambush) that is characterized by low motility and tendency to stay near the soil surface. They tend not to respond to volatile and contact host cues unless presented in an appropriate sequence. These ambush foragers (ambushers) efficiently infect mobile host species such as cutworms and mole crickets near the soil surface (e.g., *S. carpocapsae* and *S. scapterisci*). Other species are widely foraging strategists (cruisers) that are characterized by high motility and are distributed throughout the soil profile. They have the ability to orient to volatile host cues and switch to localized search after host contact. These species are well adapted to infecting sessile hosts (e.g., *S. glaseri* and *H. bacteriophora*). Most entomopathogenic nematode species, however, appear to be situated somewhere along a continuum between these two extremes using an intermediate type of foraging strategy (e.g., *S. riobrave* and *S. feltiae*) (Campbell and Gaugler, 1997; Campbell and Kaya, 1999a, 1999b).

IJs of entomopathogenic nematode species exhibit a typical behavior termed body-waving where 30 – 95% of their body is raised off the substrate for a few seconds. Some nematode species can raise > 95% of their body off the substrate, standing on a bend in their tail and assuming a straight posture or alternating periods of no motion and active waving (Campbell and Kaya, 1999a, 1999b). This behavior is termed nictation. Typical ambusher species nictate > 70% of their foraging time, whereas species with intermediate foraging strategies nictate less frequently and for shorter periods (e.g., *S. riobrave*) or cannot nictate (e.g., *S. feltiae*). Cruisers, on the other hand, cannot nictate but can body-wave. IJs of nictating species can also jump. This jumping behavior can be directed in that it appears to be used for host attachment (only in typical ambushers) or non-directed when it may play a role in dispersal (Campbell and Kaya, 1999b).

22.5.3.2 Dispersal

In addition to jumping for nematodes with nictating behavior, IJs can actively disperse in soil for up to 90 cm in both horizontal and vertical direction within 30 days (Kaya, 1990). As discussed above, this dispersal activity allows the entomopathogenic nematodes to actively seek out hosts. Nematodes can be dispersed great distances passively by water, wind, phoresis, infected hosts, human activity, etc. which may, in part, account for their widespread distribution. Factors that influence the motility of IJs are moisture, temperature, and soil texture. The most important factor is moisture because nematodes need a water film in the interstitial spaces of soil for effective propulsion. If this water film becomes too thin (in dry soil) or the interstitial spaces become completely filled with water (in saturated soil), nematode movement is restricted (Koppenhöfer, et al., 1995). Different nematode species/strains have different temperature optima and ranges (Grewal, et al., 1994; Griffin, 1993), but generally nematodes will be ineffective at low temperatures (< 10 – 15℃) and will be inactivated at higher temperatures (>30 – 40℃). Soil porosity affects nematode dispersal with less dispersal occurring as soil pores become smaller (Kaya, 1990), whereas soil salinity only affects IJ dispersal at extremely high levels (Thurston, et al., 1994b).

22.5.3.3 Survival

After field application of IJs, their persistence, in most cases, is limited. For example, when applied onto the soil surface, losses can reach 50% within hours of application because of UV radiation and desiccation (Smits, 1996). Even those IJs that enter the soil will be reduced at a rate of 5 – 10% per day until usually only around 1% of the original inoculum survives after 1 – 6 weeks. Although persistence of applied IJs is influenced by behavioral, physiological (Womersley, 1993; Wright, et al., 1998), and genetic characteristics (Gaugler, 1993), we will focus primarily on extrinsic factors. These include abiotic factors [extreme temperatures, soil moisture, osmotic stress, soil texture, RH, and UV radiation (Baur and Kaya, 2001; Glazer, 1996; Kaya, 1990; Smits, 1996)] and biotic factors [antibiosis, competition and natural enemies (Kaya, 2002; Kaya and Koppenhöfer, 1996)].

Moisture is an important factor in IJ survival. Yet, the IJs can survive low moisture conditions by adapting to lower moisture conditions and lowering their rate of metabolism. If water is gradually removed from the IJs, it gives them time to adapt to the desiccating conditions (Glazer, 1996; Patel, et al., 1997; Solomon, et al., 1999; Womersley, 1990). This is the case in natural soils allowing the IJs to actually persist longer in dry soil. On foliage and in other exposed habitats, nematode survival is short (few minutes to hours) unless the RH is close to 100%. Nematodes may survive desiccating conditions by remaining inside the host cadaver until the moisture situation improves (Brown and Gaugler, 1997; Koppenhöfer, et al., 1997).

Temperature effects on nematode survival vary with nematode species and strains (Grewal, et al., 1994; Griffin, 1993). Generally, extended exposure to temperature extremes (below 0℃ or above 40℃) is lethal to most species of entomopathogenic nematodes but the effect depends on exposure time (Brown and Gaugler, 1996). In the soil environment, IJs are normally buffered from temperature extremes. For most species, the best longevity of IJs is between 5℃ and 15℃. At higher temperatures, the IJs have increased metabolic activity and deplete their energy reserves, shortening their life span.

UV light can kill nematodes within minutes (Gaugler, et al., 1992a). Direct exposure to UV light (i. e., sunlight) can be minimized by applying IJs early in the morning or evening, or using sufficient amounts of water to wash IJs into the soil.

Soil texture has an effect on IJ survival with survival being lowest in clay soils. The poor survival rate in clay soils is probably due, in part, to lower oxygen levels in the smaller soil pores. Oxygen may also become a limiting factor in water-saturated soils and soils with high contents of organic matter. Soil pH does not have a strong effect on IJ survival. Thus, IJ survival at pH values between 4 and 8 does not vary, but survival declines at pH 10. Similarly, soil salinity has only limited negative effects on entomopathogenic nematode survival even at salinity above the tolerance levels of most crop plants (Thurston, et al., 1994b). Seawater has no negative effects on survival of several *Heterorhabditis* species/strains (Griffin, et al., 1994) and they have been frequently isolated from soils near the seashore. However, high salinity (seawater) reduced the ability of *Heterorhabditis* to infect hosts but did improve their tolerance to high temperatures (Finnegan, et al., 1999).

Kaya and Koppenhöfer (1996) and Kaya (2002) reviewed the various biotic factors (competition and natural enemies) affecting nematode survival. Intraspecific and interspecific competition can affect nematode survival. For example, if too many IJs of the same species infect a host, intraspecific competition may reduce nematode fitness. In interspecific competition, if more than one species of *Steinernema* infects a host, the prevalence of a given nematode's bacterial symbiont will exclude the other nematode's reproduction or dramatically reduce the fitness of the other nematode's progeny and eventually lead to its local extinction. If a *Steinernema* sp. and a *Heterorhabditis* sp. infect the same host, the *Steinernema* sp. will outcompete the *Heterorhabditis* sp. Interspecific competition may also occur with other insect pathogens, and the outcome of the competition will depend on the kind of competitor (e. g., entomopathogenic fungi, bacteria, or viruses), the timing of infection, and environmental factors such as temperature or soil moisture. Generally, when a nematode and an entomopathogenic fungus are applied at the same time, the nematode will outcompete the fungus for the same host. However, *Bacillus thuringiensis* can outcompete the nematodes because the host will feed on the bacterium and become infected before the nematodes can release their symbiotic bacterium. The presence of

B. thuringiensis in the host is detrimental to the development of the nematode. Among the natural enemies of nematodes, nematophagous fungi, especially *Hirsutella rhossiliensis*, have received the most attention. *H. rhossiliensis* causes a higher mortality of *S. glaseri* IJs than of *H. bacteriophora* IJs. This differential mortality is associated with the retention of the second-stage cuticle by *H. bacteriophora* IJs. Other natural enemies of IJs include invertebrate predators such as collembolans, mites, tardigrades and predatory nematodes. These natural enemies reduce IJ populations in laboratory experiments, but their impact under field conditions is poorly understood. Scavengers such as ants will also feed on insects killed by steinernematids but less so on those killed by heterorhabditids (Baur, et al., 1998). The difference in feeding activity by ants is associated with a "deterrent" factor(s) produced by *Photorhabdus*.

22.5.3.4 Recycling of Nematodes

Recycling is desirable after an application of entomopathogenic nematodes because it can provide additional and prolonged control of a pest. The abiotic and biotic factors that affect persistence, infectivity, and motility of IJs influence nematode recycling. In addition, because they are obligate pathogens, the availability of suitable hosts is a key to recycling of the nematodes. Although recycling probably is quite common (Kaya, 1990; Klein, 1993), as discussed in section 22.5.3.3, the survival of IJs is poor and nematodes have to be reapplied to obtain adequate control of soil insect pests.

In natural populations of entomopathogenic nematodes, recycling occurs in their insect hosts, but only few studies have examined the dynamics of nematode populations and the factors affecting them. Within-site distribution of nematode populations is patchy (Campbell, et al., 1997, 1998; Strong, et al., 1996; Stuart and Gaugler, 1994), and biotic and abiotic factors including seasonal fluctuations, foraging strategy of the IJs, host population dynamics, alternate hosts, etc., play a key role in nematode recycling. Entomopathogenic nematode populations probably persist as metapopulations which exhibit a "shifting mosaic" type of dynamics with asynchronous fluctuations and little migration between patches (Levins, 1970). These patches are highly vulnerable to extinction, and to persist as a metapopulation, the founding rate of local populations has to be the same as the extinction rate (Lewis, et al., 1998).

22.5.4 Genetics

22.5.4.1 Laboratory Colonization

In biological control, genetic variation for attributes affecting natural enemy success can be missed during collection or lost during importation and rearing (Roush, 1990). The three topics pertaining to preservation of genetic variation for nematodes are founder effect, inbreeding, and inadvertent selection (Stuart and Gaugler, 1996). Founder effect is a serious problem for entomopathogenic nematodes because only a limited number of insect cadavers are collected at single geographical sites resulting in reduced genetic variance. A way of maintaining genetic diversity is to collect the same species from as many geographical sites as possible and hybridize the isolates. Repositories for entomopathogenic nematodes wherein hundreds of identified isolates are stored and available, would facilitate the development of hybrid strains (Hominick and Reid, 1990).

Inbreeding depression in laboratory rearing may be a problem in small populations but should not be a major issue for entomopathogenic nematodes because they are reared on a large scale under laboratory conditions (Stuart and Gaugler, 1996). The low dispersal capability of nematodes suggests that inbreeding occurs in natural populations, and through natural selection the deleterious recessive alleles may have been eliminated.

Inadvertent selection in the laboratory can lead to the loss of field-adapted alleles important for biological control by entomopathogenic nematodes but can be reduced by minimizing generation turnover. Inadvertent selection can be significant for entomopathogenic nematodes because they are reared in large populations, especially when reared *in vitro* (Stuart and Gaugler, 1996). The development of liquid nitrogen methods allows for storage of a new field isolate into a stable stock inoculum in a few generations without risk of laboratory adaptation (Curran, et al., 1992). Laboratory adaptation of nematode strains can be corrected by outcrossing with new field isolates.

22.5.4.2 Genetic Improvement

The growing body of knowledge on all aspects of entomopathogenic nematodes and their symbiotic bacteria has presented the opportunity to improve their beneficial traits or eliminate weak ones through genetic manipulation. By understanding the traits relevant to efficacy, their biological control potential can be greatly enhanced through genetic manipulation. A necessary step is to obtain the genetic foundation on the biological basis of these traits. The sequencing of the genome of the free-living nematode *Caenorhabditis elegans* has been completed (Blaxter, 1998), and this rhabditoid nematode could serve as a useful model for improving the understanding of entomopathogenic nematode genetics.

Burnell and Dowds (1996) stated that the main targets for genetic improvement in entomopathogenic nematodes are increased efficacy, resistance to environmental extremes, development of anhydrobiotic strains, and increased suitability of *Heterorhabditis* for culture in the liquid fermentation process. For example, resistance to environmental extremes can be initiated by screening of natural isolates that may contain a desirable attribute such as a heat tolerance gene (Shapiro, et al., 1997a). This gene can then be transferred into commercial strains by cross hybridization. However, because complex behavioral and physiological traits are likely to result from the interaction of many genes, an effective means of genetically improving such traits is by selective breeding. Successful selective breeding of entomopathogenic nematodes includes the selection for cold tolerance (Grewal, et al., 1996; Griffin and Downes, 1994), improved control efficacy (Tomalak, 1994), and nematicide resistance (Glazer, et al., 1997).

Mutagenesis is another useful approach, especially in situations where key regulatory genes may control the expression of several genes. This method has been used to isolate desiccation tolerant strains (O'Leary and Burnell, 1997). Genetic engineering has also been adopted for the improvement of beneficial traits of entomopathogenic nematodes. Using microinjection and microprobes (Hashmi, et al., 1995a, b), a plasmid containing heat-shock protein genes from *C. elegans* has been introduced into *H. bacteriophora* and the resulting transgenic strain has a higher tolerance to short temperature spikes than the wild type (Hashmi, et al., 1998). Moreover, field trials showed no increased persistence of the transgenic strain compared to the wild type indicating that the transgenic form has no advantage over the wild type (Wilson, et al., 1999). The approach of conferring commercial rather than ecological advantages to the nematodes may be the best way to improve nematode performance while limiting regulatory problems for field releases and commercialization.

Burnell and Dowds (1996) considered that the main targets for genetic improvement of the bacterial symbionts are pathogenicity, host specificity, symbiont specificity, resistance to environmental extremes, and control of phase variation. A number of genes from these bacteria such as outer membrane protein genes, low-temperature induced genes, maltose metabolism genes, lux genes, extracellular enzyme genes, and crystalline protein genes have already been cloned (Forst, et al., 1997). Proteins with insecticidal activities have been isolated, the genes identified, and they show potential to be incorporated into plants for insect control (Bowen, et al., 1998).

22.5.5 Commercialization

22.5.5.1 Mass Production

Although entomopathogenic nematodes are obligate pathogens in nature, they are easily cultured either *in vivo* or *in vitro* in the laboratory (Friedman, 1990). For commercial production, both *in vivo* and *in vitro* mass production schemes have been employed. The *in vivo* method is labor intensive, more costly, and more amenable in less-developed countries, although some cottage industries in developed countries use this technology. The *in vivo* method lacks economies of scale, but it requires minimal expertise and capital investments and may be an important future sector in nematode commercialization for specific niche markets.

For large-scale production, *in vitro* methods using the three-dimensional solid media or liquid fermentation methods have been employed (Ehlers, 1996; Grewal and Georgis, 1998). The advantages of the solid media method are that capital costs are low, limited expertise is required (but more than in the *in vivo* method), and the logistics of production are flexible. Because of limited economics of scale, this method is feasible for countries with low labor costs. In contrast, the liquid fermentation method has economies of scale because the proportion of labor and capital costs decreases in scale up as operating cost increase. This technology has the lowest mass-production cost and is the method of choice for larger companies with multiple products in industrialized countries. A number of nematode species have been successfully produced in 7500 – 80,000 liter fermenters including *S. carpocapsae*, *S. riobrave*, *S. kushidai*, *S. feltiae*, *S. glaseri*, *S. scapterisci*, *H. bacteriophora*, and *H. megidis* with yield capacity as high as 250,000 IJs/ml depending on the nematode species. However, successful fermentation is only one aspect of commercialization. For example, *S. glaseri* is difficult to market because of problems with IJ induction and formulation and storage. With *Heterorhabditis*, liquid fermentation yields inconsistent numbers of IJs. The production time can be prolonged due to the variable recovery of the IJs inoculated into the cultures and the inability of the amphimictic adults to mate under liquid culture conditions (Ehlers, et al., 1998; Strauch and Ehlers, 1998).

22.5.5.2 Formulation and Storage

IJs can be stored in an aqueous suspension at 4 – 15℃ (depending on nematode species) without much loss of activity for 6 – 12 months for *Steinernema* species and 3 – 6 months for *Heterorhabditis* species. Viability at higher temperatures is considerably shorter. Many commercial nematode products are formulated on moist substrates (e.g., sponge) or held in aqueous suspensions that require continuous refrigeration to maintain nematode quality for extended periods. To improve IJ shelf life and resistance to temperature extremes, formulations that reduce IJ metabolism by immobilization or partial desiccation have been developed. These formulations contain alginate, vermiculite, clays, activated charcoals, polyacrylamide, and water dispersible granules (Georgis and Kaya, 1998; Grewal and Georgis, 1998). It is difficult to obtain an optimal formulation for all nematode species because they have different specific requirements for moisture and oxygen. One of the best formulations is the water dispersible granule that has been developed for steinernematids (e.g., *S. carpocapsae*, *S. riobrave*, and *S. feltiae*) as it combines relatively long nematode shelf life without refrigeration but with ease of handling. Anhydrobiotic IJs in water dispersible granules have a shelf life at 25℃ of 5 to 6 months for *S. carpocapsae*, 2 months for *S. feltiae*, and 1 month for *S. riobrave* (Grewal, 2000). This formulation is mixed with water prior to spray application, and the partially desiccated IJs rehydrate after application to a moist soil environment. However, to achieve optimal infectivity the IJs need to rehydrate for up to 3 days in soil (Baur, et al., 1997b). On foliage, rehydration has to occur before application making this formulation impractical for these situations.

Before and after formulation, the quality of the nematodes should be checked. At a minimum, their viability and infectivity should be monitored. In a recent study, the quality of commercial nematodes (*S. carpocapsae* and *H. bacteriophora*) from cottage industry sources was assessed (Gaugler, et al., 2000). These sources aimed largely towards the mail-order market. Although the nematodes were received in a satisfactory condition with acceptable levels of viability, consistency was a problem with each supplier having to correct some weak areas. Generally, the product did not contain the expected number of nematodes and these were not as pathogenic as the laboratory controls. In some cases, two nematode species were mixed together. The cottage industry lacked rigorous quality control, and improved reliability by the nematode industry will be most likely achieved via industry-generated agreement on standards for quality.

22.5.5.3 Regulations

Regulations on the use of entomopathogenic nematodes for insect control vary among countries (Richardson, 1996; Bedding, et al., 1996; Rizvi, et al., 1996). These nematodes, if they are indigenous to the area, are exempted from registration in many European countries (Denmark, Germany, The

Netherlands, Spain, and UK), Australia, and USA. In other countries, they are subject to similar registration procedures as a chemical pesticide, e. g., Austria, Belgium, Hungary, Japan, Switzerland, and New Zealand. The importation and use of non-indigenous and transgenic nematode species are subject to strict regulations in most countries. Some countries consider foreign strains of endemic species to be exotic, and this can be a major obstacle for the commercialization of entomopathogenic nematodes. A workshop with 15 expert participants from 10 countries made two recommendations (Ehlers and Hokkanen, 1996) that should form a strong base for future regulatory decisions concerning the use of entomopathogenic nematodes. First, because of biological and ecological features that make entomopathogenic nematodes exceptionally safe for use in biological control, they should not require registration. Second, the introduction of non-indigenous nematodes should be regulated at the species level and not at any lower taxon.

22.5.6 Efficacy

22.5.6.1 Key Target Pests

Entomopathogenic nematodes have been tested against a large number of insect pest species with results varying from poor to excellent control (Bedding, et al., 1993; Begley, 1990; Klein, 1990). Many factors can influence the successful use of nematodes as biological insecticides, but matching the biology and ecology of both the nematode and the target pest is a crucial step towards successful application. Consideration has to be given to the foraging behavior and temperature requirements of a nematode species and to accessibility and suitability of the pest to the nematode. Entomopathogenic nematodes have been most efficacious in habitats that provide protection from environmental extremes, especially in soil which is their natural habitat and in cryptic habitats. Excellent control has been achieved against plant-boring insects because their cryptic habitats are favorable for nematode survival and infectivity (e. g., no natural enemies of the nematodes and adequate moisture). Low or highly variable control has been achieved in manure because of high temperatures in animal rearing facilities and toxic effects of manure contents (ammonia) on IJs. Control of aquatic insects has been unsuccessful because the nematodes are not adapted to directed motility (host finding) in this environment. On foliage and other exposed habitats, the IJs face harsh conditions that can be only marginally remedied by adjuvants. A list of some insect pests and the commodities in which they have been successfully controlled with entomopathogenic nematodes is provided in Table 22.2.

Entomopathogenic nematodes may also provide control of plant-parasitic

Table 22.2 Target pests for commercially available entomopathogenic nematodes[1]

Pest group	Common name	Life-stage[2]	Application Site	Commodity	Nematode sp.[3]
BLATTODEA					
Blattellidae	German cockroach	A/N	Baits	apartments/structures	Sc
COLEOPTERA					
Chrysomelidae	Flea beetles	L	Soil	mint, potato, sweet potato, sugar beets, vegetables	Sc
Curculionidae	Billbugs	L	Turf	turfgrass	Sc, Hb
	Root weevils	L	Soil	berries, citrus, forest seedlings, hops, mint, ornamentals, sweet potato, sugar beet, banana[4]	Sc, Hb, Hm, Sr
	Rootworms	L	Soil	corn, peanuts, vegetables	Sc, Sr
Scarabaeidae	White grubs	L	Soil, turf	berries, field crops, ornamentals, turfgrass	Hb, Sg, Hm, Sk
DIPTERA					
Agromyzidae	Leaf miners	L	Foliage	ornamentals, vegetables	Sc
Ephydridae	Shore flies	L	Soil	ornamentals, vegetables	Sf
Sciaridae	Sciarid flies	L	Soil	mushrooms, ornamentals, vegetables	Sf
Muscidae	Filth flies	A	Baits	animal rearing facilities	Sf, Hb
Tipulidae	Crane flies	L	Turf, soil	turfgrass, ornamentals	Sc, Hm

Continued

Pest group	Common name	Life-stage[2]	Application Site	Commodity	Nematode sp.[3]
LEPIDOPTERA					
Carposinidae	Peach borer moth	L	Soil	apple	Sc
Noctuidae	Cutworms, armyworms	L/P	Soil, turf	corn, cotton, peanuts, turfgrass, vegetables	Sc
Cossidae	Carpenter worms	L	Cryptic	ornamentals, shrubs	Sc
	Leopard moth	L	Cryptic	apple, pear	Sc
Pyralidae	Webworms	L	Soil, turf	cranberries, ornamentals, turf	Sc
Sesiidae	Crown borers	L	Cryptic	berries	Sc
	Stem borers	L	Cryptic	cucurbits, ornamentals, fruit trees	Sc
ORTHOPTERA					Sc
Gryllotalpidae	Mole crickets	A/N	Turf, soil	turfgrass, vegetables	Ss, Sr
SIPHONAPTERA					
Pulcidae	Cat fleas	L/P	Soil, turf	pet/vet	Sc
NEMATODA	Plant-parasitic nematodes		Turf	Turf	Sr

[1] After Georgis and Manweiler (1994), Georgis et al. (1995), Klein (1990), Begley (1990).

[2] L = larva; P = pupa; N = nymph; A = adult.

[3] Sc = *S. carpocapsae*; Sf = *S. feltiae*; Sk = *S. kushidai*; Sr = *S. riobrave*; Ss = *S. scapterisci*; Hb = *H. bacteriophora*; Hm = *H. megidis*.

[4] Applied to residual rhizomes in a cryptic habitat (see text).

nematodes. Several laboratory and greenhouse studies have shown that inundative applications of entomopathogenic nematodes have a suppressive effect on plant-parasitic nematodes (e.g., Bird and Bird, 1986; Ishibashi and Kondo, 1987). Grewal et al. (1997) showed that *S. riobrave* can be as effective as chemical nematicides in the suppression of root knot, sting, and ring nematodes in turfgrass at economic application rates. In another study, Perry et al. (1998) observed that invasion of potato roots by juveniles of the potato cyst nematode (*Globodera rostochiensis*) in glasshouse and outdoor trials was reduced by *S. carpocapsae* and *S. feltiae*. However, there was no significant difference in the number of cysts between treated and untreated plants at senescence, and the use of *S. carpocapsae* and *S. feltiae* is unlikely to provide adequate control for the potato cyst nematode. More recent evidence suggests that allelopathic interactions occur between the plant-parasitic and the entomopathogenic nematodes (Grewal, et al., 1999). Root-penetration by root-knot nematode (*Meloidogyne incognita*) juveniles is temporarily reduced by the repellent effect of metabolites of the entomopathogenic nematodes' symbiotic bacteria. The effects are reduced egg production and egg hatch in *M. incognita* females exposed as juveniles to soil treated with entomopathogenic nematodes and their bacteria.

22.5.6.2 Application Strategies

Entomopathogenic nematodes have almost exclusively been applied using the inundative approach where high numbers of IJs are released in a uniform distribution and control of pest populations is expected to be fast and thorough. This approach is feasible for high value niche crops. Whether this is the best approach for other cropping systems has been in question for some time because of their limited shelf life, susceptibility to environmental extremes, high price, etc. Nematodes are a poor fit for an inundative approach (i.e., chemical pesticide paradigm) in many cropping systems. Augmentative (i.e., inoculative) releases and conservation and management of endemic nematode populations need to be explored, as they may be more promising and feasible in many pest situations (Lewis, et al., 1998).

Inoculative release of entomopathogenic nematodes with the expectation that they will establish new populations or augment low populations for long-term pest suppression has only been attempted a few times and little is known about the optimal approach to this strategy. *Steinernema glaseri*, isolated originally from scarab grubs in New Jersey, was released in a massive inoculative control program from 1939 to 1942 against Japanese beetle larvae in turfgrass (Gaugler, et al., 1992b). Gaugler et al. (1992b) noted that the elimination of bacterial symbionts by the use of antimicrobials in the *in vitro* rearing procedure, and possibly poor climatic adaptation of this neotropical nematode limited the success of this program. More recently, *S. scapterisci* originally isolated from Uruguay was successfully introduced into Florida for the classical biological control of mole cricket pests (Parkman and Smart, 1996).

Successful inoculative releases of entomopathogenic nematodes are dependent on long-term, multigenerational survival and recycling of the nematode populations. To achieve this goal, several conditions must be met including (1) the presence of moderately susceptible insect hosts throughout most of the year; (2) a high economic threshold level of the target insect pests; and (3) soil conditions favorable for nematode survival (Kaya, 1990). Augmentative releases into established nematode populations and/or management of the susceptibility of the host/pest populations (for example using stressors such as other control agents) are two additional approaches that may be used to boost or manage established nematode populations and warrant more attention (Koppenhöfer, et al., 1999, 2000).

22.5.6.3 Application Methods

The most common application method for entomopathogenic nematodes is to use the same type of equipment used for spraying chemical pesticides. Thus, nematodes can be applied to the target site with most commercially available spray equipment such as hand or ground sprayers, mist blowers, and aerial sprayers on helicopters (Georgis, et al., 1995). IJs can withstand pressures up to 1068 kPa and pass through common nozzle type sprayers with openings as small as 100 μm in diameter, but the screens in the nozzles should be removed to minimize damage to the nematodes. Nematodes have also been applied via irrigation systems including drip, microjet, sprinkler, and furrow irrigation (Cabanillas and Raulston, 1996; Georgis, et al., 1995). Pre- and postapplication irrigation and continued moderate soil moisture are essential for good nematode performance. If water is limited, subsurface injection of nematodes can be an efficient delivery method (Klein, 1993).

Plant-boring insects have been successfully controlled by injecting nematode suspensions directly into the borer holes or blocking the holes with sponges soaked with nematode suspensions (Yang, et al., 1993). For the banana weevil, a nematode suspension can be placed into insect-attracting cuts in residual rhizomes of bananas (Treverrow and Bedding, 1993). Baits containing IJs can offer a cost-effective way of controlling mobile insects such as adult house flies (Renn, 1998) and German cockroaches (Appel, et al., 1993) with the use of trap stations that ensure intimate IJ-pest contact and protection of the IJs from UV radiation and desication. However, these trap stations have been experimental and are not yet commercially available.

The detrimental effects of desiccation and UV radiation often can be alleviated by the addition of adjuvants to the nematode formulation/suspension. Adjuvants have been used to improve nematode performance against foliage-feeding pests. Thus, solar radiation can be filtered with stilbene brighteners (Baur, et al., 1997a; Nickle and Shapiro, 1994), whereas antidesiccants such as certain commercial oils, plant-based products, and glycerin (Baur, et al., 1997a; Broadbent and Olthof, 1995; Glazer, et al., 1992) have provided short-term protection for nematodes in exposed habitats. Surfactants (i.e., detergents) may also improve nematode speed of penetration into soil (Schroeder and Sieburth, 1997), but further studies are necessary to determine the mechanism of this interaction and whether these combinations are feasible under field conditions.

22.5.6.4 Effects of Agrochemicals and Other IPM Components

Entomopathogenic nematodes are often applied to systems/substrates that are regularly treated with many other agents, including chemical or biorational pesticides, soil amendments, and fertilizers. Depending on the agents, application timing, and physico-chemical characters of the system, the nematodes may or may not interact with these other agents. If interactions occur, they may range from antagonistic to synergistic. In addition to these agents, intraguild predation between parasitoids and nematodes may occur (Rosenheim, et al., 1995), but Sher et al. (2000) showed that a leaf miner parasitoid was compatible with *S. carpocapsae*.

Entomopathogenic nematodes appear to be compatible with many, but not all, herbicides, fungicides, acaricides, insecticides, nematicides (e.g., Georgis and Kaya, 1998; Ishibashi, 1993; Rovesti and Deseö, 1990), azadirachtin (Stark, 1996), *Bacillus thuringiensis* products (Kaya, et al., 1995), and pesticidal soap (Kaya, et al., 1995). Negative effects of various pesticides on the IJs have been documented (e.g., Patel and Wright, 1996; Rovesti and Deseö, 1990). On the other hand, synergistic interaction between entomopathogenic nematodes and other control agents has been observed for various insecticides (e.g., Koppenhöfer, et al., 2000; Nishimatsu and Jackson, 1998) and pathogens (Koppenhöfer, et al., 1999; Thurston, et al., 1994a). In view of the diversity of available chemical and biorational insecticides, a generalization on pesticide-nematode compatibility cannot be made. The compatibility of each chemical pesticide and nematode species should be evaluated on a case-by-case basis.

Inorganic fertilizer may be compatible with nematodes in inundative releases to control soil pests (Bednarek and Gaugler, 1997). Natural nematode populations, on the other hand, were negatively affected by inorganic fertilizers, but positively affected by manure (Bednarek and Gaugler, 1997). Composted manure or urea does not have negative effects on nematode virulence but fresh dung does (Shapiro, et al., 1997b).

22.6 Rhabditidae and Slugs

The only parasite that has been developed as a biological control agent of slugs is *Phasmarhabditis hermaphrodita* (Wilson, et al., 1993a) which has been available commercially in Europe since 1994. *P. hermaphrodita* is a lethal parasite capable of killing many species of pest slugs. In the slug, *Deroceras reticulatum,* the infective juvenile penetrates into the shell cavity through a natural opening that connects the shell cavity to the dorsal integumental pouch. The juvenile develops into an adult and reproduces. During this time, the mantle region of the slug swells, and 7 – 21 days after infection, the slug dies and nematode progeny move from the mantle and feed on the entire slug cadaver. When the food resources are depleted, the nematodes produce infective juveniles that leave the cadaver in search of fresh hosts.

Phasmarhabditis hermaphrodita can be mass-produced *in vitro* on the bacterium, *Moraxella osloensis* (Wilson, et al., 1993b, 1995c). It has been field tested in Europe in many crops including wheat (Wilson, et al., 1994, 1996), lettuce (Wilson, et al., 1995a), oilseed rape (Wilson, et al., 1995b), strawberries and sugar beet (see Wilson and Gaugler, 2000), and in most cases, the nematode gave control equivalent to or superior than the chemical standards. Although the nematode may take more than 7 days to kill slugs, their feeding is strongly inhibited within a few days of infection, providing rapid crop protection.

22.7 Prospects for Biological Control

Nematodes have proven to be excellent biological control agents of some insect pests. In Australia, the wood wasp *Sirex noctilio* has been suppressed initially with the nematode *Deladenus siricidicola* using classical biological control. When *Sirex* populations re-occur in high numbers in old infested areas, an augmentative biological control program can be initiated because the nematodes can be reared *in vitro* and inoculative releases can be made. However, because of its narrow host range and the only occasional need to augment natural populations, *D. siricidicola* does not attract commercial interests.

Fastidious nematodes, such as mermithids and the iotonchiids, offer greater challenges in biological control of their pests. Because they have to be cultured *in vivo* which is labor intensive and costly, they will be limited to classical, augmentative or conservation biological control programs. Although there was commercial interest with *Romanomermis culicivorax* for mosquito control, it could not compete with other biological control agents. The future of using these nematodes will more than likely depend upon government agencies providing the main support.

In contrast, the entomopathogenic nematodes (steinernematids and heterorhabditids) have attracted widespread commercial interest. These biological control agents are endowed with many advantages including host-seeking capability, high virulence, ease of production, ease of application, mammalian safety, and exemption from registration in many countries. They also possess a broad host range, are compatible with many other control agents, are widespread in distribution, and can be formulated and stored for a reasonable length of time. Currently, because of their high retail cost in comparison with other control agents, their use is restricted to high-value crops in niche markets or to homeowners. Yet, among microbial control agents of insects, they rank second in sales to *Bacillus thuringiensis*. However, the efficacy of these nematodes against insect pests has been mixed. The reasons for the varied efficacy, particularly in the soil environment, are often unknown, emphasizing the need to obtain basic information on the biology, behavior, ecology, and genetics of these nematodes. Indeed, there has been a surge in research to obtain more basic information and in exploration to discover new nematode isolates. These discoveries have resulted in the description of many more species and new genetic material.

Research in behavioral ecology has clearly demonstrated that these entomopathogenic nematodes are not generalist pathogens. Although most of them have a broad host range in the laboratory, in the field they are adapted to hosts in a particular environment depending on their foraging strategy. Their foraging strategy restricts much of their activity to a certain soil stratum eliminating many insects from infection. By understanding their behavioral patterns, the right nematode species can be matched with the insect pests, enhancing their use and efficacy in the field. Another major advancement has been in molecular engineering of a heat-shock protein into the nematodes that offers the possibility of better shipping stability. Insertion of other genes, for example desiccation tolerance, may also extend shelf life and survival in the field. Although the transgenic nematodes with the heat-shock protein gene have been field tested with no adverse effect (Wilson, et al., 1999), environmental risk assessments may still be needed for each new transgenic nematode.

These nematodes are ubiquitous in nature, but their populations usually are not sufficiently high to cause epizootic and reduce pest populations. Research is needed to better understand the factors that regulate their populations and on how their populations can be manipulated to initiate epizootic in insect pest populations. Although they have high searching capacity, they have a limited dispersal range. Can they respond in a density-dependent fashion and respond to increasing host population densities? Do their own natural enemies limit the nematodes themselves? As we learn more about nematode survival, we can address these questions and use the nematodes more effectively as inundative agents against a number of insect pests and as inoculative agents for classical biological control. However, the soil environment is a difficult medium in which to conduct some of this research. In addition, stressors can be used as pest management tools against an insect pest to increase its susceptible to the nematode. This approach has been highly successful against white grubs in turfgrass. Conservation and augmentation of natural nematode populations through proper management practices and periodic nematode releases offer further possibilities for insect pest suppression.

Significant advances are also being made with the mutualistic bacteria associated with entomopathogenic nematodes. Insecticidal compounds have been isolated and some of the genes show potential for engineering into plants for insect suppression (ffrench-Constant and Bowen, 1999). Moreover, the mutualistic bacteria (or their metabolites) show activity against plant-parasitic nematodes. This new area of research may provide growers with compounds for plant-parasitic nematode control.

Finally, the molluscicidal nematode, *P. hermaphrodita,* is commercially available in Europe for slug control. A search for other molluscicidal nematodes is being made in other countries. Currently, regulations prevent the importation of the European species into the United States as it is an exotic nematode. However, many of the pest slug species in the United States are not native, and the possibility is high that *P. hermaphrodita* or a closely related species will be found.

Significant information has been generated since the writing of this chapter and includes the following: (1) description of several new *Steinernema* species (*S. anatoliense, S. scarabaei, S. websteri,* and *S. weiseri*) and two *Heterorhabditis* species (*H. downesi* and *H. taysearae*) (Hazir, et al., 2003; Mrácek, et al., 2003; Stock and Koppenhöfer, 2003; Stock and Reid, 2003); (2) a further characterization of the ant deterrent factor by Zhou et al. (2002); (3) a recent review article on *Photorhabdus* by ffrench-Constant et al. (2003); and (4) cloning of the *Photorhabdus* genome by Duchaud et al. (2003).

References

AbuHatab, M. A., S. Selvan, and R. Gaugler. 1993. Role of proteases in penetration of insect gut by the entomopathogenic nematode *Steinernema glaseri* (Nematoda: Steinernematidae). Journal of Invertebrate Pathology 66: 125 – 130.

Adams, B. J., A. M. Burnell, and T. O. Powers. 1998. A phylogenetic analysis of *Heterorhabditis* (Nemata: Rhabditidae) based on internal spacer 1 DNA sequence data. Journal of Nematology 30: 22 – 39.

Adams, B. J., and K. B. Nguyen. 2002. Taxonomy and systematics. Pp. 1 – 33. In R. Gaugler, ed. Entomopathogenic Nematology. Wallingford, UK: CAB International.

Aguillera, M. M., N. C. Hodge, R. E. Stall, and G. C. Smart, Jr. 1993. Bacterial symbionts of *Steinernema scapterisci*. Journal of Invertebrate Pathology 62: 68 – 72.

Aguillera, M. M., and G. C. Smart, Jr. 1993. Development, reproduction, and pathogenicity of *Steinernema scapterisci* in monoxenic culture with different species of bacteria. Journal of Invertebrate Pathology 62: 289 – 294.

Akhurst, R. J. 1990. Safety to nontarget invertebrates of nematodes of economically important pests. Pp. 233 – 240. In M. Laird, L. A. Lacey, and E. W. Davidson, eds. Safety of Microbial Insecticides. Boca Raton, FL: CRC Press.

Akhurst, R. J., and N. E. Boemare. 1990. Biology and taxonomy of *Xenorhabdus*. Pp. 75 – 90. In R. Gaugler and H. K. Kaya, eds. Entomopathogenic Nematodes in Biological Control. Boca Raton, FL: CRC Press.

Akhurst, R. J., R. G. Mourant, L. Baud, and N. E. Boemare. 1996. Phenotypic and DNA relatedness between nematode symbionts and clinical strains of the genus *Photorhabdus* (Enterobacteriaceae). International Journal of Systematic Bacteriology 46: 1034 – 1041.

Appel, A. G., E. P. Benson, J. M. Ellenberger, and S. A. Manweiler. 1993. Laboratory and field evaluations of an entomogenous nematode (Nematoda: Steinernematidae) for German cockroach (Dictyoptera: Blatellidae) control. Journal of Economic Entomology 86: 777 – 784.

Babic, I., M. Fischer-Le Saux, E. Giraud, and N. Boemare. 2000. Occurrence of natural dixenic associations between the symbiont *Photorhabdus luminescens* and bacteria related to *Ochobactrum* spp. in tropical entomopathogenic *Heterorhabditis* spp. (Nematoda, Rhabditida). Microbiology 146: 709 – 718.

Barbercheck, M. E., and L. C. Millar. 2000. Environmental impacts of entomopathogenic nematodes used for biological control in soil. Pp. 287 – 308. In P. A. Follett and J. J. Duan, eds. Nontarget Effects of Biological Control. Boston, MA: Kluwer.

Bathon, H. 1996. Impact of entomopathogenic nematodes on non-target hosts. Biocontrol Science and Technology 6: 421 – 434.

Baur, M. E., and H. K. Kaya. 2001. Persistence of entomopathogenic nematodes. Bulletin from Regional Project S-265, Southern Cooperative Series Bulletin 400. http://www.agctr.lsu.edu/s265/

Baur, M. E., H. K. Kaya, R. Gaugler, and B. Tabashnik. 1997a. Effects of adjuvants on entomopathogenic nematode persistence and efficacy against *Plutella xylostella*.

Biocontrol Science and Technology 7: 513 – 525.

Baur, M. E., H. K. Kaya, and D. R. Strong. 1998. Foraging ants as scavengers on entomopathogenic nematode-killed insects. Biological Control 12: 231 – 236.

Baur, M. E., H. K. Kaya, and B. Tabashnik. 1997b. Efficacy of a dehydrated steinernematid nematode against black cutworm (Lepidoptera: Noctuidae) and diamondback moth (Lepidoptera: Plutellidae). Journal of Economic Entomology 90: 1200 – 1206.

Bedding, R. A. 1984. Nematode parasites of Hymenoptera. Pp. 755 – 795. In W. R. Nickle, ed. Plant and Insect Nematodes. New York: Marcel Dekker.

Bedding, R. A. 1993. Biological control of *Sirex noctilio* using the nematode *Deladenus siricidicola*. Pp. 11 – 20. In R. Bedding, R. Akhurst, and H. Kaya, eds. Nematodes and the Biological Control of Insects. East Melbourne, Australia: CSRIO Publications.

Bedding, R. A., and R. J. Akhurst 1974. Use of the nematode *Deladenus siricidicola* in the biological control of *Sirex noctilio* in Australia. Journal of the Australian Entomological Society 13: 129 – 137.

Bedding, R., R. Akhurst, and H. Kaya, eds. 1993. Nematodes and the biological control of insect pests. East Melbourne, Australia: CSIRO Publications.

Bedding, R. A., S. Tyler, and N. Rochester. 1996. Legislation on the introduction of exotic entomopathogenic nematodes into Australia and New Zealand. Biocontrol Science and Technology 6: 465 – 475.

Bednarek, A., and R. Gaugler. 1997. Compatibility of soil amendments with entomopathogenic nematodes. Journal of Nematology 29: 220 – 227.

Begley, J. W. 1990. Efficacy against insects in habitats other than soil. Pp. 215 – 321. In R. Gaugler and H. K. Kaya, eds. Entomopathogenic Nematodes in Biological Control. Boca Raton, FL: CRC Press.

Bird, A. F., and J. Bird. 1986. Observations on the use of insect parasitic nematodes as a means of biological control of root-knot nematodes. International Journal for Parasitology 16: 511 – 516.

Blaxter, M. 1998. *Caenorhabditis elegans* is a nematode. Science 282: 2041 – 2046.

Blaxter, M. L., P. De Ley, J. R. Garey, L. X. Liu, P. Scheldeman, A. Vierstraete, J. R. Vanfleteren, L. Y. Mackey, M. Dorris, L. M. Frisse, J. T. Vida, and W. Kelley. 1998. A molecular evolutionary framework for the phylum Nematoda. Nature 392: 71 – 75.

Boemare, N. 2002. Biology, taxonomy, and systematics of *Photorhabdus* and *Xenorhabdus*. Pp. 35 – 56. In R. Gaugler, ed. Entomopathogenic Nematology. Wallingford, UK: CABI Publishing.

Boemare, N., C. Laumond, and H. Mauleon. 1996. The entomopathogenic nematode-bacterium complex; biology, life cycle and vertebrate safety. Biocontrol Science and Technology 6: 333 – 345.

Bohan, D. A., and W. R. Hominick. 1996. Investigations on the presence of an infectious proportion amongst populations of *Steinernema feltiae* (Site 76 strain) infective stages. Parasitology 112: 113 – 118.

Bohan, D. A., and W. R. Hominick. 1997. Long-term dynamics of infectiousness in the infective-stage pool of the entomopathogenic nematode *Steinernema feltiae* (Site 76 strain) Filipjev. Parasitology 114: 301 – 308.

Bonifassi, E., M. Fischer-Le Saux, N. Boemare, A. Lanois, C. Laumond, and G. C. Smart, Jr. 1999. Gnotobiological study of infective juveniles and symbionts of *Steinernema scapterisci*: a model to clarify the concept of the natural occurrence of monoxenic associations in entomopathogenic nematodes. Journal of Invertebrate Pathology 74: 164 – 172.

Bowen, D., T. A. Rocheleau, M. Blackburn, O. Andreev, E. Golubeva, R. Bhartia, and R. H. ffrench-Constant. 1998. Insecticidal toxins from the bacterium *Photorhabdus luminescens*. Science 80: 2129 – 2132.

Broadbent, A. B., and Th. H. A. Olthof. 1995. Foliar application of *Steinernema carpocapsae* (Rhabditida: Steinernematidae) to control *Liriomyza trifolii* (Diptera: Agromyzidae) larvae in chrysanthemums. Environmental Entomology 24: 431 – 435.

Brown, I. M., and R. Gaugler. 1996. Cold tolerance of steinernematid and heterorhabditid nematodes. Journal of Thermal Biology 21: 115 – 121.

Brown, I. M., and R. Gaugler. 1997. Temperature and humidity influence emergence and survival of entomopathogenic nematodes. Nematologica 43: 363 – 375.

Burnell, A. M., and B. C. A. Dowds. 1996. The genetic improvement of entomopathogenic nematodes and their symbiont bacteria: phenotypic targets, genetic limitations and an assessment of possible hazards. Biocontrol Science and Technology 6: 435 – 447.

Burnell, A. M., and S. P. Stock. 2000. *Heterorhabditis, Steinernema* and their bacterial symbionts-lethal pathogens of insects. Nematology 2: 1 – 12.

Cabanillas, H. E., and J. R. Raulston. 1996. Effects of furrow irrigation on the distribution and infectivity of *Steinernema riobravis* against corn earworm in corn. Fundamental and Applied Nematology 19: 273 – 281.

Campbell, J. F., and R. Gaugler. 1997. Inter-specific variation in entomopathogenic nematode foraging strategy: dichotomy or variation along a continuum? Fundamental and Applied Nematology 20: 393 – 398.

Campbell, J. F., and H. K. Kaya. 1999a. How and why a parasitic nematode jumps. Nature 397: 485 – 486.

Campbell, J. F., and H. K. Kaya. 1999b. Mechanism, kinematic performance, and fitness consequences of jumping behavior in entomopathogenic nematodes (*Steinernema* spp.). Canadian Journal of Zoology 77: 1 – 9.

Campbell, J. F., A. M. Koppenhöfer, H. K. Kaya, and B. Chinnasri. 1999. Are there temporarily non-infectious dauer stages in entomopathogenic nematode populations: a test of the phased infectivity hypothesis. Parasitology 118: 499 – 508.

Campbell, J. F., E. E. Lewis, F. Yoder, and R. Gaugler. 1997. Entomopathogenic nematode (Heterorhabditidae and Steinernematidae) spatial distribution in turfgrass. Parasitology 113: 473 – 482.

Campbell, J. F., G. Orza, F. Yoder, E. Lewis, and R. Gaugler. 1998. Spatial and temporal distribution of endemic and released entomopathogenic nematode populations in turfgrass. Entomologia Experimentalis et Applicata 86: 1 – 11.

Chirico, J. 1990. Sex ratio, emergence and infection rates of *Heterotylenchus*-exposed *Musca autumnalis* (Diptera: Muscidae). Journal of Parasitology 76: 124 – 127.

Chirico, J. 1996. Transmission of *Heterotylenchus autumnalis* nematodes into field populations of *Musca autumnalis* by release of laboratory-reared specimens. Medical and

Veterinary Entomology 10: 187 – 189.
Choo, H. Y., and H. K. Kaya. 1993. Parasitism of the brown planthopper by the mermithid nematode *Agamermis* in rice. Pp. 21 – 26. In R. Bedding, R. Akhurst, and H. Kaya, eds. Nematodes and the Biological Control of Insect Pests. East Melbourne, Australia: CSIRO Publications.
Choo, H. Y., H. K. Kaya, and H. H. Kim. 1995. Biological studies on *Agamermis unka* (Nematoda: Mermithidae), a parasite of the brown plant hopper *Nilaparvata lugens*. Biocontrol Science and Technology 5: 209 – 223.
Choo, H. Y., H. K. Kaya, and J. B. Kim. 1989. *Agamermis unka* (Mermithidae) parasitism of *Nilaparvata lugens* in rice fields in Korea. Journal of Nematology 21: 254 – 259.
Choo, H. Y., H. H. Kim, and H. K. Kaya. 1998. Effects of selected chemical pesticides on *Agamermis unka* (Nematoda: Mermithidae), a parasite of the brown planthopper, *Nilaparvata lugens*. Biocontrol Science and Technology 8: 413 – 427.
Coler, R. R., and K. B. Nguyen. 1994. *Paraiotonchium muscadomesticae* n. sp. (Tylenchida: Iotonchiidae) a parasite of the house fly (*Musca domestica*) in Brazil and a key to species of the same genus *Paraiotonchium*. Journal of Nematology 26: 392 – 401.
Cook, A. G., and T. J. Perfect. 1989. The population characteristics of the brown planthopper, *Nilaparvata lugens*, in the Philippines. Ecological Entomology 14: 1 – 9.
Creighton, C. S., and G. Fassuliotis. 1981. A laboratory technique for culturing *Filipjevimermis leipsandra*, a nematode parasite of *Diabrotica balteata* larvae (Insecta: Coleoptera). Journal of Nematology 13: 226 – 227.
Creighton, C. S., and G. Fassuliotis. 1982. Mass rearing a mermithid nematode, *Filipjevimermis leipsandra* (Mermithida: Mermithidae) on the banded cucumber beetle (Coleoptera: Chrysomelidae). Journal of Economic Entomology 75: 701 – 703.
Creighton, C. S., and G. Fassuliotis. 1983. Infectivity and suppression of the banded cucumber beetle (Coleoptera: Chrysomelidae) by the mermithid nematode *Filipjevimermis leipsandra* (Mermithida: Mermithidae). Journal of Economic Entomology 76: 615 – 618.
Curran, J., C. Gilbert, and K. Buttler. 1992. Routine cryopreservation of isolates of *Steinernema* and *Heterorhabditis* spp. Journal of Nematology 24: 269 – 270.
Cuthbert, F. P., Jr. 1968. Bionomics of a mermithid (nematode) parasite of soil-inhibiting larvae of certain chrysomelids (Coleoptera). Journal of Invertebrate Pathology 12: 283 – 287.
Duchaud, E., et al. (25 other authors). 2003. The *Photorhabdus luminescens* genome reveals a biotechnological weapon to fight microbes and insect pests. Nature Biotechnology 21: 1307 – 1313.
Ehler, L. E. 1990. Some contemporary issues in biological control of insects and their relevance to the use of entomopathogenic nematodes. Pp. 1 – 19. In R. Gaugler and H. K. Kaya, eds. Entomopathogenic Nematodes in Biological Control. Boca Raton, FL: CRC Press.
Ehler, L. E. 1998. Conservation, biological control: Past, present, and future. Pp. 1 – 7. In P. Barbosa, ed. Conservation Biological Control. San Diego, CA: Academic Press.
Ehlers, R.-U. 1996. Current and future use of nematodes in biocontrol: practice and commercial aspects with regard to regulatory policy issues. Biocontrol Science and

Technology 6: 303 – 316.

Ehlers, R.-U., and H. M. T. Hokkanen. 1996. Insect biocontrol with non-endemic entomopathogenic nematodes (*Steinernema* and *Heterorhabditis* spp.): Conclusions and recommendations of a combined OECD and COST workshop on scientific and regulatory policy issues. Biocontrol Science and Technology 6: 295 – 302.

Ehlers, R.-U., S. Lunau, K. Krasomil-Osterfeld, and K. H. Osterfeld. 1998. Liquid culture of the entomopathogenic nematode-bacterium complex *Heterorhabditis megidis/ Photorhabdus luminescens*. Biocontrol 43: 77 – 86.

Ehlers, R.-U., S. Stoessel, and U. Wyss. 1990. The influence of the phase variants of *Xenorhabdus* spp. and *Escherichia coli* (Enterobacteriaceae) on the propagation of entomopathogenic nematodes of the genera *Steinernema* and *Heterorhabditis*. Revue de Nematologie 13: 417 – 424.

Eidt, D. C., and G. S. Thurston. 1995. Physical deterrents to infection by entomopathogenic nematodes in wire worms (Coleoptera: Elateridae) and other soil insects. Canadian Entomologist 127: 423 – 429.

Fassuliotis, G., and C. S. Creighton. 1982. *In vitro* cultivation of the entomogenous nematode *Filipjevimermis leipsandra*. Journal of Nematology 14: 126 – 131.

ffrench-Constant, R., and D. Bowen. 1999. *Photorhabdu*s toxins: novel biological insecticides. Current Opinion in Microbiology 2: 284 – 288.

ffrench-Constant, R., N. Waterfield, P. Daborn, S. Joyce, H. Bennett, C. Au, A. Dowling, S. Boundy. S. Reynolds, and D. Clarke. 2003. *Photorhabdus:* towards a functional genomic analysis of a symbiont and pathogen. FEMS Microbiology Reviews 26: 433 – 456.

Finnegan, M. M., M. J. Downes, M. O'Regan, and C. T. Griffin. 1999. Effect of salt and temperature stresses on survival and infectivity of *Heterorhabditis* spp. infective juveniles. Nematology 1: 69 – 78.

Fischer-Le Saux, M., V. Viallard, B. Brunel, P. Normand, and N. E. Boemare. 1999. Polyphasic classification of the genus *Photorhabdus* and proposal of new taxa: *P. luminescens* subsp. *luminescens* subsp. nov., *P. luminescens* subsp. *akhurstii* subsp. nov., *P. luminescens* subsp. *laumondii* subsp. nov., *P. temperata* sp. nov., *P. temperata* subsp. *temperata* subsp. nov. and *P. asymbiotica* sp. nov. International Journal of Systematic Bacteriology 49: 1645 – 1656.

Forst, S., and D. Clarke. 2002. Bacteria-nematode symbiosis. Pp. 57 – 77. In R. Gaugler, ed. Entomopathogenic Nematology. Wallingford, UK: CAB International.

Forst, S., B. Dowds, N. Boemare, and E. Stackebrandt. 1997. *Xenorhabdus* and *Photorhabdus* spp.: bugs that kill bugs. Annual Review of Microbiology 51: 47 – 72.

Forst, S., and K. Nealson. 1996. Molecular biology of the symbiotic-pathogenic bacteria *Xenorhabdus* spp. and *Photorhabdus* spp. Microbiological Review 60: 21 – 43.

Friedman, M. J. 1990. Commercial production and development. Pp. 153 – 172. In R. Gaugler and H. K. Kaya, eds. Entomopathogenic Nematodes in Biological Control. Boca Raton, FL: CRC Press.

Gaugler, R. 1987. Entomogenous nematodes and their prospects for genetic improvement. Pp. 457 – 484. In K. Maramorosch, ed. Biotechnology in Invertebrate Pathology and Cell Culture. New York: Academic Press.

Gaugler, R. 1993. Ecological genetics of entomopathogenic nematodes. Pp. 89 – 95. In R.

Bedding, R. Akhurst, and H. Kaya, eds. Nematodes and the Biological Control of Insects. East Melbourne, Australia: CSRIO Publications.

Gaugler, R. ed. 2002. Entomopathogenic Nematology. Wallingford, UK: CAB International.

Gaugler, R., A. Bednarek, and J. F. Campbell. 1992a. Ultraviolet inactivation of heterorhabditid and steinernematid nematodes. Journal of Invertebrate Pathology 59: 155 - 160.

Gaugler, R., J. F. Campbell, S. Selvan, and E. E. Lewis. 1992b. Large-scale inoculative releases of the entomopathogen *Steinernema glaseri*: assessment 50 years later. Biological Control 2: 181 - 187.

Gaugler, R., P. Grewal, H. K. Kaya, and D. Smith-Fiola. 2000. Quality assessment of commercially produced entomopathogenic nematodes. Biological Control 17: 100 - 109.

Gaugler, R., and H. K. Kaya, eds. 1990. Entomopathogenic Nematodes in Biological Control. Boca Raton, FL: CRC Press.

Gaugler, R., E. Lewis, and R. J. Stuart. 1997. Ecology in the service of biological control: the case of entomopathogenic nematodes. Oecologia 109: 483 - 489.

Gaugler, R., Y. Wang, and J. F. Campbell. 1994. Aggressive and evasive behaviors in *Popillia japonica* (Coleoptera: Scarabaeidae) larvae: defenses against entomopathogenic nematode attack. Journal of Invertebrate Pathology 64: 193 - 199.

Geden, C. J. 1997. Evaluation of *Paraiotonchium muscadomesticae* (Nematoda: Tylenchida: Iotonchiidae), a potential biological control agent of the house fly (Diptera: Muscidae). Biological Control 10: 42 - 47.

Georgis, R., D. B. Dunlop, and P. S. Grewal. 1995. Formulation of entomopathogenic nematodes. Pp. 197 - 205. In F. R. Hall and J. W. Barry, eds. Biorational Pest Control Agents: Formulation and Delivery. Washington, DC: ACS Symposium Series No. 595, American Chemical Society.

Georgis, R., and H. K. Kaya. 1998. Formulation of entomopathogenic nematodes. Pp. 289 - 308. In H. D. Burges, ed. Formulation of Microbial Biopesticides: Beneficial Microorganisms, Nematodes and Seed Treatments. Dordrecht, The Netherlands: Kluwer.

Georgis, R., H. K. Kaya, and R. Gaugler. 1991. Effect of steinernematid and heterorhabditid nematodes (Rhabditida: Steinernematidae and Heterorhabditidae) on nontarget arthropods. Environmental Entomology 20: 815 - 822.

Georgis, R., and S. A. Manweiler. 1994. Entomopathogenic nematodes: a developing biological control technology. Agricultural Zoology Reviews 6: 63 - 94.

Glazer, I. 1996. Survival mechanisms of entomopathogenic nematodes. Biocontrol Science and Technology 6: 373 - 378.

Glazer, I., M. Klein, A. Navon, and Y. Nakache. 1992. Comparison of efficacy of entomopathogenic nematodes combined with antidesiccants applied by canopy sprays against 3 cotton pest (Lepidoptera: Noctuidae). Journal of Economic Entomology 85: 1636 - 1641.

Glazer, I., L. Salame, and D. Segal. 1997. Genetic enhancement of nematicide resistance in entomopathogenic nematodes. Biocontrol Science and Technology 6: 499 - 512.

Grewal, P. S. 2000. Anhydrobiotic potential and long-term storage of entomopathogenic nematodes (Rhabditida: Steinernematidae). International Journal for Parasitology 30:

995 – 1000.
Grewal, P. S., R. Gaugler, H. K. Kaya, and M. Wusaty. 1993. Infectivity of the entomopathogenic nematode *Steinernema scapterisci* (Nematoda: Steinernematidae). Journal of Invertebrate Pathology 62: 22 – 28.
Grewal, P. S., R. Gaugler, and Y. Wang. 1996. Enhanced cold tolerance of the entomopathogenic nematode *Steinernema feltiae* through genetic selection. Annals of Applied Biology 129: 335 – 341.
Grewal, P. S., and R. Georgis. 1998. Entomopathogenic nematodes. Pp. 271 – 299. In F. R. Hall and J. Menn, eds. Methods in Biotechnology, Vol. 5: Biopesticides: Use and Delivery. Totowa, NJ: Humana Press.
Grewal, P. S., E. E. Lewis, and S. Venkatchari. 1999. Allelopathy: A possible mechanism of suppression of plant-parasitic nematodes by entomopathogenic nematodes. Nematology 1: 735 – 743.
Grewal, P. S., W. R. Martin, R. W. Miller, and E. E. Lewis. 1997. Suppression of plant-parasitic nematode populations in turfgrass by application of entomopathogenic nematodes. Biocontrol Science and Technology 7: 393 – 399.
Grewal, P. S., S. Selvan, and R. Gaugler. 1994. Thermal adaptation of entomopathogenic nematodes: Niche breadth for infection, establishment, and reproduction. Journal of Thermal Biology 19: 245 – 253.
Griffin, C. T. 1993. Temperature responses of entomopathogenic nematodes: implications for the success of biological control programmes. Pp. 115 – 126. In R. Bedding, R. Akhurst, and H. Kaya, eds. Nematodes and the Biological Control of Insects. East Melbourne, Australia: CSRIO Publications.
Griffin, C. T. 1996. Effects of prior storage conditions on the infectivity of *Heterorhabditis* sp. (Nematoda: Heterorhabditidae). Fundamental and Applied Nematology 19: 95 – 102.
Griffin, C. T., and M. J. Downes. 1994. Recognition of low temperature active isolates of the entomopathogenic nematode *Heterorhabditis* spp. (Rhabditida: Heterorhabditidae). Nematologica 40: 106 – 115.
Griffin, C. T., M. M. Finnegan, and M. J. Downes. 1994. Environmental tolerances and the dispersal of *Heterorhabditis*: Survival and infectivity of European *Heterorhabditis* following prolonged immersion in seawater. Fundamental and Applied Nematology 17: 415 – 421.
Griffin, C. T., K. M. O'Callaghan, and I. Dix. 2001. A self-fertile species of *Steinernema* from Indonesia: Further evidence of convergent evolution amongst entomopathogenic nematodes? Parasitology 122: 181 – 186.
Hashmi, G., and R. Gaugler. 1998. Genetic diversity in insect-parasitic nematodes (Rhabditida: Heterorhabditidae). Journal of Invertebrate Pathology 72: 185 – 189.
Hashmi, S., G. Hashmi, and R. Gaugler. 1995a. Genetic transformation of an entomopathogenic nematode by microinjection. Journal of Invertebrate Pathology 66: 293 – 296.
Hashmi, S., G. Hashmi, I. Glazer, and R. Gaugler. 1998. Thermal response of *Heterorhabditis bacteriophora* transformed with the *Caenorhabditis elegans hsp*70 encoding gene. Journal of Experimental Zoology 281: 164 – 170.
Hashmi, S., P. Ling, G. Hashmi, M. Reed, R. Gaugler, and W. Trimmer. 1995b.

Genetic transformation of nematodes using arrays of micromechanical piercing structures. Biotechnique 19: 166 – 770.

Haugen, D. A. 1990. Control procedures for *Sirex noctilio* in the Green Triangle: Review from detection to severe outbreak (1977 – 1987). Australian Forestry 53: 24 – 32.

Haugen, D. A., and M. G. Underdown. 1990. *Sirex noctilio* control program in response to the 1987 Green Triangle outbreak. Australian Forestry 53: 33 – 40.

Hazir, S., S. P. Stock, and N. Keskin. 2003. A new entomopathogenic nematode, *Steinernema anatoliense* n. sp. (Rhabditida: Steinernematidae), from Turkey. Systematic Parasitology 55: 211 – 220.

Hominick, W. M. 2002. Biogeography. Pp. 115 – 143. In R. Gaugler, ed. Entomopathogenic Nematology. Wallingford, UK: CABI Publishing.

Hominick, W. M., B. R. Briscoe, F. G. del Pino, J. Heng, D. J. Hunt, E. Kozodoy, Z. Mráček, K. B. Nguyen, A. P. Reid, S. Spiridonov, S. P. Stock, D. Sturhan, C. Waturu, and M. Yoshida. 1997. Biosystematics of entomopathogenic nematodes: Current status, protocols and definitions. Journal of Helminthology 71: 271 – 298.

Hominick, W. M., and A. P. Reid. 1990. Perspectives on entomopathogenic nematology. Pp. 327 – 345. In R. Gaugler and H. K. Kaya, eds. Entomopathogenic Nematodes in Biological Control. Boca Raton, FL: CRC Press.

Hominick, W. M., A. P. Reid, D. A. Bohan, and B. R. Briscoe. 1996. Entomopathogenic nematodes: Biodiversity, geographical distribution and the convention on biological diversity. Biocontrol Science and Technology 6: 317 – 331.

Hominick, W. M., and G. A. Tingley. 1984. Mermithid nematodes and the control of insect vectors of human disease. Biocontrol News and Information 5: 7 – 20.

Hu, K., J. Li, and J. M. Webster. 1999. Nematicidal metabolites produced by *Photorhabdus luminescens* (Enterobacteriaceae), bacterial symbiont of entomopathogenic nematodes. Nematology 1: 457 – 469.

Ishibashi, N. 1993. Integrated control of insect pests by *Steinernema carpocapsae*. Pp. 105 – 113. In R. Bedding, R. Akhurst, and H. Kaya, eds. Nematodes and the Biological Control of Insects. East Melbourne, Australia: CSRIO Publications.

Ishibashi, N., and E. Kondo. 1987. *Steinernema feltiae* (DD – 136) and *S. glaseri*: Persistence in soil and bark compost and their influence on native nematodes. Journal of Nematology 18: 310 – 316.

Jackson, T. J., H. Wang, M. T. Nugent, C. T. Griffin, A. M. Burnell, and B. C. A. Dowds. 1995. Isolation of insect pathogenic bacteria, *Providencia rettgeri*, from *Heterorhabditis* spp. Journal of Applied Bacteriology 78: 237 – 244.

Kaiser, H. 1991. Terrestrial and semiterrestrial Mermithidae. Pp. 899 – 965. In W. R. Nickle, ed. Manual of Agricultural Nematology. New York: Marcel Dekker.

Kaya, H. K. 1990. Soil ecology. Pp. 93 – 115. In R. Gaugler and H. K. Kaya, eds. Entomopathogenic Nematodes in Biological Control. Boca Raton, FL: CRC Press.

Kaya, H. K. 2002. Natural enemies and other antagonists. Pp. 189 – 203. In R. Gaugler, ed. Entomopathogenic Nematology. Wallingford, UK: CAB International.

Kaya, H. K., T. M. Burlando, H. Y. Choo, and G. S. Thurston. 1995. Integration of entomopathogenic nematodes with *Bacillus thuringiensis* or pesticidal soap for control of insect pests. Biological Control 5: 432 – 441.

Kaya, H. K., and R. Gaugler. 1993. Entomopathogenic nematodes. Annual Review of

Entomology 38: 181 – 206.

Kaya, H. K., and A. M. Koppenhöfer. 1996. Effects of microbial and other antagonistic organism and competition on entomopathogenic nematodes. Biocontrol Science and Technology 6: 333 – 345.

Kaya, H. K., and R. D. Moon. 1978. The nematode *Heterotylenchus autumnalis* and face fly, *Musca autumnalis*: A field study in northern California. Journal of Nematology 10: 333 – 341.

Kaya, H. K., R. D. Moon, and P. L. Witt. 1979. Influence of the nematode, *Heterotylenchus autumnalis*, on the behavior of the face fly, *Musca autumnalis*. Environmental Entomology 8: 537 – 540.

Kaya, H. K., and S. P. Stock. 1997. Techniques in insect nematology. Pp. 281 – 324. In L. Lacey, ed. Manual of Techniques in Insect Pathology. San Diego, CA: Academic Press.

Kerwin, J. L., and R. K. Washino. 1985. Recycling of *Romanomermis culicivorax* (Mermithidae: Nematoda) in rice fields in California, USA. Journal of Medical Entomology 22: 637 – 643.

Klein, M. G. 1990. Efficacy against soil-inhabiting insect pests. Pp. 195 – 214. In R. Gaugler and H. K. Kaya, ed. Entomopathogenic Nematodes in Biological Control. Boca Raton, FL: CRC Press.

Klein, M. G. 1993. Biological control of scarabs with entomopathogenic nematodes. Pp. 49 – 57. In R. Bedding, R. Akhurst and H. Kaya, eds. Nematodes and the Biological Control of Insects. East Melbourne, Australia: CSRIO Publications.

Koppenhöfer, A. M. 2000. Nematodes. Pp. 283 – 301. In L. A. Lacey and H. K. Kaya, eds. Field Manual of Techniques in Invertebrate Pathology. Dordrecht, The Netherlands: Kluwer.

Koppenhöfer, A. M., M. E. Baur, S. P. Stock, H. Y. Choo, B. Chinnasri, and H. K. Kaya. 1997. Survival of entomopathogenic nematodes within host cadavers in dry soil. Applied Soil Ecology 6: 231 – 240.

Koppenhöfer, A. M., H. Y. Choo, H. K. Kaya, D. W. Lee, and W. Gelernter. 1999. Improved field and greenhouse efficacy with a combination of entomopathogenic nematode and *Bacillus thuringiensis* against scarab grubs. Biological Control 14: 37 – 44.

Koppenhöfer, A. M., I. M. Brown, R. Gaugler, P. S. Grewal, H. K. Kaya, and M. G. Klein. 2000. Synergism of imidacloprid and entomopathogenic nematodes: greenhouse and field evaluation. Biological Control 19: 245 – 251.

Koppenhöfer, A. M., H. K. Kaya, and S. Taormino. 1995. Infectivity of entomopathogenic nematodes (Rhabditida: Steinernematidae) at different soil depths and moistures. Journal of Invertebrate Pathology 65: 193 – 199.

Krafsur, E. S., C. J. Church, M. K. Elvin, and C. M. Ernst. 1983. Epizootiology of *Heterotylenchus autumnalis* (Nematoda) among face flies (Diptera: Muscidae) in central Iowa, USA. Journal of Medical Entomology 3: 318 – 324.

Kung, S. P., R. Gaugler, and H. K. Kaya. 1991. Effects of soil temperature, moisture, and relative humidity on entomopathogenic nematode persistence. Journal of Invertebrate Pathology 57: 242 – 249.

Lacey, L. A., and H. K. Kaya, eds. 2000. Field manual of techniques in invertebrate pathology. Dordrecht, The Netherlands: Kluwer.

Levins, R. 1970. Extinction. Pp. 77 – 107. In M. Gerstenhaber, ed. Lectures on Mathematics in the Life Sciences, Vol. 2. Providence, RI: American Mathematical Society.

Lewis, E. E. 2002. Behavioural ecology. Pp. 205 – 223. In R. Gaugler, ed. Entomopathogenic Nematology. Wallingford, UK: CAB International.

Lewis, E. E., J. F. Campbell, and R. Gaugler. 1998. A conservation approach to using entomopathogenic nematodes in turf and landscapes. Pp. 235 – 254. In P. Barbosa, ed. Conservation Biological Control. San Diego, CA: Academic Press.

Liu, J., G. O. Poinar, Jr., and R. E. Berry. 2000. Control of insect pests with entomopathogenic nematodes: the impact of molecular biology and phylogenetic reconstruction. Annual Review of Entomology 45: 287 – 306.

Lysenko, O., and J. Weiser. 1974. Bacteria associated with the nematode *Neoaplectana carpocapsae* and the pathogenicity of this complex for *Galleria mellonella* larvae. Journal of Invertebrate Pathology 24: 332 – 336.

Mamiya, Y. 1989. Comparison of infectivity of *Steinernema kushidai* (Nematoda: Steinernematidae) and other steinernematid and heterorhabditid nematodes for three different insects. Applied Entomology and Zoology 24: 302 – 308.

Mrácek, Z., D. Sturhan, and A. Reid. 2003. *Steinernema weiseri* n. sp. (Rhabditida: Steinernematidae), a new entomopathogenic nematode from Europe. Systematic Parasitology 56: 37 – 47.

Nappi, A. J. 1973. Effects of parasitization by the nematode, *Heterotylenchus autumnalis*, on mating and oviposition in the host, *Musca autumnalis*. Journal of Parasitology 59: 963 – 969.

Nickle, W. R., and M. Shapiro. 1994. Effects of eight brighteners as solar radiation protectants for *Steinernema carpocapsae*, All strain. Supplement Journal of Nematology 26: 782 – 784.

Nickle, W. R., and H. E. Welch. 1984. History, development, and importance of insect nematology. Pp. 627 – 653. In W. R. Nickle, ed. Plant and Insect Nematodes. New York: Marcel Dekker.

Nishimatsu, T., and J. J. Jackson. 1998. Interaction of insecticides, entomopathogenic nematodes, and larvae of the western corn rootworm (Coleoptera: Chrysomelidae). Journal of Economic Entomology 91: 410 – 418.

Nishimura, Y., A. Hagiwara, T. Suzuki, and S. Yamanaka. 1994. *Xenorhabdus japonicus* sp. nov. associated with the nematode *Steinernema kushidai*. World Journal of Microbiology and Biotechnology 10: 207 – 210.

O'Leary, S. A., and A. M. Burnell. 1997. The isolation of mutants of *Heterorhabditis megidis* (Strain UK211) with increased desiccation tolerance. Fundamental and Applied Nematology 20: 197 – 205.

Parkman, J. P., and G. C. Smart, Jr. 1996. Entomopathogenic nematodes, a case study: introduction of *Steinernema scapterisci* in Florida. Biocontrol Science and Technology 6: 413 – 419.

Patel, M. N., R. N. Perry, and D. J. Wright. 1997. Desiccation survival and water contents of entomopathogenic nematode *Steinernema* spp. (Rhabditida: Steinernematidae). International Journal for Parasitology 27: 61 – 70.

Patel, M. N., and D. J. Wright. 1996. The influence of neuroactive pesticides on the behaviour of entomopathogenic nematodes. Journal of Helminthology 70: 53 – 61.

Peng, Y., J. Song, G. Tian, Q. Xue, F. Ge, J. Yang, and Q. Shi. 1998. Field evaluations of *Romanomermis yunanensis* (Nematoda: Mermithidae) for control of Culicinae mosquitoes in China. Fundamental and Applied Nematology 21: 227 – 232.

Perry, R. N., W. M. Hominick, J. Beane, and B. Briscoe. 1998. Effect of the entomopathogenic nematodes, *Steinernema feltiae* and *S. carpocapsae*, on the potato cyst nematode, *Globodera rostochiensis*, in pot trials. Biocontrol Science and Technology 8: 175 – 180.

Peters, A. 1996. The natural host range of *Steinernema* and *Heterorhabditis* spp. and their impact on insect populations. Biocontrol Science and Technology 6: 389 – 402.

Peters, A., and R.-U. Ehlers. 1994. Susceptibility of leatherjackets (*Tipula paludosa* and *T. oleracea*; Tipulidae: Nematocera) to the entomopathogenic nematode *Steinernema feltiae*. Journal of Invertebrate Pathology 63: 163 – 171.

Peters, A., D. H. Gouge, R.-U. Ehlers, and N. G. M. Hague. 1997. Avoidance of encapsulation by *Heterorhabditis* spp. infecting larvae of *Tipula oleracea*. Journal of Invertebrate Pathology 70: 161 – 164.

Petersen, J. J. 1984. Nematode parasitism of mosquitoes. Pp. 797 – 820. In W. R. Nickle, ed. Plant and Insect Nematodes. New York: Marcel Dekker.

Petersen, J. J. 1985. Nematodes as biological control agents: Part I. Mermithidae. Advances in Parasitology 24: 307 – 344.

Petersen, J. J., and O. R. Willis. 1975. Establishment and recycling of a mermithid nematode for the control of mosquito larvae. Mosquito News 35: 526 – 532.

Poinar, G. O., Jr. 1979. Nematodes for Biological Control of Insects. Boca Raton, FL: CRC Press.

Popiel, I., and W. M. Hominick. 1992. Nematodes as biological control agents: Part II. Advances in Parasitology 31: 381 – 433.

Remillet, M., and C. Laumond. 1991. Sphaerularioid nematodes of importance in agriculture. Pp. 967 – 1024. In W. R. Nickle, ed. Manual of Agricultural Nematology. New York: Marcel Dekker.

Renn, N. 1998. The efficacy of entomopathogenic nematodes for controlling housefly infestations of intensive pig units. Medical and Veterinary Entomology 12: 46 – 51.

Richardson, P. N. 1996. British and European legislation regulating rhabditid nematodes. Biocontrol Science and Technology 6: 449 – 463.

Rizvi, S. A., R. Hennessey, and D. Knott. 1996. Legislation on the introduction of exotic nematodes in the US. Biocontrol Science and Technology 6: 477 – 480.

Rosenheim, J. A., H. K. Kaya, L. E. Ehler, J. J. Marois, and B. A. Jaffee. 1995. Intraguild predation among biological-control agents: theory and practice. Biological Control 5: 303 – 335.

Roush, R. T. 1990. Genetic variation in natural enemies: critical issues for colonization in biological control. Pp. 263 – 288. In M. Mackauer, L. E. Ehler, and J. Roland, eds. Critical Issues in Biological Control. Andover, UK: Intercept.

Rovesti, L., and K. V. Deseö. 1990. Compatibility of chemical pesticides with the entomopathogenic nematodes, *Steinernema carpocapsae* Weiser and *S. feltiae* Filipjev (Nematoda: Steinernematidae). Nematologica 36: 237 – 245.

Schroeder, W. J., and P. J. Sieburth. 1997. Impact of surfactants on control of the root weevil *Diaprepes abbreviatus* larvae with *Steinernema riobravis*. Journal of Nematology 29: 216 – 219.

Shapiro, D. I., I. Glazer, and D. Segal. 1997a. Genetic improvement of heat tolerance in *Heterorhabditis bacteriophora* through hybridization. Biological Control 8: 153 – 159.

Shapiro, D. I., G. L. Tylka, and L. C. Lewis. 1997b. Effects of fertilizers on virulence of *Steinernema carpocapsae*. Applied Soil Ecology 3: 27 – 34.

Sher, R. B., M. P. Parrella, and H. K. Kaya. 2000. Biological control of the leafminer *Liriomyza trifolii* (Burgess): Implications for intraguild predation between *Diglyphus begini* Ashmead and *Steinernema carpocapsae* Weiser. Biological Control 17: 155 – 163.

Simões, N., and J. S. Rosa. 1996. Pathogenicity and host specificity of entomopathogenic nematodes. Biocontrol Science and Technology 6: 403 – 411.

Smith, K. A., R. Miller, and D. H. Simser. 1992. Entomopathogenic nematode bibliography – heterorhabditid and steinernematid nematodes. Southern Cooperative Service Bulletin 370. Arkansas Agricultural Experiment Station: Fayetteville.

Smits, P. 1996. Post-application persistence of entomopathogenic nematodes. Biocontrol Science and Technology 6: 379 – 387.

Solomon, A., I. Paperna, and I. Glazer. 1999. Desiccation survival of the entomopathogenic nematode *Steinernema feltiae*: Induction of anhydrobiosis. Nematology 1: 61 – 68.

Song, J., Y. Song, Y. Peng, and E. G. Platzer. 2001. Cryopreservation and low temperature storage of the pre-parasitic juveniles of *Romanomermis culicivorax*. Nematology 3: 417 – 421.

Stark, J. D. 1996. Entomopathogenic nematodes (Rhabditida: Steinernematidae): Toxicity of neem. Journal of Economic Entomology 89: 68 – 73.

Stock, S. P., and A. M. Koppenhöfer. 2003. *Steinernema scarabaei* n. sp. (Rhabditida: Steinernematidae), a natural pathogen of scarab beetle (Coleoptera: Scarabaeidae) from New Jersey. Nematology 5: 191 – 204.

Stock, S. P., and A. P. Reid. 2003. Biosystematics of entomopathogenic nematodes (Steinernematidae, Heterorhabditidae): Current status and future directions. In R. Cook and D. J. Hunt, eds. Proceedings of the Fourth International Congress of Nematology, 8 – 13 June 2002, Tenerife, Spain. Nematology (Monographs and Perspectives) 2: 435 – 446.

Stoffolano, J. G., Jr. 1967. The synchronization of the life cycle of diapausing face flies, *Musca autumnalis*, and of the nematode, *Heterotylenchus autumnalis*. Journal of Invertebrate Pathology 9: 395 – 397.

Stoffolano, J. G., Jr. 1968. Distribution of the nematode *Heterotylenchus autumnalis*, a parasite of the face fly, in New England with notes of its origin. Journal of Economic Entomology 61: 861 – 863.

Strauch, O., and R.-U. Ehlers. 1998. Food signal production of *Photorhabdus luminescens* inducing the recovery of entomopathogenic nematodes *Heterorhabditis* spp. in liquid culture. Applied Microbiology and Biotechnology 50: 369 – 374.

Strong, D. R., H. K. Kaya, A. V. Whipple, A. L. Child, S. Kraig, M. Bondonno, and J. L. Maron. 1996. Entomopathogenic nematodes: natural enemies of root-feeding caterpillars on bush lupine. Oecologia 108: 167 – 173.

Stuart, R. J., and R. Gaugler. 1994. Patchiness in populations of entomopathogenic nematodes. Journal of Invertebrate Pathology 64: 39 – 45.

Stuart, R. J., and R. Gaugler. 1996. Genetic adaptation and founder effect in laboratory populations of the entomopathogenic nematode *Steinernema glaseri*. Canadian Journal of Zoology 74: 164 – 170.

Talbot, P. H. B. 1977. The *Sirex-Amylostereum-Pinus* association. Annual Review of Phytopathology 15: 41 – 54.

Tanada, Y., and H. K. Kaya. 1993. Insect pathology. San Diego, CA: Academic Press.

Thurston, G. S., H. K. Kaya, and R. Gaugler. 1994a. Characterizing the enhanced susceptibility of milky disease-infected scarabaeid grubs to entomopathogenic nematodes. Biological Control 4: 67 – 73.

Thurston, G. S., Y. Ni, and H. K. Kaya. 1994b. Influence of salinity on survival and infectivity of entomopathogenic nematodes. Journal of Nematology 26: 345 – 351.

Tomalak, M. 1994. Selective breeding of *Steinernema feltiae* (Filipjev) (Nematoda: Steinernematidae) for improved efficacy in control of a mushroom fly, *Lycoriella solani* Winnertz (Diptera: Sciaridae). Biocontrol Science and Technology 4: 187 – 198.

Treverrow, N. L., and R. A. Bedding. 1993. Development of a system for the control of the banana weevil borer, *Cosmopolites sordidus* with entomopathogenic nematodes. Pp. 41 – 47. In R. Bedding, R. Akhurst, and H. Kaya, eds. Nematodes and the Biological Control of Insects. East Melbourne, Australia: CSRIO Publications.

Volgyi, A., A. Fodor, A. Szentirmai, and S. Forst. 1998. Phase variation in *Xenorhabdus nematophilus*. Applied Environmental Microbiology 64: 1188 – 1193.

Wang, Y., J. F. Campbell, and R. Gaugler. 1995. Infection of entomopathogenic nematodes *Steinernema glaseri* and *Heterorhabditis bacteriophora* against *Popillia japonica* (Coleoptera: Scarabaeidae) larvae. Journal of Invertebrate Pathology 66: 178 – 184.

Wang, Y., and R. Gaugler. 1998a. Host and penetration site location by entomopathogenic nematodes against Japanese beetle larvae. Journal of Invertebrate Pathology 72: 313 – 318.

Wang, Y., and R. Gaugler. 1998b. *Steinernema glaseri* surface coat protein suppresses the immune response of *Popillia japonica* (Coleoptera: Scarabaeidae) larvae. Biological Control 14: 45 – 50.

Wilson, M. J., and R. Gaugler. 2000. Terrestrial mollusc pests. Pp. 787 – 804. In L. A. Lacey and H. K. Kaya, eds. Field Manual of Techniques in Invertebrate Pathology. Dordercht, The Netherlands: Kluwer.

Wilson, M. J., D. M. Glen, and S. K. George. 1993a. The rhabditid nematode Phasmarhabditis hermaphrodita as a potential biological control agent for slugs. Biocontrol Science and Technology 3: 503 – 511.

Wilson, M. J., D. M. Glen, S. K. George, and R. C. Butler. 1993b. Mass cultivation and storage of the rhabditid nematode Phasmarhabditis hermaphrodita, a biocontrol agent for slugs. Biocontrol Science and Technology 3: 513 – 521.

Wilson, M. J., D. M. Glen, S. K. George, and L. A. Hughes. 1995a. Biocontrol of slugs in protected lettuce using the rhabditid nematode Phasmarhabditis hermaphrodita. Biocontrol Science and Technology 5: 233 – 242.

Wilson, M. J., D. M. Glen, S. K. George, J. D. Pearce, and C. W. Wiltshire. 1994. Biological control of slugs in winter wheat using the rhabditid nematode, Phasmarhabditis hermaphrodita. Annals of Applied Biology 125: 377 – 390.

Wilson, M. J., D. M. Glen, J. D. Pearce, and P. B. Rodgers. 1995c. Monoxenic culture of the slug parasite Phasmarhabditis hermaphrodita with different bacteria in solid and liquid phase. Fundamental and Applied Nematology 18: 159 – 166.

Wilson, M. J., L. A. Hughes, and D. M. Glen. 1995b. Developing strategies for the nematode, Phasmarhabditis hermaphrodita, as a biocontrol agent for slugs in integrated crop management systems. Pp. 33 – 40. In R. G. McKinlay and D. Atkinson, eds. Integrated Crop Protection: Towards Sustainability. Farnham, UK: British Crop Protection Council.

Wilson, M. J., L. A. Hughes, G. M. Hamacher, L. D. Barahona, and D. M. Glen. 1996. Effects of soil incorporation on the efficacy of the rhabditid nematode Phasmarhabditis hermaphrodita as a biological control agent for slugs. Annals of Applied Biology 128: 117 – 126.

Wilson, M., W. Xin, S. Hashmi, and R. Gaugler. 1999. Risk assessment and fitness of a transgenic entomopathogenic nematode. Biological Control 15: 81 – 87.

Womersley, C. Z. 1990. Dehydration survival and anhydrobiotic potential. Pp. 117 – 137. In R. Gaugler and H. K. Kaya, eds. Entomopathogenic Nematodes in Biological Control. Boca Baton, FL: CRC Press.

Womersley, C. Z. 1993. Factors affecting physiological fitness and modes of survival employed by dauer juveniles and their relationship to pathogenicity. Pp. 79 – 88. In R. Bedding, R. Akhurst, and H. Kaya, eds. Nematodes and the Biological Control of Insect Pests. East Melbourne, Australia: CSIRO Publications.

Wright, D. J., S. Piggott, and M. N. Patel. 1998. Physiological aspects of survival by entomopathogenic nematodes. Proceedings of the VIIth International Colloquium in Invertebrate Pathology and Microbial Control, Sapporo, Japan, Aug 23 – 28, 1998. Pp. 20 – 23.

Yan, S., H. Qian, K. Xia, and L. Ling. 1986. Life cycle of *Amphimermis* sp., a nematode parasite of the brown planthopper in China. Acta Phytophylacica Sinica 13: 79 – 84. (In Chinese).

Yang, H., G. Zhang, S. Zhang, and J. Heng. 1993. Biological control of tree borers (Lepidoptera: Cossidae) in China with the nematode *Steinernema carpocapsae*. Pp. 33 – 40. In R. Bedding, R. Akhurst, and H. Kaya, eds. Nematodes and the Biological Control of Insects. East Melbourne, Australia: CSRIO Publications.

Zondag, R. 1969. A nematode infection of *Sirex noctilio* (F.) in New Zealand. New Zealand Journal of Science 12: 732 – 747.

Zhou, X., H. K. Kaya, K. Heungens, and H. Goodrich-Blair. 2002. Response of ants to a deterrent factor(s) produced by the symbiotic bacteria of entomopathogenic nematodes. Applied and Environmental Microbiology 68: 6202 – 6209.

23 Cost-benefits of Nematode Management through Regulatory Programs

Paul S. Lehman

23.1 Introduction

Damage caused by nematodes may be overlooked by growers because crop losses are small, but the additive cost over time is substantial. Usually, when nematodes cause serious crop losses, growers have different options to minimize the damage caused by these pests. Among the control measures, chemical approaches are the most popular among farmers because of impressive yield increase following their application. Among non-chemical approaches, cropping systems, cultural practices, organic amendments, plant resistance, and solarization are the most common methods recommended for nematode management. Pest exclusion strategies such as, regulatory certification, sanitation procedures, and quarantine are often overlooked as cost-effective methods for nematode management, although it is well documented in the literature that management and eradication of nematode pests accidentally introduced into new environments are extremely difficult and expensive (Koenning, et al., 1999; Lee, et al., 1999; Poucher, et al., 1967).

The history, cost-benefits, and future opportunities and challenges for nematode management through regulatory programs are discussed in this chapter.

23.2 History and Development of Regulatory Management Programs

From the earliest times when humans began cultivating plants, seeds and vegetative plant propagative materials were moved to new regions of production. For most crops, for thousands of years new regions of production expanded slowly around the centers of origin, where pests and diseases coevolved with the crops and other plants in natural ecosystems. For a few crops, most of the world's production remains near the region where cultivation began. For example, olive pits dug from Levantine Soils suggest that olive cultivation began in the eastern Mediterranean 6000 years ago, and today 99 percent of the world's olive oil is produced around the rim of the Mediterranean (Zwingle, 1999). During the past 500 years however, the production of many crops expanded worldwide coupled with exploration and the expansion of travel and trade. Inadvertently, hitchhiking pests were introduced to new areas of production aided by explorers, governments, agricultural industries, and scientists. This is especially true for vegetatively propagated crops. As an example, nematodes and other pests were carried with banana plants when they were introduced to different geographical regions throughout the world. From the 7th to the 10th century, traders carried banana cultivars to Madagascar from the Indo-Malay region in Southeast Asia, where cultivation probably began millennia earlier. From there bananas were spread to southern and western Africa, and eventually Portuguese explorers in the late 1400s carried them from Guinea to the Canary Islands. In the early 1500s, Spanish missionaries carried them from the Canary Islands to the Dominican Republic, and from the 16th to the 19th centuries explorers, traders and settlers carried bananas all over tropical America (Marin, et al., 1998). In the 19th century when the banana became established in tropical America, germplasm from the Gros Michel and Dwarf Cavendish group was introduced into the region from Southeast Asia. It is likely that burrowing nematode, *Radopholus similis* (Cobb) Thorne, hitchhiked on an introduction of Gros Michel to Jamaica from Southeast Asia. The eminent nematologist N. A. Cobb first studied the burrowing nematode in Fiji during his visit in 1890 to 1891, and it is probable that the population he observed there had nearly completed a trip around the world a few years earlier when introductions of the Gros Michel group originally from Asia were sent from Jamaica to Fiji. Although this link is not mentioned in the taxonomic description of *R. similis*, the nematode origin was probably recognized by Cobb, because he received infected Gross Michel corms from Jamaica to complete his description of this nematode in 1915 (Cobb, 1915). Burrowing nematodes, however, were of little concern to the banana industry because the Gros Michel group is moderately resistant to this nematode. But in the 1950s, a mutant of this Gros Michel group, known as Cocos, which is very susceptible to this nematode, was distributed by the United Fruit Company to their plantations in at least seven countries in Central America, the Caribbean, and South America. Unfortunately the seed beds in Panama, from which new Cocos germplasm was distributed, were infested with burrowing nematode, thus virtually guaranteeing that a serious problem would erupt throughout the region (Marin, et al., 1998).

A similar saga to that of the burrowing nematode problem on banana could be told for many crops and the spread of pests around the world. As Europeans began colonizing North America they brought with them the food crops that were a part of their cultural heritage. Of the major food crops grown in the USA today, only a few, e. g., blueberries, cranberries, strawberries, pecans, raspberries, and some grapes, are native (Foster, 1991). Even the corn that Native Americans were cultivating, had been introduced earlier from Native Americans migrating from Mexico. On his second voyage Columbus started the trend to bring new plants to the new world. In the late 1700s, American diplomats were ordered to send home rare plant material (Foster, 1991). By the late 1800s, plant exploration was actively pursued by the government. In 1862, the United States Department of Agriculture (USDA) was founded, and one of its objectives was to introduce new plants to America. An office of plant and seed introductions was established within the USDA about 1887, and was directed by David Fairchild (Fairchild, 1906). It was estimated that 50,000 kinds of plants and seeds had been introduced by the USDA up to 1922 (Crawford, 1922). In an article published in Scientific American in 1922, Crawford described how scientists "combed the world" for new crops for America and reflected the mood of the times when he wrote: "there are few things more romantic than the work of an agricultural explorer." Among the explorers Crawford revered were Barbour Lathrop and David Fairchild, who made a three-year agricultural exploration, visiting every continent and one-half of the countries of the world; H. L. Shantz, who traveled from Capetown to Egypt, "exploring the wild jungle-country in the upper Congo and journeyed into the wild-animal country of British East Africa" and brought back 1600 specimens of African plants which were tried out in America; and Frank Meyer, who "walked 10,000 miles through the heart of China, Manchuria, Korea, and parts of Tibet and Russian Turkestan looking for plants that might be of value in America." Crawford warned that the life of an agricultural explorer was not always an easy one, citing as an example the experiences of Frank Myers who "was once attacked by ruffians in Harbin and one time stood up against a wall to be shot, but managed to talk himself out of this uncomfortable situation." Myers is credited with introducing alfalfas, almonds, apricots, barleys, bamboos, citrus, chestnuts, dwarf cherries, fruited oleasters, jujubes, peaches, pears, persimmons, strawberry trees, sorghums, wheats, wild apples and hundreds of other plants from this region of the world. Agricultural explorers not only impacted agriculture but also changed the face of the urban landscape, as David Fairchild indicated when he eulogized the work of Meyers:

"His hardy yellow rose peers in on me through my study window, and up in the border his scarlet lily is in bud, while the perfume of his lilac has barely passed away. His white-barked pine is dusting its pollen in the air, his Euonymous and his bamboo are growing in the corners of the house, and his dryland elm with its delicate branches shades the entrance. So much of China has he successfully transformed to this country" (Crawford, 1922).

Unfortunately, new plants were not the only things that agricultural explorers brought to America. New pests were inadvertently introduced and became established along with these early plant introductions. In addition, some pests were introduced accidentally in cargo, baggage, ship stores, domesticated animals, and ship ballast. Undoubtedly, many non-native nematodes were inadvertently introduced with some of these new crops and the contaminated soil that accompanied them. In the late 1800s and early 1900s, scientists in the USA imported soil from China and Japan to insure adequate inoculation of newly imported legumes with their nitrogen-fixing bacteria (Fairchild, 1948). Soils from these fields were then used to inoculate other fields, and recommendations actually were given to farmers as to how much soil they should obtain from other fields and distribute with the seed at planting when a field was planted to a legume for the first time (Hopkins, 1904). Fortunately some scientists recognized that this practice was shortsighted. In 1916, a scientist from a Georgia Experiment Station warned that seed and soil inoculations had serious drawbacks since "troublesome plant maladies, such as cotton wilt, pea wilt, melon wilt, and nematodes" were being distributed (Temple, 1916). It has been speculated that the soybean cyst nematode may have been introduced with soil that was imported to provide nodulating bacteria (Noel, 1992).

Today it is difficult to imagine that 150 years ago there were no government agencies, anywhere in the world, that specifically focused on regulating plant pests. But when one considers that certain fundamental biological concepts emerged only in the last 150 years, it is not surprising that when new plants and crops were imported, adequate precautions were not taken to prevent the importation of these hitchhiking pests. At that time it was not well understood that in their native habitats plants have their own group of endemic pests that coevolve with them, and these pests and pathogens also have their own set of biological or environmental antagonists that limit the damage. Nor was it understood that many times when pests and disease organisms are inadvertently introduced with imported plants without the native fauna and flora that limit damage, the consequences to native plants and crops that originated from other areas can sometimes be devastating. The role of insects in destroying crops had been recognized for thousands of years, but an understanding of the role of microscopic organisms such fungi, bacteria and nematodes as causal agents of plant disease only developed in the mid-1800s. About this time, the devastating epiphytotic of the potato blight in Europe caused by *Phytophthora infestans* (Mont.) De Bary, which was probably introduced from the Andean regions of South America, resulted in the death by famine of one million Irish peasants. The human misery and social upheaval were followed by mass immigration of Europeans to America. A few years later, in 1859, the importance of plant nematodes began to be recognized when Schacht discovered the sugar-beet cyst nematode, *Heterodera schachtii* Schmidt, in Germany.

During this same time period, a series of events began to unfold in the French vineyards that should have left an indelible impression that introduced pests from America might threaten French culture as much as any invading army. The powdery mildew fungus, *Uncinula necator* (Schw.) Burr., probably introduced from North America, reduced French wine production by 80%, but extensive application of sulfur allowed the production to rise again. To select grape cultivars resistant to powdery mildew, American rootstocks were imported, but unfortunately along with a hitchhiking root aphid, *Viteus vitifoliae* (Fitch), that did little damage to grapes in America, but devastated French rootstocks in the 1860s. In response, the French imported resistant American rootstocks, but this time introductions brought the downy mildew fungus, *Plasmopara viticola* (Burk. & Curt.) Berl. & de Toni in Sacc., which in the 1870s became the third and most devastating grape plague. The monetary loss to downy mildew amounted to nearly US$50 billion, with European wine production in the 1880s being far less than it was in the 1840s (Schumann, 1991).

In response to these events, for the first time European governments began enacting legislation to exclude or quarantine plant pests that were known to be causing serious economic losses in other countries. In 1873 Germany enacted legislation to prevent establishment of grape phylloxera and in 1875 prohibited the importation of Irish potatoes from the USA to prevent introduction of the Colorado potato beetle, *Leptinotarsa decemlineata* (Say). In the next few years other European countries enacted similar legislation (Dowling, et al., 1982).

Meanwhile in the USA, growers also were experiencing the negative impact of introduced pests with the introduction of the cottony cushion scale, *Icerya purchasi* Maskell, in 1869 from Australia, and the San Jose scale, *Quadraspidiotus perniciosus* (Comstock), from China in 1879. In response, the state California enacted the first plant quarantine legislation in the USA in 1881 in an attempt to exclude new pests and diseases of grape. In 1885 California passed legislation authorizing the inspection of all incoming plant material (Anonymous, 1933; Foster, 1991). In 1889 a law was enacted in Florida that stated: "It is unlawful for any person to knowingly sell or give away any diseased nursery stock or seeds in the State of Florida. Any person violating this section shall be fined not more than $500 or imprisoned not more than six months."

By 1908, 39 of the 48 states in the USA had enacted quarantine legislation (Dowling, et al., 1982). The federal government, however, was slower and more reluctant to introduce quarantine legislation. But with the introduction of the boll weevil, *Anthonomus g. grandis* Boheman, the white pine blister rust *Cronartium ribicola* J. C. Fisch., and the chestnut blight fungus, *Cryphonectria parasitica* (Murrill) Barr, the pressure mounted for unified national effort to exclude such pests (Foster, 1991). James Wilson, who was U. S. Secretary of Agriculture from 1897 – 1913, supported the concern of some of the Department's scientists that plant introductions carried with them a risk of devastating pest introductions. An effort was made to pass a national plant quarantine law, but Congress failed to pass it (Jefferson and Fusonie, 1977). But at least by 1906, the USDA began to develop its own inspection process for the plants they were importing. This was important because Secretary of Agriculture Wilson also actively supported plant introductions, since one of his primary goals was to work toward the day when the USA would become agriculturally self-sufficient. To further this objective he established the Office of Foreign Seed Plant Introduction, headed by David Fairchild, a plant explorer who later established the renowned Fairchild Botanical Gardens in Miami. And it was Fairchild's efforts to import ornamental flowering cherry trees that would ultimately test the integrity of the Department's inspection process (Jefferson and Fusonie, 1977).

At that time, one of the key projects of the President of the United States' wife, Mrs. William Howard Taft, was to beautify what was then known as the speedway, the several miles open corridor that stretches from the presidential memorials to the capital building. It was David Fairchild who convinced the President's wife that Japan's "Royal Flower", the flowering cherry trees, should grace this area, which has each spring for decades attracted thousands of visitors who enjoy their beauty. Because of Fairchild's earlier contacts with Japanese officials when he imported 30 varieties of cherry trees, the word spread in Japan of the American's project to plant Japan's "Royal Flower". Before long the news arrived through diplomatic channels that the mayor of Tokyo decided to donate 2000 trees to the project. The newspapers in Washington D. C. added to the local excitement by covering this event, which now involved the heads of State of both governments. The trees arrived in

Seattle, Washington, on December 10, 1909 and were sent by train to Washington, D. C. When they arrived in early January, the Department of Agriculture's inspection team, which included the eminent nematologist N. A. Cobb, was sent to examine them. They observed several scales, wood-boring lepidopterus larvae, and root-knot nematodes on the trees. In their report to the Secretary of Agriculture Wilson, the head of the inspection team stated that this shipment of trees had the "worst infestation of insects and root galls he had ever encountered", and courageously recommended that the trees be "destroyed by burning". In final report it is stated that Cobb found that about 72% of the lots were infected with the "root gall worm" and he is quoted as concluding: "I have no hesitation in saying that in a country where a proper inspection of disease material was legally in force with the object of protecting agriculture, the importation of these trees would not be permitted." These recommendations resulted in a series of consultations among highest levels of government including the Secretary of State, Secretary of War, and the President, who was likely well informed since this was his wife's project. In the end, the mayor of Tokyo was diplomatically informed that because of the experiences with destructive foreign pests, the Department of Agriculture had no choice, "but the painful duty of destroying the trees". The following year the Japanese government involved their best scientists in preparing a new pest-free gift of 6000 trees that had been fumigated twice with hydrocyanic acid gas to assure that the embarrassing incident could never be repeated (Jefferson and Fusonie, 1977). After this second shipment arrived the first trees were planted by the Mrs. Taft and the wife of the Japanese Ambassador. In 1923 the mayor of Tokyo and his daughters visited Washington D. C. to enjoy the cherry blossoms, indicating the success of both science and diplomacy 13 years earlier. But in the end it may have been root-knot nematodes and other pests that accompanied the mayor's first gift, that finally helped to convince the US Congress to pass the Plant Quarantine Act of 1912, which is considered the watershed for all subsequent federal quarantine legislation in the USA (Lehman, 1995).

World War II left the economies of many European and Asian countries in shambles, but the USA, with intact massive production facilities, emerged as the most powerful economic power on the earth. In 1947, the US Marshall Plan and other economic rescue efforts were initiated for the recovering nations. As part of the recovery plan the General Agreement on Tariffs and Trade (GATT) was negotiated in 1947 to open up markets for increased global trade. With rapidly changing conditions in trade and travel, the need for effective international cooperation in preventing the spread of pests and diseases became urgent. In 1952 the International Plant Protection Convention (IPPC) was established as a subsidiary of the United Nations Food and Agricultural Organization (FAO). There are around 100 signatory nations to the IPPC agreement, which has provided a framework for national, regional, and global effort in plant protection. Member countries are required to establish a national plant protection organization to: survey crops for pests and diseases and report the outbreak and spread of pests and diseases; inspect plant and plant products moving in international trade and issue phytosanitary certificates; publish its plant quarantine regulations and distribute them to others; and, strengthen advisory services and research capabilities (Van der Graaff, 2001). Under Article VIII of the convention, countries that share similar agricultural production systems and pest problems are encouraged to establish Regional Plant Protection Organizations (RPPOs). Nine RPPOs have been established with the overall goal of a stronger coordinated approach to resolving plant protection and quarantine problems, and worldwide 186 countries belong to one or more of these regional organizations (Table 23. 1). The European Plant Protection Organization (EPPO) was the first to be established in 1951 and has been the strongest and most active of the RPPOs. The EPPO has provided a forum for scientifically based Pest Risk Assessment (PRA) and currently ten nematodes are listed among the pests recommended to be regulated among member countries. It has also provided strong leadership in developing programs that certify the production of healthy plant propagative materials (Van der Graaff, 2001). Cooperative agreements have been reached on certification schemes for 28 crops, and these programs are important in preventing the spread of nematodes and other pests to new areas (Anonymous, 2001). Nematodes that have been targeted in these certification programs are *Aphelenchoides* spp., *Ditylenchus* spp., *Meloidogyne* spp., *Pratylenchus* spp., and virus vector nematodes, i. e., species of *Longidorus*, *Paratrichodorus*, *Trichodorus*, and *Xiphinema* (McNamara, 1995).

Table 23. 1 Member countries of the nine regional plant protection organizations and date established under the International Plant Protection Convention (IPPC).*

REGIONAL PLANT PROTECTION ORGANIZATIONS (RPPOs)
ASIA AND PACIFIC PLANT PROTECTION COMMISSION (APPPC) (1956) Australia, Bangladesh, Cambodia, China, Fiji, France (for French Polynesia), India, Indonesia, Laos, Malaysia, Myanmar, Nepal, New Zealand, Pakistan, Papua New Guinea, Philippines, Portugal (for Macau), Republic of Korea, Samoa (Western), Solomon Islands, Sri Lanka, Thailand, Tonga, Viet Nam
CARIBBEAN PLANT PROTECTION COMMISSION (CPPC) (1967) Barbados, Colombia, Costa Rica, Cuba, Dominica, Dominican Republic, France (for Guadeloupe, French Guiana, Martinique), Grenada, Guyana, Haiti, Jamaica, Mexico, Netherlands (for Aruba, Netherlands Antilles), Nicaragua, Panama, Saint Kitts and Nevis, Saint Lucia, Suriname, Trinidad and Tobago United Kingdom (for British Virgin Islands), United States of America (for United States Virgin Islands, Puerto Rico), Venezuela

Continued

REGIONAL PLANT PROTECTION ORGANIZATIONS (RPPOs)
COMITE REGIONAL DE SANIDAD VEGETAL PARA EL CONO SUR (COSAVE) (1980) Argentina, Brazil, Chile, Paraguay, Uruguay
COMUNIDAD ANDINA (CA) (1969) Bolivia, Colombia, Ecuador, Peru, Venezuela
EUROPEAN AND MEDITERRANEAN PLANT PROTECTION ORGANIZATION (EPPO) (1951) Albania, Algeria, Austria, Belgium, Bulgaria, Croatia, Cyprus, Czechia, Denmark, Estonia, Finland, France, Germany, Greece, Guernsey, Hungary, Ireland, Israel, Italy, Jersey, Jordan, Kyrgyzstan, Latvia, Lithuania, Luxembourg, Macedonia FYR, Malta, Morocco, Netherlands, Norway, Poland, Portugal, Romania, Russia, Slovakia, Slovenia, Spain, Sweden, Switzerland, Tunisia, Turkey, Ukraine, United Kingdom of Great Britain and Northern Ireland
INTERAFRICAN PHYTOSANITARY COUNCIL (IAPSC) (1954) Algeria, Angola, Benin, Botswana, Burkina Faso, Burundi, Cameroon, Cape Verde, Central African Republic, Chad, Comoros, Congo (Democratic Republic of), Congo (Republic of), Côte d'Ivoire, Djibouti, Egypt, Equatorial Guinea, Ethiopia, Gabon, Gambia, Ghana, Guinea, Guinea-Bissau, Kenya, Lesotho, Liberia, Libyan Arab Jamahiriya, Madagascar, Malawi, Mali, Mauritania, Mauritius, Mozambique, Namibia, Niger, Nigeria, Rwanda, Sãoé Tom and Principe, Senegal, Seychelles, Sierra Leone, Somalia, South Africa, Sudan, Swaziland, Togo, Tunisia, Uganda, United Republic of Tanzania, Zambia, Zimbabwe
NORTH AMERICAN PLANT PROTECTION ORGANIZATION (NAPPO) (1976) Canada, Mexico, United States of America
ORGANISMO INTERNACIONAL REGIONAL DE SANIDAD AGROPECUARIA (OIRSA) (1953) Belize, Costa Rica, El Salvador, Guatemala, Honduras, Mexico, Nicaragua, Panama
PACIFIC PLANT PROTECTION ORGANIZATION (PPPO) (1995) Australia (Including Norfolk Island), Cook Islands, Fiji, France (for French Polynesia, New Caledonia, Wallis and Futuna Islands), Kiribati, Marshall Islands, Micronesia (Federated States of), Nauru, New Zealand, Niue, Northern Mariana Islands, Palau, Papua New Guinea, Samoa (Western), Solomon Islands, Tokelau, Tonga, Tuvalu, United Kingdom (for Pitcairn), United States of America (for American Samoa and Guam), Vanuatu

* Source: IPPC homepage, http://www.fao.org/ag/AGPP/PQ/, August 3, 2000.

23.3 Exclusion Methods and Procedures

The primary objective of plant quarantine or regulatory pest management programs is to reduce the spread of harmful organisms either within or between countries. These programs primarily focus on the ways that humans aid in the dispersal of pests and pathogens, and the implementation of regulatory exclusion procedures to prevent pests and pathogens from entering the areas where the host is grown. Exclusion efforts may be direct, i.e., the movement of plant materials or articles is only restricted if a pest or pathogen is observed. If risks of pest introduction and subsequent economic risks are high, indirect exclusion procedures may be established and plant materials or articles are restricted regardless of the absence or presence of a pest or pathogen.

23.3.1 Direct Exclusion Procedures

Exclusion of pests and pathogens at the port of entry Plant materials or other imported articles are routinely inspected at ports of entry in most countries. This is primarily done by visual inspections or by sampling. If excluded pests are encountered, sometimes an eradicative treatment is applied to eliminate the pest or pathogen, but more often the consignment is denied entry or destroyed. Nematodes listed for exclusion are usually not established in the importing country. But established nematode pests are sometimes excluded if: (1) an eradication or containment program is underway against the pest of concern; (2) external introduction of a nematode would threaten domestic certification programs for establishing clean planting stock; or (3) there are exotic races of a nematode present in other countries.

Exclusion of pests and pathogens at origin Countries which lack well-developed quarantine services frequently rely solely on the services of the country of origin to declare in a phytosanitary certificate that the pest of concern is not present in the country or in some cases a geographical region within the country of origin. Pre-clearance programs are another means of reducing risks of importing unwanted pests by excluding them in the country of origin. Through a cooperative agreement with the importing country, growers in the country of origin produce plants under strict phytosanitary regimes. Prior to shipment, the regulatory officials in the exporting country certify that the plant material meets the plant production health standards specified by the importing country. In addition the exporting country is usually required to sample or index the plant material prior to export to assure apparent freedom of pests and pathogens. In return there is minimal inspection of the plant material at the port of entry, primarily sampling for quality assurance.

Domestic programs for certification of plant propagative material One of the major ways nematodes have been spread worldwide is through infested planting stock. Even if a nematode is widely distributed in a country, domestic regulatory certification programs of seed or plant propagative material are often one of the most economical means of managing nematodes, especially because multiple species of nematodes and other pests can be effectively excluded with the same sanitation practices used in a certification program (Anonymous, 2001). Excluding nematodes through certification programs is often by far the most cost-effective means of managing nematodes, especially for perennial fruit and nut crops and field crops which are transplanted from seedbeds.

23.3.2 Indirect Exclusion Procedures

Because nematodes are microscopic in size, they can rarely be detected by visual inspection. Trained nematologists can sometimes detect *Meloidogyne* females or cysts of *Heterodera* or *Globodera* with the naked eye or the aid of a hand lens, but most plant-parasitic nematodes cannot be detected by this means. Few countries have the laboratory facilities or trained personnel to sample and extract nematodes from soil and plant material or to identify the nematodes using a microscope. Furthermore, if large volumes of perishable plant material are imported, there are time constraints for a diagnosis. So even with unlimited resources, a procedure that requires days to complete an analysis for nematodes is often prohibitive. For this reason many countries use indirect methods to exclude nematodes. In fact these methods are probably more important in preventing the dispersal of nematodes than direct visual inspection for galls or cysts to exclude nematodes at ports of entry.

Exclusion of plants if accompanied by prohibited articles Most nematodes may be soil borne, and indirectly they are excluded because most countries prohibit importation of soil, or plants in soil. Many countries also prohibit importation of plants established in artificial growing media or bare-rooted plants in packing materials (hay, straw, forest litter, etc.) that may harbor nematodes and other pests.

Exclusion of host plants If a pest or pathogen has potential to cause severe losses to a major crop, and if the pest is difficult to eradicate or contain once it is established, then even a small number of failures to detect a pest at the port of entry can threaten the economic well being of growers and in some cases the nation as a whole. For this reason many countries often indirectly exclude pests by simply prohibiting all known plants that may serve as a carriers of major pests or hosts of major pathogens of economically important crops (Table 23.2). In some cases, only plants or seeds are excluded, but in many cases both are excluded. Excluding a crop plant or plant propagative material is often a multi-targeted attempt to exclude several major organisms. A collateral benefit is that even nematodes that cause minor economic losses are also excluded. For example, South Africa excludes 270 hosts to indirectly exclude nematodes such as *Aphelenchoides besseyi* Christie, *Ditylenchus dipsaci* (Kuhn) Filipjev, and *Radopholus similis*.

Table 23.2 Exclusion of nematodes pests by indirect quarantine regulations, which prohibit entry of crops or genera.[1]

Crops/Genera	Number of Countries in which crops/genera are prohibited	Percentages of countries prohibiting		
		Plants only	Seeds only	Plants and seeds
FRUIT CROPS				
Citrus	62	55	0	45
Cocos	28	29	7	64
Fragaria	20	65	0	35
Musa	39	54	0	46
Pome Fruits	37	85	0	15
Prunus	37	85	0	15
Ribes	16	69	0	31
Vitis	41	90	0	10
VEGETABLE CROPS				
Ipomoea	23	61	0	39
Solanum	48	77	0	23
FOREST CROPS				
Acer	14	100	0	0
Castanea	34	76	0	24
Conifers	27	100	0	0
Crategus	14	100	0	0
Juglans	21	100	0	0
Populus	27	93	0	7
Quercus	25	92	0	8
Salix	22	100	0	0
Sorbus	24	96	0	4
Uimus	32	84	0	16
OTHER CROPS				
Coffea	49	24	18	57
Gossypium	52	25	14	61

Continued

Crops/Genera	Number of Countries in which crops/ genera are prohibited	Percentages of countries prohibiting		
		Plants only	Seeds only	Plants and seeds
Helianthus	15	20	0	80
Hevea	28	29	0	71
Nicotiana	26	35	7	58
Palm, oil	16	56	0	44
Rice	33	21	18	61
Rosa	22	100	0	0
Theobroma	43	19	2	79
Tea	20	45	0	55
Saccharum	40	63	0	37

[1] Data from survey of regulations of 125 countries (Kahn, 1982).

Exclusion of host plants by added declaration on phytosanitary certificates

Declarations on phytosanitary certificates may require that specified pests and pathogen do not occur in the country or region of origin. In many cases nematodes are indirectly excluded by default because the country of origin cannot meet all the requirements for other types of pests and pathogens listed on the phytosanitary certificate.

23.4 Nematodes in International Quarantine Legislation

The nematodes that countries most frequently regulate in 2000 and the ten nematodes listed most frequently in the survey of Kahn (1982) two decades earlier are listed in Table 23.2. In general, in the past two decades the number of countries regulating nematodes has increased. For example, the number of countries regulating *Globodera rostochiensisis* (Wollenweber) Behrens increased from 51 in 1982 to 106 countries in 2000. The ranking of the nematodes in the top ten list has also shifted in the past two decades. For example, in 1982 *Aphelenchoides besseyi* ranked 10th and was regulated by nine countries, but it ranked 2nd in 2000 with 70 countries regulating it, primarily to protect rice crops. This indicates that regulations are not static, and there are many reasons for these changes over the past decades. New political entities and shifts in trade relations after the breakup of the Soviet Union in the 1990s increased the number of countries in Europe regulating nematodes. Changes have been influenced by the increase in the Regional Plant Protection Organizations that have provided a more efficient means for countries to conduct pest risk analyses and prepare regulatory pest lists. There has been an increase in the number of trained nematologists in the past several decades in many countries, especially in South America, Asia, and Africa. The surveys and applied research these nematologists have conducted, along with their expertise, has enabled many countries to develop regulations for nematodes that are biologically based. In some cases, the change in rank reflects nematological research during the past two decades. For example, the increase in the number of countries regulating *Bursaphelenchus xylophilus* (Steiner & Buhrer) Nickle, *Globodera pallida* (Stone) Behrens, *Nacobbus aberrans* (Thorne) Thorne & Allen and *Meloidogyne chitwoodi* Golden, O'Bannon, Santo & Finley primarily, is a result of research during the past two decades, which has resulted in increased international awareness of these nematode pests. The four major root-knot nematodes, *Meloidogyne arenaria* (Neal) Chitwood , *M. hapla* Chitwood, *M. incognita* (Kofoid & White) Chitwood, and *M. javanica* (Treub) Chitwood, which worldwide may cause greater total crop loss than any other pest pathogen, are actually regulated by fewer countries than 20 years ago. For example, in 1980, 11 countries had regulations to exclude *M. javanica*, but in 2000 no country listed it as a prohibited nematode. Surveys conducted as a part of the International *Meloidogyne* Project indicated that the common root-knot nematodes are widely distributed in many countries. Most countries have removed these nematodes from their prohibited pest lists because they are aware, according to the rules of international trade, a country could be subject to prosecution if it regulates the pest externally but does not regulate this same pest internally, especially if it is widely distributed domestically, because in this case the pest is being used as a trade barrier.

Even though a long list of countries considers a given nematode a quarantine pest, the enforcement of its exclusion is not always successful. The federal regulatory inspectors at most ports of entry, including those in the USA, rely primarily on visual symptoms or signs such as galls and cysts to indicate the presence of quarantine nematodes. These regulatory port introduction centers are not equipped to sample, extract, and identify nematodes and other pests that do not cause visible symptoms and cannot be seen with the aid of a hand lens. Additional personnel trained in nematode identification are needed.

The last decades of the 20th century were marked by multilateral trade agreements, and the emergence of multinational trading blocks such as the European Union and the North American Free Trade Agreement (NAFTA). Upgraded GATT agreements lowered trade restriction further, and from world trade negotiations in Uruguay and Kyoto Japan, the World Trade Organization (WTO) emerged in 1995 with 135 member countries accounting for more than 90 percent of international trade. As a result, the value of world trade in agriculture rose from about US$40 billion in 1970 to US$200 billion in the 1990s (USDA, 1994). Consequently, the task of terminal inspections to monitor agricultural imports for hitchhiking exotic pests has become challenging and in some countries overwhelming. For example, agricultural inspections at United States borders increased 40-fold during this period. The task of excluding microscopic pests such as nematodes is especially difficult, considering that in the absence of plant symptoms, most nematodes will be missed during visual inspections. With the substantial changes in the patterns of trade and travel it is appropriate to ask: "How efficient and cost effective are regulatory efforts in pest exclusion and management, and do regulatory agencies continue to have a role as we begin the 21st century?" Several examples of past successful regulatory efforts are presented which illustrate that investments in well-managed regulatory programs yield high returns. With fewer chemical options to manage nematodes under field conditions, and growing environmental and energy conservation concerns, the cost avoidance benefits of excluding nematodes will increase.

23.5 Examples of Regulatory Management Programs with High Returns

Regulatory pest exclusion programs, especially programs for perennial crops, require long-term commitments. The cost-benefits from regulatory programs that exclude nematodes from perennial crops are compounded with time. The cost-benefit analysis from a 40-year investment in the nematode program in Florida demonstrates this increase in cost-benefits over time.

23.5.1 Citrus Regulatory Programs in Florida

Cost-benefit analysis of the burrowing nematode regulatory program during 1960 – 2000 In the early 1950s, the Florida citrus industry was threatened with a serious problem known as spreading decline caused by the burrowing nematode (BN), *Radopholus similis*. One of the principal ways that the nematode was being spread was by growers planting seedlings from nurseries that were infected with burrowing nematodes. The impact of planting infected seedlings was not evident to many growers at that time because the disease progresses slowly at first causing minimal damage for the first four to five years. But because citrus is a perennial crop with many trees producing for more than 30 years, the long term impact of this nematode can be financially devastating to the grower. This is illustrated in Figure 23. 1 that compares production data for citrus in two 25 hectare groves with the same cultivar, soil type and cultural practices, but in one healthy seedlings were used and in the other seedlings infected with burrowing nematodes. In the central Florida where citrus is grown in deep sandy soils, growers expect that citrus trees will begin producing about 124 boxes of fruit per hectare (50 boxes per acre) 4 years after a new grove is established and that production will increase for the next seven to eight years until it reaches around 1400 to 1500 boxes per hectare (600 boxes per acre). With good management practices, this level of production may continue for 30 to 40 years (Fig. 23. 1, certified seedlings). For example, if a grower had planted a hectare of citrus with certified nematode-free plants in 1960, by 1995 the cumulative on-tree yield dollar value for the 35-year period would have been $137,000 based on average

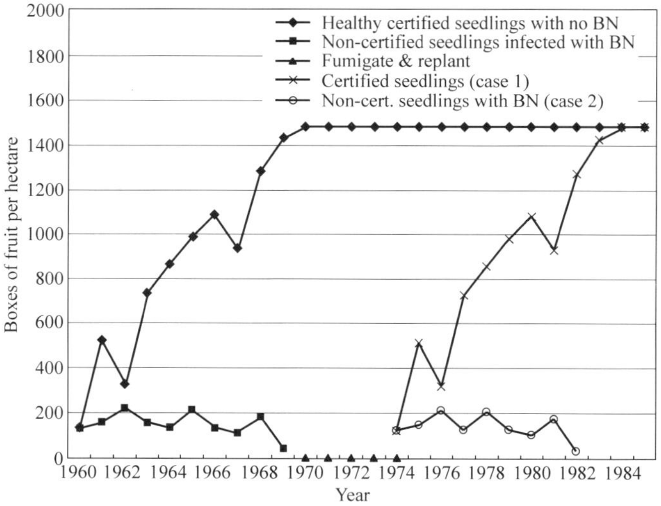

Figure 23. 1 Predicted production that growers would have experienced during a 25-year period (1960 – 1985) after planting healthy seedlings versus burrowing nematode infected seedlings (adapted from grove data, Poucher et al., 1967).

published values per acre indexed to 1995 dollar values (Fig. 23.2). However, if this grower had planted BN-infested plants, after several years citrus production would have begun to decline and fluctuate at a low level until losses due to BN required tree removal, fumigation, and replanting of the grove. After replanting with seedlings certified free of BN, the grower would wait another four years for the trees to begin producing, and then another 10 years for the trees to reach peak production, or the level of profit the grower could have experienced 14 years earlier by planting certified BN-free seedlings the first time (Fig. 23.1, case 1). The total on-tree value of the hectare of citrus after 35 years would have been US$86,500 (Fig. 23.2). If after fumigation, however, this grower would have carelessly replanted with BN-infected seedlings (Fig. 23.1, case 2) then this grower is destined to repeat the same experience, i.e., fumigation and replant after ten years. In this case the total on-tree value of citrus produced from the hectare of citrus in the 35-year period from 1960 to 1995 would have been only US$59,600, a 57% reduction of return over 35 years compared to the grower who planted healthy certified seedlings that were not infected with BN (Fig. 23.2).

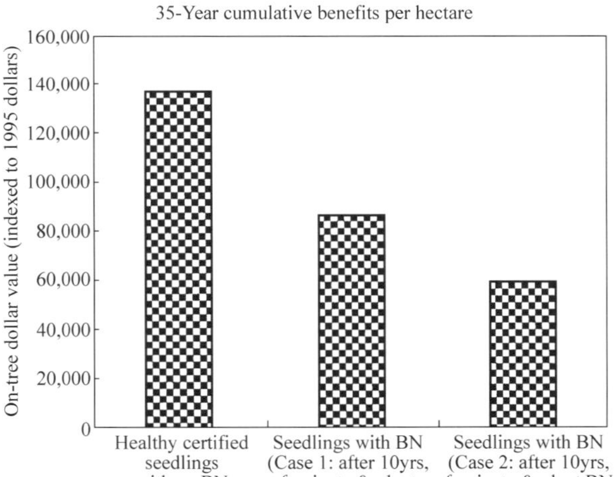

Figure 23.2 Economic impact to citrus growers planting healthy certified versus burrowing nematode infected seedlings during a 35-year period. Data for healthy grove based on average on published on tree values for citrus in Florida from 1960 – 1995. Predicted yields for infected seedlings based on data from Poucher et al. (1967).

Due to the severe losses that Florida citrus growers were experiencing at that time, the Florida Department of Agriculture, in cooperation with citrus growers, launched regulatory programs to prevent the further spread of this serious pest. These included: (1) *surveys* more than three million survey samples were taken from 1955 to 1984 to define infested areas; (2) an *eradication program* a total of 15, 197 acres (6153 hectares) of infested trees were removed and burned, and the infested areas were fumigated and kept fallow for two years before replanting was allowed; (3) a *barrier program* the perimeter of an infested area was fumigated and kept fallow to prevent the growth of citrus roots and movement of nematodes to healthy trees; and (4) *certification programs* regulations were established to ensure that the citrus seedlings which commercial nurseries sold to growers were free of burrowing, citrus, and coffee lesion nematodes. Citrus nursery sites were only approved if they were sampled and found free of these regulated nematodes. Nursery sites were required to meet strict standards as to drainage and set-back distances from roadways or plants known to be burrowing nematode hosts. Approved nurseries were inspected and sampled annually to ensure that growers used good sanitation practices and that citrus seedlings continued to be free of regulated nematodes. In addition, soil pits (sites providing soil for fill) were also required to be certified and found free of the nematodes regulated for citrus nurseries to ensure that the groves in which certified seedlings were planted would not be contaminated with regulated nematodes when roads and power lines were constructed in or near groves.

These programs greatly reduced the rate of spread of the burrowing nematode in Florida citrus groves. Because of the implementation of nematode eradication, barrier, and nursery certification programs in the 1950s, the nematode-infested area in Florida citrus groves was only 3800 hectares (9385 acres) by 1984. The citrus nursery certification program has been the keystone for the success of regulatory programs on citrus in Florida. In the mid 1950s when the threat of BN to Florida's citrus production was first recognized, surveys revealed that there were 278 citrus nurseries infested with BN (Esser, at al., 1988). With the implementation of nursery certification programs the number of BN-infested nurseries rapidly declined and finally declined to zero detection in 1970 (Table 23. 3). From 1970 to 2000, commercial citrus nurseries were sampled annually by the state as required by law and burrowing nematodes were not detected in any commercial citrus nursery in Florida (Table 23.3).

Table 23.3 Incidence of *Radopholus similis* in Florida citrus nurseries (1953 – 2000).

Years	Nurseries infested
1953 – 1958	278
1958 – 1962	93
1962 – 1966	59
1966 – 1970	14
1970 – 1900	0

Although it is difficult to estimate the area that would have become nematode-infested to date without these regulatory management programs, it was predicted in 1958, based on observations of the spread of the disease at that time, that within a 10-year period without the implementation of regulatory control programs, approximately 18,000 additional hectares (44,500 acres) would become infested. If only the most conservative estimate for rate of spread of the disease is used for a cost-benefit analysis, i.e., using the area predicted to become nematode infested within 10 years (18,000 hectares), the benefits of these regulatory programs to Florida growers and the Florida economy were much greater than the costs of implementing the regulatory programs. Taking into consideration the average financial loss growers experience when their groves are infested with BN (Fig. 23.2) and using a conservative estimate of 18,000 hectares, which did not become infested with BN because of the regulatory programs, the cumulative benefits for Florida citrus growers during this 35-year period was almost US$1.4 billion. The actual cost of the regulatory programs from 1954 to 1994 was about US$100 million. Program costs include the budget for required pushing of trees and fumigating 6104 hectares, maintaining buffers around 2227 infested hectares, and the costs of 3 million survey and certification samples. Thus, the regulatory management costs were only about 7% of the benefits to the growers, or for every US $70,000 invested in the programs there was a return of US $1 million.

Citrus certification annual cost-benefits An additional benefit of the nematode certification program has been the reduction of the percentage of the citrus area in Florida that has become infested with other nematodes of citrus. Records from the citrus groves sampled for BN in the mid-1950s and early 1960s show that about half of them were infested with the citrus nematode (CN), *Tylenchulus semipenetrans* Cobb (Lee, et al., 1999). To prevent further dissemination of this nematode through nursery stock, it also was included as a regulated nematode when the nursery certification programs for

BN were established. In the 1980s, many groves in the northern region of citrus production in Florida, which had been infested with *T. semipenetrans* prior to the establishment of nursery certification programs, were not replanted after several severe freezes, and citrus production shifted to southern regions. Because nematodes have not infested the new citrus areas planted over the past 40 years, only about 20 – 30% of the citrus acreage in Florida is now infested (Lee, et al., 1999). This is much lower than in other states and countries without certification programs where it is not uncommon for the 90% of the citrus area to be infested with citrus nematode. Citrus production has increased threefold worldwide in the past 35 years, mostly as a result of an increase of the area planted, but unfortunately most of the area is now infested with the citrus nematode because infected seedlings were used to establish new areas of production.

Thus, the collateral benefits for exclusion of *T. semipenetrans* also should be included in the calculations for the annual benefits that Florida citrus growers received from regulatory certification programs. In 2000 the benefits from excluding *T. semipenetrans* was estimated to be 33 million dollars based on average losses of 5% where susceptible rootstocks are planted. About 50% of the citrus area planted after the freezes of the 1980s was on nematode-resistant rootstocks such as Swingle citrumelo rootstock and these areas were excluded for the cost-benefit analysis. However, as with all perennial crops, if a rootstock is continually exposed to a pathogen, selection pressure may result in the development of resistance-breaking biotypes that threaten the continued usefulness of a rootstock. In 1995 a resistance-breaking biotype was found on Swingle rootstock in Florida. A secondary benefit of excluding nematodes from new and established groves through the certification program is a reduction of selection pressure against host resistance and thus lowering the development and spread of resistance-breaking biotypes.

An estimate of the combined benefit that Florida citrus growers received in 2000 from nursery certification programs is 33 million for exclusion of citrus nematode and 17 million from burrowing nematode for a total of 50 million dollars. The estimate for BN nematode is based on 25% losses in infested areas and assumes that growers used practices recommended to manage BN. This estimate is considered conservative because based on the predicted rates for the spread of burrowing nematode without the establishment of regulatory programs in 1958, by 2000 the BN infested area would have likely been twice that used in this estimate.

Total annual costs to the Plant Inspection and Nematology Sections for nematode certification programs for citrus nurseries was US$75,000, compared to the benefit from the certification program which excludes BN and CN which were estimated to be US$50 million in 2000. Thus, in 2000 for every dollar invested in the administration of the nematode certification program in Florida, there was a return of approximately US$666 to growers.

Additional benefits were obtained by excluding the coffee lesion nematode and other nematodes that were not included in the calculation. The total economic impact of the Florida citrus industry on the general economy through jobs and revenue from processing, transportation, and sales is about seven times the on-tree fruit value. Thus, the positive economic impact of this citrus nematode certification program was far greater for the general public than just the increase of on-tree fruit value for growers. The public also benefits indirectly from these programs because of lower food costs. Because pests are excluded by certification programs, there are also environmental and economic benefits to growers because there is less need for pesticide applications.

23.5.2 Cooperative Nematode Certification Programs for Ornamental Plants

To prevent the introduction of nematodes and other pests, many countries have quarantine regulations that prohibit importation of plants grown in natural soils. Many countries, including the USA, allow the importation of plants grown in certain types of approved pest-free artificial soil mixes if the exporting country's regulatory agency provides a phytosanitary certificate which states that the plants are free of pests that are of concern to the importing country. Because the race of the burrowing nematode that parasitizes citrus is only known to occur on citrus in Florida and it also has a wide range of ornamental hosts, Florida's ornamental plant growers face additional restrictions. Citrus producing states and countries require that any nursery that produces ornamental plants for export is sampled annually by the Florida Department of Agriculture and found free of burrowing nematode.

The state of California has stringent regulations not only for the burrowing nematode, but for 22 additional nematode species which might cause damage to the state's agricultural crops if they are introduced and become established in the state. Ornamental growers from certain states, including Florida, where the market value of ornamentals exceeds one billion dollars, must adhere to strict sanitation requirements in order to export ornamental plants to California. These standards include growing plants on benches at least 18 inches above the ground and ensuring that soil, water, and plant propagation materials are free of regulated plant nematodes. State regulatory agencies are required to annually sample nurseries that export plants and monitor nursery sanitation. California state regulatory officials inspect and sample each shipment of plants at the state border prior to allowing their entry into the state. If regulated nematodes are found in a nursery during the annual sampling or in plants that are shipped, then the nursery's shipping privileges are revoked. If requested, state regulatory scientists help nursery managers discover where they may not have followed good sanitation practices and suggest measures to ensure that future plant shipments are not infested with regulatory nematodes. Before the shipping privileges of these nurseries are reinstated, production areas in the nursery where plants are grown for California export must be sampled and found free of any nematodes regulated by the state of California.

Although many ornamental plant growers consider these requirements too restrictive, the cooperative certification programs required by California have been effective in excluding many nematodes, which might cause serious economic losses to California growers. California regulatory officials estimate that the potential damage that exotic nematodes might cause to California's diverse agricultural crops is equal to US$ 600 million, about 3% of the total crop value of 19 billion dollars (California DFA, 2000). The reniform nematode, *Rotylenchulus reniformis* Linford & Oliveira, is an example of a nematode that is not present in California and is a serious pest that has the potential of damaging many crops, since there are 314 plant species in 74 families that are reported as hosts of this nematode (Robinson, et al., 1997). The cost-benefits of excluding reniform nematode from cotton were analyzed as follows. In 1997 the total crop value of the cotton crop in California was US$1.1 billion (USDA, 1997). In other states, where the reniform nematode is established in approximately 10% of the cotton production areas, losses average 7%. To date regulatory practices have been successful in preventing reniform nematode from becoming established in California. Assuming that reniform nematode would cause similar damage to cotton in California as occurs in other states, in 1997 the benefits that growers received from regulatory efforts that excluded the reniform was estimated to be US$7.2 million. Excluding reniform nematodes also benefits California's 4 billion dollar melon and vegetable industry, because this nematode is known to cause approximately 10% crop losses cantaloupe, squash, and snap beans (Robinson, et. al., 1997). The benefits to California growers by excluding reniform nematode far exceeds the cost of implementing the regulatory programs that prevented this nematode from becoming established in California. The costs of Florida's nematode certification programs for all nematodes regulated by the state of California is approximately US$100,000 annually, which is only about 1.3% of the benefits calculated for excluding reniform nematode from cotton in California.

23.5.3 Regulatory Programs for Potato Cyst Nematodes

Potato cyst nematodes (PCN), *Globodera rostochiensis* and *G. pallida*, are recognized as important quarantine pests worldwide, with more countries regulating PCN than any other phytoparasitic nematode. In addition to countries where PCN are not known to occur, the list of countries regulating PCN also includes countries where PCN are known to occur, but are not widely distributed. Generally, if a pest is widely distributed within a country it is not regulated, but in the case of PCN some countries impose regulations on these pests because virulence groupings vary considerably between regions and countries.

South and Central America Potato cyst nematodes occur in Bolivia, northern Chile, Ecuador, Colombia, Peru, and Venezuela in South America and Panama in Central America (Brodie, 1998; Franco, et al., 1998). In spite of reports of the occurrence of PCN in Argentina and Costa Rica recent surveys failed to confirm the presence of PCN in these countries (Brodie, 1998; Doucet, et al., 1997). About 40% of the potato production in South America occurs in Brazil and Argentina and these countries which are currently considered free from potato cyst nematodes have exterior quarantines and domestic seed certification programs to prevent the introduction and spread of PCN (Franco, et al., 1998). Other countries that have import quarantines for both species of PCN are Chile, Paraguay, and Uruguay. *Globodera rostochiensis*, but not *G. pallida*, is officially regulated by Venezuela, Costa Rica, El Salvador, Guatemala, and Honduras. Argentina, Brazil and Chile have potato seed certification programs. Due to the fact that PCN occurs only in the northern area, Chile has also established plant quarantine inspection points to ensure that potato cyst nematodes do not spread to other regions of the country.

North America *Globodera rostochiensis* is present in confined areas of Canada and USA, and in many states in Mexico. In North America *G. pallida* only is known to occur in Canada. To prevent the spread of the PCN, all three countries include potato cyst nematodes in their potato seed certification programs and have domestic quarantines to regulate the movement of host plants, infested soil, contaminated equipment, etc. These NAPPO member countries also have phytosanitary regulations that are directed at preventing the introduction of these nematodes from Europe, South America, or other regions where it is known to occur.

In Canada, PCN are only known to occur on islands in the most eastern and western extremes of the country. *Globodera rostochiensis* occurs on Vancouver Island and Newfoundland. The PCN was discovered in Newfoundland in 1962 and was probably brought to the island before it became a province of Canada in 1949. Both *G. rostochiensis and G. pallida* have been detected in Newfoundland, and they occur primarily in home gardens, as there is very little area suitable for commercial potato farming in this province. The nematodes were likely introduced to Newfoundland from Europe because prior to 1949 trade with Europe was unrestricted. The infestation on Vancouver Island in British Columbia is believed to have resulted from importation of flower bulbs from The Netherlands, since fields where the bulbs were grown were later planted to potato. The Canadian government has established regulatory programs that have been successful in preventing the spread of PCN to potato production areas in other parts of the country. Planting host crops is not permitted in any area of Vancouver Island where the PCN has been detected in surveys. To prevent the spread of the nematode from Newfoundland to the large commercial potato growing areas on Prince Edward Island and mainland Canada, before boarding the ferry to leave the island every vehicle is thoroughly washed in a high-pressure wash facility. The efforts to prevent the spread of PCN in Canada have been aided by the fact that the principal area of potato production, Prince Edward Island, and the areas infested with PCN are physically isolated from the mainland by natural barriers of water.

In Mexico, surveys indicate that *G. rostochiensis* occurs in 46 counties involving nine states: Coahuila, Distrito Federal, Guanajuato, Hidalgo, Mexico, Nuevo Leon, Puebla, Tlaxcala, and Veracruz (Brodie, 1998). Mexico bans the importation of seed tubers, and certifies and maintains areas free of the nematode through phytosanitary control.

In the USA, the golden nematode, *G. rostochiensis,* was first detected on Long Island in New York State in 1941. The nematode is believed to have been introduced on contaminated military equipment returning from Europe after World War I. The temporary airfield where military equipment returned to the USA was later abandoned and the site became a potato field (Brodie, 1998). Subsequent detection has been made several times in five counties in Upstate New York. The nematode was also found at two sites in Newcastle County, Delaware in the 1960s, but the quarantine of this area was removed in 1970 because eradication treatments were effective according to subsequent surveys (S. Poe, pers. comm.). Except for this episode, for more than 40 years the spread of this nematode was restricted to limited areas in New York State by the quarantine coupled with surveys to detect new infestations, and the application of nematicides in infested fields. In recent years, the management program of APHIS has focused on requiring resistant varieties to be planted in infested fields for two out of every four years instead of using chemical management approaches. This has been even more effective in reducing potato cyst nematode densities below detectable levels than the programs that used nematicides to achieve this objective. This management program that requires less survey and nematicides is less costly. In 1998, APHIS spent about US$580,000 on the golden nematode eradication and control programs, a savings of US$1.5 million annually compared to costs of nematicide management programs in prior years. However, the use of resistant varieties promotes the selection of resistant-breaking biotypes. Until 1994, when a new biotype (Ro2) was discovered, only one biotype (Rol) was known to be present in infested areas in New York. The selection of this biotype is a cause for considerable concern, since it appears that it is not controlled by the *H1* gene, the principle source of resistance in all resistant varieties available in the USA. The efforts and programs of state and federal regulatory agencies, which have been successful in preventing the spread of the golden nematode to other commercial areas where potatoes are grown in the USA, will become more difficult and challenging in the future.

Europe Potatoes were first domesticated in the Andean region of South America. Thousands of years later, they were introduced to Europe in the sixteenth century from where they were distributed throughout the world. Following the devastation of the potato blight disease in the mid-1800s, plant breeders made several expeditions in South America to collected wild and cultivated potato tubers in order to breed varieties resistant to the late blight fungus, and it is speculated that the PCN was introduced from South America to Europe with breeding material. Cyst nematodes were first noticed on potatoes by Kühn in Germany in 1881. In the early part of the 20th century there was renewed worldwide interest in potato as a food crop and many expeditions were made to collect wild potatoes from South America. Improved varieties were developed in Europe, and seed tubers of these varieties and adhering soil were distributed throughout Europe and the rest of the world. This played an important role in the dissemination of the PCN to potato-growing areas worldwide (Franco, et al., 1998).

Both species of PCN are widely distributed in western Europe, occurring in Austria, Belgium, France, Germany, Greece and Crete, Iceland, Ireland, Italy, Luxemburg, Malta, The Netherlands, Norway, Portugal, Spain, Sweden, Switzerland, and the UK (Turner and Evans, 1998; Zasada and Gatt, 2000). *Globodera rostochiensis* but not *G. pallida* is reported from Finland and mainland Denmark (EPPO, 1997). In central and eastern Europe, *G. rostochiensis* is also widely distributed, occurring in Albania, Belorussia, Bulgaria, Czech Republic, Estonia, Hungary, Latvia, Lithuania, Poland, Romania, Russian Federation, Slovakia, and Ukraine. Except for Bulgaria, Poland, and Romania, *Globodera pallida* has not been reported in the countries of central and eastern Europe, however, in some countries surveys have not been conducted to detect this nematode (EPPO, 1997; van Riel and Mulder, 1998).

Potato cyst nematodes are considered A2 quarantine organisms by the EPPO and consequently most countries require fields in which seed potatoes are grown to be inspected and sampled according to approved methods and found free of PCN. Crop rotation is enforced in the EC countries in seed potato production areas. The EC requires that other crops must be rotated with seed potatoes in two of 3 years, but many countries increase this, with some requiring 7-year rotations. In some countries land in which PCN is detected is not eligible for seed potatoes production for many years. For example, in Northern Ireland infested land is not even eligible to be sampled until at least 15 years from the last potato crop, and if it is still infested thereafter it is only eligible to be resampled every five years until declared free from PCN (Whitehead and Turner, 1998). All EPPO countries require a phytosanitary certificate (or a plant passport between EU countries) for the import of seed potatoes. These phytosanitary regulations have helped to reduce the spread of PCN and their pathotypes to new areas. The benefits from regulatory programs is illustrated in Northern Ireland, where surveys indicate that over the last 30 years the level of PCN-infested land remains at between 1% and 3%, and after 20 years a distinctive pathotype was contained in a 20 km radius of where it was first detected (Whitehead and Turner, 1998).

Asia and Africa Similar efforts to contain the spread of the PCN have been made in Asia and Africa. In Asia, both PCN species are reported from Cyprus, India, and Pakistan. Only *G. rostochiensis* has been reported from Japan, Lebanon, The Philippines, Sri Lanka, and Tajikistan. In Africa, *G. rostochiensis* occurs in Algeria, Egypt, Libya Morocco, Sierra Leon, South Africa and Tunisia, but *G. pallida* has only been reported from the North African countries of Algeria and Tunisia (Turner and Evans, 1998). Collaborative programs with the German government in India and Pakistan have largely focused on containing the spread of the PCN through intensive applications of nematicides, although the collaborative project in Pakistan also aims to establish a seed certification program (Zaheer, 1998). In South Africa the Department of Agriculture enforces the Agricultural Pests Act that allows areas infested with PCN to be placed in quarantine, and subject to control measures that include prohibition of cultivating host crops and the movement of plants, plant products, and soil from infested areas (Kleynhans, 1998).

New Zealand and Australia Pest exclusion programs are usually implemented by governmental agencies, but in some cases, such as the PCN programs in New Zealand, they are funded and administered by growers. Extensive infestations of both species of potato cyst nematodes were found south of Auckland, New Zealand in 1972. From 1972 to 1988 extensive national surveys revealed that although the sites of PCN infestation were widely spread throughout New Zealand, only about 2.2% of the production area surveyed was infested (Marshal, 1998). From the time PCN was first detected until 1988, the New Zealand government controlled PCN through a policy of "detect, control and eradicate." If PCN was detected on a grower's property, a survey team determined the extent of the infestation, and all land occupied by the grower within a radius of 80 km was declared to be infested. Infested areas were subject to annual sampling and quarantine restrictions on the movement of machinery, potatoes, other solanaceous plants, and soil. Infested fields were fumigated, replanting of solanaceous hosts was prohibited for a minimum of 3 years, and the growers' uninfested lands were subject to close monitoring and subsequently surveyed at least once every 5 years. Quarantines on infested land were lifted for seed potato production only if no solanaceous crops had been grown on the infested lands for a period of 15 years from the time of the last positive identification of PCN on the land. These regulations greatly reduced the rate of spread of PCN in New Zealand, and by 1988 the total infested area began to decline. In 1988, the New Zealand government adopted a "user pay policy" for all services provided by government departments, and the concept that regulatory programs should be funded for the national good was abandoned. Growers were required to meet the full cost of quarantine management, approximately NZ$ 300,000 per year. Though willing to pay, unfortunately growers were unable to meet this cost, and all government-funded surveys, fumigations and monitoring of infested areas ceased. Regulations and legal restrictions on farm titles were revoked and PCN management became the growers' problem. The Ministry of Agriculture continues PCN surveys for seed potato certification, but this program is funded by the growers (Marshal, 1998).

In 1988, *G. rostochiensis* was detected in southwestern Australia near Perth and in 1991 it was found in southeastern Australia near in Victoria. *Globodera pallida* has not been found in Australia. Infested fields were harvested under supervision and fumigated. The government continues to support extensive surveys for PCN throughout the production areas. In the seed production, areas additional surveys are supported by growers. Potato cyst nematode has not been found in any other state, and these regulatory management programs appear to be successful in reducing the risk of PCN spread. Australia has benefited from the experience and the collaboration of New Zealand nematologists. The efforts of regulatory scientists in both countries have ensured that farmers could continue to grow potatoes crops with minimal damage due to PCN and meet the phytosanitary obligations necessary for international trade (Marshal, 1998).

Cost-benefits of PCN exclusion programs It is difficult to determine precisely how widely distributed the potato cyst nematode would be if regulatory exclusion efforts had not been implemented in the past 50 years. But a general comparison of North American and European experiences with this pest provides little doubt that the efforts to exclude PCN throughout the world have been beneficial and cost effective. In the USA the 1998 eradication and control budget was about US$580,000 and the annual PCN research budget was approximately US$500,000 (Brodie, pers. comm.). Although these programs have been costly, they have been cost effective in protecting the entire US potato crop. Because for many years there was limited effort to prevent the spread of *G. rostochiensis* and *G. pallida* in many areas of Europe, these nematodes are widely distributed and annual yield losses are estimated to be 9% (Evans and Brodie, 1980). In 1998 the value of the potato crop in the USA was US$2.6 billion (USDA, 1999) and losses of 9% would have cost the growers US$207 million, an amount approximately 200 times greater than the annual budget for the exclusion of this pest. Potato cyst nematodes are generally included in the list of pests and disease organisms regulated in potato seed certification programs, and this has been a very important factor in reducing the spread of PCN.

23.5.4 *Nacobbus aberrans*

In the past several decades, research by nematologists in countries of the South American Andean region has demonstrated that *Nacobbus aberrans* (Thorne) Thorne & Allen is a pest that warrants special effort in regulatory exclusion efforts. This pest, also known by the common names of false root knot nematode and rosario de la papa, may severely reduce yields of major food crops such as potato, sugarbeet, and some vegetables. Because it is not present in many parts of the world where it has potential to reduce crop yields, 38 countries impose quarantine restrictions against *N. aberrans* (Table 23.4). In the USA, *N. aberrans* has been reported on sugarbeet in five western states. Populations in the USA differ from those in South America in that they do not parasitize potato. Even though the race that causes severe losses to potato in the Andean region of South America is not known to occur in the USA, the federal regulatory agencies in the USA do not routinely monitor this pest at ports of entry.

Table 23.4 Nematodes regulated by twenty or more countries in international quarantine legislation in 2000 compared to number of countries Kahn listed as regulating these nematodes in 1982.[1]

Nematode species	Number of countries regulating	
	in 2000	in 1982
Globodera rostochiensis	106	51
Aphelenchoides besseyi	70	9
Ditylenchus dipsaci	58	23
Radopholus similis[2]	55	11
Globodera pallida	55	_*
Ditylenchus destructor	53	12
Heterodera glycines	52	_*
Aphelenchoides fragariae	47	13
Bursaphelenchus xylophilus	46	_*
Xiphinema index	42	_*
Nacobbus aberrans	38	_*
Xiphinema americanum	30	_*
Anguina tritici	24	_*
Heterodera schachtii	22	16
Bursaphelenchus cocophilus	21	_*

[1] Data compiled from APHIS/USDA data base for countries regulating pests in September 2000 (http://excerpt.ceris.purdue.edu) and from Kahn's 1982 list of ten most frequently regulated nematodes.
_* = Nematode not listed in Kahn's list of ten most regulated nematodes.
[2] 2000 data for *R. similis* includes countries regulating *R. citrophilus,* which Valette et al. (1998) proposed as a junior synonym of *R. similis.* In 2000, 36 countries only list *R. similis* in their regulations, 7 countries list *R. citrophilus* only, and 12 countries list both.

Nacobbus aberrans should be included along with other pests and diseases that are monitored in potato seed certification programs. The potato race is easily disseminated because active second-stage juveniles (J2) are able to invade tubers, where without feeding they molt to inactive J3 and J4, and vermiform adults under favorable environmental conditions (Souza and Baldwin, 1998). These inactive life-stages are a source of inoculum when infected tubers are planted in clean soil (Jatala and de Scurrah, 1975). Most of the nematodes are concentrated in the lenticels and are found in the first millimeters below the epidermis. In small tubers males and females with egg masses are rarely observed (Costilla, 1985). Slight protuberances can be observed around the lenticels, but usually there are no galls, lesions or other infection symptoms on the tubers. Nematodes can survive in the tuber for at least 10 months (Costilla, 1985). Third and fourth stage juveniles may survive in the tubers for many months a quiescence state. Infected tubers are a major means of nematode dissemination that has allowed to be dispersed great distances from the Andean regions to Finland and Russia (Kirjanova and Lobanova, 1975).

The production of certified seed-potatoes free of *N. aberrans* and other nematode pests should be encouraged by local governments in South America to limit the spread of this pest inside and outside those countries. In South America, only Bolivia has adopted a nematode seed certification program for the production of potato tubers free of *N. aberrans* and cyst nematodes (Franco, 1994; Franco, et al., 1998).

More attention needs to be given to the other hosts that may serve as avenues for the dissemination of this pest. *Nacobbus abberans* has many hosts which produce fleshy roots or tubers such as: *Beta vulgaris* L. (table beets and sugarbeets), *Brassica napobrassica* Mill (rutabaga), *Brassica rapa* L. (turnip), *Daucus carota* L. (carrot), *Raphanus sativus* (radish), *Oxalis tuberosa* Molina (oca), *Tropaeolum tuberosum* Ruiz & Pav. (Mashua or anu), and *Ullucus tuberosus* Caldas (olluco) (Canto, 1992). As a result of new trade agreements during the past decade between the USA and countries where *N. aberrans* is found, many vegetable hosts of this nematode, including five fleshy root hosts, are marketed commercially between countries in the trading block with minimal regulatory inspection. In the next decades many regional trading blocks will likely merge and expand. It is likely that a unified trading block will emerge that will include most countries in North, Central and South America. If the trends of the past decade are indicative of the future, this will increase the risk of dissemination of *N. aberrans* because more food crop hosts will be commercially exchanged between countries in these large trading blocks.

Benefits of excluding this pest can be illustrated for potatoes in the USA where both *N. dorsalis* and *N. aberrans* occur, but populations of these species are not known to parasitize potatoes. If races of *N. aberrans* that parasitize potatoes were introduced and became established on potatoes in the USA and losses averaged only 1%, potato growers would lose US$2.6 million annually. Similar losses could be calculated for European potato growers. But in addition, European sugarbeet growers would benefit from efforts to prevent the introduction *N. aberrans* from both South and North America where races of the nematode from both regions parasitize sugarbeet. The introduction of this nematode could have a very negative impact because European countries produce about 85% beet sugar that represents approximately one third of the world's total sugar production from cane and beet.

23.6 Principles for Establishing Successful Cost-beneficial Regulatory Management Programs

Efforts to prevent the introduction of exotic nematodes are interdependent as are links in a chain, and many times a single weakness in a regulatory program may lead its demise or failure. Before any regulatory pest exclusion program is initiated by any agency or country, it is important to evaluate if there are weak links or factors that may limit the program's success. For example, a regulatory program may be eventually doomed to failure if some of the pest's important survival mechanisms or dispersal pathways cannot be regulated due to economic and political considerations.

An understanding of pest dissemination and survival mechanisms is one of the most critical factors regulatory agencies must consider in developing strategies for successful pest exclusion programs. For example, certification and regulatory programs that attempt to prevent the dispersal of *R. reniformis* with ornamental crops must consider that this nematode has a wide host range, including many native plants and weeds that can serve as reservoirs for contamination of plants. In addition, consideration must be given to the fact that the reniform nematode has an exceptional ability to survive extended periods, up to 27 months or longer, in a dehydrated state. The long persistence of this nematode in the soil in the absence of a host complicates any attempt to manage or eradicate it. If these types of key biological factors are overlooked by regulatory agencies, then their pest exclusionary efforts may fail or be of limited value.

Attempts to limit the spread of soybean cyst nematode, *Heterodera glycines*, in the USA illustrate how incomplete biological knowledge in a non-integrated system can result in ineffective regulatory efforts. State and federal quarantines in the USA focused on attempts to prevent the spread of this pest in the soil peds that contaminated seed or on contaminated equipment. Although these efforts slowed down the spread of the pest, the value of these regulatory efforts was diminished because regulatory agencies were unable to address other means of nematode dissemination, such as with river flooding and dispersal by migrating birds. It is not difficult visualize the ironic image a regulatory agent rigorously patrolling ground transportation at inspection stations, while at the same time, birds flying overhead are dropping on the agent's head viable cyst nematodes they have ingested with the worms in nearby infested fields.

Pest distribution and correct pest identification are critical factors. Accurate knowledge of a pest distribution is essential before any regulatory effort is initiated and program strategies must be continually revised as new information of pest distribution becomes available. In 1926 a Federal Domestic Narcissus Bulb Quarantine was enacted in an effort to prevent the spread of *Ditylenchus dipsaci* (Kuhn) Filipjev, the bulb and stem nematode, because of its potential to damage alfalfa and clover crops in western USA. In 1935 this quarantine was revoked when it became evident that this nematode pest was widely distributed and impossible to eradicate. If adequate knowledge of the pest distribution and host range had been available in 1926, the quarantine likely would not have been established. Root-knot nematode quarantines were considered in 1936, but were not invoked for similar reasons (O'Bannon and Esser, 1987).

As new taxonomic information on pests becomes available, regulatory agencies must consider the impact and revise regulations to reflect changes in nomenclature. Failure to do this leads to considerable confusion as to which pest is actually being regulated. As an example, in the late 1990s *Heterodera marioni* was still listed as a regulated pest in Montana, USA and it was impossible to know which species of nematode was being regulated. Some regulatory agencies might have concluded that Montana prohibits a species of cyst nematode, but it would have been more correct to conclude that the regulations might actually refer to any root-knot nematode species, because the linage of the specific epithet *marioni* can be traced to *Anguillula marioni*, a name first used by Cornu in 1879 to refer to nematodes that caused galls, and then later was placed in the genus *Heterodera* by Marcinowski in 1909 where it remained until Chitwood in 1949 recognized that this taxon consisted of many species and described them in the genus *Meloidogyne*.

In some cases, taxonomic research on regulatory nematodes is essential to provide regulatory agencies a sound basis for revising regulatory policy. This is illustrated by regulatory research on *Tylenchulus* species in Florida which was supported by citrus growers. In the 1950s the Florida Department of Agriculture began regulating *Tylenchulus semipenetrans* in citrus nurseries. Prior to establishing a new commercial citrus nursery, the potential site had to be sampled and certified to be free of burrowing and citrus nematodes. In the 1960s and the decades that followed many requests by growers to establish citrus nurseries in uncultivated lands with no history of citrus production were not approved by the Florida Department of Agriculture because a *Tylenchulus* species was found when these areas were sampled. At that time, it was assumed that these *Tylenchulus* populations found in uncultivated areas were strains or races of the citrus nematode. Suspicion arose that these "wild" citrus nematode populations occurring in native areas should not be subjected to regulatory restrictions; however, more information was needed on the distribution, host preference, and morphological characteristics of these populations to justify any change in regulatory policy. In the mid-1980s, the citrus industry supported research by Florida Department of Agriculture nematologists who studied *Tylenchulus* populations from areas that had no previous history of citrus cultivation. It was found that *Tylenchulus* populations from native areas actually consist of two distinct types, and that each type has its own distinct group of hosts. It was also confirmed that neither type was able to infect citrus, and that it was possible to distinguish morphologically these two types of the "wild" populations from each other and from the *Tylenchulus semipenetrans* that damages citrus (Inserra, et al., 1988). The "wild" populations that parasitize climbing hempweed and several other dicot hosts were described as *T. palustris*. The populations that were first found on *Andropogon virginicus* and other grasses were described as *T. graminis* (Inserra, et al., 1988). Because these two new nematode species are not subject to regulatory restrictions, it has resulted in fewer citrus nursery site certification failures. From 1977 to 1986, before the two new *Tylenchulus* species were described a total of 154, or one of every 13 locations failed because the citrus nematode or its "wild" race was detected in samples taken at these locations, but there have been no certification failures in these areas after 1986 (Lehman, et al., 1996). This was very beneficial to Florida's citrus growers because at the time when this regulatory research was conducted there was a critical need for new approved citrus nursery sites that would meet nematode certification standards. From 1978 to 1985, 22 million trees were planted in Florida's citrus groves, but from 1986 to 1993 three times this number, 66 million citrus seedlings were set. This increased demand for citrus seedlings was primarily due to replacement needs following freezes and changes in cultural practices which led to almost twice as many tree being set per acre.

It is important to monitor plant introductions and germplasm banks, especially when new crops are first introduced into a country or region. This is especially difficult because most nematode infections cannot be detected through visual inspection. Most plant introduction stations, including those of APHIS in the USA, do not have an adequate number of trained personnel and equipment to detect nematodes in plant tissues.

Cost-benefits from investments in nursery certification programs that exclude nematodes from perennial crops, even those that cause minor losses, are compounded with time. The data presented in this chapter on Florida's nursery certification program for nematodes of citrus illustrates this principle.

Grower support is essential for success, especially for certification programs. If growers are not convinced that regulatory programs are beneficial to them, they will find ways to get around regulations. In Brazil, for example, nematode certification requirements were established for coffee nurseries by the state of Minas Gerais, but many growers illegally purchased nematode-infested seedlings in the nearby state of Sao Paulo, because the seedlings were lower in price and because they considered the regulations a nuisance restricting their freedom rather than a necessity. In some cases socioeconomic factors have been major obstacles to the success of regulatory certification programs. In Sicily for centuries citrus seedlings have been traditionally grown in the same area in sandy river bottoms that were infested with *Tylenchulus semipenetrans*. Efforts to establish certified citrus nurseries failed because there was a shortage of uninfested land for new nurseries and because growers resisted producing seedlings on raised benches.

In Egypt the government is reluctant to establish new citrus nurseries in the new irrigated areas where citrus is currently being established in nematode-free areas. Instead, citrus seedlings are produced in nurseries that are in the traditional area of citrus production that is infested with citrus nematodes. One of the primary reasons nurseries are not established in the new areas of production is that this would have a very negative economic impact on nurseries in the region where citrus seedlings have been grown for many years.

There must be continuity of legal and financial support of programs and research. Regulatory programs, especially certification programs for perennial crops, require long term commitments. There must be uninterrupted and adequate funding for these regulatory programs. Growers, and regulatory scientists must cooperate in conducting these programs with integrity and consistency.

23.7 Future Perspectives: Opportunities and Challenges

23.7.1 Opportunities

There are still many opportunities to manage nematodes through regulatory exclusion programs. Many nematode pests that are capable of damaging major crops still have a very limited distribution worldwide. For example, the sting nematode, *Belonolaimus longicaudatus* is distributed primarily in southeastern USA. Widely cultivated crops, such as soybean, cotton, corn strawberries, small grains, forage crops, peanut, and vegetables are severely damaged by sting nematodes. Fortunately, sting nematodes are limited in distribution by soil type, requiring soils with at least 80% sand for survival. Nevertheless, most countries have some soils meeting this criteria, sometimes cultivated by growers with limited resources and for whom the introduction of sting nematodes could cause devastating losses that would threaten the survival of those who are dependent on subsistence agriculture. Some countries import turf grasses from southeastern USA and many grasses are also good hosts of the sting nematode. Yet only two countries, the Republic of Korea and New Zealand, have regulations to prohibit the introduction of sting nematodes.

Worldwide there are still many opportunities for growers to prevent nematode damage in perennial fruit crops and annual vegetable crops by using sanitation practices which exclude nematodes from seedlings produced in nurseries and seed beds. But unfortunately, too few of these programs have been implemented. Examples of nematodes that are frequently distributed on root stocks are: *Hemicriconemoides mangiferae* Siddiqi on mango and lychee; *Heterodera fici* Kirjanova on fig; *Meloidogyne* species on citrus, cocoa, coffee, fig, guava, kiwi, papaya, passion fruit, pistachio, and tea; *Pratylenchus coffeae* (Zimmerman) Filipjev & Schuurmans Stekhoven on citrus and coffee; *Pratylenchus loosi* Loof on tea; *Pratylenchus vulnus* Allen & Jensen on fig, peach and olive; *Radopholus* species on citrus and tea; *Rotylenchulus reniformis* on coffee, papaya, and tea; *Tylenchulus semipenetrans* on citrus, olive and persimmon; and *Xiphinema index* Thorne & Allen on grape (Campos, et al., 1990; Cohn and Duncan, 1990; Duncan and Cohn, 1990).

Many times governments promote and support the planting of certain crops in new regions. For example, in recent years the government of Egypt has had reclamation projects and promoted the planting of citrus and other crops in new regions. In such cases there is a window of opportunity to educate growers, and develop sanitation and certification programs that are effective in preventing the introduction of nematodes on contaminated equipment or infected planting stock. These efforts are especially important when perennial crops are established in a new region, because growers would realize the benefits of excluding nematodes for many years or even decades.

There are many opportunities for regional cooperation to determine the distribution of races and species of nematodes using both classical and molecular techniques. Accurate information on the genetic diversity and nematode distribution is essential for plant breeders to develop resistant crops that are of practical use to growers. This type of regional cooperative project to determine the distribution of the races and root-knot species of coffee in order to develop durable nematode resistance has been initiated for Latin America by the major coffee research agencies in Latin America and the European Union. Whenever possible, regulatory nematologists should provide cooperative support for such effort because such systematic surveys often reveal opportunities to establish seed or planting stock certifications programs because the common nematode pests on a crop may have a disjunct distribution. For example, preliminary results from the Latin American survey project of coffee nematodes indicate that the three most important root-knot species, *Meloidogyne exigua*, *M. incognita*, and *M. paranaensis* each occur in separate regions where coffee is grown in Brazil (R. Carneiro, pers. comm.).

The science of nematology witnessed an exponential growth of knowledge in the last half of the 20th century and this has been coupled with an opportunity for scientifically based regulatory nematode exclusion programs. In West Africa, surveys revealed that the two most damaging nematodes of banana and plantains, *Pratylenchus goodeyi* and *R. similis*, have distinct regional distributions associated with temperature and altitude (Bridge, et al., 1995). *Radopholus similis* is adapted to higher temperatures and is consistently found below 900 m where most commercial production occurs. *Pratylenchus goodyei*, which has a lower temperature preference than *R. similis*, is also a pest of bananas and plantains growing at higher altitudes where these crops are primarily grown by small landholders. Because *P. goodeyi* is a pest exclusively of small holder cultivation, this may explain why it has not been widely distributed in contrast to *R. similis*, which has been disseminated on commercial planting stock worldwide. Although *P. goodeyi* populations from the highlands do not appear to be adapted to the higher temperatures in lowland areas, however, *P. goodeyi* has the potential to become an important pest of bananas where they are grown in cooler climatic zones such as in the Mediterranean and Middle East countries. These examples illustrate the importance of understanding the edaphic and climatic requirements for nematode pests and how they influence nematode distribution, in order to evaluate the costs and benefits for establishing programs that promote the use of nematode-free planting stock.

Even if nematodes are widely distributed in a geopolitical region, regulatory management to exclude them from areas where they do not occur, still may be beneficial. These opportunities are often overlooked. An example is citrus nematode, *Tylenchulus semipenetrans,* which occurs in all major citrus producing countries of the world. Certainly in the strictest definition, this nematode would not be classified as an exotic pest in any of these countries. This nematode pest, however, illustrates an important principle. In a world with expanding multinational trade agreements, national regulatory agencies need to rethink pest exclusion policies that are based primarily on the organism distribution in relation to geopolitical boundaries. Even if this nematode occurs within a country, based on citrus nematode biology, it should be considered regionally exotic and excluded from any new area where citrus is planted for the first time, or even in areas where citrus has been planted previously, if sampling indicates the absence of citrus nematode. The citrus nematode probably coevolved with citrus or citrus relatives in southeast Asia. This nematode is a highly specialized parasite, in that it develops very high densities on its obligate host with relatively low damage. Physiological races of this nematode have evolved, but it parasitizes only a few other crops, such as, olive, persimmons and grapes, and it has few reservoir hosts. Even within infested groves or citrus orchards the distribution and densities of the citrus nematode is generally not very uniform, reflecting that this nematode has evolved a very specialized form of parasitism that has minimized the selection pressure for specialized survival and dispersal modes outside its host. Therefore, the citrus nematode can be excluded from very limited areas, if sanitation practices are established for groves and nurseries.

Regulatory agencies have an opportunity to shift their thinking from pest exclusionary policies that are geopolitically based to a more cooperative multinational biologically based approach that includes exclusion policies for nematode races that are not present in a region, even though other races of the same species have already been introduced into a country or region.

23.7.2 Future Challenges

Regulatory agencies need to respond to the following changes and challenges associated with increased travel and emergence of large trade organizations:

Regulatory agencies must continue to strengthen efforts leading to greater regional cooperation Regulatory efforts should not be competitive, but must be cooperative and international in scope. There is a need to strengthen regional regulatory organizations and practice greater transparency in conducting pest risk analyses. Information must be obtained cooperatively and shared freely utilizing information technology. The principles of the International Plant Protection Convention (IPPC), which were adopted in 1952 and have been ratified by most nations, provide for international cooperation in which each country agrees to report on the existence, outbreak, and spread of economically important pests and to share information on means of control of pests (Van der Graaff, 2001). Too often, however, the absence of evidence is considered evidence of absence. If countries certify that a pest is not present, this should be supported by adequate survey data. It is counterproductive, if countries benefit because they have a paucity of information on nematode distribution. Scientists who find new or introduced pests in some cases may hesitate to share and publish this information because there are few positive incentives to do so, but instead they expect from past experience that they will receive a negative response from administrators and growers because such news will result in economic losses due to regulatory restrictions on trade. One of the greatest challenges facing regulatory agencies in the new millennium is harmonization, or developing biologically based international regulatory standards and guidelines that are as similar as possible, and which assure that any measures that restrict trade are commensurate with risk (Van der Graaff, 2001). The Agreement on Sanitary and Phytosanitary Measures adopted by member countries of the World Trade Organization require that each countries' regulations are developed from pest risk analyses that is scientifically based and in line with international standards. Any member country may challenge another country to justify its actions if it believes the other country's phytosanitary regulations contravenes these principles and unjustifiably serve as a barrier to trade. This means that in the future regulatory nematologists will be asked to provide more precise information with regard to nematode detection and identification, ecoclimatic requirements, geographic distribution, host economic damage, etc. In many cases this will require regional cooperative efforts.

There are many ways that efforts to exclude exotic nematodes need to be strengthened based on current biological knowledge. In Europe, for example, the EPPO quarantine lists emphasize the need to protect temperate crops, probably reflecting that strong trade blocks and extensive understanding of nematological problems first developed in northern Europe. Nematodes such as *Rotylenchulus reniformis*, *Pratylenchus coffeae*, *Belonolaimus* and *Hoplolaimus* species are not present in southern European countries. Biological studies indicate these nematodes may cause substantial losses to many warm-climate crops, which are grown in southern Europe, so this raises the question of whether these nematodes also need to be included on EPPO lists of exotic pests that warrant exclusionary efforts.

The increased movement of plant propagative material must be consistently monitored A weak link or inconsistency in efforts to exclude exotic nematodes relates to rapid changes in patterns of commerce and trade that have influenced the ornamental plant industry worldwide during the past 15 years. Many countries, including the USA, for many years have prohibited the importation of soil or plants in soil in order to exclude exotic soil-borne pests, however, the importation of plant cuttings and bare-root plants has been permitted. This policy was established when few bare-rooted plants were imported and when it was thought that most root pests could be detected by visual inspection. To reduce labor costs, it is becoming more and more common for nurseries to have multinational operations and airfreight planting material from other countries. For example, ornamental nurseries in Florida import about 500 million cuttings and bare-rooted plants annually from more than 40 different countries. This exchange of propagative plant material also occurs in other states and countries that market their ornamental nursery products throughout their country or export them worldwide. Studies of plant-parasitic nematodes such as *Radopholus*, *Pratylenchus*, and *Rotylenchulus* species may migrate into stem tissue or may remain attached to very minute roots at the stem base. Furthermore, careful examination of many plants that are shipped bare-root frequently reveal some soil adhering to some of the roots, so even ectoparasitic nematodes may be shipped with bare-rooted plants. In this case, United States federal regulatory policy does not reflect current understanding of nematode biology and that ornamental plants may serve as a "Trojan horse" from which exotic nematodes are introduced to food crops. This is not consistent with the stronger sanitation requirements of not permitting importation of plants in soil. This is a weak link in current pest exclusion efforts. Federal inspection standards at the ports of entry, or sanitation standards at the site where bare-rooted plants and cuttings originate, need to be modified and strengthened.

Regulatory agencies need to cooperatively establish centers of excellence for training and research in new diagnostics techniques Correct identification and knowledge of pest distribution are essential for development of regulatory policy and management. As world trade continues to expand, this will increase the volume of plant propagative materials and vegetables that regulatory agencies need to monitor for nematode pests. This will require rapid inexpensive diagnostic techniques, not only for nematode species but also races and pathotypes. Regulatory agencies will need to support long-term research that will allow them to integrate classical taxonomic and molecular diagnostic techniques.

As an example, molecular and morphological analysis of isolates of *Pratylenchus coffeae* and closely related species has provided a biological basis for regulating field sites for citrus nurseries in Florida (Duncan, et al., 1999; Inserra, et al., 2001). For many years certain populations of lesion nematodes found on weeds in uncultivated areas were thought to be *P. coffeae,* and consequently these sites were not approved for citrus nurseries. This research has clarified that populations on weeds are different species that are distinct from populations that attack citrus and this made it possible to modify regulatory approval of citrus nursery sites.

Molecular studies can assist regulatory scientists when taxonomic actions, such as has occurred for the anguinids, complicate regulatory decisions. In addition to taxonomic changes, morphometric identification and regulatory decisions for anguinids have been further complicated by fact that often, only juveniles are found in galls recovered from seed shipments. Recent studies have shown that DNA molecular analysis of PCR-RFLP patterns for the ITS1 region is useful to identify closely related anguinid nematodes (Powers, et al., 2001). *Anguina* species that had been synonymized in the past, such as, *Anguina agrostis* (Steinbuch) Filepjev, *A. funesta* Price, Fisher & Kerr, and *A. wevelli* (*Afrina wevelli*) van den Berg are readily differentiated with this technique. Molecular research has indicated that *A. funesta*, the vector of the notoriously toxic bacteria, *Rathayibacter toxius*, should be the primary target of regulatory action. In this case, new molecular diagnostic tools should allow regulatory agencies to revise their regulations and standardize identification, because misidentification of anguinid species can result in significant economic and political impact for countries that market grain and seeds in international trade.

Another example is the clarification of the relationship of *Radopholus similis* and *R. citrophilus*, which was possible only after many years of research, and without which there would be continuing confusion as to which of these nematodes are regulated by some countries. Following the description of *R. citrophilus* the regulations of some countries included both species. Other countries, because of practical considerations, did not accept or include *R. citrophilus* in their regulations and continued to treat both as one species *R. similis, sensu lato*. In some cases, this was a source of confusion when regulatory agencies issued phytosanitary certificates. Molecular genetic studies have provided information that provides a scientific basis for the synonymization of *R. similis* and *R. citrophilus* (Kaplan and Opperman, 1997, 2000; Kaplan, et al., 2000; Valette, et al., 1998).

23.8 Conclusion

The past decades have been characterized by greater environmental awareness and concern, recognition of need for less dependence on chemical pesticides, and increased emphasis on alternative methods of control. In coming decades, the expanding world population will increase the worldwide need for food and fiber, and the desire to share in the benefits of technology will continue to spread globally. As these needs conflict with the planets fragile ecosystems and global environmental concerns increase, there will be a growing need for cooperative regional biologically based regulatory policy. Sanitation and certification programs, which exclude nematodes lower pesticide use and thus impact favorably on the environment. Exclusion of nematodes from new areas of production is the most effective and cost-beneficial means of managing nematodes. The rapid expansion of international trade and travel that has characterized the end of the past millennium will continue, as will the need for cost-efficient regulatory management programs.

References

Anonymous. 1933. A brief history of plant quarantine in California. Pp. 109 – 114. In The Efficacy and Economic Effects of Plant Quarantines in California. University of California Agricultural Experiment Station, Berkeley, California Bulletin 553.

Anonymous. 2001. Schemes for the production of healthy plants for planting. OEPP/EPPO Bulletin 31: 435 – 478.

Bridge, J., N. S. Price, and P. Kofi. 1995. Plant parasitic nematodes of plantain and other crops in Cameroon, West Africa. Fundamental and Applied Nematology 18: 251 - 260.

Brodie, B. B. 1998. Potato cyst nematodes (*Globodera* species) in Central and North America. Pp. 317 - 331. In R. J. Marks and B. B. Brodie, eds. Potato Cyst Nematodes Biology, Distribution and Control. Wallingford, UK: CAB International.

California DFA (Department of Food and Agriculture). 2000. Nematology laboratory. http//www. cdfa. ca. gov/plant/ppd/nematolo. htm

Campos, V. P., P. Sivapalan, and C. Gnanapragasam. 1990. Nematode parasites of coffee, cocoa, and tea. Pp. 387 - 430. In M. Luc, R. A. Sikora, and J. Bridge, eds. Plant Parasitic Nematodes in Subtropical and Tropical Agriculture. Wallingford, UK: CAB International.

Canto, M. 1992. Life cycle and pathogenicity of *Nacobbus aberrans* (Thorne, 1935) Thorne and Allen 1994. Pp. 113 - 127. In F. J. Gommers and P. W. Th. Maas, eds. Nematology from Molecule to Ecosystem. Wildervank, The Netherlands: Dekker and Huisman.

Cobb, N. A. 1915. *Tylenchus similis,* the cause of a root disease of sugarcane and bananas. Journal of Agricultural Research 4: 561 - 568.

Cohn, E., and L. Duncan. 1990. Nematode parasites of tropical and subtropical fruits. Pp. 347 - 362. In M. Luc, R. A. Sikora, and J. Bridge, eds. Plant Parasitic Nematodes in Subtropical and Tropical Agriculture. Wallingford, UK: CAB International.

Costilla, M. A. 1985. El falso nematode del nudo *Nacobbus aberrans* (Thorne, 1935) Thorne and Allen, 1944 ans su relaciôn con el cultivo de papa en el noreste argentino. Revista Industrial y Agricola de Tucuman 62: 79 - 97.

Crawford, R. P. 1922. World crops of America. Scientific American 126: 226 - 227.

Doucet, M. E., and M. M. A. de Doucet. 1997. Nematodes and agriculture in continental Argentina: an overview. Fundamentals of Applied Nematology 20: 521 - 539.

Dowling, C. F., A. E. Graham, and S. A. Alfieri. 1982. Plant inspection and certification. Plant Disease 66: 345 - 351.

Duncan, L. W., and E. Cohn. 1990. Nematode parasites of citrus. Pp. 321 - 346. In M. Luc, R. A. Sikora, and J. Bridge, eds. Plant parasitic nematodes in subtropical and tropical agriculture. Wallingford, UK: CAB International.

Duncan, L. W., R. N. Inserra, W. K. Thomas, D. Dunn, I. Mustika, L. M. Frisse, M. L. Mendes, K. Morris, and D. Kaplan. 1999. Molecular and morphological analysis of isolates of *Pratylenchus coffeae* and closely related species. Nematropica 29: 61 - 80.

EPPO. 1997. Data sheets on quarantine organisms: *Globodera rostochiensis* and *Globodera pallida*. Pp. 601 - 606. In I. M. Smith, D. G. McNamara, P. R. Scott, M. Holderness, and B. Burger, eds. Quarantine Pests for Europe. Wallingford, UK: CAB International.

Esser, R. P., J. H. O'Bannon, and C. C. Riherd. 1988. The citrus site approval program for the burrowing nematode and its beneficial effect on the citrus industry of Florida. OEPP/EPPO Bulletin 18: 579 – 586.

Evans, K., and B. B. Brodie. 1980. The origin and distribution of the golden nematode and its potential in the USA. American Potato Journal 57: 79 – 89.

Fairchild, D. 1948. Early experiences with the soybean. Soybean Digest 9: 14 – 15.

Fairchild, D. 1906. Our plant immigrants. National Geographic Magazine 17: 179 – 291.

Foster, J. 1991. Exclusion of plant pests by inspections, certification and quarantines. Pp. 311 – 338. In A. A. Hanson and D. Pimental, eds. CRC Handbook of Pest Management in Agriculture, 2nd edition, Vol I. Roca Raton, FL: CRC Press.

Franco, J. 1994. Problemas de nematodos en la producción de papa en climas templados en la región andina. Nematropica 24: 179 – 195.

Franco, J., R. Oros, G. Main, and N. Ortuno. 1998. Potato cyst nematodes (*Globodera* species) in South America. Pp. 239 – 269. In R. J. Marks and B. B. Brodie, eds. Potato Cyst Nematodes Biology, Distribution and Control. Wallingford, UK: CAB International.

Hopkins, C. G. 1904. Nitrogen bacteria and legumes (with special reference to red clover, cowpeas, soybeans, alfalfa, and sweet clover) on Illinois soils. Illinois Agricultural Experiment Station Bulletin 94: 307 – 328.

Inserra, R. N., N. Vovlas, J. H. O' Bannon, and R. P. Esser. 1988. *Tylenchulus graminis* n. sp. and *T. palustris* n. sp. (Tylenchulidae), from native flora of Florida, with notes on *T. semipenetrans* and *T. furcus*. Journal Nematology 20: 266 – 287.

Inserra, R. N., L. W. Duncan, A. Troccoli, D. Dunn, J. Maia dos Santos, N. Kaplan, and N. Vovlas. 2001. *Pratylenchus jaehni* sp. n. from citrus in Brazil and its relationship with *P. coffeae* and *P. loosi* (Nematoda: Pratylenchidae). Nematology 3: 653 – 665.

Jatala, P., and M. M. De Scurrah. 1975. Mode of dissemination of *Nacobbus* spp. in certain potato – growing areas of Peru and Bolivia. Journal of Nematology 7: 324.

Jefferson, R. M., and A. E. Fusonie. 1977. The Japanese flowering cherry trees of Washington D. C. – a living symbol of friendship. National Arboretum Contribution No. 4, ARS, USDA.

Kahn, R. P. 1982. The host as a vector: exclusion as control. Pp. 123 – 149. In K. F. Harris and K. Marmorsch, eds. Pathogens, Vectors and Plant Diseases: Approaches to Control. New York: Academic Press.

Kaplan, D. T., and C. H. Opperman. 1997. Genome similarity implies that citrus-parasitic burrowing nematodes do not represent a unique species. Journal of Nematology 29: 430 – 440.

Kaplan, D. T., and C. H. Opperman. 2000. Reproductive strategies and karyotype of the burrowing nematode, *Radopholus similis*. Journal of Nematology 23: 126 – 133.

Kaplan, D. T., W. T. Thomas, L. M. Frisse, J. L. Sarah, J. M. Stanton, P. R. Speijer, D. H. Marin, and C. H. Opperman. 2000. Phylogenetic analysis of geographically diverse *Radopholus similis* via rDNA sequence reveals a monomorphic motif. Journal of Nematology 32: 134 – 142.

Kirjanova, E. S., and N. A. Lobanova. 1975. [A potato parasite] Zashchita Rastenii, Moscow 9, 49. (From Helminthological Abstracts, Series B, 1979, 48 (4): 150 # 1429.)

Kleynhans, K. P. N. 1998. Potato cyst nematodes (*Globodera* species) in Africa. Pp. 347 – 351. In R. J. Marks and B. B. Brodie, eds. Potato Cyst Nematodes Biology, Distribution and Control. Wallingford, UK: CAB International.

Koenning, S. R., C. Overstreet, J. W. Noling, P. A. Donald, J. O. Becker, and B. A. Fortnum. 1999. Survey of crop losses in response to Phytoparsitic nematodes in the United States for 1994. Supplement to Journal of Nematology 31: 587 – 618.

Lee, R. F., P. S. Lehman, and L. Navarro. 1999. Nursery practices and certification programs for budwood and rootstocks. Pp. 35 – 46. In L. W. Timmer and L. W. Duncan, eds. Citrus Health Management. St. Paul, MN: APS Press.

Lehman, P. S. 1995. Role of plant protection organizations in nematode management. Pp. 137 – 148. In Programa e anais Congresso International de Nematologia Tropical, June 4 – 9, 1995, RioQuente, Brasil.

Lehman, P. S., W. W. Smith, and R. N. Inserra. 1996. A ten year assessment of benefits to the citrus nursery and soil pit industries in Florida from regulatory research on *Tylenchulus* species. Florida Department of Agriculture and Consumer Services, Division of Plant Industry Nematology Circular 215.

McNamara, D. G. 1995. The role of European an Mediterranean Plant Protection Organization in nematode management. Pp. 122 – 127. In Programa e Anais Congresso International de Nematologia Tropical, June 4 – 9, 1995, Rio Quente, Brasil.

Marin, D. H., T. B. Sutton, and K. R. Barker. 1998. Dissemination of bananas in Latin America and the Caribbean and its relationship to the occurrence of *Radopholus similis*. Plant Disease 82: 964 – 974.

Marshall, J. W. 1998. Potato cyst nematodes (*Globodera* species) in New Zealand and Australia. Pp. 353 – 394. In R. J. Marks and B. B. Brodie, eds. Potato cyst nematodes biology, distribution and control. Wallingford, UK: CAB International.

Noel, G. R. 1992. History, distribution and economics. Pp. 1 – 13. In R. D. Riggs and J. A. Wrather, eds. Biology and Management of the Soybean Cyst Nematode. St. Paul, MN: APS Press.

O'Bannon, J. H., and R. P. Esser. 1987. Regulatory perspectives in nematology. Pp. 38 – 46. In J. A. Veech and D. W. Dickson, eds. Vistas on Nematology. Hyattsville, MD: Society of Nematologists.

Poucher, C., H. W. Ford, R. F. Suit, and E. P. DuCharme. 1967. Burrowing Nematode in Citrus. Florida Department of Agriculture, Division of Plant Industry Bulletin 7.

Powers, T. O., A. L. Szalanski, P. G. Mullin, T. S. Harris, T. Bertozzi, and J. A. Griesbach. 2001. identification of seed gall nematodes of agronomic and regulatory concern with PCR-RFLP of ITS1. Journal of Nematology 33: 191 – 194.

Robinson, A. F., R. N. Inserra, E. P. Caswell-Chen, N. Vovlas, and A. Troccoli. 1997. *Rotylenchulus* species: identification, distribution, host ranges, and crop plant resistance. Nematropica 27: 127 – 180.

Schumann, G. L. 1991. Plant Diseases: Their Social Impact. St. Paul, MN: APS Press.

Souza, R. M., and J. G. Baldwin. 1998. Changes in the esophageal gland activity during the life cycle of *Nacobbus abberans* (Nemata: Pratylenchidae). Journal of Nematology 30: 275 – 290.

Temple, J. C. 1916. Studies in *Bacillus radicicola*. Georgia Experiment Station Bulletin 120.

Turner, S. J., and K. Evans. 1998. The origins, global distribution and biology of potato cyst nematodes (*Globodera rostochiensis* (Woll.) and *Globodera pallida* Stone). Pp. 27 – 49. In R. J. Marks and B. B. Brodie, eds. Potato Cyst Nematodes Biology, Distribution and Control. Wallingford, UK: CAB International.

USDA. 1994. Desk reference Guide to U. S. Agricultural Trade. Agriculture handbook No. 683, Foreign Agricultural Service, Washington, D. C.

USDA. 1997. National Agricultural Statistics Service 1997 census of agriculture. http://www.nass.usda.gov/census

USDA. 1999. National Agricultural Statistics Service 1999 potato summary. http://USDA.mannlib.cornell.edu:80/USDA/

Valette, C. D., D. Mounport, M. Nicole, J. L. Sarah, and P. Baujard. 1998. Scanning electron microscope study of two African populations of *Radopholus similis* (Nematoda: Pratylenchidae) and proposal of *R. citrophilus* as a junior synonym of *R. similis*. Fundamentals and Applied Nematology 21: 139 – 146.

Van der Graaff, N. A. 2001. International cooperation in plant protection 1951 – 2001. OEPP/EPPO Bulletin 31: 341 – 343.

van Reil, H. R., and A. Mulder. 1998. Potato cyst nematodes (Globodera species) in western Europe. Pp. 271 – 298. In R. J. Marks and B. B. Brodie, eds. Potato Cyst Nematodes Biology, Distribution and Control. Wallingford, UK: CAB International.

Whitehead, A. G., and S. J. Turner. 1998. Management and regulatory control strategies for potato cyst nematodes (*Globodera rostochiensis* and *Globodera pallida*). Pp. 135 – 152. In R. J. Marks and B. B. Brodie, eds. Potato Cyst Nematodes Biology, Distribution and Control. Wallingford, UK: CAB International.

Zaheer, K. 1998. Potato cyst nematodes (*Globodera* species) in Asia. Pp. 333 – 345. In R. J. Marks and B. B. Brodie, eds. Potato Cyst Nematodes Biology, Distribution and Control. Wallingford, UK: CAB International.

Zasada, I. A., and M. Gatt. 2000. The first report of *Globodera pallida* in Malta. Nematologia Mediterranea 28: 101 – 102.

Zwingle, E. 1999. Olive oil, elixir of the gods. National Geographic 123: 66 – 81.

24 Nematicides: Past and Present Uses *

J. R. Rich, R. A. Dunn and J. W. Noling

24.1 Introduction

The discovery that certain chemicals possessed nematicidal properties and their subsequent use in agriculture greatly impacted crop production by improving both yield and quality of crops worldwide. Equally important was the recognition from the use of nematicides that nematodes were causing previously unidentified damage to plants. The resulting increase in nematology positions during the early 1950s helped mold the Science of Nematology as we know it today. Subsequently, nematologists have made tremendous strides in basic nematode biology, problem diagnosis, and nematode management methods. Nematicides, however, shaped the face of phytonematology and continue to be important in production of many crops. The importance of nematicides to nematology is evidenced by a number of excellent reviews published over the years (Bunt, 1987; Castro and Thomason, 1971; Johnson and Feldmesser, 1987; Lembright, 1990; Newhall, 1955; Taylor, 1959; Thorne, 1961; Van Berkum and Hoestra, 1979; Van Gundy and McKenry, 1977; Whitehead, 1978; Winfield, 1978; Wright, 1981). Since extensive reviews are available, information in this chapter will be presented as a general overview of current nematicide status and uses. For more detailed information on specific topics, please refer to the listed references.

24.2 History

Concepts, ideas and theory generally precede discoveries in science, and this was true with the development of nematicidal controls for plant-parasitic nematodes (Taylor, 1979). The stage for nematode control by chemicals was

---

* This chapter is dedicated to the memory and many achievements of J. R. Christie and A. L. Taylor who worked extensively in the early development of nematicides.

set when Garreau (1854) recognized the insecticidal value of carbon disulfide (bisulfide), and later, soil application of carbon disulfide by Thenard in 1872 to control grape phylloxera (Fleming and Baker, 1935; Wilhelm, 1966). Phylloxera had destroyed over 1,000,000 ha of French vineyards (Metcalf and Flint, 1962), and carbon disulfide became extensively used to control this soilborne insect (Newhall, 1955). The successful control of phylloxera probably led Kuhn (1881) to initiate experiments to control the sugar beet cyst nematode, *Heterodera schachtii,* with carbon disulfide but these experiments apparently were inconclusive (Thorne, 1961). Later work by Bessey (1911), however, documented root-knot nematode control with carbon disulfide in the field. Few records indicate use by farmers of carbon disulfide for nematode control although its nematicidal efficacy has been shown on numerous occasions over the years (Taylor, 1959). Little early use probably was due to limited knowledge of nematode-induced disease problems, and in addition, the disadvantages of high cost, high rates needed, and flammability of the chemical posed risks. In recent years, there has been renewed interest in use of carbon disulfide by application of sodium tetrathiocarbonate (Enzone[R]), which generates carbon disulfide as it hydrolyzes in soil (Matheron and Matejka, 1988).

The next chemical found to possess nematicidal properties was chloropicrin (trichloronitromethane), first tested by Matthews (1920) in England for nematode and wireworm control. It was tested, among other reasons, to find a use for large stockpiles of chloropicrin accumulated during World War I (Taylor, 1979). Chloropicrin tear "gas" was used during the war to disable enemy troops and still is used widely in riot control. In pot tests, Matthews found that chloropicrin increased tomato yields by 73% and was "most effective" in control of nematodes. Later, Johnson and Godfrey (1932) reported that chloropicrin controlled root-knot nematodes in California. Further work confirmed efficacy of chloropicrin in Hawaiian pineapple production where negative correlations between chloropicrin rates and nematode infection levels were shown (Godfrey, 1935). Yields of pineapple were improved by as much as 57%. Subsequently, chloropicrin fumigation was widely adapted in Hawaiian pineapple production as long as surplus supplies lasted (Thorne, 1961). Other soil uses of chloropicrin included seedbeds, nurseries, and specialty crops (Johnson and Feldmesser, 1987). Like carbon disulfide, chloropicrin is a broad-spectrum biocide (Stark, 1948) and continues to be used over 80 years after its discovery as a nematicide. Chloropicrin, however, is most useful as a soil fungicide and is applied only secondarily as a nematicide due to the availability of more efficacious and economical nematicidal products (Lembright, 1990; Rich and Whitty, 1999; Wilhelm, et al., 1961).

The decade between 1940 and 1950 was profoundly important for the Science of Nematology. Nematicidal properties were discovered for three chemicals: methyl bromide (bromomethane), D-D mixture (1, 3-dichloropropene, 1, 2-dichloropropane) and EDB (1, 2-dibromoethane; commonly called ethylene dibromide). Methyl bromide was first reported by Christie and Cobb (1940) to have activity against the chrysanthemum foliar nematode. However, it was phytotoxic so could not be used for their purposes. Pot trials by Hawkins (1939) prompted Taylor and McBeth (1940) to test the material as a soil nematicide with some success. Taylor and McBeth (1941) increased the efficacy of methyl bromide by covering treated soil with a gas impervious paper to help hold the gas in soil longer. These experiments were the forerunners of the plastic mulch system of methyl bromide application that still is used (Noling and Becker, 1994). Methyl bromide has been shown on numerous occasions to be an excellent biocide with broad-spectrum activity on various weeds, insects, nematodes, and fungi. It is widely utilized on seedbeds and high value crops but less on other crops due to specialized application techniques and expense. Unfortunately, recent suggestions that methyl bromide may deplete the atmospheric ozone (Watson, et al., 1992) have led to a scheduled phaseout of the chemical in industrialized countries by 2005 and remaining countries by 2015 (Noling and Gilreath, 2000).

The major discovery of D-D for nematode control and its subsequent importance to plant nematology commenced with work of Carter (1943) at the Pineapple Research Institute in Hawaii. He found that D-D mixture controlled nematodes in pineapple more efficiently than chloropicrin. Almost concurrently with the discovery of nematicidal activity of D-D, good nematode control with EDB was reported (Christie, 1945; Thorne and Jensen, 1946). It was equally effective to D-D mixture in controlling root-knot nematodes. Both D-D and EDB, unlike previously identified fumigants, were primarily nematicidal chemicals, easier to apply, and more economical to use. These properties plus the combined efforts of Shell Chemical Company, Dow Chemical Company, The Pineapple Research Institute, and state and federal agencies stimulated widespread acceptance and use of these products within several years (Taylor, 1979; Thorne, 1961). In later years, the 1, 3-dichloropropene (1, 3-D) component of the D-D mixture was shown to represent approximately 98% of the nematicidal activity of the mixture (Youngson and Goring, 1970). As a result of these findings and presence of 1, 2-dichloropropane (1, 2-D) contaminants in drinking water, 1, 2-D was subsequently removed from the mixture. After extensive re-registration review of 1, 3-D by the US Environmental Protection Agency (EPA), the product was approved, with some restrictions, for continued use in the USA. EDB was canceled by EPA in 1983 due to possible carcinogenicity and widespread groundwater contamination. In subsequent years, this action was followed by most countries in the world.

Another widely used fumigant nematicide was DBCP (1, 2-dibromo-3-chloropropane) (McBeth, 1954; McBeth and Bergeson, 1955; Raski, 1954). The material was an effective nematicide and could be used at low rates on many plants, thus providing economical control of nematodes in crops generally considered low value. Unlike other fumigants, DBCP was relatively non-toxic to living plants, so it found wide use on perennial crops as well as by applications at planting time for many annual crops (Taylor, 1979). DBCP generally was applied by soil injection but also could be mixed into soil in a granular formulation or applied in irrigation water. A limitation of DBCP was phytotoxicity to some members of the plant family Solanaceae and a few other crops, so it was not recommended for use on such crops as tobacco, tomatoes, and potatoes (Johnson and Feldmesser, 1987). The chemical is no longer used in most countries due to documentation of male sterility, possible carcinogenesis, and presence in groundwater (Feldmesser and Smart, 1982).

Metham (sodium N-methyl dithiocarbamate dihydrate) was the last fumigant nematicide/biocide introduced (Lear, 1956) and has been shown to control various nematodes, weeds, some fungi and insects. This material hydrolyzes in soil to form a volatile gas, methyl isothiocyanate (MIT), which is the toxic entity. Metham can be applied as a drench, in irrigation water or injected into the soil. Over the years, metham has been shown to be useful as a soil fungicide but like chloropicrin, is less effective as a nematicide and more expensive than EDB, DD, and DBCP. The material still is used widely today for control of soilborne fungi on a large number of crops, but generally only secondarily as a nematicide.

A second era of nematicide chemistry and mode of action was initiated with tests using V-C 13 (0-2, 4-dichlorophenyl 0, 0-diethyl phosphorothioate) (Christie and Perry, 1958; Manzelli, 1955). V-C 13 was the first non-fumigant nematicide, and movement was in soil water rather than the soil air to water interface of fumigants. The discovery of nematicidal activity of this chemical led to testing and development of several other non-fumigant nematicides such as aldicarb, carbofuran, ethoprop, and fenamiphos which are still in production today (Tables 24.1 – 24.2).

Table 24.1 Chronological history of the development of nematicides.

Common name	Chemical name(s)[1]	Author(s) & date(s)[2]
Carbon disulfide	Carbon disulfide, Carbon bisulfide	Kuhn, 1881; Bessey, 1911
Chloropicrin	Trichloronitromethane	Matthews, 1920
Methyl bromide	Bromomethane	Christie and Cobb, 1940
D-D mixture	1,2-dichloropropane; 1,3-dichloropropene	Carter, 1943
Ethylene dibromide	1,2-dibromoethane	Christie, 1945; Thorne and Jenson, 1946
DBCP	1,2-dibromo-3-chloropropane	McBeth, 1954; Raski, 1954
Metham[3]	Sodium N-methyl dithiocarbamate	Lear, 1956
V-C 13	0-2,4-dichlorophenyl 0, 0-diethyl phosphorothiate	Manzelli, 1955
Thionazin	0, 0-diethyl 0-pyrazinyl phosphorothioate	Jenkins and Guengerich, 1959
Fensulfothion	0,0-diethyl 0-4-(methylsulfinyl) Phenylphosphorothioate	Good, 1963; Reed, 1963
Aldicarb	2-methyl-2-(methylthio) propionaldehyde 0-(methyl-carbamoyl) oxime	Miller, 1966
Fenamiphos (Phenamiphos)	Ethyl 3-methyl-4-(methylthio) phenyl (1-methyl) phosphoramidate	O'Bannon and Taylor, 1967
Ethoprop(hos)	O-ethyl S, S-dipropyl phosphorodithioate	Osborne, et al., 1969
Oxamyl	S-methyl N'N'-dimethyl-N-[(methyl Carbamoyl) oxy]-1-thio-oxamimidate	Radewald, et al., 1970
Terbufos	S-[[(1,1-dimethylethyl) thio] methyl] 0,0-di-ethylphosphoro-dthioate	Gordon, 1977
Carbofuran	2,3-dihydro-2,2-dimethyl-7-benzofuranyl methylcarbamate	Gowen, 1978; DiSanzo, 1981

[1] Source of chemical names, Farm Chemicals Handbook (Curran, 2001).

[2] Represents the first recognized scientific publications and dates.

[3] Metham is only one of several methyl isothiocyanate-generating compounds.

Table 24.2 Major nematicides, past and present, used worldwide for plant-parasitic nematode control.

Common name	Trade names in the USA[1]	Regulatory status in the USA[2]	Fumigant (F)/ Non-fumigant(NF)
Carbon tetrathiocarbonate	Enzone	R	F
Chloropricrin	Chlor-O-Pic Others	R	F
Methyl bromide	Many	C[3]	F
D-D mixture	Shell D-D	C	F
	Vidden-D	C	F
1,3-D	Telone II	R	F
Ethylene dibromide	Soilbrom Dowfume W-40, Dowfume W-85	C	F
DBCP	Nemagon Fumazone Many Others	C	F
Metham[4]	Vapam Busan Others	R	F
Fensulfothion	Dasanit	C	NF
Aldicarb	Temik	R	NF
Phenamiphos (Fenamiphos)	Nemacur	R	NF
Ethoprop(hos)	Mocap	R	NF
Oxamyl	Vydate, Oxamyl	R	NF
Carbofuran	Furadan	R	NF
Terbufos	Counter	R	NF

[1] Trade names vary worldwide; those listed are sold in the USA.
[2] Indicates regulatory status in the United States, R = registered and C = canceled.
[3] Methyl bromide is scheduled for cancellation in developing countries in 2005 and in remaining countries by 2015.
[4] One of several methyl isothiocyanate-generating compounds.

24.3 Fumigants

24.3.1 General Properties

Nematicidal chemicals are divided into two main groups, fumigants (volatile) and non-fumigants (non-volatile), according to their movement in soil. Fumigant nematicides received their name from the property of gaseous dispersion after application to soil. Non-fumigants are not volatile and disperse in the soil water phase. Nematicides most commonly are applied to soil and must dissolve in moisture surrounding soil particles to be effective since this is the location of target nematode species (Wright, 1981). Other modes of application such as foliar sprays (Westerdahl, et al., 1991) and seed treatments (Gray and Soh, 1989) may be used in a few cases.

Chemicals with fumigant properties frequently are divided into the two general groups of multipurpose fumigants and nematicidal fumigants. Multipurpose fumigants have good activity against plant-parasitic nematodes as well as fungi, weeds, and insects (Dunn and Noling, 1997; Lembright, 1990). These fumigants are general soil biocides and normally are applied to control more than one soilborne pest (Wright, 1981). Some of the most commonly or previously used multipurpose fumigants include carbon disulfide, chloropicrin, dazomet, metham, and methyl bromide (Tables 24.1 – 2). Nematicidal fumigants (D-D, 1,3-D, EDB, and DBCP) are primarily active against nematodes and soil arthropods at normal use rates, but may be active against other soilborne pests at higher rates. Both fumigant types are applied to soil (usually injected) and subsequent movement is by gaseous diffusion. If sufficient concentration × time ratios are achieved in the soil solution, nematodes are killed (Goring, 1962; McKenry and Thomason, 1974).

Multipurpose soil fumigants seldom are used solely for nematode control since they are generally expensive and can only be used economically on crops of high value. The most widely used multipurpose fumigant is methyl bromide due to its spectrum of activity on nematode, fungi, weed, and insect pests (Methyl Bromide Technical Options Committee, 1995). Methyl bromide is an excellent nematicide and as evidence of such, it has been used as the sole nematode management tool in high value vegetable production systems in Florida for over 30 years (Noling, 1996; Rich and Olson, 1999). Chloropicrin and metham are more efficacious than methyl bromide to control soilborne fungi, however, so higher proportions of these materials (e.g., 67% methyl bromide-33% chloropicrin) rather than the common formulation of 98% methyl bromide-2% chloropicrin were often chosen when significant fungicidal activity was needed (Methyl Bromide Technical Options Committee, 1995; South, et al., 1997; Wilhelm, et al., 1961). With the phaseout of methyl bromide soil fumigation in the United States, all formulations of methyl bromide will have progressively higher proportions of chloropicrin to achieve the gradual reductions of total poundage of methyl bromide demanded by EPA.

Three alkyl halides, EDB, DBCP, and 1,3-D, are the active ingredients found in nematicidal fumigants and have been the fumigants of choice when nematode control was the primary purpose for fumigation. These materials were comparatively inexpensive, more efficacious, and easier to apply than multipurpose fumigants (Dunn and Noling, 1997; Taylor, 1959). As stated earlier, EDB and DBCP are no longer in use worldwide. The remaining nematicidal fumigant, 1, 3-D, continues to be available after a series of regulatory reviews in the United States and still is widely used. The 1,3-D has traditionally been available as a single active ingredient, but it has been increasingly formulated in combination with chloropicrin or metham to increase spectrum of activity. The worldwide phaseout of methyl bromide will prompt more use of these combinations for dual nematode and fungal control.

Both multipurpose and nematicidal fumigants may be used on a wide range of crops due to both their generally good pest control efficacy and lack of residues in plants. With the exception of DBCP, these materials are phytotoxic to germinating seeds and living plants, so must be used as preplant treatments. Recent work with 1,3-D has shown that crops such as soybean and cotton may tolerate at-planting treatments (Rich and Kinloch, 2000; Rodriguez-Kabana, et al., 1979). Post-planting applications of 1, 3-D on perennial fruit crops and perennial bermudagrass golf course plantings also have shown promise (Zehr, 1994; Dunn, pers. obs.). Widespread use of 1, 3-D for such applications, however, are not expected due to the likelihood of damage to living crop plants.

24.3.2 Application Techniques

Fumigants generally are injected 25 – 45 cm deep into soil with chisel spacings between 30 – 45 cm (Dunn and Noling, 1997; Lembright, 1990). Depth of application and horizontal spacings are related to specific fumigant vapor pressure, characteristics of the soil, and depth of soil profile desired to be treated. Deeper fumigant placement is necessary under conditions such as tree planting sites or nurseries for deep-rooted plant species (Nyczepir, 1991). The point of fumigant application must be sealed by pressing or bedding soil behind the injection knives to prevent escape of the fumigant through the chisel tunnels. Soil injection is most common, but alternative applications have been made with certain fumigants. For example, DBCP often was applied through irrigation systems (flood, overhead sprinkler, and drip) for crops as widely different as banana, peanut, and turf (Dunn, pers. comm.). Metham may be applied through overhead irrigation systems to manage soilborne fungal and nematode diseases (Johnson, et al., 1992; Noling, 1991). Recently, emulsifiable 1, 3-D formulations have been tested for efficacy through drip irrigation systems, with good results in finer-textured soil types and mixed results in sandier soils (Melichar, et al., 1996).

Unlike non-fumigant nematicides, existing multipurpose or nematicidal fumigants have no significant residual effects on nematodes. The exception, DBCP, no longer is available for use. Most fumigants break down rapidly in soil, and not enough is left after the waiting period to affect nematodes should they reenter the treated area (Lembright, 1990). Therefore, it is important that nematodes not be reintroduced into treated soil, e. g., by mixing with untreated soil or using infested planting stock. However, the use of fumigants can lead to a "biological vacuum" in the soil, thus allowing any remaining nematode populations to increase rapidly (Van Gundy, 1985).

Land preparation and correct soil moisture are critical for maximum nematicidal activity of soil fumigants. Fumigants are adsorbed to organic matter; if organic matter is excessive, application rates must be increased to offset loss of efficacy (Lembright, 1990; Van Gundy and McKenry, 1977). Nematodes also may be harbored in roots and protected from sufficient fumigant concentrations for adequate kill. The correct moisture level is important: The gaseous phase must move through pore spaces for distribution within the treated area, but the active ingredient must be dissolved in soil water to reach the target nematodes, which are entirely aquatic organisms. Thus, soil pore spaces should not be occluded by water (Goring, 1962; Lembright, 1990), but soil that is too dry will allow fumigant gases to escape before the fumigant can dissolve in the soil moisture in sufficient concentrations to kill nematodes. Waiting periods for fumigant dissipation from soil to avoid phytotoxicity to crop plants may range from 3 days to more than 3 weeks. Fine textured soils, low temperatures and excessive soil moisture retard escape of the fumes from the soil, requiring longer planting delays (Thomason, 1987). Unlike non-fumigants, fumigant nematicides kill nematodes and may be applied months in advance of planting, provided that treated soil is not recontaminated with nematodes.

24.4 Non-fumigants

24.4.1 General Properties

These products may be formulated as either liquid or granular materials and include widely used materials such as aldicarb, ethoprop(hos), fenamiphos, and oxamyl (Tables 24.1–24.2). Non-fumigant nematicides are represented by two main classes of chemical compounds, organophosphates and carbamates. These nematicides must move in the soil water phase, and subsequent movement is accomplished by rainfall or irrigation (Smelt and Leistra, 1992). The mode of action of both non-fumigant nematicide classes is that of acetylcholinesterase inhibitors (Opperman and Chang, 1990; Wright, 1981). As such, the non-fumigant nematicides are mostly nemastatic rather than nematicidal. Thus, seldom do economically applied dosages of these materials directly kill nematodes, and nematodes recover and may become infective after removal from contact with non-fumigant nematicides. The non-fumigants must remain in contact with nematodes for a suitable period of time (4 –8 weeks) to prevent infection and allow for good early plant root growth. In addition to activity against nematodes, all non-fumigant nematicides show toxicity to one or more groups of insects (Johnson and Feldmesser, 1987; Taylor, 1959). Conversely, a number of insecticides are also mildly nematicidal and many commercial companies have routinely tested new chemistry developed for insect control for nematicidal activity as well.

24.4.2 Application Techniques

Non-fumigant nematicides are predominately applied to soil as in-furrow, band or broadcast applications (Dunn and Noling, 1997). Band and broadcast applications usually are followed by tilling into the soil for further uniformity of distribution and to prevent runoff. Distribution in soil is entirely in the soil solution, so ingredients move as the soil water moves (Smelt and Leistra, 1992). With excessive rainfall or irrigation, there is a risk of prematurely losing the active ingredient from the root zone and water contamination by leaching or run-off. Organic matter and clay content of soil also affect movement of non-fumigant nematicides. In sandy soils with low organic matter, low cation exchange properties may permit excessive leaching and rapid loss of the material from the root zone (Johnson and Feldmesser, 1987; Rahi, et al., 1992). Therefore, these products often are more effective in fine-textured soil types. Soil and water pH affect how long an active ingredient remains in the soil. Active ingredients in the organophosphate and carbamate families are acidic, so generally degrade more rapidly in soils at pH greater than 7.0.

Unlike fumigant nematicides, many non-fumigants are absorbed systemically and may translocate in living plant tissue either upwardly (apoplastic), downwardly (symplastic), or both (ambimobile) (Wright, 1981). The knowledge of chemical movement and the direction of movement offer flexibility in application strategies for the individual nematicides. For example, oxamyl is ambimobile; it moves symplastically when applied to plant foliage and has shown activity against soilborne nematodes, resulting in crop yield increase (Westerdahl, et al., 1991). Alternately, aldicarb moves apoplastically in plants when applied to soil and is used to suppress both nematode populations in soil and foliar-feeding thrip insect populations (Rhone-Poulenc, 1998). The disadvantage of a chemical being mobile in plants may outweigh advantages since chemical residue may persist in plants, limiting crop registrations. This is one reason that most non-fumigants do not have wide crop use registrations compared to fumigants. Ethoprop, a product of long standing that is not absorbed systemically, is registered on far more crops and target pests than most other non-fumigant nematicides.

24.5 Other "Nematicides"

Initial successes with chemicals to control nematodes spawned numerous attempts by industry and university scientists to identify other nematicidal chemicals, a process ongoing to the present day. Over 40 years ago, Taylor (1959) stated "Undoubtedly, thousands of chemical compounds have been given some sort of testing for this purpose". Over the years, few of these "thousands" have been placed on the market or generally available for use.

A large group of chemicals reported to have nematicidal properties have been organophosphate insecticides such as phorate, parathion, disulfoton, and diazinon (Johnson and Feldmesser, 1987). The fungicides benomyl and pentachloronitribenzene also have been reported to possess nematicidal properties (Wright, 1981). A listing of various other chemicals includes sodium selenate, benzene hexachloride, chlorophenyl mercury chloride, sodium azide, formaldehyde, and ammonia (Taylor, 1959). More recently, some chemicals are being tested that were essentially abandoned many years ago. These include methyl iodide, propargyl bromide, sodium azide, and chemicals in the avermectin group. These latter chemicals have proven effective as nematicides but registration and use were limited by factors such as cost, movement in soil, or handling characteristics. As conditions have changed, however, a number of previously identified nematicidal chemicals are being subjected to retesting. For example, the search for chemical alternatives for methyl bromide has focused heavily although not exclusively on older chemicals such as 1,3-D, chloropicrin, and metham (Noling and Becker, 1994).

24.6 Current Challenges

Considerable optimism was evident during the 1950s and 1960s for expanded use and discovery of nematicidal chemicals. This optimism was reasonable since discoveries in the 1940s gave birth to a rapidly growing Science of Nematology and expanding use of nematicides to manage nematodes in agriculture. For example, in the USA only a "small hectarage" was treated with nematicides in 1948 while approximately 200,000 hectares received nematicide application 10 years later (Taylor, 1959). Additionally by 1970, nematicide sales in the USA had reached US$51 million (Hodges, 1973). Review articles during this time indicated the need for more nematicide use, and general information or expectations did not include those problems that would appear later (Thomason, 1987).

Nematicides are a distinctly 20th-century phenomenon since they were first discovered, developed and widely used during that century. With nematicide technology, as in most scientific advances, a large body of knowledge was created which eventually not only showed the many advantages but also limitations and disadvantages of the technology. As a result, several nematicides were canceled or use restrictions placed upon them to mitigate problems not recognized when they first were used. Wright (1981) clearly alluded to problems of nematicides and many of his comments were prophetic of future events.

24.6.1 Present Uses

Despite product cancellations and use restrictions, and the lesser specter of enhanced biodegradation, nematicides are widely used, particularly in developed countries and on higher value crops. For example, over 80% of flue-cured tobacco hectarage in Canada, USA, and Zimbabwe receive annual nematicide applications (Rich, et al., 1989). In Florida, almost 100% of the 16,000 hectares of fresh market tomatoes is also treated with multipurpose fumigants, with nematode control as a major element in choosing this treatment (Noling, 1996). Production of these high-value crops is very risky economically, so many growers have used the most effective broad-spectrum fumigants possible to limit even small losses. Multipurpose fumigants have given this assurance to growers in the past and were readily adopted since other management techniques were less reliable. For many vegetable crops, for example, plant resistance is not available, is limited to only a few potential nematode pests, may be limited by soil temperature, and/or is subject to resistance breaking biotypes. Crop rotation with less profitable crops often is not an economical option or if possible, only shortened rotations are practical. Thus, agriculture continues to need and demand nematicides.

24.6.2 Biodegradation

Repeated use of most soil-applied pesticides leads to enhanced rates of biodegradation by microbial adaptation (Kaufman, et al., 1985; Mojtahedi, et al., 1991; Racke and Coats, 1990). Exposure of a soil microbial population to the same or similar active ingredient has led to an increase in populations that are able to utilize the chemical and its metabolites. Nematicides as a group are applied only once or twice annually and are less subject to enhanced biodegradation pressures than multiple use materials such as insecticides. However, enhanced degradation rates of several nematicides have been shown.

After three annual applications, biodegradation of 1, 3-D was shown to progressively increase (Chung, et al., 1999). Once application ceased, the enhancement lasted two years. Nematode control, however, was not monitored in this test or a similar one (Ou, et al., 1994), so any loss of nematicidal efficacy was not determined.

A prominent example of enhanced biodegradation among nematicides is that of fenamiphos which has been found to be subject to enhanced microbial degradation in many locations and in association with many different crops around the world. Rapid degradation of fenamiphos first came to serious attention in banana plantations (a crop for which applications were made two or more times annually) in locations as widely separated as Ecuador, the Ivory Coast, and Costa Rica (Anderson, 1989). In a field survey conducted in 1991 – 1992, enhanced biodegradation was found to have developed widely in commercial turf plantings in Florida, including golf courses, sports fields, and turf research grounds where fenamiphos had been applied repeatedly (Dunn, pers. obs.). Of 30 sites sampled throughout Florida, 16 were found to have developed enhanced microbial degradation of fenamiphos such that the compound degraded 10 to 30 times as rapidly in soil that had been treated repeatedly as in soil that had never been treated with fenamiphos. In only five sites did the rate of microbial degradation of fenamiphos not differ significantly from that of soil that never had been treated with that chemical. Degradation of fenamiphos and its two principal oxidation products, fenamiphos sulfoxide and fenamiphos sulfone, was related to continuous application (Chung and Ou, 1996).

In addition to that of fenamiphos, enhanced biodegradation of ethoprop, carbofuran, aldicarb and oxamyl after repeated soil treatments also have been reported (Karpouzas, et al., 1999; Smelt, et al., 1987). Although little published research has been conducted on the practical importance of enhanced biodegradation of non-fumigant nematicides on nematode control, the well-documented need for prolonged exposure of nematodes to these products for them to be effective implies that enhanced microbial degradation directly reduces their efficacy. As with other pesticides, enhanced biodegradation of nematicides is expected to increase and impact nematode management programs.

24.7 Trends

The emergence of the 21st century brings a crucial time for both nematicides and by default for the broader future of Nematology. Choices of chemicals for efficacious nematode control have become highly restricted and if the past is an indication, will become fewer in the future. Environmental fate, toxicology, and carcinogenicity became critical issues during the last part of the 20th century. Existing nematicides and possible new products may be excellent in all nematicidal respects but must pass even more stringent tests for human and environmental safety.

Commercial companies and nematologists have continued to develop and test potential nematicide candidates. For example, the senior author has field tested over 30 compounds for companies over the past two decades. Several were effective and worthy of development as nematicides. However, product registration costs, manufacturing costs/unit, potential for groundwater or other environmental contamination and/or human health effects prohibited development. As noted, the discovery that a material is nematicidal sometimes is only a minor part of the process for governmental registration and subsequent widespread use.

Terbufos was the last major nematicide registered (1974) in the USA (Rich Michell of EPA, pers. comm.). A few materials have been registered and/or used on specific crops or in regions of the world since then. With society's newfound resolution that pest control agents must be safe first and efficacious second, most recent efforts to identify new commercial nematicides have focused on products of natural or organic origin and deemed to be extremely safe to man and the environment.

Loss of many nematicides from use in the USA by regulation or marketing decisions has left nematode problems for which there is no established nematicidal product, and yet for which other management tactics are inadequate (Dunn and Noling, 1997). As a result, many products have entered the marketplace with claims to reduce in some way the effects of nematodes on, for instance, public turf plantings, landscape ornamental plantings, home lawns, and home vegetable gardens. These materials often lack registration and labeling as nematicides even as they claim to provide nematode control in some way. Sales promotion of most of these products depends heavily on testimonials based on lay trials without objective controls. Hopeful producers and managers with unsolved nematode problems continue to buy such products even though there are no objective data by which to assess their nematicidal efficacy. Most have none. For example, Dunn (pers. obs.) tested 15 materials in 1998 and 1999, most offered for sale to the public for nematode control, with only two (neither yet available commercially as nematicides in late 2001) showing a degree of nematicidal efficacy against nematodes associated with turfgrasses in the field.

Many of these "non-traditional" products have been offered to solve nematode problems and include detergents and wetting agents, botanical products, and products of animal orgin. Another array of products has claimed to enable the plant to offset injurious effects of nematodes by strengthening the plant through various combinations of enzymes, growth hormones, nutritional supplements, and other special stimulants; these usually are said to come from completely safe "natural" sources.

A few products known to be effective wetting agents and/or detergents have been tested extensively and some have been sold for years for nematode control. Nematodes contain water under relatively high tonic pressure that is retained in the body by highly effective lipid membranes. Solvents, detergents, and other chemical agents that break down those membranes can readily kill nematodes. These products often cause rapid death of nematodes in vitro, but have not been consistently effective when applied to the chemically and physically complex soil medium. They often are very injurious to plant membranes as well, limiting the safety of their application to living plantings.

Many kinds of plants and animal products have been known for centuries to be able to reduce damage from root-knot and other nematodes when incorporated into the soil as discussed in detail elsewhere in this book. Most, however, require very high rates of application to be effective. For example, up to 1 mt/ha of raw crab scrap was necessary to significantly reduce root-gall ratings on tomato in one test, while 20 mt/ha of crab scrap compost was needed to achieve the same result (Rich and Hodge, 1993). The mass of material to be moved and applied, and the difficulty of incorporating that much (often bulky) material has made it difficult to keep these products economically practical for most uses. Some products offered to the lay public have depended on the popularity of "natural" products for horticultural uses, but have recommended that their products be used at small fractions of the rates proven to actually be effective. Despite setbacks suffered because of sometimes misleading marketing strategies, it seems likely that the near future will see successful commercial development of nematicidal products based on various plant materials, including but not limited to sesame (*Sesamum indicum*), neem (*Azadirachta indica*), and oriental mustard (*Brassica juncea*).

Chemicals and organic materials often improve plant growth as a result of plant growth regulator effects or improved fertility. Separating nutritional effects of organic/naturally sourced materials from any directly nematicidal effects can be difficult. Thus, scientists and users should always be wary of testimonials for products that have not undergone thorough testing. Similarly, trials of efficacy for such products should include nutritional controls that provide similar levels of nitrogen and other major nutrients without the presumed active ingredients of the putative nematicide under trial.

Nematicide development over the past 30 years has not paralleled that of chemical controls in other pest disciplines. Numerous herbicides, fungicides, and insecticides have been developed over that time. These materials have become increasingly safe, effective, and environmentally friendly and used at steadily reducing rates. Several factors may be given for lack of comparable nematicide development during this period. They include the location of nematodes in the complex soil environment, thus making contact difficult (Van Gundy and McKenry, 1977). Probably more important has been the limited resources devoted to nematicide development because there are far fewer nematologists than counterparts in other disciplines and because nematicides are considered a substantially smaller potential market compared to insecticides, fungicides, and herbicides.

As indicated at the beginning of this chapter, nematicides have been very important to the Science of Nematology. Ironically, cancellations of nematicides and lack of developing new ones may have been a positive stimulus as well. In many cases, it has necessitated research into other nematode management options so as not to depend solely upon chemical control for nematodes. The progress in non-chemical nematode management is evidenced by other chapters in this book. However, nematicides have and for the foreseeable future will remain important components of integrated nematode management programs. Continued research with currently available nematicides is important, particularly to find ways to reduce applications and rates where possible. An emerging technology of precision agriculture may prove beneficial for using nematicidal products more efficiently. Candidate nematicides also should be vigorously studied so we can improve and maintain nematicides in our arsenal of nematode management techniques well into the 21st century.

Chapter Appendix

Physical Characteristics of Common Nematicides, Technical Products

- aldicarb Toxicity oral (rats) LD_{50} 1 mg/kg, vapor pressure 2.9×10^{-5} mm Hg (25℃), non-flammable, solubility in H_2O 6000 mg/kg (25℃).
- carbofuran Toxicity oral (rats) LD_{50} 8 mg/kg, vapor pressure 2×10^{-6} mm Hg (25℃), non-flammable, solubility in water 700 ppm.
- carbon disulfide Toxicity oral (rats) LD_{50} 3188 mg/kg, flammable, vapor pressure 352.6 mm Hg (25℃), solubility in water 0.2 gm/100 ml.
- chloropicrin Toxicity oral (rats) LD_{50} (inhalation LC_{50} 150 mg/kg in 15 min), vapor pressure 18.3 mm Hg (20℃), non-flammable, solubility in water 2000 mg/kg.

- 1,3-D Toxicity oral LD_{50} (rats) 224 – 300 mg/kg (inhalation LC_{50} 904 mg/kg), vapor pressure 28 mm Hg (20℃), flammable, solubility in water 2700 mg/kg.
- DBCP Toxicity oral (rats) 170 – 300 mg/kg, vapor pressure 0.58 mm Hg, solubility in water 1230 mg/kg.
- ethylene dibromide Toxicity oral (rats) LD_{50} 146 mg/kg, vapor pressure 7.69 mm Hg, solubility in water 3370.
- ethoprop(hos) Toxicity oral (rats) LD_{50} 61.5 mg/kg, vapor pressure 3.5×10^{-4} mm Hg (26℃), non-flammable, solubility in water 760 mg/kg.
- fensulfothion Toxicity oral LD_{50} 5 mg/kg.
- methyl bromide Toxicity oral (rats) LD_{50} 214 mg/kg (inhalation LC_{50} 3120 mg/kg (15 min.), vapor pressure 1400 mm Hg (20℃), solubility in water 17,500 mg/kg.
- oxamyl Toxicity oral (rats) LD_{50} 5.4 mg/kg, solubility in water 28 g/100 g.
- fenamiphos Toxicity oral (rats) LD_{50} 6 mg/kg, vapor pressure 0.12 Pa (20℃), non-flammable, solubility in water 700 mg/kg.
- terbufos Toxicity oral (rats) LD_{50} 4.5 mg/kg, non-flammable, solubility in water 15 mg/kg.

References

Anderson, J. P. E. 1989. Accelerated microbial degradation of nematicides and other plant-protection chemicals in soils. Nematropica 19:1 (abstr.).

Bessey, E. A. 1911. Root-knot and its control. Bureau of Plant Industry, United States Department of Agriculture, Washington, D. C., Bulletin 217.

Bunt, J. A. 1987. Modes of action of nematicides. Pp. 461 – 468. In J. A. Veech and D. W. Dickson, eds. Vistas on Nematology, Hyattsville, Maryland: Society of Nematologists.

Carter, W. 1943. A promising new soil amendment and disinfectant. Science 97: 383 – 384.

Castro, C. E., and I. J. Thomason. 1971. Mode of action of nematicides. In B. M. Zuckerman, W. F. Mai, and R. A. Rohde, eds. Plant Parasitic Nematodes, Vol. II, New York: Academic Press.

Christie, J. R., and G. S. Cobb. 1940. The inefficiency of methyl bromide fumigation against the chrysanthemum foliar nematode. Proceedings of the Helminthological Society of Washington 7:62.

Christie, J. R. 1945. Some preliminary tests to determine the efficacy of certain substances when used as soil fumigants to control the root-knot nematodes *Heterodera marioni* (cornu) Goodey. Proceedings of the Helminthological Society of Washington 12: 14 – 19.

Christie, J. R., and V. G. Perry. 1958. A low-phytotoxic nematicide of the organic phosphate group. Plant Disease Reporter 42: 74 – 75.

Chung, K. -Y., D. W. Dickson, and L. -T. Ou. 1999. Differential enhanced degradation of cis- and trans-1, 3-D in soil with a history of repeated field application of 1, 3-D. Journal of Environmental Science and Health B34: 749 – 768.

Chung, K. -Y., and L. -T. Ou. 1996. Degradation of fenamiphos sulfoxide and fenamiphos sulfone in soil with a history of continuous applications of fenamiphos. Archives of Environmental Contamination and Toxicology 30: 452 – 458.

Curran, B., ed. 2001. Farm Chemicals Handbook Volume 87. Willoughby, OH: Miester Publishing.

DiSanzo, C. P. 1981. Effects of foliar application of carbofuran and a related compound on plant-parasitic nematodes under greenhouse and growth chamber conditions. Journal of Nematology 13: 20 – 24.

Dunn, R. A., and J. W. Noling. 1997. Florida Nematode Management Guide. Publication SP-54, Florida Cooperative Extension Service, University of Florida, Gainesville, Florida.

Feldmesser, J., and G. C. Smart, Jr., co-leaders. 1982. The biologic and economic assessment of DBCP. United States Department of Agriculture, Washington D. C., Technical Bulletin 1666.

Fleming, W. E., and F. E. Baker. 1935. The use of carbon disulfide against the Japanese beetle. United States Department of Agriculture, Washington D. C., Technical Bulletin 478: 1 – 92.

Garreau, M. 1854. Notice sur la destruction des charcons du blé. Arch. Agr. Nord. France 2: 195 – 198.

Godfrey, G. H. 1935. Experiments on the control of root-knot nematode in the field with chloropicrin and other chemicals. Phytopathology 25: 67 – 90.

Good, J. M., 1963. Root-knot and free-living nematodes on tobacco. Fungicide – Nematocide Tests of 1963 19: 11.

Gordon, F. M. 1977. U. S. Patent Office No. 4065558, granted for a new nematicide. American Cyanamid Company, Trenton, NJ.

Goring, C. A. I. 1962. Theory and principles of soil fumigation. Advances in Pest Control Research 5: 47 – 84.

Gowen, S. R. 1978. Efecto de diferentes nematicides sobre el rendimento del banano en las Islas de Barlavento. Nematropica 8: 9 – 13.

Gray, F. A., and D. H. Soh. 1989. A nematicide seed treatment to control *Ditylenchus dipsaci* on seedling alfalfa. Journal of Nematology 21: 184 – 188.

Hawkins, L. A. 1939. Bureau of Entomology and Plant Quarantine Circular E-84. United States Department of Agriculture, Washington, D. C.

Hodges, L. R. 1973. Nematodes and their control. Union Carbide Corporation, Salinas, California.

Jenkins, L., and H. W. Guengerich. 1959. Chemical dips for control of nematodes on bare root nursery stock. Plant Disease Reporter 43: 1095 – 1097.

Johnson, A. W., A. S. Csinos, A. M. Golden, and N. C. Glaze. 1992. Chemigation for control of black shank-root-knot complex and weeds in tobacco. Journal of Nematology 24: 648 – 655.

Johnson, A. W., and J. Feldmesser. 1987. Nematicides – a historical review. Pp. 448 – 454. In J. A. Veech and D. W. Dickson, eds. Vistas on Nematology. Hyattsville, MD: Society of Nematologists.

Johnson, M. O., and G. H. Godfrey. 1932. Chloropicrin for nematode control. Industrial Engineering Chemistry 24: 311 – 313.

Karpouzas, D. S. H., D. G. Walker, A. Froud-Williams, and R. J. Drennan. 1999. Evidence for enhanced biodegradation of ethoprophos and carbofuran in soils from Greece and the U. K. Pesticide Science 55: 300 – 311.

Kaufman, D. D., Y. Katan, D. F. Edwards, and E. G. Jordan. 1985. Microbial adaptation and metabolism of pesticides. Pp. 437 – 451. In J. L. Milton, ed. Agricultural Chemicals of the Future. Totowa, NJ: Rowman and Allanheld.

Kuhn, J. 1881. Die Ergebnisse de Versuche zur Ermittelung der Ursache der Rubenmudigkeit und zur Erforschung der Natur der Nematoden. Ber. Physiol. Lab. Landwirtsh. Inst. Univ. Halle 3153 pp.

Lear, B. 1956. Results of laboratory experiments with Vapam for control of nematodes. Plant Disease Reporter 40: 847 – 852.

Lembright, H. W. 1990. Soil fumigation: Principles and application technology. Supplement to Journal of Nematology 22: 632 – 644.

Manzelli, M. A. 1955. A residual organophosphorus nematocide. Plant Disease Reporter 39: 400 – 404.

Matheron, M. E., and J. C. Matejka. 1988. In vitro activity of sodium tetrathiocarbonate on sporulation and growth of six *Phytophthora* species. Phytopathology 78: 1234 – 1237.

Matthews, J. D. 1920. Report of the work of the W. B. Randall Research Assistant from October 1st, 1918 to December 31st, 1919. Experiment Station Chestnut, Herts., U. K., 1919 Annual Report 5: 18 – 21.

McBeth, C. W. 1954. Some practical aspects of soil fumigation. Plant Disease Reporter Supplement 227: 95 – 97.

McBeth, C. W., and G. B. Bergeson. 1955. 1, 2-dibromo-3-chloropropane – a new nematicide. Plant Disease Reporter 39: 223 – 225.

McKenry, M. V., and I. J. Thomason. 1974. 1, 3-dichloropropene and 1, 2-dichloropropane compounds: I. Movement and fate as affected by various soils. II. Organism-dosage-response studies in the laboratory with several nematode species. Hilgardia 42: 392 – 438.

Melichar, M. W., J. P. Mueller, and J. M. Richardson. 1996. Pre-plant drip irrigation application of 1, 3-dichloropropene for nematode control in annual crops. In Proceedings of the Annual International Research Conference on Methyl Bromide Alternatives and Emissions Reductions, Orlando. Florida.

Metcalf, C. L., and W. P. Flint. 1962. Destructive and Useful Insects: Their Habits and Control. 4th Edition. New York: McGraw-Hill Book Company.

Methyl Bromide Technical Options Committee-United Nations Environment Program. 1995. 1994 report of the methyl bromide technical options committee for the 1995 assessment of the Montreal protocol on substances that deplete the ozone layer. United Nations Environment Program, Australia.

Miller, P. M. 1966. Control of *Heterodera tabacum* with volatile and nonvolatile nematocides. Plant Disease Reporter 50: 506 – 509.

Mojtahedi, H., G. S. Santo, and J. N. Pinkerton. 1991. Efficacy of ethoprop on *Meloidogyne hapla* and *M. chitwoodi* and enhanced biodegradation in soil. Journal of Nematology 27:372 – 379.

Newhall, A. G. 1955. Disinfectation of soil by heat, flooding and fumigation. Botanical Review 21:189 – 250.

Noling, J. W. 1991. Metham sodium application via drip irrigation systems. Florida Department of Agriculture and Consumer Services, Division of Plant Industry, Nematology Circular 196, Gainesville, Florida.

Noling, J. W. 1996. The role of soil fumigants in Florida agriculture. Pp. 14 – 24. In J. N. Seiber, J. A. Knuteson, J. E. Woodrow, N. L. Wolfe, M. V. Yates, and S. R. Yates, eds. Fumigants: Environmental Fate, Exposure, and Analysis. American Chemical Society, Symposia Series 652.

Noling, J. W., and J. O. Becker. 1994. The challenge of research and extension to define and implement alternatives to methyl bromide. Supplement to the Journal of Nematology 26:573 – 586.

Noling, J. W., and J. P. Gilreath. 2000. Methyl bromide: Progress and problems identifying alternatives. Citrus and Vegetable Magazine 64: A1 – A15.

Nyczepir, A. P. 1991. Nematode management strategies in stone fruits in the United States. Journal of Nematology 23:334 – 341.

O'Bannon, J. H., and A. L. Taylor. 1967. Control of nematodes on citrus seedlings by chemical bare-root dip. Plant Disease Reporter 51:995 – 998.

Opperman, C. H., and S. Chang. 1990. Plant-parasitic nematode acetylcholinesterase inhibition by carbamate and organophosphate nematicides. Journal of Nematology 22: 481 – 488.

Osborne, W. W., C. Harris, L. M. Harrison, W. F. Brown, R. L. Shaw, and H. S. Adams. 1969. The efficacy of certain chemical soil treatments for the control of *Meloidogyne incognita acrita* in tobacco. Journal of Nematology 1:22 (Abstr.).

Ou, L.-T, J. E. Thomas, and D. W. Dickson. 1994. Degradation of fenamiphos in soil with a history of continuous fenamiphos applications. Soil Science Society of America Journal 58:1139 – 1147.

Racke, K. D., and J. R. Coates, eds. 1990. Enhanced biodegradation of pesticides in the environment. ACS Symposium Series 426. Washington, D. C.: American Chemical Society.

Radewald, J. D., F. Shibuya, J. Nelson, and J. Bivens. 1970. Nematicide control with 1410, an experimental nematicide-insecticide. Plant Disease Reporter 54:187 – 190.

Rahi, G. S., J. R. Rich, and C. H. Hodge. 1992. Ethoprop depletion from soil as influenced by simulated rainfall. Supplement to the Journal of Nematology 24: 642 – 647.

Raski, D. J. 1954. Soil fumigation for the control of nematodes on grape replants. Plant Disease Reporter 38:811 – 817.

Reed, H. E. 1963. Root-knot nematode on tobacco. Fungicide-Nematocide Tests, Results of 1963, 19:118.

Rhone-Poulenc Ag Company. 1998. Temik briefing book. Research Triangle Park, Raleigh, North Carolina.

Rich, J. R., J. D. Arnett, J. A. Shepherd, and M. C. Watson. 1989. Chemical control of nematodes of flue-cured tobacco in Brazil, Canada, United States, and Zimbabwe. Supplement to the Journal of Nematology 21:609 – 611.

Rich, J. R., and C. H. Hodge. 1993. Utilization of blue crab scrap compost to suppress *Meliodogyne javanica* on tomato. Nematropica 23:1 – 5.

Rich, J. R., and R. A. Kinloch. 2000. Comparative responses of cotton yield to preplant and at-plant applications of 1,3-dichloropropene. Nematropica 30: 223 – 229.

Rich, J. R., and S. M. Olson. 1999. Utility of Mi gene resistance in tomato to manage *Meloidogyne javanica* in north Florida . Supplement to the Journal of Nematology 31: 715 – 718.

Rich, J. R., and E. B. Whitty. 1999. Efficacy of chloropicrin and ditera to manage *Meloidogyne javanica* in north Florida. Tobacco Science 43:18 – 22.

Rodriguez-Kabana, R., H. W. Penick, P. S. King, F. A. Gray, E. L. Carden, N. R. McDaniel, and F. B. Selman. 1979. Efficacy of planting-time applications of 1, 3-dichloropropene for control of nematodes on soybeans. Nematropica 9:66 – 71.

Smelt, J. H., S. J. H. Crum, W. Teunisson, and M. Leistra. 1987. Accelerated transformation of aldicarb, oxamyl and ethoprophos after repeated soil treatments. Crop Protection 6:295 – 303.

Smelt, J. H., and M. Leistra. 1992. Availability, movement and (accelerated) transformation of soil-applied nematicides. Pp. 281 – 293. In F. J Gommers and P. W. Th. Maas, eds. Nematology from Molecule to Ecosystem. Wildervank, The Netherlands: Dekker and Huisman.

South, D. B., W. A. Carey, and S. A. Enebak. 1997. Chloropicrin as a soil fumigant in forest nurseries. The Forestry Chronicle 73:489 – 494.

Stark, F. L., Jr. 1948. Investigations of chloropicrin as a soil fumigant. New York Agricultural Experiment Station Memoirs 278:1 – 61.

Taylor, A. L. 1959. Progress in chemical control of nematodes. Pp. 427 – 434. In C. S. Holton, G. W. Fischer, R. W. Fulton, Helen Hart, and S. E. A. McCallan, eds. Plant pathology – problems and progress, Madison, Wisconsin: University of Wisconsin Press.

Taylor, A. L. 1979. Nematocides and nematicides-history. Florida Department of Agriculture and Consumer Services, Division of Plant Industry Memo Series, 11 pp.

Taylor, A. L., and C. W. McBeth. 1940. Preliminary tests of methyl bromide as a nematocide. Proceedings of the Helminthological Society of Washington 7:93 – 96.

Taylor, A. L., and C. W. McBeth. 1941. A practical method of using methyl bromide as a nematocide in the field. Proceedings of the Helminthological Society of Washington 8: 26 – 28.

Thomason, I. J. 1987. Challenges facing nematology: Environmental risks with nematicides and the need for new approaches. Pp. 469 – 476. In J. A. Veech and D. W. Dickson, eds. Vistas on Nematology. Hyattsville, Maryland: Society of Nematologists.

Thorne, G. 1961. Principles of Nematology. New York: McGraw-Hill Book Company.

Thorne, G., and V. Jensen. 1946. A preliminary report on the control of the sugarbeet nematode with two chemicals, D-D and Dowfume W-15. Pp. 322 – 326. In Proceedings 4th General Meeting American Society Sugarbeet Technologists.

Van Berkum, J. A., and H. Hoestra. 1979. Practical aspects of the chemical control of nematodes in soil. Pp. 53 – 134. In D. Mulder, ed. Soil Disinfestation. Amsterdam,

The Netherland: Elsevier.
Van Gundy, S. D. 1985. Biological control of nematodes: status and prospects in IMP systems. Pp. 467 – 478. In M. A. Hoy and D. C. Herzog, eds. Biological Control in Agricultural IMP Systems. Orlando, Florida: Academic Press.
Van Gundy, S. D., and M. V. McKenry. 1977. Action of nematicides. Pp. 263 – 280. In J. G. Horsfall and E. B. Cowling, eds. Vol. 1, Plant Disease: an Advanced Treatise. New York: Academic Press.
Watson, R. T., D. L. Albritton, S. O. Anderson, and S. Lee-Bapty. 1992. Methyl bromide: Its atmospheric science, technology and economics. Montreal Protocol Assessment Supplement, United Nations Environmental Programme on Behalf of the Contracting Parties to the Montreal Protocol, Nairobi, Kenya.
Westerdahl, B. B., D. Giraud, J. D. Radewald, and C. A. Anderson. 1991. Management of *Ditylenchus dipsaci* in daffodils with foliar applications of oxamyl. Supplement to the Journal of Nematology 23: 706 – 711.
Whitehead, A. G. 1978. Chemical control, (a) Soil treatment. Pp. 283 – 296. In J. F. Southey, ed. Plant Nematology. London: Her Majesty's Stationary Office.
Wilhelm, S. 1966. Chemical treatments and inoculum potential of soil. Annual Review of Phytopathology 4: 53 – 77.
Wilhelm, S., R. C. Storkan, and J. E. Sagen. 1961. Verticillium wilt of strawberry controlled by fumigation of soil with chloropicrin and chloropicrin-methyl bromide mixtures. Phytopathology 51: 744 – 748.
Winfield, A. L. 1978. Chemical control (b) Treatments of plants, planting material and seeds. Pp. 296 – 301. In J. F. Southey, ed. Plant Nematology. London: Her Majesty's Stationary Office.
Wright, D. J. 1981. Nematicides: Mode of action and new approaches to chemical control. Pp. 421 – 449. In B. M. Zuckerman and R. A. Rhodes, eds. Plant Parasitic Nematodes. New York: Academic Press.
Youngston, C. R., and C. A. I. Goring. 1970. Nematicidal activity of 1, 3-dichloropropene and 1, 2-dichloropropane to three types of plant-parasitic nematodes. Plant Disease Reporter 54: 196 – 199.
Zehr, E. I. 1994. 1, 3-dichloropropene as a postplanting nematicide for peach trees. United States Department of Agriculture National Technical Information Service Publication 122: 66 – 71.

25 Irradiation Effects on Plant-parasitic Nematodes

D. W. Dickson

25.1 Introduction

There is interest in irradiation as an alternative to fumigation for disinfestation of plants, plant products, and other materials of pests, and pathogens including plant-parasitic nematodes. With the proposed suspension of methyl bromide (Thomas, 1996) irradiation takes on added dimensions as an important alternative disinfestation method in the future. Unfortunately, there have been relatively few studies done on effects of irradiation on nematodes, although there are numerous plant-parasitic nematodes that are of importance in quarantine and regulatory programs. For example, the North American Plant Protection Organization list the following nematodes as principals of regulatory concern: *Ditylenchus dipsaci*, *Globodera pallida*, *G. rostochiensis*, *Heterodera glycines*, *H. goettingiana*, *H. trifoli*, *H. zeae*, *Meloidogyne chitwoodi*, and *Radopholus similis* (Lehman, 1995). Other nematodes that are likely to appear on regulatory lists in other countries of the world include: *Anguina tritici*, *Aphelenchoides* spp., *Belonolaimus longicaudatus*, *Bursaphelenchus xylophilus*, *B. cocophilus*, *Ditylenchus destructor*, *Dolichodorus* spp., *Hemicycliophora arenaria*, *Longidorus* spp., *Meloidogyne* spp., *Paralongidorus* spp., *Paratrichodorus* spp., *Pratylenchus* spp., *Rotylenchulus reniformis*, *Trichodorus* spp., and *Xiphinema* spp. Use of irradiation to kill, sterilize, or otherwise render nematodes harmless would be a relatively safe option to the use of fumigation.

The earliest reports of the effects of radiation on nematodes dates back to the early 1900s, when X-ray radiation was evaluated for its deleterious effects on trichinae (*Trichinella spiralis*) in meat. It was found that encysted trichinae subjected to radium radiation failed to develop in mice, however the radiation failed to destroy sexually mature trichinae (Tyzzer and Honeji, 1916). Since that time irradiation has been applied to all groups of nematodes, animal parasitic, free-living, plant-parasitic, and entomopathogenic, with much mixed results. A review of important studies on irradiation of plant-parasitic nematodes is presented to bring together pertinent data. For convenience and comparative purposes all irradiation dosages reported herein were converted to kGy. Conversion of dosages were based on the following: 1 roentogens = 0. 93 rad; 100 rad = 0. 1 kilorad = 1 Gray = 0. 001 kilogray.

25. 2 Radiation Effects on Plant-parasitic Nematodes

Cobb (1920), who is considered the father of American nematology, was the first to report on the use of ionizing radiation on plant-parasitic nematodes. His tests, applying unspecified dosages of X-rays on root-knot nematodes, had no observable effects since root galls developed normally. The next report on the use of irradiation on plant-parasitic nematodes was not published until 35 years later. Fassuliotis and Sparrow (1955) treated the golden nematode, *Globodera rostochiensis* with X-rays with dosage ranging from 0. 02 to 0. 72 kGy. The life cycle was interrupted by X-rays applied at dosages of 0. 19 kGy and above. This work was followed up by studies using both X-rays and gamma rays from a ^{60}Co source (Fassuliotis, 1958a; 1958b). Hatching tests from cyst exposed to both types of radiations revealed a delay in onset of hatching and a reduction in the total number of juveniles emerging. After exposure of cysts to 5. 95 kGy, no juveniles emerged. Irradiation with 0. 19 kGy, 0. 37 kGy, and 0. 74 kGy sterilized 84%, 98%, and 100%, of juveniles, respectively. Viability was not altered in developing females following post-irradiation storage of irradiated cysts for 6 months at 4. 4℃. Chromosome aberrations in the form of fragments and bridges at anaphase were found in maturing eggs recovered from females that developed from irradiated juveniles.

A summary from research conducted on effects of irradiation on plant-parasitic nematodes over a 41-year period is presented in Table 25. 1. An effort was made to include data that shows a positive detrimental effect on some life stage of the various nematodes tested. The minimum dosages of irradiation rendering plant-parasitic nematodes noninfective or unable to reproduce are summarized in Table 25. 2.

Ditylenchus destructor and *Rhabditis* spp. are an important nematode pathogen and pest, respectively, of cultivated mushroom beds. Wood and Goodey (1957) applied gamma radiations from ^{60}Co at doses of 0. 56 to 0. 89 kGy to infested mushroom compost in order to control these nematodes. The nematode inactivation dose for the two species was between 0. 45 and 0. 89 kGy, which was over four times as great as the 0. 09 and 0. 19 kGy found to inactivate the golden nematode (Fassuliotis and Sparrow, 1955).

Table 25.1 Dosages of irradiation and their effects on various developmental stages of plant-parasitic nematodes. Dosages presented herein were selected because they gave a specific detrimental effect on some life stage of the nematode. Other dosages were sometimes tested but are not presented.

Nematode species	Dose (kGy)[a]	Developmental stage treated[b]	Specific effects	Reference
Aphelenchoides sp. (in infested soil and chopped roots)	0.37	Mixed	No reproduction in bioassay.	Myers, 1960.
Bursaphelenchus xylophilus (in wood chips)	8.0	J2	No nematodes found. (length of time required for death to occur not given).	Eichholz, et al., 1991.
B. xylophilus (in water)	7.0	J2	No nematodes found. (length of time required for death to occur not given).	Smith, 1991.
Ditylenchus destructor (in potato)	2.0	Mixed	Increased juvenile mortality by 10%; increased female to male ratio 2.3 times.	Karnkowski and Ignatowicz (pers. comm.).
D. destructor (in infested compost)	0.89	Mixed	No reproduction.	Wood and Goodey, 1957.
Ditylenchus dipsaci (in onion)	0.50	Mixed	Decreased rate of development to adulthood by 14%; increased male to female by 1.14 times.	Karnkowski and Ignatowicz (pers. comm.).
D. dipsaci	0.50	Mixed	Increased juvenile mortality (in garlic) by 10%.	
D. dipsaci (aqueous suspension of axenic nematodes)	2.88	J4	Eliminate apical swellings typically induced by the nematode in alfalfa.	Siddiqui and Viglierchio, 1970.
	0.48		Stopped reproduction when inoculated on alfalfa.	

Continued

Nematode species	Dose (kGy)[a]	Developmental stage treated[b]	Specific effects	Reference
D. myceliophagus (in infested soil or chopped roots)	1.12	Mixed	No reproduction in bioassay.	Myers and Dropkin, 1959.
Globodera rostochiensis (*cysts*)	5.95	Cysts	No J2 emerged.	Fassuliotis, 1958a.
	0.74		Caused 100% sterility.	
	0.19		F_1 females produced no viable J2.	
G. rostochiensis (air dried, wet, or root-diffusate soaked cyst)	0.16	Cysts (eggs, J1, and J2)	Preirradiation treatment of cysts affected response. Hatch rate of dry cyst was stimulated 1.6 fold, whereas wet cyst or root-diffusate soaked cyst were not affected. Death rate of eggs was not affected by any treatment. F_1 generation from radiated cysts – 98% to 99% of radiated eggs were dead, hatch was decreased to 0% to 2% in dry or root-diffusate soaked cyst, respectively.	Evans, 1970.
	0.64		F_1 generation from radiated cysts – no cyst developed. F_1 generation from radiated J2 – decreased cyst production 99.6%.	
G. rostochiensis	4.40	Cysts	No J2 emerged.	Van de Woestijne and Van den Brande, 1960.
Helicotylenchus buxophilus (in infested soil and chopped roots)	2.98	Mixed	No reproduction in bioassay.	Myers, 1960.

Continued

Nematode species	Dose (kGy)[a]	Developmental stage treated[b]	Specific effects	Reference
Heterodera schachtii (in infested soil)	5.12	J2	No F_1 cysts developed on beet after 45 days.	Townshend, 1967.
	1.28		No J2 emerged from F_1 cysts.	
	0.32		No F_2 cysts developed on beet after 180 days.	
	5.12	Cysts	Reduced emergence of J2 by 76%.	
Meloidogyne arenaria (in infested soil or chopped roots)	1.67	Mixed	No reproduction in bioassay.	Myers and Dropkin, 1959.
Meloidogyne graminicola (in rice roots)	0.03	Egg masses (eggs, J1, and possibly J2)	Decreased the number of hatched J2 reaching adulthood by 29%; F_1 J2 generation was normal.	Prasad, et al., 1982.
Meloidogyne hapla	6.40	Egg mass	No root-galling occurred.	Van de Woestijne and Van den Brande, 1960.
Meloidogyne incognita (in infested soil or chopped roots)	1.67	Mixed	No reproduction in bioassay.	Myers and Dropkin, 1959.
M. incognita (in infested soil or chopped roots)	1.67	Mixed	No reproduction in bioassay.	Myers and Dropkin, 1959.
M. incognita (in water)	0.39	White egg masses (eggs, J1, and J2) Yellow-brown egg masses	Emerged J2 failed to mature and produce eggs. A few emerged J2 did mature and produced egg masses with eggs.	Ishibashi and Taguchi, 1965.

Continued

Nematode species	Dose (kGy)[a]	Developmental stage treated[b]	Specific effects	Reference
M. incognita (medium not known)	0.40	Eggs (50% exceeding 4-cell stage)	Decreased hatch but the J2 that did hatch were infective.	Cox, et al., 1976.
		Eggs (50% equal to or less than 4-cell stage)	Percentage decrease in hatch unknown.	
Meloidogyne javanica (in sterile distilled water)	7.50	J2	Killed within 1 day after treatment.	Chinnasri, et al., 1997.
	7.0		Killed within 5 days after treatment.	
	6.25	Egg masses (eggs, J1, and J2)	Hatch inhibited completely.	
Mesocriconema axeste (in infested soil and chopped roots).	1.49	Mixed	No reproduction in bioassay.	Myers, 1960.
Paratrichodorus minor (in infested soil and chopped roots)	2.98	Mixed	No reproduction in bioassay.	Myers, 1960.
Pratylenchus thornei (in soil, 150 g)	7.50	Mixed	Killed within 2 days after treatment. Nematodes were likely in an anhydrobiotic condition at time of treatment.	Thompson, 1990.
Pratylenchus vulnus (in infested soil and chopped roots)	0.37	Mixed	No reproduction in bioassay.	Myers, 1960.

Continued

Nematode species	Dose (kGy)[a]	Developmental stage treated[b]	Specific effects	Reference
Pratylenchus zeae (in tap water)	0.50	Eggs	45% reduction in hatch in 24 h.	Nath, et al., 1986.
		J2	30% killed in 24 h.	
		J3 and J4	23% killed in 24 h.	
		Adults	14% killed in 24 h.	
Scutellonema bradys (in yam tubers)	0.05	Mixed	Eliminated signs of tuber rotting.	Adesiyan, 1977.
	0.15		Decreased number of nematodes extracted from tubers by 88%.	
(in yam peels)	0.30		Killed 80% of nematodes in peels.	
Scutellonema bradyurum (in infested soil and chopped roots)	2.98	Mixed	No reproduction in bioassay.	Myers, 1960.
Tylenchorhynchus claytoni (in infested soil and chopped roots)	0.74	Mixed	No reproduction in bioassay.	Myers, 1960.
Xiphinema diversicaudatum (in infested soil and chopped roots)	0.74	Mixed	No reproduction in bioassay.	Myers, 1960.

[a] Conversion of dosages were based on the following: 1 roentogen = 0.93 rad; 100 rad = 0.1 kilorad = 1 Gray.

[b] Mixed = eggs, second-, third-, fourth-stage juveniles, and adults; J1, J2, J3, J4 = first-, second-, third-, fourth-stage juveniles; and cysts = dead mature female stage of *Heterodera* and *Globodera* spp. that serves as a protective shell for eggs.

Table 25.2 Minimum dosages of irradiation rendering plant-parasitic nematodes noninfective or unable to reproduce.

Nematode species	Dose (kGy)[a]	Developmental stage treated[b]	Reference
Aphelenchoides sp.	0.37	Mixed	Myers, 1960.
Ditylenchus destructor	0.89	Mixed	Wood and Goodey, 1957.
D. dipsaci	0.48	J4	Siddiqui and Viglierchio, 1970.
D. myceliophagus	1.12	Mixed	Myers and Dropkin, 1959.
Globodera rostochienesis	0.64	Cysts	Evans, 1970.
G. rostochienesis	5.95	Cysts	Fassuliotis, 1958a.
G. rostochienesis	4.40	Cysts	Van de Woestijne and Van den Brande, 1960.
Helicotylenchus buxophilus	2.98	Mixed	Myers, 1960.
Heterodera schachtii	5.12	J2	Townshend, 1967.
Meloidogyne arenaria	1.67	Mixed	Myers and Dropkin, 1959.
M. hapla	6.40	Egg masses	Van de Woestijne and Van den Brande, 1960
M. incognita	0.39	White egg masses	Ishibashi and Taguchi, 1965.
M. incognita	1.67	Mixed	Myers and Dropkin, 1959.
M. incognita	5.95	Cysts	Fassuliotis, 1958a.
M. javanica	4.25	J2 and eggs	Chinnasri et al., 1997.
Mesocriconema axeste	1.49	Mixed	Myers, 1960.
Paratrichodorus minor	2.98	Mixed	Myers, 1960.
Pratylenchus vulnus	0.37	Mixed	Myers, 1960.
Scutellonema brachyurum	2.98	Mixed	Myers, 1960.
Tylenchorhynchus claytoni	0.74	Mixed	Myers, 1960.
Xiplinema diversicaudatus	0.74	Mixed	Myers, 1960.

[a] Conversion of dosages were based on the following: 1 roentogen = 0.93 rad; 100 rad = 0.1 kilorad = 1 Gray.

[b] Mixed = eggs, second-, third-, fourth-stage juveniles, and adults; J2, J4 = second-, fourth-stage juveniles; and cysts = dead mature female stage of *Heterodera* and *Globodera* spp. that serves as a protective shell for eggs.

25.2 Radiation Effects on Plant-parasitic Nematodes

Weischer (1957; 1960) reported on the effects of X-rays on the golden nematode and also on the sugarbeet cyst nematode, *Heterodera schachtii*. Treatment at 0.37 kGy/hour for less than 72 h had no apparent effect, whereas longer exposure decreased hatch of the sugarbeet cyst nematode but the golden nematode juveniles were stimulated to hatch. A dose of 0.27 kGy decreased the reproduction rate of the golden nematode and soaking the cysts in potato root-diffusate made them much more sensitive to radiation.

Three non-parasitic nematode species and 13 plant-parasitic nematodes in 10 genera, both ectoparasites, and migrating and sedentary endoparasites, were tested in soil or plant media for their response to radiation with ^{60}Co (Myers, 1960; Myers and Dropkin, 1959). The dose required for complete sterilization was variable. Reproduction in the plant parasites was completely stopped with a dose of 0.37 kGy in only three species. Seven species required a dose of over 1.5 kGy to stop reproduction, although a large reduction in reproduction of most species was attained with a dose of 0.74 kGy. The non-parasitic nematodes also were variable in their response to irradiation. Myers and Dropkin (1959) pointed out that irradiation of field soil to combat plant-parasitic nematodes would not be practical because of the length of time required to irradiat even a small area in a field. Also, they pointed out that the dosages required to interrupt the life cycle of plant-parasitic nematodes is generally well above that known to be lethal to living plant tissue, thus controlling nematodes in or on plant roots or tubers did not appear feasible. Based on their studies, they calculated that more than 0.71 kGy of gamma radiation would have to be delivered 0.61 m deep in soil to control nematodes in a manner similar to that attained by application of a nematicide at a moderate dose.

The high dosage of radiation of 7.5 kGy was required to eliminate nematodes from soil (Thompson, 1990). Approximately 1% of *Pratylenchus thornei* survived 5 kGy of gamma irradiation, whereas none survived 7.5 kGy or more. Some of the nematodes in these treatments may have been in an anhydrobiotic state, which greatly increases the nematodes ability to withstand adverse conditions and this maybe makes nematodes more resistant to the impact of radiation.

Myers (1960) points out that once embryos were past the gastrulation stage of embryogenesis they were much more resistant to irradiation, which is probably related to the fact that all the basic tissue lines are essentially completed and few additional cellular divisions are necessary to complete juvenile development. This result agreed with findings by Seide (1925), who showed that the rate of abnormalities caused by irradiation in the animal parasitic nematode, *Ascaris* sp., was related to the maturity of the embryo.

Recent studies on the bacterial feeding nematode, *Caenorhabditis elegans*, also confirm that nematodes in advanced stages of development are more resistant to radiation (Ishii and Suzuki, 1990). In *C. elegans* radiation resistance decreased slightly throughout the first proliferative phase of embryogenesis. It was suggested this might be due to the increase in target size, since most cells in *C. elegans* are autonomously determined. Nematodes irradiated in the second half of embryogenesis were about 40-fold more resistant to the lethal effects of X-rays, and this may be due to the absence of cell divisions during this time. The radiation resistance increased still more with advancing juvenile stages. A radiation hypersensitive mutant, rad-1, irradiated in the first half of embryogenesis, was about 30-fold more sensitive than wild-type, but in the second half of embryogenesis it was the same as the wild-type.

Cox et al. (1976) divided *M. incognita* eggs into two groups (group 1 = 50% exceeded the 4-cell stage, and group 2 = 50% equal to or less than the 4-cell stage). The effects of irradiation (0, 0.01, 0.02, 0.03, and 0.04 kGy) and temperature (7℃, 22℃, and 32℃) were evaluated on rate of hatch and infectivity of juveniles that did hatch from the two groups of *M. incognita* eggs. Interestingly, the younger eggs were not clearly more susceptible to radiation damage. In the case of group 1 the rate of decreased hatch was proportional to the dosages applied, but was independent of exposure temperatures. However, the hatch of the second group of eggs was not proportional to the radiation dose and was affected by temperature. Unfortunately, the percentage reduction of hatch was not given, but it was clear that viable juveniles hatched from both groups of treated eggs and that they produced viable progeny when inoculated onto host plants. The F_1 second-stage juveniles from some radiation treatments were infective on a resistant cultivar of cowpea. Whether this implies a mutation in the nematode induced by radiation whereby the nematode becomes infective on a resistant host plant is not clear and is deserving of further study.

Research by Van de Woestijne and Van den Brande (1960) confirmed the earlier work by Fassuliotis (1955, 1958a, 1958b), Myers and Dropkin (1959), Myers (1960), and others that large doses of radiation were required to be lethal to plant-parasitic nematodes. These workers determined that juveniles from cysts of the golden nematode could still infect potato plants after treatment with 1.6 kGy, whereas 3.2 kGy were lethal. However, this dose was not lethal to all juveniles of the northern root-knot nematode, *Meloidogyne hapla*. Some slight infection of tomato plants occurred after treatment at 3.2 kGy, but infection by juveniles treated with 0.8 kGy was not impaired.

One of the few studies done on the effect of irradiation on nematode development was conducted by Ishibashi (1965). Plant roots were irradiated with 0.93 kGy and 0.19 kGy of ^{60}Co 3 and 6 days following inoculation with root-knot nematode juveniles. The timing of irradiation affected nematode development. Irradiation of *Meloidogyne incognita* J2 before sexual differentiation caused retardation of gonad growth in the adults of the same generation. Early irradiation (3rd day after inoculation) with 0.09 or 0.19 kGy increased the number of males in the population from 1.5% in the control to 18% at 0.09 kGy to 27% at 0.19 kGy in irradiated treatments. Late irradiation (6th day after inoculation) increased males with two testis, which indicated sex reversal, from 1.5% in the control to 25% at 0.09 kGy to 34% at 0.19 kGy.

Townshend (1967), who applied irradiation directly to nematode infested soil and to the cyst stage of the sugarbeet cyst nematode, was interested in testing the effects of irradiation at less then a sterilizing dose. Earlier workers (Myers and Dropkin, 1959; Myers, 1960) tested dosages of irradiation required for complete sterilization, however, they did not take into account the effect of these high doses of irradiation on the succeeding generation of nematodes. Townshend's approach was to apply non-lethal doses of irradiation to nematodes in soil to determine their effects on the succeeding generation. Samples of 450 grams of soil each infested with *H. schachtii* were placed in polyethylene bags and exposed to 0.07, 0.30, 1.2, and 4.8 kGy of gamma irradiation. The lower doses up to 1.2 kGy increased the recovery of juveniles compared with the control, which was speculated to be caused by an inactivation of the natural hatch inhibitor contained in eggs of this nematode. The highest dose of irradiation almost eliminated all recovery of juveniles, which depended on nematode mobility. When the irradiated juveniles were placed on beet the number of F_1 cyst was increased over the control by only the two lowest dosages, but unexpectedly 1.2 kGy greatly suppressed the formation of cyst and 4.8 kGy completely eliminated any cyst. Fewer juveniles emerged from the F_1 cysts at 0.07 kGy compared with the control and were almost eliminated at 0.32 kGy. The only F_2 cyst formed was from the juveniles recovered from F_1 cyst in the 0.07 kGy treatment and viable juveniles were recovered from these F_2 cyst. This work clearly demonstrated that fully embryonated eggs within cyst are more resistant to irradiation than eggs not fully embryonated. Thus, eggs with embryonating tissues appeared to be more susceptible to radiation damage.

A comprehensive study of the effects of gamma radiation on *G. rostochiensis* was made by Evans (1970). His objectives were to determine the effects of radiation on newly-hatched juveniles, males and females, cyst exposed to wetting, dryness, and hatching factor (potato root diffusate), and the resulting F_1 generation. Dosages of exposure included 0.02, 0.04, 0.08, and 0.16 kGy in some tests, and other test included 0. 32 and 0. 64 kGy. Hatching from air-dried cysts was stimulated by the lower doses (0. 02, 0. 08, and 0. 16 kGy), whereas hatching from cysts exposed to root diffusate before irradiation was decreased slightly. In the F_1 generation irradiating doses of 0. 02, 0. 08, and 0. 16 kGy reduced the number of embryonated eggs per cyst, with the latter two doses causing substantial reductions. However, when cysts that had been stored for an extra year were irradiated the impact on the F_1 was much greater than on cysts that had not been stored. All doses, except 0. 02 kGy reduced the number of new cysts on plants and embryonated eggs per cyst. No progeny formed from cysts exposed to 0. 64 kGy. Irradiated J2, however, produced an F_1 generation at all doses. The number of cyst formed from J2 was reduced by doses of 0. 04 kGy and greater, however even at 0. 64 kGy three cysts were formed. Adults were more resistant to irradiation than hatched J2 or J2 within cysts. The fertility of females was more affected by irradiation than males, apparently because oocytes continue dividing after mating. Males on the other hand have mature sperm and leave roots seeking females thus they are not as vulnerable to radiation.

The only study of gamma irradiation effects on a monoxenic culture of a plant-parasitic nematode (*D. dipsaci*) was conducted by Siddiqui and Viglierchio (1970). They recorded morphological abnormalities of treated nematodes and their subsequent generations, and the effects of radiation on nematodes in an oxygen and nitrogen atmospheres were compared. Fifteen doses of radiation ranging from 0 to 4. 0 kGy were tested. *Ditylenchus dipsaci* motility was lower in treatments of 0. 48, 0. 96, and 1. 92 kGy compared with the control, but some nematodes regained their motility after 2 days. Increasing the rate of radiation decreased motility, but 40% remained motile 2 weeks after irradiation. No immobilization or rupturing of nematode bodies was observed even at the highest dosages tested. Doses of 2. 88 kGy or above inhibited the nematode's ability to induce characteristic seedling symptoms following infection. An irradiation dose of 0. 48 kGy appeared to cause sterility in the treated nematodes. The ratio of adults to juveniles increased with increased radiation beginning at 0. 32 kGy. Nematodes irradiated in a nitrogen atmosphere compared with those in an oxygen atmosphere were better protected from the radiation effects.

Adesiyan (1977) applied gamma irradiation to white yams infected with yam spiral nematode, *Scutellonema bradys* in storage. Dosages of radiation applied were 0. 05, 0. 08, 0. 1, 0. 13, and 0. 15 kGy. All doses reduced the number of nematodes recovered, but none eradicated the nematode from the tubers. Dosages of 0. 2 to 0. 3 kGy reduced 70% to 80%, respectively, of the nematodes within yam peels. All doses reduced rotting caused by the nematode when checked at 12 weeks following treatment, but even the lowest dose inhibited sprouting thereby making the treatment impractical for seed yams. Irradiation at 0.08 and 0.15 kGy did not materially affect yam appearance or taste.

The effect of radiation on the growth and development of root-knot nematode, *Meloidogyne graminicola* in rice roots was determined by Prasad, et al. (1982). A dosage of 0.03 kGy applied to the nematode's egg masses for 10 minutes reduced nematode infection by hatched J2 by more than half. Root penetration took longer for irradiated nematodes, irradiated females were smaller, and the incidence of male nematodes was greater compared with control nematodes.

Bursaphelenchus xylophilus, the pinewood nematode, is an obstacle to the export of raw softwood products and green lumber from North America to some foreign countries. Irradiation as a deinfestation alternative to fumigation for softwood products and green lumber has been investigated (Eichholz, et al., 1991). Exposure of pinewood nematode infested woodchips were exposed for periods varying from 1 h to 2 weeks to gamma ray doses up to 12 kGy. Lethal doses were found to lie in the range above 6 to 8 kGy, which is considered too high to make irradiation an economically attractive means of deinfestation for commercial woodchips. Smith (1991) reported a similar dosage of 7 kGy was required to kill pinewood nematode in aqueous solution, which supports the contention that a higher dosage is necessary to eliminate pinewood nematode in vivo than in vitro.

Recent studies of irradiation effects on nematodes confirmed that relative high dosages are required to cause mortality (Chinnasri, et al., 1997; Karnkowski and Ignatowicz, pers. comm.). Chinnasri et al. (1997) showed that with root-knot nematodes (*M. javanica*) a dose of 7.5 kGy was required to kill all J2 within 1 day following treatment and egg hatch was completely inhibited at 6.25 kGy. Dosages of 2 and 3 kGy were moderately successful in reducing egg hatch; dosages of 4 and 5 kGy reduced hatch to 15% or less compared with the control. Doses of 4 kGy or less had little effect on J2. At 6 kGy, it took 5 days before mortality was observed, and at 15 days after treatment this increased to 80% mortality compared with the control. With a dose of 7 kGy, some mortality was observed by the second day following treatment and 100% were dead by day 5. At dosages of 7.5 and 8 kGy, 100% of J2 were dead the day following treatment. Mortality was based on the assumption that J2 were dead if they were lying motionless in a straightened position, swollen, or appeared darker in color.

These workers followed up on the mortality test with an infectivity test (root galling and reproduction), which is more conclusive of the effects of irradiation on plant-parasitic nematodes. Obviously, if root-knot nematode infection is stopped then you have a successful treatment. A dosage of 2 kGy significantly reduced the rate of galling of J2 and 4.25 kGy completely prevented any galling and egg production. Chinnasri et al. (1997) also evaluated sublethal dosages of irradiation (0.005, 0.01, and 0.015 kGy) in combination with stress induced by heat (43℃ and 49℃ for 10 min each, and 49℃ for 20 min), but heat at 49℃ was sufficient in itself to cause nematode mortality, thus failing to increase nematode sensitivity to irradiation.

Karnkowski and Ignatowicz's (pers. comm.) recent irradiation studies on *D. dipsaci*, the stem and bulb nematode, and *Ditylenchus destructor*, the potato rot nematode confirmed earlier reports that high dosages of irradiation are needed to kill these nematode. These workers did not test any dosages that were lethal. They recovered live juveniles and adults of the stem and bulb nematode from onion and garlic samples 3 weeks following treatment with doses up to 0.5 kGy and live potato rot nematodes from potato tubers treated with doses up to 2 kGy. Although the dosages applied did not kill nematodes they did inhibit or delay development during the post-treatment period. Sex ratios were not altered except at the highest dosage of 0.5 kGy, which favored a slight increase in males. More *D. dipsaci* emerged from onion than from garlic, but this was caused by the lower moisture levels in stored garlic than in stored onion.

25.3 Summary

It is apparent that all groups of nematodes and their developmental growth stages vary in their sensitivity to irradiation. One fact appears clear – nematodes require relatively high dosages of radiation to cause immediate mortality or the inhibition of hatch. The dose that kills nematodes appears to be between 4 and 8 kGy. Fortunately, lower dosages appear effective in causing nematode sterility or loss of infectivity, however, even these lower dosages appear to be beyond the limit of tolerance for most plants that might harbor plant-parasitic nematodes. In most cases sublethal doses of radiation do not appear to affect nematode viability. Some nematodes do not reproduce following radiation at 0.37 kGy, whereas other nematodes required up to 3.0 kGy to stop reproduction. However, up to this time relatively few nematodes have been evaluated. Many more species of nematodes must be tested in order to confirm earlier work and set thresholds for untested species.

Nematodes appear to be most sensitive to radiation while in the egg stage, especially when the eggs are in early phases of embryogenesis. Most nematodes, unfortunately, cannot be accessed only in the egg stage of development and even when eggs can be assessed, e. g., root-knot nematode egg masses or nematodes with their cyst stage filled with eggs, they are likely to contain eggs in varying stages of embryological development, as well as J1 and J2. Many nematodes will exist in plant material or soil in mixed life stages, thus radiation must be effective for the different life forms, e. g., eggs, J1, J2, J3, J4, and adults. Several important plant-parasitic nematodes are capable of developing survival stages, e. g., cysts, and juvenile stages that enter a quiescent stage, which are likely to impact the effects of radiation. From the very few radiation studies done on nematode where survival stages existed it is apparent that they do impact the effects of radiation. Cysts treated with "hatching factor" were more susceptible to radiation than dry cysts. Actually radiation treatment stimulated hatch in dry cysts.

In many of the studies conducted on nematodes the impact of radiation on infectivity and reproduction of plant-parasitic nematodes is not always clear. Only few workers have looked at the F_1 and F_2 generations following irradiation treatments. Much lower dosages can be used to cause losses in infectivity or losses in reproduction potential than are required to kill nematodes. From the few studies done it does appear that each nematode has its own radiation threshold.

There is great need for future studies on effects of irradiation on plant-parasitic nematodes. First order of work needs to be collaboration of previous studies. There have relatively few genera of nematodes tested and where the same species has been tested by more than one researcher, the results are not always in agreement. This is likely because of the use of different methodologies, however different populations of nematodes from the same species may react differently. It is not unusual to find differences in virulence within a species, which might impact their response to radiation. Also, there is lack of any standardization in protocols used to test radiation effects on nematodes. Often the medium containing nematode life stages being treated with radiation is not specified. In fact, this is one area that needs critical work. For example, do nematodes show the same response if treated in sterile distilled water vs. being contained in plant roots, tubers, bulbs, or soil? Another area that needs work is repeated treatment of nematodes. All studies done to date have been with the application of a single dosage of irradiation. Would applications applied two, three, or more times over a period of days improve the effects of radiation vs. a single dosage?

Relatively few of the plant-parasitic nematodes that are important in quarantine and regulatory programs have been tested for their response to irradiation. The burrowing nematode, *Radopholus similis*, which is likely to appear on all regulatory lists, has never been tested for its response to irradiation. Establishing thresholds to irradiation for survival and sterility of the numerous nematodes important to international trade should be a high priority for future studies.

References

Adesiyan, S. O. 1977. Studies on the effect of gamma radiation (from Cobalt 60 source) on storage life of white yam (*Dioscorea rotundata* var. *efon*) infected with *Scutellonema bradys*. Annals of Applied Biology 86: 213 – 218.

Chinnasri, B., J. H. Moy, B. S. Sipes, and D. P. Schmitt. 1997. Effect of gamma-irradiation and heat on root-knot nematodes, *Meloidogyne javanica*. Journal of Nematology 29: 30 – 34.

Cobb, N. A. 1920. Proceedings of the 42nd meeting of Helminthological Society of Washington. Journal of Parasitiology 7: 101.

Cox, R. J., E. H. Moody, and C. J. Southards. 1976. Effects of neutron irradiation on *Meloidogyne incognita*. Journal of Nematology 8: 282 (Abstr.).

Eichholz, G. G., A. A. Bogdanov, and L. D. Dwinell. 1991. Radiation sensitivity of pine wood nematodes in woodchips. Applied Radiation and Isotopes 42: 177 – 179.

Evans, K. 1970. The effects of gamma radiation on *Heterodera rostochiensis*. Nematologica 16: 284 – 294.

Fassuliotis, G. 1958a. Effects of ionizing radiations on the golden nematode, *Heterodera rostochiensis*. Radiation Research 9: 112 – 113 (Abstr.).

Fassuliotis, G. 1958b. Observations on the biological effects of ionizing radiations on the life cycle of *Heterodera rostochiensis* Wollenweber, 1923. Ph. D. dissertation. New York University, NY.

Fassuliotis, G., and A. H. Sparrow. 1955. Preliminary report on X-ray studies on the golden nematode. Plant Disease Reporter 39: 572.

Ishibashi, N. 1965. The increase in male adults by gamma-ray irradiation in the root-knot nematode, *Meloidogyne incognita* Chitwood. Nematologica 11: 361 – 369.

Ishibashi, N., and S. Taguchi. 1965. Studies on the hatching of the root-knot nematode, *Meloidogyne incognita* Chitwood. II. Radio-sensitivity of embryos and infectivity of larvae emerging from irradiated egg masses. Annotationes Zoologicae Japonenses 38: 12 – 19.

Ishii, N., and K. Suzuki. 1990. X-ray inactivation of *Caenorhabditis elegans* embryos or larvae. International Journal of Radiation Biology 58: 827 – 833.

Lehman, P. S., 1995. Role of plant protection organizations in nematode management. Regional, federal, and state organizations in North America. Congresso Internacional de Nematologia Tropical. Program and Proceedings. 4 – 9 June 1995. Rio Quente, Brazil.

Myers, R. F. 1960. The sensitivity of some plant-parasitic and free-living nematodes to gamma and X-irradiation. Nematoalogica 5: 56 – 63.

Myers, R. F., and V. H. Dropkin. 1959. Impracticability of control of plant parasitic nematodes with ionizing radiations. Plant Disease Reporter 43: 311 – 313.

Nath, R. P., G. Swarup, and M. S. Chatrath. 1986. Effect of gamma radiations on the survival and multiplication of *Pratylenchus zeae*. Journal of Nuclear Agriculture and Biology 15: 214 – 216.

Prasad, K. S. K., Y. S. Rao, and K. Krishnappa. 1982. Effect of ^{60}Cobalt radiation on the growth and development of *Meloidogyne graminicola* in rice roots. Indian Journal of Nematology 12: 171 – 173.

Seide, J. 1925. Zur kenntnis der biologischen strahlenwirkung. Untersuchungen an Ascaris-ei mit ultravioletten, rontgen- und radiumstrahlen. Z. Wiss. Zool. 124: 252 – 304.

Siddiqui, I. A., and D. R. Viglierchio. 1970. The effects of gamma irradiation on the motility, infectivity, reproduction and morphology of *Ditylenchus dipsaci*. Nematologica 16: 459 – 469.

Smith, R. S., ed. 1991. The use of heat treatment in the eradication of the pinewood nematode and its insect vectors in softwood lumber. Report of the Task Force on Pasteurization of Softwood Lumber. Forintek Canada Corp., Vancouver, B. C., Canada.

Thomas, W. B. 1996. Methyl bromide: Effective pest management tool and environmental threat. Supplement to the Journal of Nematology 28: 586 – 589.

Thompson, J. P. 1990. Treatments to eliminate root-lesion nematode (*Pratylenchus thornei* Sher & Allen) from a vertisol. Nematologica 36: 123 – 127.

Townshend, J. L. 1967. Gamma irradiation of *Heterodera schachtii*. Nematologica 13: 586 – 592.

Tyzzer, E. E., and J. A. Honeij. 1916. The effects of radiation on the development of *Trichinella spiralis* with respect to its application to the treatment of other parasitic diseases. Journal of Parasitology 3: 43 – 56.

Van de Woestijne, N., and J. Van den Brande. 1960. Gevoeligheid van plantenparasiterende aaltjes voor radioaktieve straling. Overdruk uit Mededelingen van de landbouwhogeschool en de Opzoekingsstations van de staat te Gent, Belgium. Deel 25: 1057 – 1064.

Weischer, B. 1957. Die Wirkung ionisierender Strahlen auf die Entwicklung von *Heterodera rostochiensis* und *H. schachtii*. Nematologica 2: 300 – 305.

Weischer, B. 1960. Aktivitatszustand und Strahlenempfindlichkeit beim Kartoffelnematoden (*Heterodera rostochiensis* Wollenweber). Mitt. biol. BundAnst. Ld-u. Forstw. 99: 59 – 65.

Wood, F. C., and J. B. Goodey. 1957. Effects of gamma-ray irradiation on nematodes infesting cultivated mushroom beds. Nature 180: 760 – 761.

Index

Index